Desert Plants

K.G. Ramawat
Editor

Desert Plants

Biology and Biotechnology

Springer

Editor
Prof. Dr. K.G. Ramawat
M.L. Sukhadia University
Department of Botany
Udaipur 313001, India
e-mail: kg_ramawat@yahoo.com

ISBN: 978-3-642-02549-5 e-ISBN: 978-3-642-02550-1
DOI 10.1007/978-3-642-02550-1
Springer Heidelberg Dordrecht London New York

Library of Congress Control Number: 2009938013

Cover illustration: Capparis decidua (Photo K.G. Ramawat)

Cover design: WMXDesign GmbH, Heidelberg, Germany

Printed on acid-free paper

Springer is part of Springer Science+Business Media (www.springer.com)

Preface

Deserts appear very fascinating during our short visits. However, the lives of plants and animals are very difficult under the harsh climatic conditions of high temperature and scant water supply in deserts, sometimes associated with high concentrations of salt. The editor of this book was born and brought up in the Great Indian Desert, and has spent much of his life studying the growth and metabolism of desert plants. It is very charming on a cool summer evening to sit at the top of a sand dune listening only to blowing air and nothing else. It has been my dream to prepare a volume on desert plants encompassing various aspects of desert plant biology. In this book, I have tried to present functional and useful aspects of the vegetation resources of deserts along with scientific input aimed at understanding and improving the utility of these plants. The scant vegetation of deserts supports animal life and provides many useful medicines, timber and fuel wood for humans. Therefore, there are chapters devoted to medicinal plants (Chap. 1), halophytes (Chaps. 13, 14), and fruit plants (Chaps. 17, 20). Desert plants have a unique reproductive biology (Chaps. 9–11), well-adapted eco-physiological and anatomical characteristics (Chap. 7), and specialised metabolism and survival abilities. These plants are difficult to propagate and pose many problems to researchers developing biotechnological approaches for their amelioration (Chaps. 18–20). Covering all the above aspects, this book provides an excellent amalgam of the morphology, physiology and biotechnology of desert plants.

Finally, I would like to acknowledge my contributors, who have made serious efforts to ensure high scientific quality of the book. I also would like to thank my colleagues at Springer.

October 2009 K.G. Ramawat

Contents

Contributors

M. Alonso-Pedano Laboratorio de Investigaciones Botánicas (LABIBO), Facultad de Ciencias Naturales, Universidad Nacional de Salta, Av. Bolivia 5150, 4400 Salta, Argentina

J.A. Alvarez Departamento de Dendrocronología e Historia Ambiental, Instituto Argentino de Nivología, Glaciología y Ciencias Ambientales, Centro Científico Tecnológico, CONICET, Mendoza, CC. 330–5500 Mendoza, Argentina

M.E. Apple Department of Biological Sciences, Montana Tech of the University of Montana, Butte, Montana, MT 59701, USA, mapple@mtech.edu

J. Arora Laboratory of Bio-Molecular Technology, Department of Botany, M.L. Sukhadia University, Udaipur 313001, India

S. Bravo Facultad de Ciencias Naturales, Laboratorio de Investigaciones Botánicas (LABIBO), Universidad Nacional de Salta, Av. Bolivia 5150, 4400 Salta, Argentina

I. Carrillo-Angeles Departamento de Ecología de la Biodiversidad, Laboratorio de Dinámica de Poblaciones y Evolución de Historias de Vida, Instituto de Ecología, Universidad Nacional Autónoma de México, Apartado Postal 70-275, Ciudad Universitaria UNAM, C.P., 04510 México, D.F., México

E. Cherif Laboratory of Molecular Genetics, Immunology and Biotechnology, Tunis Faculty of Sciences, Campus University, 2092 El Manar Tunis, Tunisia

J. Chikara Central Salt and Marine Chemicals Research Institute (CSIR), G.B. Marg, Bhavnagar, Gujarat 364 021, India

S. Dakhlaoui-Dkhil Laboratory of Molecular Genetics, Immunology and Biotechnology, Tunis Faculty of Sciences, Campus University, 2092 El Manar Tunis, Tunisia

Ma. del Coro Arizmendi FES Iztacala UNAM, Laboratorio de Ecología UBIPRO, Av. De los Barrios 1, Los Reyes Iztacala, Tlalnepantla, Edo. México, CP 54090, México, coro@servidor.unam.mx

I. Furukawa Faculty of Agriculture, Tottori University, Koyama, Tottori 680-8553, Japan

C. Giordano Instituto Argentino de Investigaciones de las Zonas Áridas, CONICET, Mendoza, Argentina

H. Godínez-Alvarez UBIPRO, FES-Iztacala, UNAM, Av. de los Barrios 1, Los Reyes Iztacala, Tlalnepantla, 54090 Estado de México, Ap. Postal 314, Iztacala, México

J. Golubov Laboratorio de Ecología, Sistemática y Fisiología Vegetal, Departamento El Hombre y Su, Ambiente-CBS-Universidad Autónoma Metropolitana-Xochimilco, Calzada del Hueso 1100, Col. Villa Quietud, 04960 México D.F., México, jgolubov@gmail.com

S. Goyal Laboratory of Bio-Molecular Technology, Department of Botany, M.L. Sukhadia University, Udaipur 313001, India

N. Jain Laboratory of Bio-Molecular Technology, Department of Botany, M.L. Sukhadia University, Udaipur 313001, India

Z. Jiang Research Institute of Wood Industry, Chinese Academy of Forestry, Beijing 100091, China

D. Kalita Cellulose, Pulp & Paper Division, North East Institute of Science and Technology (CSIR), Jorhat 785006, Assam, India, dipkalita_2001@yahoo.co.in, kalita_d@rrljorhat.res.in

P.K. Kasera Laboratory of Plant Ecology, Department of Botany, Jai Narain Vyas University, Jodhpur 342 033, India, jnvupkkasera@yahoo.co.in

V.V. Kuznetsov Timiryazev Institute of Plant Physiology, Russian Academy of Sciences, Botanicheskaya str 35, 127276 Moscow, Russia, vlkuzn@ippras.ru

X. Li Institute of Applied Ecology, Chinese Academy of Sciences, Shenyang 110016, People's Republic of China, lixuehua@iae.ac.cn

U. Lüttge Institute of Botany, Department of Biology, Technical University of Darmstadt, Schnittspahn-Strasse 3–5, 64287 Darmstadt, Germany, luettge@bio.tu-darmstadt.de

M.C. Mandujano Departamento de Ecología de la Biodiversidad, Laboratorio de Dinámica de Poblaciones y Evolución de Historias de Vida, Instituto de Ecología,

Universidad Nacional Autónoma de México, Apartado Postal 70-275, Ciudad Universitaria UNAM, C.P., 04510 México, D.F., México, mcmandu@miranda.ecologia.unam.mx

J. Márquez-Guzmán Facultad de Ciencias UNAM, Laboratorio de Desarrollo en Plantas, Circuito Exterior s/n, Ciudad Universitaria, Avenida Universidad 3000, CP 04510 México, México, jmg@fciencias.unam.mx

M. Marrakchi Tunis Faculty of Sciences, Laboratory of Molecular Genetics, Immunology and Biotechnology, Campus University, 2092 El Manar Tunis, Tunisia

C. Martínez-Peralta Departamento de Ecología de la Biodiversidad, Laboratorio de Dinámica de Poblaciones y Evolución de Historias de Vida, Instituto de Ecología, Universidad Nacional Autónoma de México, Apartado Postal 70-275, Ciudad Universitaria UNAM, C.P., 04510 México, D.F., México

S.K. Mehar Department of Biological and Earth Sciences, Shri Venkateshwara University, Tirupathi, Andra Pradesh, India

S. Mohammed Department of Botany, Government Lohia (PG) College, Churu 331 001, India, smdchuru@gmail.com

H.Y. Mohan Ram 194, SFS DDA Flats, Mukherji Nagar, Delhi 110 009, India, hymohanram@gmail.com

H. Nakano National Agricultural Research Center for Kyushu Okinawa Region, Chikugo, Fukuoka 833-0041, Japan, nakanohr@affrc.go.jp

P. Ortega-Baes Laboratorio de Investigaciones Botánicas (LABIBO), Facultad de Ciencias Naturales, Universidad Nacional de Salta, Av. Bolivia 5150, 4400 Salta, Argentina, portega@miranda.ecologia.unam.mx, ortega_baes@yahoo.com.ar

A. Othmani Laboratory of Molecular Genetics, Immunology and Biotechnology, Tunis Faculty of Sciences, Campus University, 2092 El Manar Tunis, Tunisia

A. Ould Mohamed Salem Tunis Faculty of Sciences, Laboratory of Molecular Genetics, Immunology and Biotechnology, Campus University, 2092 El Manar Tunis, Tunisia

D.V.N.S. Pamidimarri Central Salt and Marine Chemicals Research Institute (CSIR), G.B. Marg, Bhavnagar, Gujarat 364 021, India

R. Raj Bhansali Division of Plant Sciences and Biotechnology, Central Arid Zone Research Institute, Jodhpur 342003, India, anjalirikhab@gmali.com

K.G. Ramawat Laboratory of Bio-Molecular Technology, Department of Botany, M.L. Sukhadia University, Udaipur 313001, India, kg_ramawat@yahoo.com

M.F. Ramos-Ordoñez FES Iztacala UNAM, Laboratorio de Ecología UBIPRO, Av. De los Barrios 1, Los Reyes Iztacala, Tlalnepantla, Edo. México, CP 54090 México, México, mafyx@correo.unam.mx

M.P. Reddy Plant Stress Genomics and Technology Center, 4700 King Abdullah University of Science and Technology, Mail Box # 2053, Thuwal 23955-6900, Kingdom of Saudi Arabia, reddymuppala@gmail.com

S. Rhouma Laboratory of Molecular Genetics, Immunology and Biotechnology, Tunis Faculty of Sciences, Campus University, 2092 El Manar Tunis, Tunisia

J. Sajama Laboratorio de Investigaciones Botánicas (LABIBO), Facultad de Ciencias Naturales, Universidad Nacional de Salta, Av. Bolivia 5150, 4400 Salta, Argentina

N.I. Shevyakova Timiryazev Institute of Plant Physiology, Russian Academy of Sciences, Botanicheskaya str 35, 127276 Moscow, Russia

K.R. Shivanna Ashoka Trust for Research in Ecology and the Environment, Royal Enclave, Srirampura, Jakkur Post, Bangalore 560 064, India, shivanna@atree.org

E. Sotola Laboratorio de Investigaciones Botánicas (LABIBO), Facultad de Ciencias Naturales, Universidad Nacional de Salta, Av. Bolivia 5150, 4400 Salta, Argentina

P. Su Laboratory of Plant Stress Ecophysiology and Biotechnology, Linze Inland River Basin Research Station, Cold and Arid Regions Environmental and Engineering Research Institute, Chinese Academy of Sciences, Lanzhou 730000, China, supx@lzb.ac.cn

S. Sühring EEA-Salta, Instituto Nacional de Tecnología Agropecuaria (INTA), Ruta Nac. 68 km 172, (4403) Cerrillos, Salta, Argentina

S. Sundaramoorthy Department of Botany, JNV University, Jodhpur 342001 Rajasthan, India, jnvusundar@rediffmail.com

M.S. Suthar Department of Botany, JNV University, Jodhpur 342001, Rajasthan, India

R. Tandon Department of Botany, University of Delhi, Delhi 110 007, India, rjtnd@rediffmail.com

M. Trifi Laboratory of Molecular Genetics, Immunology and Biotechnology, Tunis Faculty of Sciences, Campus University, 2092 El Manar Tunis, Tunisia, mokhtar.t@fst.rnu.tn, trifimokhtar@yahoo.fr

A. Vilela Museo Egidio Feruglio-CONICET, Av. Fontana 140, 9100 Trelew, Chubut, Argentina

P.E. Villagra Departamento de Dendrocronología e Historia Ambiental, Instituto Argentino de Nivología, Glaciología y Ciencias Ambientales, Centro Científico Tecnológico, CONICET, Mendoza, CC. 330–5500 Mendoza, Argentina, villagra@mendoza-conicet.gov.ar

S. Yang Research Institute of Wood Industry, Chinese Academy of Forestry, Beijing 100091, China, Shangke620@hotmail.com

S. Zehdi-Azouzi Laboratory of Molecular Genetics, Immunology and Biotechnology, Tunis Faculty of Sciences, Campus University, 2092 El Manar Tunis, Tunisia

Part A
General Biology

Chapter 1
Biodiversity, Biology and Conservation of Medicinal Plants of the Thar Desert

Jaya Arora, Shaily Goyal, and Kishan Gopal Ramawat

Abstract The Great Indian Desert known as the Thar Desert occupies about 60% of the area of Rajasthan – the largest state of India. It is one of the most heavily populated (in terms of both people and cattle) deserts of the world. The animal and human populations exert tremendous pressure on the scant vegetation of the region, making several plants vulnerable to becoming endangered. Inherent biological problems associated with these plants make their survival difficult and have forced adaptation to the harsh environment. The biological activities of these plants range from analgesic, antifungal, antimicrobial, hypolipidemic to hepatoprotective and anticancerous. This chapter reviews the biological problems faced by the medicinal plants of this region, their bioactive molecules, as well as biotechnological approaches aimed at improving and conserving these plants.

1.1 Introduction

Deserts have played a special role in human evolution and adaptation. They appear to be the major terrestrial habitat that channelled early human dispersal, representing barriers at some times, corridors at others (Gamble 1993). Studies of desert societies have also provided some of the most fertile ground for debate regarding human adaptability and how societies cope with marginal – often precarious – environmental circumstances, and about the effects of these environmental conditions on human land use, mobility, and dispersal (Kelly 1995).

J. Arora, S. Goyal, and K.G. Ramawat (✉)
Laboratory of Bio-Molecular Technology, Department of Botany, M.L. Sukhadia University, Udaipur-313001, India
e-mail: kg_ramawat@yahoo.com

K.G. Ramawat (ed.), *Desert Plants*,
DOI 10.1007/978-3-642-02550-1_1, © Springer-Verlag Berlin Heidelberg 2010

1.2 Deserts of the World

Deserts are large bands of dry lands along the tropics in both the Northern and Southern hemispheres (Mares 1999; Middleton and Thomas 1997). The United Nations Environment Program (UNEP) has prepared a map of the extent of world deserts (Middleton and Thomas 1997). Deserts cover around 25,500,000 km^2, approximately 20% of the land area of the world. The boundaries of these deserts, which are constantly changing due to various climatic and human factors, are likely to drift over the next century as human-induced global warming takes effect. The defining characteristic of world deserts is aridity. The current UNEP definition of desert is a moisture deficit under normal climatic conditions where P/PET <0.20, i.e. where rainfall is less than 20% of potential moisture loss through evaporation (Smith et al. 1995).

1.2.1 The Thar Desert

Rajasthan is the largest state in India, and is located in the northwestern part of the country. The state is rich in floral diversity, with 911 wild species belonging to 780 genera and 154 families growing here (Shetty and Singh 1987–1993; Bhandari 1999). Geographically, Rajasthan lies between 23°3′ to 30°12′ longitude and 69°30′ to 78°17′ latitude. It occupies 342,239 km^2 land area, which is 10.41% of the total land area of India. The desert in northwestern India is known as the Thar Desert, and is one of the most heavily populated deserts in the world. The Thar Desert lies between 24° to 28° N latitude and 68° to 71° E longitude, occupying an area of about 200,000 km^2. Physically the desert stretches as far as Delhi to the east, south to the Run of Kutch and the Arabian Sea, to the arid rocky mountains of Baluchistan in the west, and is bounded in the north by the irrigated plains of Punjab (Fig. 1.1). The Aravalli hills divide the state of Rajasthan into two parts: (1) north-western desert, and (2) south-eastern hilly semi-arid forest. The altitude of the Thar Desert

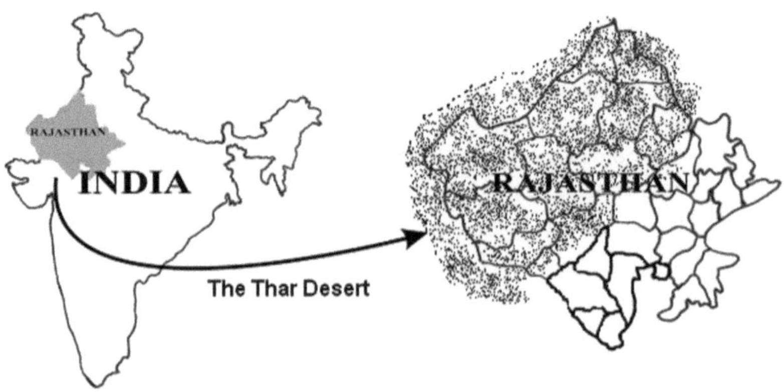

Fig. 1.1 Map of India showing location of the Thar Desert

ranges from 61 m a.s.l. near Run of Kutch to 457 m in the lower reaches of Aravalli (where the highest peak Guru Shikhar in Mt. Abu is at 1,722 m a.s.l.). These geographical conditions provide extreme habitat for a wide range of flora, including bryophytes, pteridophytes, a lone gymnosperm – *Ephedra foliata* – and angiosperms including hydrophytes, halophytes and xerophytes.

About 720,000 ha desert area is saline and is used for production of table salt through open pits (subsoil) or wells (underground). Due to the high salt conditions, plants in this region have adapted to withstand high salt concentrations. The mechanism of salt tolerance differs in different species (Ramani et al. 2006). Some of the plants commonly found in saline habitats include *Cressa cretica, Haloxylon recurvum, Haloxylon salicornicum, Portulaca oleracea, Salsola baryosma, Sesuvium sesuvioides, Suaeda fruticosa, Tamarix aphylla, Trianthema triquetra, Zaleya redimita, and Zygophyllum simplex.*

1.2.2 Climate

Dry hot summers and pleasant dry winters are prominent features of the Thar Desert. The mean daily maximum temperature in summer ranges from 41°C to 46°C, and temperatures can reach up to 53°C in the shade during the hot summer noon. Rainfall is sparse, ranging from 127 mm to 254 mm annually, and is confined mainly to the rainy season (July–September).

1.2.3 Topographical Features

The topography of the Thar Desert is distinctly marked with sand, scattered rocky ridges and steep slopes. Topography and climatic factors play a significant role in determining the type of vegetation. Most regions consist of dry undulating plains of hardened sand, with the rest consisting largely of a rolling plain of loose sand that form shifting sand dunes 2–10 km long and 20–30 m in height.

1.2.4 Phytogeography

Phytogeographically, most of the Thar Desert area lies within the Saharo-Sindhian region. The desert area west of the Aravalli Hills is floristically very poor, comprising 682 species belonging to 351 genera and 87 families of flowering plants (Bhandari 1999), representing only 5% of the flora of India, which has ~17,500 flowering plants (Rao 2006). The flora east of Aravalli harbours about 8% of the flora of India, with 1,378 species belonging to 126 families (Tiagi and Aery 2007). Permanent features of the vegetation of the Thar Desert include trees and shrubs like *Acacia jacquemontii*, *Acacia nilotica*, *Calligonum polygonoides*, *Capparis decidua*, *Commiphora wightii*, *Leptadenia phytotechnica*, *Lycium barbarum*,

Prosopis cineraria, *Salvadora oleoides*, *Salvadora persica*, *Tamarix aphylla*, and *Zizyphus nummularia*. Herbs and shrubs like *Aerva persica*, *Blepharis scindica*, *Calotropis procera*, *Crotalaria burhia*, *Cymbopogon javarancusa*, *Euphorbia caducifolia*, *Grewia tenax* and *Tephrosia purpurea*, can generally be observed on the rocks and sandy ridges.

1.3 Ethnobotanical Studies

In India, traditional folklore medicine has a long history and is very deep rooted in rural and tribal populations. It was practiced long before the beginning of the Christian era and perhaps even in the "Pre-vedic" periods of the Mohanjodaro and Harrapan civilisations. Indeed, knowledge of plant species producing medicines, essential oils and insecticides dates back to the beginning of civilisation. The traditional health care practices of indigenous people pertaining to human health is termed ethnomedicine (Ramawat et al. 2009). Several tribes lead a nomadic life in Rajasthan, and movement of such tribes and their cattle causes destruction of vegetation. In addition, several tribes living in East Rajasthan obtain their livelihood from plants, with these minor forest products being purchased by cooperatives. Such produce includes various types of gums (gum arabic from *Acacia senegal*, gum karaya from *Sterculia urens*, dhawda gum from *Anogeissus latifolia*, salai gum from *Boswellia serrata*, oleogum resin of *Commiphora wightii*), catha from *Acacia catechu*, dyes (red colour from *Bixa orellana* and *Mallotus philippensis*), several types of fruits, roots or root tubers from plants like *Chlorophytum borivilianum*, *Curculigo orchioides*, several *Dioscorea* species, leaves for making bidi (a local cigarette containing tobacco) from *Diospyros melanoxylon*, seeds and leaves of *Datura* species, flowers of *Madhuca indica* for making country liquor, and fibre for various usages from plants like *Calotropis procera* and *Crotalaria burhia*, etc. Several works have described the usage of plants by the tribes and local people of Rajasthan in detail (Bhandari 1974; Sebastian and Bhandari 1984a, 1984b; Jain 1991; Joshi 1995; Katewa and Sharma 1998; Katewa et al. 2003; Jain et al. 2005, 2008; Katewa 2009).

1.4 Biology of Desert Plants

The Indian desert is one of the most heavily populated (human and cattle) deserts of the world. The resulting biotic interference exerts tremendous pressure on 84 economically important species, due to which 31 species have become either vulnerable or endangered (Singh 2004). Of these, 17 species and 8 botanical varieties are endemic to The Great Indian Desert. Biological irregularities like poor seed set and production can be caused by reproductive problems, e.g. *Commiphora wightii* (Kumar et al. 2003) and *Anogeissus pendula* (Joshi et al. 1991); low seed viability, e.g. *Anogeissus pendula* (Joshi et al. 1991), *Tecomella undulata* (Arya et al. 1992), and *Azadirachta indica* (Anonymous 1980), or due to flower, fruit and seed infestation by insects, e.g. *Acacia senegal* and *Prosopis cineraria* (Sharma and Ramawat 2005). Xerophytic habit is an adaptation of plants to survive in harsh conditions (high temperature and low water availability) by modifying their requirements. However, several of these plants are affected by insect and termite infestation of stems, and by various fungal pathogens, which affects the growth and wood quality of these species (Anonymous 1980). Infestation of flower, fruit, and seed by insects causes flower abnormalities, poor flower and seed set, and abnormal physiological changes in the plants themselves (Purohit et al. 1979; Ramawat et al. 1979). For example, *Withania coagulans* and *Ephedra foliata* are unable to produce a sufficient quantity of seed because of an imbalance in the ratio of male to female plants / flowers and the predomination of androecious plants/flowers (Singh 2004). *Anogeissus pendula* produces predominantly sterile seeds (Joshi et al. 1991), and *Salvadora* exhibits very poor seed germination. Therefore, efforts to study the reproductive biology and seed physiology of these plants are required in order to be able to select and propagate resistant plants. Since it would be difficult to exploit all the available germplasm immediately, conventional (seed, plantations, pollen) and non-conventional (embryo, callus, shoot tips by cryo-preservation) methods should be used to conserve and preserve the germplasm for future use.

With the exception of a few members of Fabaceae, most of these species, being outbreeders, produce heterozygous progeny, which results in variation in the natural population of these plants, e.g. *Agele marmelos*, *Prosopis* species and *T. undulata*. This variation is expressed in both morphological (fruit size, absence or presence of thorns, crown size, etc) and physiological (sugar, protein and chlorophyll content, isozymes patterns, etc.) characters within the species, e.g. in *Ziziphus mauritiana* (Muchuweti et al. 2005; Pareek 2001). Most of these tree species are grown from seed from a wild population with intraspecific variation. So far, except for a few species like *Z. mauritiana*, no detailed procedures have been adopted to select superior material with the aim of cloning and propagating such species.

1.5 Medicinal and Biological Activities

Traditional medicine is the mainstay of primary health care in virtually all developing countries. The use of herbal medicine in developed countries is also expanding rapidly, with many people turning towards alternative treatments that they hope will be less harmful and have fewer side effects than western medicine. The World Health Organisation (WHO) estimated that $\sim$ 80% of the developing world relies on traditional medicine, and that 85% of this usage relies on plants or their extracts as the active substances. Desert areas harbour a high diversity of medicinal plants. Modern scientific validation methods have confirmed the strong analgesic, antiarthritic, antifungal, antimicrobial, antiparasitic, hepatoprotective, hypolipidemic, insecticidal and anticancerous activities of several of these species (Table 1.1). Out of 700 species known to occur in extreme desert conditions (Bhandari 1999) about three dozen have potential biological activity. Some of these, such as *Achyranthes*

Table 1.1 Medicinal plants of arid regions, plant parts used, their bioactive molecules and biological activities

Plant species (Family)	Bioactive compounds	Plant part used	Biological activities	References
Abutilon indicum (Malvaceae)	Abutilin A, (R)-N-(1′ methoxycarbonyl-2′ phenylethyl)-4-hydroxybenzamide, β-sitosterol, eugenol	Whole plant	Mosquito larvicidal	Abdul Rahuman et al. 2008
			Hepatoprotective	Porchezhian and Ansari 2005
			Hypoglycemic	Seetharam et al. 2002
			Analgesic	Ahmed et al. 2000
Acacia nilotica (Mimosaceae)	Kaempferol (AN-5), D-pinitol, a sex hormone, viz 3 β-acetoxy-17-hydroxy-androst-5-ene	Stem bark, gum, flower, leaves	Antioxidant	Singh et al. 2008
			Immunosuppressive	Aderbauer et al. 2008
			Anticancer and antimutagenic	Meena et al. 2006; Arora et al. 2003
			Antiinflammatory	Chaubal et al. 2003, 2006
			Antifungal	Hamza et al. 2006
			Antiplasmodial	Kirira et al. 2006
			Larvicidal	Chaubal et al. 2005
			Anti-leishmanial	Fatima et al. 2005
			Antidiaorrheal	Agunu et al. 2005
			Moderate antimicrobial activity against multi-drug resistant *Salmonella typhi*	Rani and Khullar 2004
			Inhibitory effect on hepatitis C virus (HCV) protease	Hussein et al. 2000
			Antihypertensive and antispasmodic activities	Gilani et al. 1999
Achyranthes aspera (Amaranthaceae)	Ecdysterone, betaine	Root, leaves	Treatment of leprosy, fistula-in-ano, bronchial asthma	Goyal et al. 2007
			Post coital antifertility activity	Vasudeva and Sharma 2006
			Immunity enhancement	Chakrabarti and Vasudeva 2006
			Anti-inflammatory	Vetrichelvan and Jegadeesan 2003
			Antiarthritic	Gokhale et al. 2002
			Cancer chemopreventive	Chakraborty et al. 2002
			Prothyroidic, antiperoxidative	Tahiliani and Kar 2000
Aerva persica (Amaranthaceae)	Persinol, persinosides A and B	Whole plant	Antioxidative	Ahmed et al. 2006a

Aristolochia bracteolata (Asclepiadaceae)	Aristolochic acid	Leaves	Wound healing activity	Shirwaikar et al. 2003
			Antiplasmodial activity	El-Tahir et al. 1999
Balanites aegyptiaca (Simaraubaceae)	Balanitin-6 and-7: diosgenyl saponins	Kernel, fruit mesocarp, root, bark	Antitumor activity	Gnoula et al. 2008
			Larvicidal	Chapagain et al. 2008
			Antiinflammatory, antinociceptive, antioxidant	Speroni et al. 2005
			Fasciolicidal	Koko et al. 2000
Barleria prionitis (Asclepiadaceae)	Iridoid glycoside, barlerin, verbascoside	Whole plant	Hepatoprotective	Singh et al. 2005
			Significant reduction in spermatogenesis	Verma et al. 2005; Gupta et al. 2000
			Antiinflammatory and antiarthritic	Singh et al. 2003
			Potent activity against respiratory syncytial virus	Chen et al. 1998
Boerhaavia diffusa (Nyctaginaceae)	Nonprenylated rotenoids viz boeravinones G(1), H(2), I(10), J(11), punarnavoside, liriodendrin	Whole plant	Cell-mediated immune response	Manu and Kuttan 2008
			Breast cancer resistance protein inhibiting activity	Ahmed-Belkacem et al. 2007
			Radioprotective	Manu et al. 2007
			Spasmolytic effects	Borrelli et al. 2005, 2006
			Immunosuppressive	Pandey et al. 2005
			Antifungal activity	Agrawal et al. 2004
			Antidiabetic activity with improvement in antioxidant status	Satheesh and Pari 2004
			Cancer chemopreventive	Bharali et al. 2003
Capparis decidua (Capparidaceae)	Triacontanol (C1), 2-carboxy-1,1dimethylpyrrolidine (C2)	Stem, flower, fruit	Insecticidal and oviposition inhibitory activity	Upadhyay et al. 2006
			Hypolipidemic	Purohit and Vyas 2005; Goyal and Grewal 2003
			Antidiabetic, antioxidative	Yadav et al. 1997
Calotropis procera (Asclepiadaceae)	Laticifer proteins, calotropagenin, calotropin, rutin	Dried latex, leaves,	Hepatoprotective by acting as antioxidants	Olaleye and Rocha 2008; Padhy et al. 2007

(*continued*)

Table 1.1 (continued)

Plant species (Family)	Bioactive compounds	Plant part used	Biological activities	References
		flowers, root, bark	Anticancer and cytotoxic	Soares de Oliveira et al. 2007; Choedon et al. 2006
			Larvicidal	Singh et al. 2005
			Anthelmintic activity	Iqbal et al. 2005
			Spasmolytic effect	Iwalewa et al. 2005
Cassia tora (Caesalpiniaceae)	Emodin, chrysophanol, chryso-obtusin, obtusifolin, physcoin, cassiaside, aloe-emodin, emodin, torachrysone, toralactone	Seeds, leaves	Improvement of skin viscoelastic properties	Ahshawat et al. 2008
			Inhibitory activity on protein glycation and aldose reductase	Jang et al. 2007, Lee et al. 2006
			Oestrogenic and anti-oestrogenic	El-Halawany 2007
			Hypolipidemic	Cho et al. 2007; Patil et al. 2004
			Improve serum lipid status in type II diabetic subjects	Cho et al. 2005
			Antifungal	Kim et al. 2004
			Used in treatment of plaque and caries	Hebbar et al. 2004
			Antinociceptive	Chidume et al. 2002
Citrullus colocynthis (Cucurbitaceae)	Cucurbitacins	Fruit	Low glycemic index food	Robert et al. 2008
			Larvicidal	Rahuman and Venkatesan 2008
			Antimicrobial efficiency	Paul 2008
			Hypolipidemic	Daradka et al. 2007
Commiphora wightii (Burseraceae)	E- and Z- guggalsterone	Gum resin	Hypolipidemic agent in clinical practice, potential anti-dementia drug	Saxena et al. 2007
			Inhibits tumour cell proliferation, used traditionally to treat obesity, diabetes, atherosclerosis and osteoarthritis	Shishodia et al. 2007
Ephedra foliata (Ephedraceae)	Ephedrine, pseudoephedrine	Stem	Neuropharmacological	Caveney et al. 2001
			Used in anti-asthmatic compound	Rogers et al. 1997

Haloxylon recurvum (Chenopodiaceae)	Halosterols A and B, Haloxysterols A-D	Whole plant	Chymotrypsin enzyme inhibitory	Hussain et al. 2006
			Cholinesterase inhibition	Ahmed et al. 2006b
Haloxylon salicornicum (Chenopodiaceae)	Haloxylines A and B	Whole plant	Antifungal and cholinesterase enzymes inhibitory potentials	Ferheen et al. 2005
Lepidagathis cristata (Acanthaceae)	Cristatin A, cycloartenol and stigmasta-5,11(12)-diene-3-β-ol	Whole plant	Immunosuppressive	Ravikanth et al. 2001
Malva parvifolia (Malvaceae)	CW1, CW2	Leaves, roots	Antimicrobial	Tadeg et al. 2005; Wang et al. 2001; Wang and Bunkers 2000
			Anti-inflammatory	Shale et al. 2005
			Active against multidrug resistant *Mycobacterium tuberculosis*	Jimenez-Arellanes et al. 2003
Maytenus emarginata (Celastraceae)	Emarginatine F [1] and emarginatine G [2]	Entire plant	Strong cytotoxicity against several experimental cancer lines	Kuo et al. 1994, 1990
Phyllanthus amarus (Euphorbeaceae)	Phyllanthin	Leaves, seeds	Hepatoprotective	Faremi et al. 2008; Khatoon et al. 2006
			Larvicidal	Rahuman et al. 2008
			Treatment of drug induced nephrotoxicity	Adeneye and Benebo 2008
			Hypotensive	Amaechina and Omogbai 2007
			Antimicrobial	Okigbo and Igwe 2007
			Radioprotective	Harikumar and Kuttan 2007
			Antinociceptive	Santos et al. 2000
Phyllanthus emblica syn Emblica officinalis (Euphorbeaceae)	Emblicanin-A, B, gallic acid, ellagic acid, pyrogallol, apigenin 7–0-(6"butyryl-β-glucopyranoside), quercetin, putranjivain A	Fruit	Antimicrobial, virucidal action against HIV-1NL4.3 and HPV infections	Talwar et al. 2008; Srikumar et al. 2007
			Hepatoprotective	Panchabhai et al. 2008
			Cancer chemopreventive	Pinmai et al. 2008; Arulkumaran et al. 2007; Sandhya and Mishra 2006; Deep et al. 2005
			Potent antioxidant in various conditions like prevention of	Saito et al. 2008; Yokozawa et al. 2007; Mythilypriya et al. 2007

(continued)

Table 1.1 (continued)

Plant species (Family)	Bioactive compounds	Plant part used	Biological activities	References
			age-related renal disease and in arthritis	
			Memory improvement and reversal of memory deficits	Vasudevan and Parle 2007a, 2007b
			Preventive role in prefibrogenesis of liver	Mir et al. 2007
			Healing activity on infected wound in form of TRIPHALA	Kumar et al. 2008
			Radioprotective effect	Singh et al. 2006
			Effective for hypercholesterolemia and prevention of atherosclerosis	Kim et al. 2005
Phyllanthus fraternus (Euphorbeaceae)	E,E-2,4-octadienamide, E,Z-2,4-decadienamide, niruriside, phyllanthin	Whole plant	Hepatoprotective	Sailaja and Setty 2006; Khatoon et al. 2006; Ahmed et al. 2002
			Antinociceptive	Catapan et al. 2000; Santos et al. 2000
			Antiplasmodial	Sittie et al. 1998
Salvadora persica (Salvadoraceae)	Four benzylamides of which N-benzyl-2-phenylacetamide is pharmacologically important	Stem	Caries prevention	Sofrata et al. 2007; Khalessi et al. 2004; Darmani et al. 2006
			Antiplasmodial	Ali et al. 2002
			Anticonvulsant and sedative effects	Monforte et al. 2002
			Antiulcer	Sanogo et al. 1999
			Hypolipidemic	Galati et al. 1999
Sida cordifolia (Malvaceae)	Indoloquinoline alkaloid-cryptolepine, 1,2,3,9-tetrahydro-pyrrolo [2,1-β] quinazolin-3-ylamine	Aerial parts, mainly leaves	Antipyretic and antiulcerogenic	Philip et al. 2008
			Chemotherapeutic agent for treatment of osteorsarcoma	Matsui et al. 2007
			Antiinflammatory and analgesic	Sutradhar et al. 2007; Franzotti et al. 2000
			Liver regeneration	Silva et al. 2006
			Cardiovascular activity, cause hypotension and bradycardia	Medeiros et al. 2006
			Depressive activity on CNS	Franco et al. 2005

			Effects in neurodegenerative diseases such as Parkinson's, Alzheimer's, loss of memory	Auddy et al. 2003
			Hyperglycemic	Kanth and Diwan 1999
Solanum xanthocarpum (Solanaceae)	Solasodine, carpesterol, steroidal glycosides, diosgenin, β-sitosterol	Fruits, root	Antifungal	Singh et al. 2007; Dabur et al. 2004
			Antimalarial	Mohan et al. 2007
			Hypoglycemic	Kar et al. 2006
			Inhibits growth of Reo virus	Jabbar et al. 2004
			Potential larvicide	Mohan et al. 2005
			Used in bronchial asthma	Govindan et al. 2004
			Antinociceptive activity	Rahman et al. 2003
			Potential molluscicidal	Wei et al. 2002
Suaeda fruticosa (Chenopodeaceae)	Coumarins, histamine	Aerial part	Hypoglycemic effect	Benwahhoud et al. 2001
			Hypolipidemic activity	Bennani-Kabchi et al. 1999
Tephrosia purpurea (Fabaceae)	Glabratephrin, (+)-tephrorins A and B, tephrsone	Leaves	Antihyperglycemic and antihyperlipidemic	Pavana et al. 2007
			Wound healing potential, hepatoprotective, antiulcer, antibacterial	Lodhi et al. 2006
			Immunomodulatory	Damre et al. 2003
			Inhibits haemolysis of erythrocytes	Gokhale et al. 2000
			Cancer chemopreventive	Khan et al. 2001; Saleem et al. 2001; Chang et al. 2000
Tribulus terrestris (Zygophyllaceae)	Protodioscin, hecogenin-3-O-β-d-glucopyranosyl (1→4)- β -D-galactopyranoside, eight steroid saponins TTS-8 to TTS-15, tribulosin, β-sitosterol-d-glucoside	Fruits, leaves, root	Antimicrobial	Al-Bayati and Al-Mola 2008
			Male erectile dysfunction	Gauthaman and Ganesan 2008
			Attenuates apoptosis in cardiocyte	Sun et al. 2008
			Hypoglycemic and hypolipidemic	El-Tantaway and Hassanin 2007; Li et al. 2002
			Protective role in induced nephrotoxicity	Kavitha and Jagadeesan 2006
			Analgesic	Heidari et al. 2007

(*continued*)

Table 1.1 (continued)

Plant species (Family)	Bioactive compounds	Plant part used	Biological activities	References
			Increase melanocyte stimulating hormone (MSH) expression	Yang et al. 2006; Deng et al. 2002
			Antifungal	Zhang et al. 2005
			75% inhibition of PPR and Reo virus	Jabbar et al. 2004
			Antihypertensive and vasodilator	Phillips et al. 2006; Sharifi et al. 2003
			Anticancerous	Neychev et al. 2007; Yang et al. 2005; Sun et al. 2004, 2003
			Anthelmintic	Deepak et al. 2002
Tridax procumbens (Fabaceae)	Lupeol, sitosterol	Leaves	Antiinflammatory	Margaret et al. 1998
			Exerts a direct pro-healing effect through release of adrenal steroid	Diwan et al. 1983
Withania coagulans (Solanaceae)	Withaferin, withanolide, withcoagin	Fruits	Hypolipidemic	Hemalata et al. 2006
Withania somnifera (Solanaceae)	Withaferin A, withanolide A, withanoside IV, withanoside VI, sominone, ashwagandhanolide, sterole glycosyltransferases	Root, leaves	Cancer chemopreventive	Stan et al. 2008; Malik et al. 2007; Widodo et al. 2007; Senthil et al. 2007; Subbaraju et al. 2006
			Useful in several neurodegenerative diseases	Srinivasan et al. 2007; Senthilnathan et al. 2006; Ichikawa et al. 2006
			Cardioprotection	Tohda 2008; Kulkarni and Dhir 2008; Kuboyama et al. 2006
			Mood stabliser	Mohanty et al. 2008
			Normalise hyperglycemia in diabetes mellitus type II	Gupta and Rana 2007
			Useful in anxiety and insomnia	Anwer et al. 2008
			Useful in arthritis treatment	Kumar and Kalonia 2007; Khan et al. 2006; Khanna et al. 2007; Rasool and Varalakshmi 2006a; 2007
			Antimalarial	Dikasso et al. 2006
			Prevention of glycation induced pathogenesis in diabetes mellitus and aging	Babu et al. 2007

			Stress management	Madina et al. 2007
			Antifungal activity by inhibiting aflatoxin B production	Krishnamurthy and Shashikala 2006; Girish et al. 2006
			AchE inhibitory	Vinutha et al. 2007
			Immunomodulatory	Spelman et al. 2006; Rasool and Varalakshmi 2006b
			Hypocholesteremic and antioxidant	Visavadiya and Narasimhacharya 2007
			Treatment of osteoporosis	Nagareddy and Lakshmana 2006
			Antidote	Machiah et al. 2006
Zizyphus mauritiana (Rhamnaceae)	Betulinic acid	Fruits	Hepatoprotective and immunomodulatory	Adhvaryu et al. 2007
			Anticacerous activity	Mukherjee et al. 2006

Fig. 1.2 Fruits and leaves of *Phyllanthus emblica*

aspera, *Balanites aegyptiaca*, *Barleria prionitis*, *Boerhaavia diffusa*, *Commiphora wightii*, *Phyllanthus emblica syn. Emblica officinalis*, *Phyllanthus amarus*, *Tribulus terrestris* and *Withania somnifera* are used in the Indian system of traditional medicine (Ramawat and Goyal 2008). Plants like *C. wightii* and *W. somnifera* are used in hundreds of tons, while fruits of *P. emblica* (Fig. 1.2) are used in thousands of tons and rank first in consumption in Indian traditional medicine (Ramawat and Goyal 2008; Anonymous 2001).

The diverse biological activities are due to the presence of a wide array of bioactive molecules (Fig. 1.3), including simple alkaloids, anthraquinones (physcion, emodin, chrysophanol), naphthopyrone glucosides (cassiaside, rubrofusarin-6–0-β-D-gentiobioside), phenolics, rotenoids [boeravinones G (1) and H (2)], saponins, steroids (β-sitosterol, carpesterol, ecdysterone), and terpenes (Table 1.1). Exploration of the chemical constituents of the plants and pharmacological screening may provide the basis for lead compounds in the development of novel agents. Indeed, herbs have already provided us with some of the most important life saving drugs used in modern medicine (Goyal et al. 2008).

1.6 Methods of Propagation

Conventional vegetative propagation methods have been developed mostly in woody plants like *Commiphora wightii*, *Ziziphus mauritiana*, *Prosopis cineraria*, etc. Some of these plants, e.g. *P. cineraria* (Ramawat and Nandwani 1991), are very

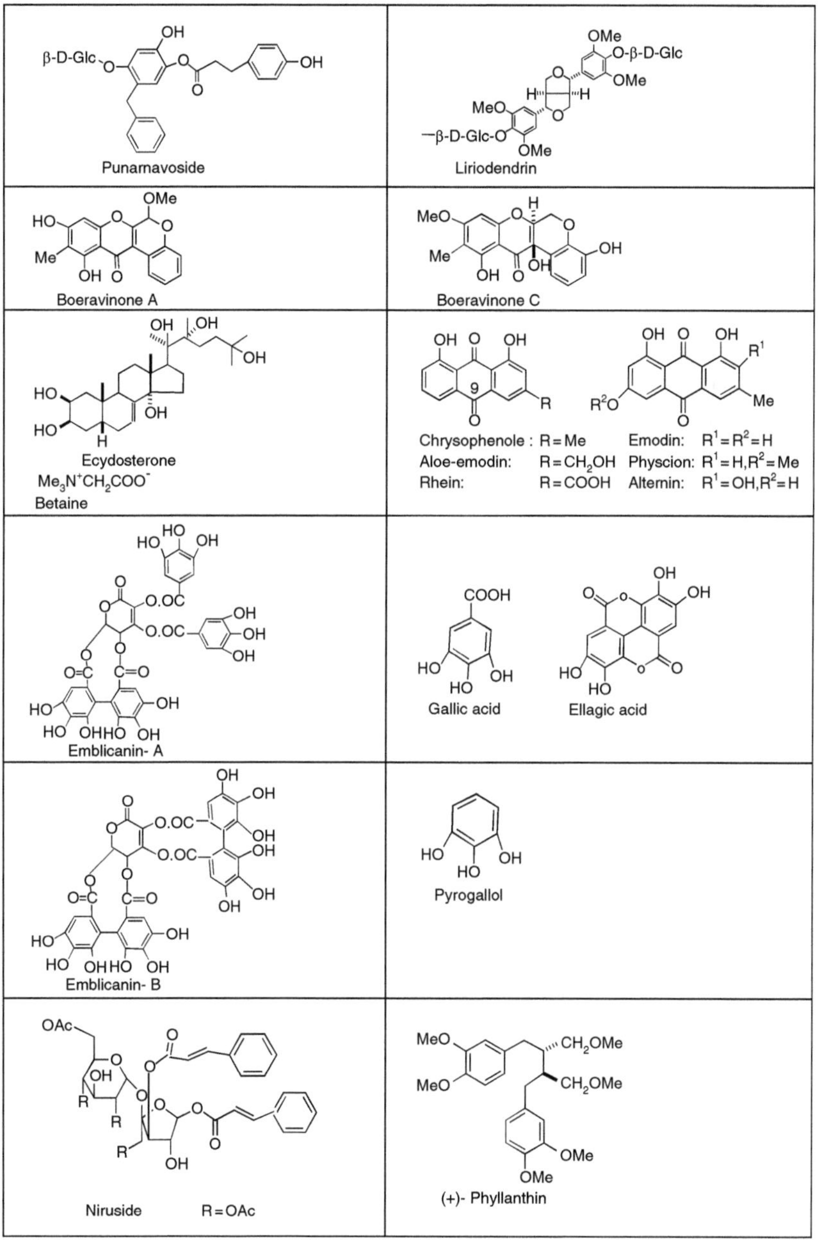

Fig. 1.3 Chemical structures of bioactive molecules of selected medicinal plants

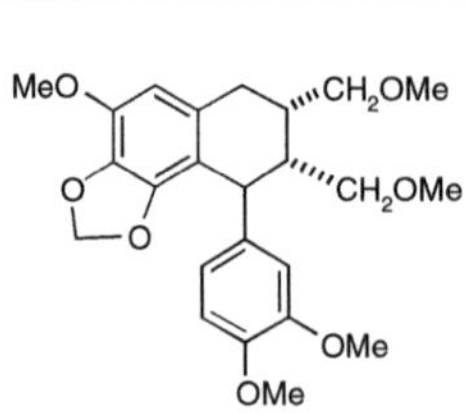

Hypophyllanthin

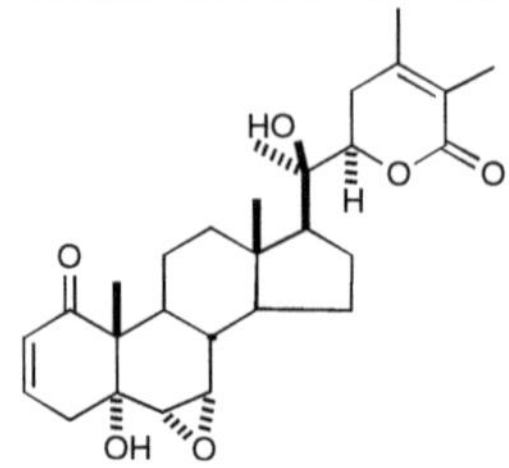

Withanolide- A

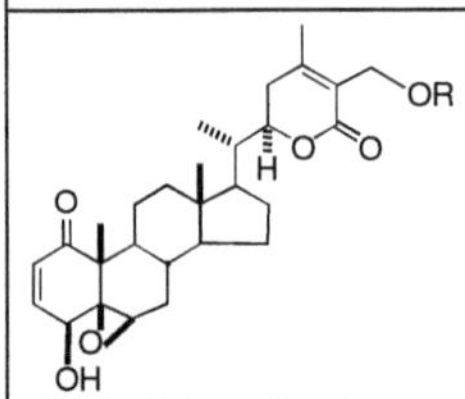

Withaferin-A: R = H
Sitoinodoside-IX: R = Glc
Sitoinodoside-X: R = Glc (6-o-palmitoyl)

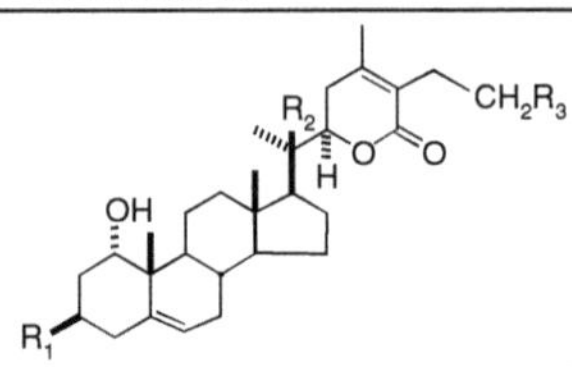

	R1	R2	R3
Withanoside-IV:	Glc(1→6)-GLc	H	OH
Withanoside–VI:	Glc(1→6)-GLc	OH	H
Coagulin	Glc	OH	H

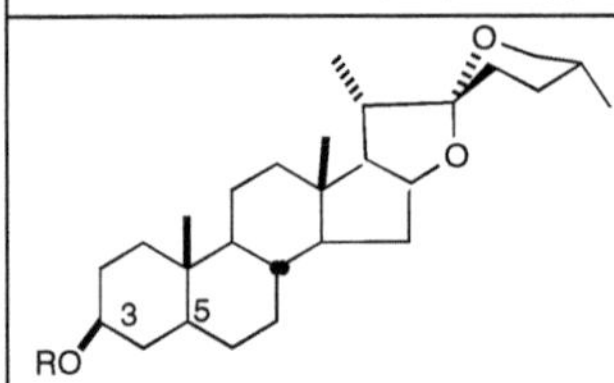

Tribulosin: 5αH, R = Gal-[(2→ 1) Rha-]-(4→1) - Glc-(2→1) –Xyl-(3→ 1)-Xyl
Isoterrestrosin-B: 5βH, R = Rha-(1→ 2) -[Glc - (1→4)]- Gal

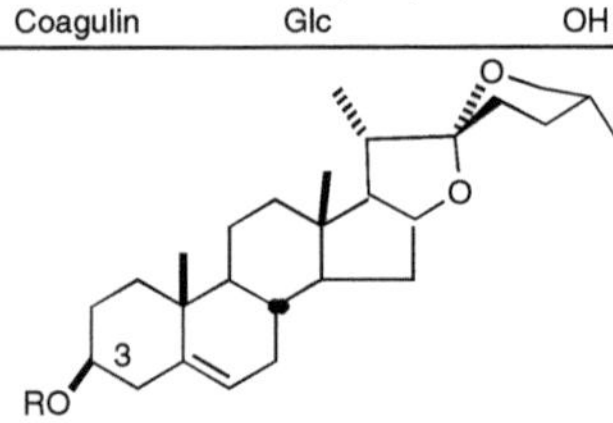

Diosgenin: R = H
Dioscin : R = Glc -(4 →1) -Man -(6- deoxy) -(2→ 1) - Man - (6- deoxy)

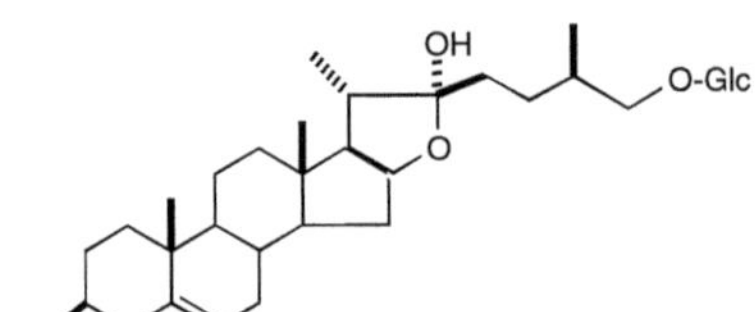

Protodioscin: R = Glc- (1→4) -Man -(6- deoxy) - (2 →1) -Man-(6- deoxy)

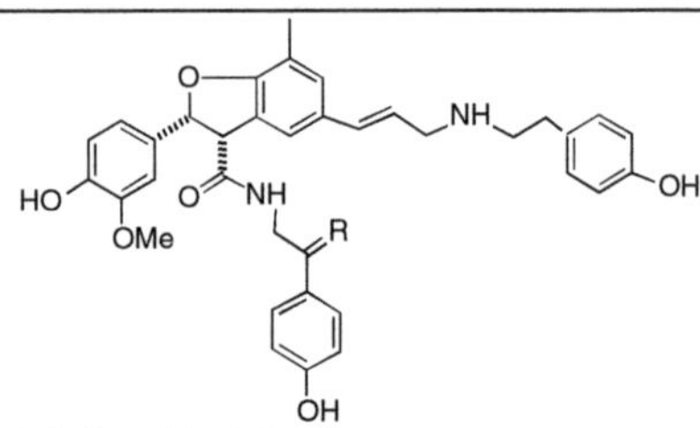

Tribulusamide- A: R = H2
Tribulusamide- B: R = O

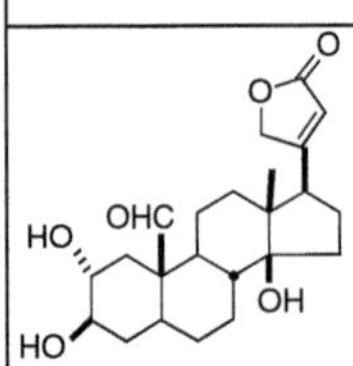

Calotropagenin

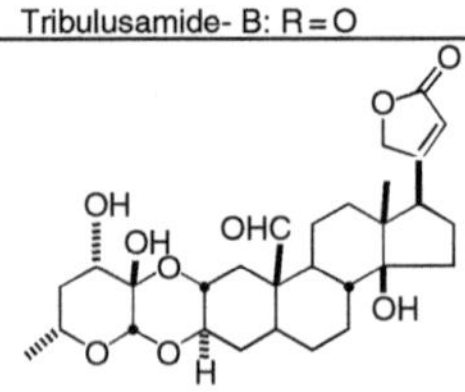

Calotropin

Fig. 1.3 (continued)

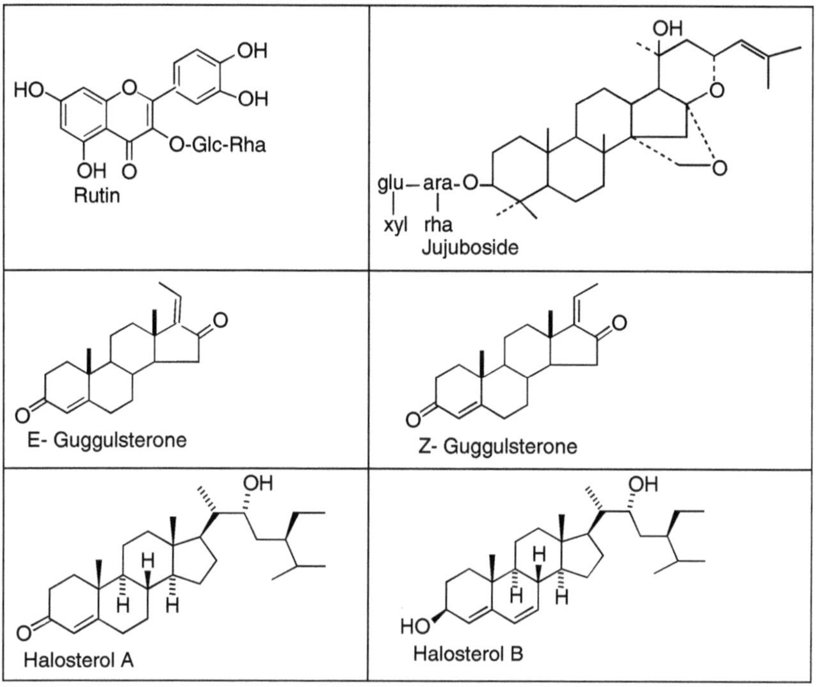

Fig. 1.3 (continued)

difficult to root. In medicinally important plants like *C. wighti*, rooting of up to 70% of stem cuttings can be obtained with the application of auxin and selection of suitable material (Singh et al. 1989). *Z. mauritiana* improved for fruit quality is propagated by grafting (Pareek 2001). Propagation of improved stock is a very difficult and challenging task in arid regions. In the case of *Z. mauritiana*, wild stock with in situ grafting was found to be more successful than nursery grafting and transplanting. The seed-produced plants are not true to the parent stock, and stem cutting of superior selections do not establish properly, leaving grafting on wild stock as the only means of propagating the selected material (Pareek 2001). Other methods, such as air-layering and grafting, are successful with only a few species, and then only on a limited scale. These techniques are adversely affected by variation between and within species, making biotechnological methods an attractive alternative.

1.7 Biotechnological Approaches

Medicinal plant biotechnology offers many novel opportunities and techniques to conserve, propagate, improve and utilise medicinal plants and herbs for the welfare of human beings. Since the emergence of the concept of cellular

Table 1.2 Plant species and their in vitro response on different media

Plant species	Medium	Plant part used	Regeneration	Reference
Achyranthes aspera	MS + 2,4-D (1.0–2.0 mg/l) + NAA (0.5 mg/l)	Leaf	Callus	Kayani et al. 2008
Aristolochia bracteolata	MS + BA (4.4–35.5 μM) + kinetin (0.45–9.2 μM) + IAA (0.57–5.8 μM)	Nodes	Shoot regeneration	Rameshree et al. 1994
	MS + kinetin (0.45–23 μM) + NAA (0.46–9.2 μM) + IAA (0.5–5.7 μM)	Leaves, node	Callus	Rameshree et al. 1994
Boerhaavia diffusa	MS + BAP (1.5 mg/l) + NAA (0.5 mg/l)	Shoot tip and nodal explants	Multiple shoot regeneration	Roy 2008
	MS + NAA (1.0 mg/l) + BAP (1.0 mg/l)	Callus	Shooting	Gupta et al. 2004
Balanites aegyptiaca	MS + BAP (2.5 mg/l) + NAA (0.1 mg/l)	Axillary bud	Shooting	Ndoye et al. 2003
	MS + IBA (20 mg/l)		Rooting	Ndoye et al. 2003
	MS + BAP (0.45 μM)	In vitro shoots axillary meristems	Shooting	Rathore et al. 2004
Capparis decidua	MS + NAA (0.1 mg/l) + BAP (5.0 mg/l) + additives	Node	Multiple shoots	Deora and Shekhawat 1995
Citrullus colocynthis	MS + BA (25 μM)	Cotyledon	Organogenic calli	Dabauza et al. 1997
	MS + BA (0.5 μM)	Organogenic calli	Shooting	Dabauza et al. 1997
	MS + IBA (2.5/5.0 μM)	In vitro shoots	Rooting	Dabauza et al. 1997
Cocculus pendulus	$^1/_4$ MS + IAA (0.1 mg/l) + kinetin (2.97 mg/l)	Callus	Shoots	Gaur et al. 1995
Commiphora wightii	MS modified + 2,4,5-T (0.25 mg/l) + kinetin (0.1 mg/l)	Zygotic embryo	Somatic embryogenesis	Kumar et al. 2003
Haloxylon recurvum	MS + BAP (4.0 μM) + additives	Nodal explant and leaves	Multiple shoots	Dagla and Shekhawat 2005
	$^1/_2$ MS + IBA (4.0 μM)+ activated charcoal (100 mg/l)	In vitro shoots	Rooting	Dagla and Shekhawat 2005
Maytenus emarginata	MS + IAA (0.1 mg/l) + BAP (2.5 mg/l)	Shoot segments	Multiple shoots	Rathore et al. 1992a
	MS + IBA (25 mg/l)	In vitro shoots	Rooting	Rathore et al. 1992a

Phyllanthus amarus	MS + kinetin (0.1 mg/l) + IAA (0.1 mg/l)	Shoot tips	Direct shoots regeneration	Bhattacharyya and Bhattacharya 2001
	MS only	Alginate-encapsulated shoot tips	Plant regeneration	Singh et al. 2006
	$^{1}/_{4}$ liquid MS	Leaves	Hairy roots	Bhattacharyya and Bhattacharya 2004
Phyllanthus emblica syn Emblica officinalis	MS + BAP (4.44 μM/l) + IBA (2.46 μM/l)	Nodal explants	Shooting	Goyal and Bhadauria 2008
	BMS + 2,4-D (1–4 mg/l) + kinetin (0.05 mg/1) and BMS + NAA (1–4 mg/l) + kinetin (0.05 mg/l)	Mature embryo explants	Callus	Tyagi and Govil 1999
Salvadora persica	MS + BAP (4.0 mg/l) + IAA (0.5 mg/l) + adenine sulphate (40 mg/l) + glutamine (100 mg/l) + thiamine HCl (10 mg/l)	Cotyledonary nodes	Shooting	Mathur et al. 2002
	$^{1}/_{2}$ MS + IBA (3.0 mg/l)	Shoots	Rooting	Mathur et al. 2002
Sida cordifolia	MS + BAP (2.0 mg/l) + NAA (0.5 mg/l) + adenine sulphate (1.0 mg/l) + 10% (v/v) coconut milk	Nodal explants	Multiple shoots	Sivanesan and Jeong 2007
	$^{1}/_{2}$ MS + IBA (2.0 mg/l)	In vitro regenerated shoots	Rooting	Sivanesan and Jeong 2007
Tephrosia purpurea	MS + PGR	Root, leaves, stem	Callus	Zafar and Mujeeb 2002
Tribulus terrestris	MS + NAA (0.2 mg/l) + BAP (0.5 mg/l) + glutamine (0.5 mg/l)	Cotyledonary leaves along with epicotyl segment	Shoots and roots	Ali et al. 1997
Withania somnifera	MS + 2,4-D (2 mg/l) + kinetin (0.2 mg/l)	Root, cotyledonary leaf segments	Callus	Rania et al. 2003
	MS + 2,4-D (2 mg/l) + kinetin (0.2 mg/l)	Callus	Shoot regeneration	Rania et al. 2003
	MS + IBA (2.0 mg/l)	Shoots	Rooting	Rania et al. 2003
	MS + BA (4.4 M) + 2,4-D (2.3 M)	Shoot tips	Shoot multiplication	Sen and Sharma 1991
		Hypocotyl	15–20 shoots	Mathur et al. 1995

(*continued*)

Table 1.2 (continued)

Plant species	Medium	Plant part used	Regeneration	Reference
Ziziphus mauritiana	MS + BAP (2.5 mg/l) + IAA (0.1 mg/l) + KNO_3 (3,800 mg/l)			
	$^1/_2$ MS + IBA (0.5 mg/l) + kinetin (0.05 mg/l)	Shoot	Rooting	Goyal and Ayra 1985
Ziziphus nummularia	MS + kinetin (2.5 mg/l)	Cotyledonary hypocotyls	8–18 adventitious shoot induction	Mathur et al. 1994
	MS + kinetin (10 mg/l) + IAA (0.1–0.5 mg/l) + KNO_3 (3,800 mg/l)	Hypocotyl	18 shoots	Mathur et al. 1993
	White's liquid + IBA (25 mg/l) + 48 h pulse then White's HF semi-solid	Regenerated shoots	Rooting	Rathore et al. 1992b

totipotency, contemporary developments in the area of plant tissue culture to devise methods allowing rapid and year-round multiplication of desired genotypes, production of pathogen-free stock, raising uniform clones from highly heterozygous plants, monitoring production of useful natural products in vitro, propagating plants (including genetically transformed plants) with changed/altered genotypes have added new dimensions to this field, pushing this science towards the realm of technological application by providing a basis for modern biotechnology (Ramawat et al. 2004; Ramawat and Goyal 2008). Micropropagation methods have been reported for around 18 species (Table 1.2), while many more have been cultured for various purposes including production of secondary metabolites, e.g. *Commiphora wightii* cell cultures (Mathur and Ramawat 2007), *Pueraria tuberosa* (Goyal and Ramawat 2008a, 2008b) and *Cayratia trifolia* (Roat and Ramawat 2009). Of these, only six are woody trees and the rest are herbs. These methods still need improvement to make them commercially viable programmes.

1.8 Conclusions

It is evident from the account above that the flora of Rajasthan is rich in medicinal plants. The recent surge in publications describing the biological activities of these plants demonstrates the potential for their application in many different areas of human medicine. This is the first compilation of its kind to describe the biological activities of such a large number of desert plants together with biotechnological input. Some of these plants are already well established in the Indian system of medicine, therefore it is desirable to study their reproductive biology and biological activities, and to develop methods for their conservation. Although these plants can be very difficult to work with in terms of propagation and improvement, they possess a gene pool representing some very useful characteristics, such as drought resistance, salinity resistance and the ability to cope with high temperatures. Thus, there is much scope to explore the genetic and molecular biology of these plants to better understand the mechanisms underlying desert plant survival.

Acknowledgements This work was supported by financial assistance from UGC-DRS under the special assistance programme for medicinal plant research and the DST-FIST programme for infrastructure development to K.G.R. J.A. and S.G. thank CSIR, New Delhi for granting of a JRF and RA fellowship, respectively.

References

Abdul Rahuman A, Gopalakrishnan G, Venkatesan P, Geetha K (2008) Isolation and identification of mosquito larvicidal compound from *Abutilon indicum* (Linn.) Sweet. Parasitol Res 102:981–988

Adeneye AA, Benebo AS (2008) Protective effect of the aqueous leaf and seed extract of *Phyllanthus amarus* on gentamicin and acetaminophen-induced nephrotoxic rats. J Ethanopharmacol 23:318–323
Aderbauer B, Clausen PH, Kershaw O, Melzig MF (2008) In vitro and in vivo trypanocidal effect of lipophilic extracts of medicinal plants from Mali and Burkina Faso. J Ethnopharmacol 119:225–231
Adhvaryu MR, Reddy N, Parabia MH (2007) Effects of four Indian medicinal herbs on isoniazid-, rifampicin- and pyrazinamide-induced hepatic injury and immunosuppression in guinea pigs. World J Gastroenterol 13:3199–3205
Agrawal A, Srivastava S, Srivastava MM (2004) Antifungal activity of *Boerhavia diffusa* against some dermatophytic species of *Microsporum.* Hindustan Antibiot Bull 45–46:1–4
Agunu A, Yusuf S, Andrew GO, Zezi AU, Abdurahman EM (2005) Evaluation of five medicinal plants used in diarrhoea treatment in Nigeria. J Ethnopharmacol 101:27–30
Ahmed B, al-Howiriny TA, Mathew R (2002) Antihepatotoxic activity of *Phyllanthus fraternus.* Pharmazie 57:855–856
Ahmed-Belkacem A, Macalou S, Borrelli F, Capasso R, Fattorusso E, Taglialatela-Scafati O, Di Pietro A (2007) Nonprenylated rotenoids, a new class of potent breast cancer resistance protein inhibitors. J Med Chem 50:1933–1938
Ahmed E, Imran M, Malik A, Ashraf M (2006a) Antioxidant activity with flavonoidal constituents from *Aerva persica.* Arch Pharm Res 29:343–347
Ahmed E, Nawaz SA, Malik A, Choudhary MI (2006b) Isolation and cholinesterase-inhibition studies of sterols from *Haloxylon recurvum.* Bioorg Med Chem Lett 16:573–580
Ahmed M, Amin S, Islam M, Takahashi M, Okuyama E, Hossain CF (2000) Analgesic principle from *Abutilon indicum.* Pharmazie 55:314–316
Ahshawat MS, Saraf S, Saraf S (2008) Preparation and characterization of herbal creams for improvement of skin viscoelastic properties. Int J Cosmet Sci 30:183–193
Al-Bayati FA, Al-Mola HF (2008) Antibacterial and antifungal activities of different parts of *Tribulus terrestris* L. growing in Iraq. J Zhejiang Univ Sci B 9:154–159
Ali G, Mughal MH, Srivastava PS, Iqbal M (1997) Micropropagation of *Tribulus terestris* L., an important medicinal plant. J Plant Biol 40:202–205
Ali H, König GM, Khalid SA, Wright AD, Kaminsky R (2002) Evaluation of selected Sudanese medicinal plants for their in vitro activity against hemoflagellates, selected bacteria, HIV-1-RT and tyrosine kinase inhibitory, and for cytotoxicity. J Ethnopharmacol 83:219–228
Amaechina FC, Omogbai EK (2007) Hypotensive effect of aqueous extract of the leaves of *Phyllanthus amarus* Schum and Thonn (Euphorbiaceae). Acta Pol Pharm 64:547–552
Anonymous (1980) Firewood crops: shrubs and tree species for energy production. National Academy of Science, Washington DC
Anonymous (2001) Demand study for selected medicinal plants. Centre for Research, Planning and Action, New Delhi, India
Anwer T, Sharma M, Pillai KK, Iqbal M (2008) Effect of *Withania somnifera* on insulin sensitivity in non-insulin-dependent diabetes mellitus rats. Basic Clin Pharmacol Toxicol 102:498–503
Arora S, Kaur K, Kaur S (2003) Indian medicinal plants as a reservoir of protective phytochemicals. Teratog Carcinog Mutagen 1:295–300
Arya S, Toky OP, Harris SM, Harris PJC (1992) *Tecomella undulata* (Rohida) A valuable tree of Thar desert. Int Tree Corps J 7:141–147
Arulkumaran S, Ramprasath VR, Shanthi P, Sachdanandam P (2007) Alteration of DMBA-induced oxidative stress by additive action of a modified indigenous preparation – Kalpaamruthaa. Chem Biol Interact 167:99–106
Auddy B, Ferreira M, Blasina F, Lafon L, Arredondo F, Dajas F, Tripathi PC, Seal T, Mukherjee B (2003) Screening of antioxidant activity of three Indian medicinal plants, traditionally used for the management of neurodegenerative diseases. J Ethnopharmacol 84:131–138
Benwahhoud M, Jouad H, Eddouks M, Lyoussi B (2001) Hypoglycemic effect of *Suaeda fruticosa* in streptozotocin-induced diabetic rats. J Ethnopharmacol 76:35–38

Babu PV, Gokulakrishnan A, Dhandayuthabani R, Ameethkhan D, Kumar CV, Ahamed MI (2007) Protective effect of *Withania somnifera* (Solanaceae) on collagen glycation. Comp Biochem Physiol B Biochem Mol Biol 147:308–313

Bhandari MM (1974) Native resources used as famine foods in Rajasthan. Econ Bot 28:73–81

Bhandari MM (1999) Flora of the Indian desert. MPS Repros, Jodhpur, Rajasthan, India

Bharali R, Azad MR, Tabassum J (2003) Chemopreventive action of *Boerhaavia diffusa* on DMBA-induced skin carcinogenesis in mice. Indian J Physiol Pharmacol 47:459–464

Bhattacharya R, Bhattacharya S (2001) High frequency in vitro propagation of *Phyllanthus amarus* Schum. & Thom. by shoot tip culture. Indian J Exp Biol 39:1184–1187

Bhattacharya R, Bhattacharya S (2004) Development of a potent in vitro source of *Phyllanthus amarus* roots with pronounced activity against surface antigen of hepatitis B virus. In Vitro Cell Dev Biol Plant 40:504–509

Bennani-Kabchi N, El Bouayadi F, Kehel L, Fdhil H, Marquié G (1999) Effect of *Suaeda fruticosa* aqueous extract in the hypercholesterolaemic and insulin-resistant sand rat. Therapie 54:725–730

Borrelli F, Milic N, Ascione V, Capasso R, Izzo AA, Capasso F, Petrucci F, Valente R, Fattorusso E, Taglialatela-Scafati O (2005) Isolation of new rotenoids from *Boerhaavia diffusa* and evaluation of their effect on intestinal motility. Planta Med 71:928–932

Borrelli F, Ascione V, Capasso R, Izzo AA, Fattorusso E, Taglialatela-Scafati O (2006) Spasmolytic effects of nonprenylated rotenoid constituents of *Boerhaavia diffusa* roots. J Nat Prod 69:903–906

Catapan E, Otuki MF, Viana AM, Yunes RA, Bresciani LF, Ferreira J, Santos AR, Calixto JB, Cechinel-Filho V (2000) Pharmacological activity and chemical composition of callus culture extracts from selected species of *Phyllanthus*. Pharmazie 55:945–946

Caveney S, Charlet DA, Freitag H, Maier-Stolte M, Starratt AN (2001) New observations on the secondary chemistry of world *Ephedra* (Epheraceae). Am J Bot 88:1199–1208

Chaubal R, Mujumdar AM, Puranik VG, Deshpande VH, Deshpande NR (2003) Isolation and X-ray study of an anti-inflammatory active androstene steroid from *Acacia nilotica*. Planta Med 69:287–288

Chaubal R, Pawar PV, Hebbalkar GD, Tungikar VB, Puranik VG, Deshpande VH, Deshpande NR (2005) Larvicidal activity of *Acacia nilotica* extracts and isolation of D-pinitol – a bioactive carbohydrate. Chem Biodivers 2:684–688

Chaubal R, Mujumdar AM, Misar A, Deshpande VH, Deshpande NR (2006) Structure–activity relationship study of androstene steroids with respect to local anti-inflammatory activity. Arzneimittelforschung 56:394–398

Chakrabarti R, Vasudeva RY (2006) *Achyranthes aspera* stimulates the immunity and enhances the antigen clearance in *Catla catla*. Int Immunopharmacol 6:782–790

Chakraborty A, Brantner A, Mukainaka T, Nobukuni Y, Kuchide M, Konoshima T, Tokuda H, Nishino H (2002) Cancer chemopreventive activity of *Achyranthes aspera* leaves on Epstein-Barr virus activation and two-stage mouse skin carcinogenesis. Cancer Lett 177:1–5

Chang LC, Chávez D, Song LL, Farnsworth NR, Pezzuto JM, Kinghorn AD (2000) Absolute configuration of novel bioactive flavonoids from *Tephrosia purpurea*. Org Lett 2:515–518

Chapagain BP, Saharan V, Wiesman Z (2008) Larvicidal activity of saponins from *Balanites aegyptiaca* callus against *Aedes aegypti* mosquito. Bioresour Technol 99:1165–1168

Chen JL, Blanc P, Stoddart CA, Bogan M, Rozhon EJ, Parkinson N, Ye Z, Cooper R, Balick M, Nanakorn W, Kernan MR (1998) New iridoids from the medicinal plant *Barleria prionitis* with potent activity against respiratory syncytial virus. J Nat Prod 61:1295–1297

Chidume FC, Kwanashie HO, Adekeye JO, Wambebe C, Gamaniel KS (2002) Antinociceptive and smooth muscle contracting activities of the methanolic extract of *Cassia tora* leaf. J Ethnopharmacol 81:205–209

Cho IJ, Lee C, Ha TY (2007) Hypolipidemic effect of soluble fiber isolated from seeds of *Cassia tora* Linn. in rats fed a high-cholesterol diet. J Agric Food Chem 55:1592–1596

Cho SH, Kim TH, Lee NH, Son HS, Cho IJ, Ha TY (2005) Effects of *Cassia tora* fiber supplement on serum lipids in Korean diabetic patients. J Med Food 8:311–318

Choedon T, Mathan G, Arya S, Kumar VL, Kumar V (2006) Anticancer and cytotoxic properties of the latex of *Calotropis procera* in a transgenic mouse model of hepatocellular carcinoma. World J Gastroenterol 12:2517–2522

Dabur R, Singh H, Chhillar AK, Ali M, Sharma GL (2004) Antifungal potential of Indian medicinal plants. Fitoterapia 75:389–391

Dabauza M, Bordas M, Salvador A, Roig L-A, Moreno V (1997) Plant regeneration and Agrobacterium-mediated transformation of cotyledon explants of *Citrullus colocynthis* (L.) Schrad. Plant Cell Rep 16:888–892

Dagla HR, Shekhawat NS (2005) In vitro multiplication of *Haloxylon recurvum* (Moq.) – a plant for saline soil reclamation. J Plant Biotechnol 3:155–160

Damre AS, Gokhale AB, Phadke AS, Kulkarni KR, Saraf MN (2003) Studies on the immunomodulatory activity of flavonoidal fraction of *Tephrosia purpurea*. Fitoterapia 74:257–261

Daradka A, Almasad MM, Qazan WS, El-Banna NM, Samara OH (2007) Hypolipidaemic effects of *Citrullus colocynthis* L. in rabbits. Pakistan J Biol Sci 10:2768–2771

Darmani H, Nasayr T, AL-Hiyasat AS (2006) Effects of extracts of miswak and derum on proliferation of Balb/C 3T3 fibroblasts and viability of cariogenic bacteria. Int J Dent Hyg 4:62–66

Deep G, Dhiman M, Rao AR, Kale RK (2005) Chemopreventive potential of Triphala (a composite Indian drug) on benzo(a)pyrene induced forestomach tumorigenesis in murine tumor model system. J Exp Clin Cancer Res 24:555–563

Deepak M, Dipankar G, Prashanth D, Asha MK, Amit A, Venkataraman BV (2002) Tribulosin and beta-sitosterol-D-glucoside, the anthelmintic principles of *Tribulus terrestris*. Phytomedicine 9:753–756

Deora NS, Shekhawat NS (1995) Micropropagation of *Capparis decidua* (Forsk.) Edgew. – a tree of arid horticulture. Plant Cell Rep 15:3–4

Deng Y, Yang L, An SL (2002) Effect of *Tribulus terrestris* L. decoction of different concentrations on tyrosinase activity and the proliferation of melanocytes. Di Yi Jun Yi Da Xue Xue Bao 22:1017–1019

Dewan A, Khana N, Gupta SC (1992) In vitro micropropagation of *Acacia nilotica* subsp. indica Brenan via cotyledonary nodes. Plant Cell Rep 12:18–21

Dikasso D, Makonnen E, Debella A, Abebe D, Urga K, Makonnen W, Melaku D, Kassa M, Guta M (2006) Anti-malarial activity of *Withania somnifera* L. Dunal extracts in mice. Ethiop Med J 44:279–285

Diwan PV, Tilloo LD, Kulkarni DR (1983) Steroid depressed wound healing and *Tridax procumbens*. Indian J Physiol Pharmacol 27:32–36

El-Halawany AM (2007) Estrogenic and antiestrogenic activities of *Cassia tora* phenolic constituents. Chem Pharm Bull Tokyo 55:1474–1482

El-Tahir A, Satti GM, Khalid SA (1999) Antiplasmodial activity of selected Sudanese medicinal plants with emphasis on *Acacia nilotica*. Phytother Res 13:474–478

El-Tantawy WH, Hassanin LA (2007) Hypoglycemic and hypolipidemic effects of alcoholic extract of *Tribulus alatus* in streptozotocin-induced diabetic rats: a comparative study with *T. terrestris* (Caltrop). Indian J Exp Biol 45:785–790

Fatima F, Khalid A, Nazar N, Abdalla M, Mohomed H, Toum AM, Magzoub M, Ali MS (2005) In vitro assessment of anti-cutaneous leishmaniasis activity of some Sudanese plants. Turkiye Parazitol Derg 29:3–6

Ferheen S, Ahmed E, Afza N, Malik A, Shah MR, Nawaz SA, Choudhary MI (2005) Haloxylines A and B, antifungal and cholinesterase inhibiting piperidine alkaloids from *Haloxylon salicornicum*. Chem Pharm Bull (Tokyo) 53:570–572

Faremi TY, Suru SM, Fafunso MA, Obioha UE (2008) Hepatoprotective potentials of *Phyllanthus amarus* against ethanol-induced oxidative stress in rats. Food Chem Toxicol 46:2658–2664

Franco CI, Morais LC, Quintans-Júnior LJ, Almeida RN, Antoniolli AR (2005) CNS pharmacological effects of the hydroalcoholic extract of *Sida cordifolia* L. leaves. J Ethnopharmacol 98:275–279

Franzotti EM, Santos CV, Rodrigues HM, Mourão RH, Andrade MR, Antoniolli AR (2000) Anti-inflammatory, analgesic activity and acute toxicity of *Sida cordifolia* L. (Malva-branca). J Ethnopharmacol 72:273–277

Galati EM, Monforte MT, Forestieri AM, Miceli N, Bader A, Trovato A (1999) *Salvadora persica* L.: hypolipidemic activity on experimental hypercholesterolemia in rat. Phytomedicine 6:181–185

Gamble C (1993) Timewalkers: the prehistory of global colonization. Sutton, Phoenix Mill

Gaur A, Singh AK, Suri SS, Ramawat KG (1995) Bud cultures and regeneration of plantlets in *Cocculus pendulus*, a woody medicinal plant. Gartenbau 60:69–72

Gauthaman K, Ganesan AP (2008) The hormonal effects of *Tribulus terrestris* and its role in the management of male erectile dysfunction – an evaluation using primates, rabbit and rat. Phytomedicine 15:44–54

Gilani AH, Shaheen F, Zaman M, Janbaz KH, Shah BH, Akhtar MS (1999) Studies on antihypertensive and antispasmodic activities of methanol extract of *Acacia nilotica* pods. Phytother Res 13:665–669

Girish KS, Machiah KD, Ushanandini S, Harish Kumar K, Nagaraju S, Govindappa M, Vedavathi M, Kemparaju K (2006) Antimicrobial properties of a non-toxic glycoprotein (WSG) from *Withania somnifera* (Ashwagandha). J Basic Microbiol 46:365–374

Gnoula C, Mégalizzi V, De Nève N, Sauvage S, Ribaucour F, Guissou P, Duez P, Dubois J, Ingrassia L, Lefranc F, Kiss R, Mijatovic T (2008) Balanitin-6 and -7: diosgenyl saponins isolated from *Balanites aegyptiaca* Del. display significant anti-tumor activity in vitro and in vivo. Int J Oncol 32:5–15

Gokhale AB, Dikshit VJ, Damre AS, Kulkarni KR, Saraf MN (2000) Influence of ethanolic extract of *Tephrosia purpurea* Linn. on mast cells and erythrocytes membrane integrity. Indian J Exp Biol 38:837–840

Gokhale AB, Damre AS, Kulkami KR, Saraf MN (2002) Preliminary evaluation of anti-inflammatory and anti-arthritic activity of *S. lappa, A. speciosa* and *A. aspera*. Phytomedicine 9:433–437

Govindan S, Viswanathan S, Vijayasekaran V, Alagappan R (2004) Further studies on the clinical efficacy of *Solanum xanthocarpum* and *Solanum trilobatum* in bronchial asthma. Phytother Res 18:805–809

Goyal BR, Goyal RK, Mehta AA (2007) Phyto-pharmacology of *Achyranthes aspera*: a review. Pharmacogn Rev 1:143–150

Goyal D, Bhadauria S (2008) In vitro shoot proliferation in *Emblica officinalis* var. Balwant from nodal explants. Ind J Biotechnol 7:394–397

Goyal R, Grewal RB (2003) The influence of teent (*Capparis decidua*) on human plasma triglycerides, total lipids and phospholipids. Nutr Health 17:71–6

Goyal S, Ramawat KG (2008a) Synergistic effect of morphactin on cytokinin-induced production of isoflavonoids in cell cultures of *Pueraria tuberosa* (Roxb. ex. Willd.) DC. Plant Growth Regul 55:175–181

Goyal S, Ramawat KG (2008b) Marked increase in isoflavonoids accumulation by ethrel in cell suspension cultures of *Pueraria tuberosa*. Acta Plant Physiol 30:849–853

Goyal S, Roat C, Ramawat KG (2008) Production of polyphenols through cell cultures: a new class of bioactive molecules. In: Arya ID, Arya S (eds) Utilization of biotechnology in plant sciences, FRI, Dehradun, India, pp 75–94

Goyal Y, Arya HC (1985) Tissue culture of desert trees: II Clonal multiplication of *Ziziphus* in vitro. J Plant Physiol 119:399–404

Girish KS, Machiah KD, Ushanandini S, Harish Kumar K, Nagaraju S, Govindappa M, Vedavathi M, Kemparaju K (2006) Antimicrobial properties of a non-toxic glycoprotein (WSG) from *Withania somnifera* (Ashwagandha). J Basic Microbiol 46:365–374

Gupta GL, Rana AC (2007) Protective effect of *Withania somnifera* dunal root extract against protracted social isolation induced behavior in rats. Indian J Physiol Pharmacol 51:345–353

Gupta RK, Srivastava A, Verma HN (2004) Callus culture and organogenesis in *Boerhaavia diffusa*: a potent antiviral protein containing plant. Physiol Mol Biol Plants 10:263–268

Gupta RS, Kumar P, Dixit VP, Dobhal MP (2000) Antifertility studies of the root extract of the *Barleria prionitis* Linn in male albino rats with special reference to testicular cell population dynamics. J Ethnopharmacol 70:111–117
Hamza OJ, Van den Bout-Van den Beukel CJ, Matee MI, Moshi MJ, Mikx FH, Selemani HO, Mbwambo ZH, Van der Ven AJ, Verweij PE (2006) Antifungal activity of some Tanzanian plants used traditionally for the treatment of fungal infections. J Ethnopharmacol 108:124–132
Harikumar KB, Kuttan R (2007) An extract of *Phyllanthus amarus* protects mouse chromosomes and intestine from radiation induced damages. J Radiat Res (Tokyo) 48:469–476
Hebbar SS, Harsha VH, Shripathi V, Hegde GR (2004) Ethnomedicine of Dharwad district in Karnataka, India-plants used in oral health care. J Ethnopharmacol 94:261–266
Heidari MR, Mehrabani M, Pardakhty A, Khazaeli P, Zahedi MJ, Yakhchali M, Vahedian M (2007) The analgesic effect of *Tribulus terrestris* extract and comparison of gastric ulcerogenicity of the extract with indomethacine in animal experiments. Ann N Y Acad Sci 1095:418–427
Hemalata S, Wahi AK, Singh PN, Chansouria JP (2006) Hypolipidemic activity of aqueous extract of *Withania coagulans* Dunal in albino rats. Phytother Res 20:614–617
Hussein G, Miyashiro H, Nakamura N, Hattori M, Kakiuchi N, Shimotohno K (2000) Inhibitory effects of Sudanese medicinal plant extracts on hepatitis C virus (HCV) protease. Phytother Res 14:510–516
Hussain S, Ahmed E, Malik A, Jabbar A, Ashraf M, Lodhi MA, Choudhary MI (2006) Halosterols A and B, chymotrypsin inhibitory sterols from *Haloxylon recurvum*. Chem Pharm Bull (Tokyo) 4:623–625
Ichikawa H, Takada Y, Shishodia S, Jayaprakasam B, Nair MG, Aggarwal BB (2006) Withanolides potentiate apoptosis, inhibit invasion, and abolish osteoclastogenesis through suppression of nuclear factor-kappaB (NF-kappaB) activation and NF-kappaB-regulated gene expression. Mol Cancer Ther 5:1434–1445
Iqbal Z, Lateef M, Jabbar A, Muhammad G, Khan MN (2005) Anthelmintic activity of *Calotropis procera* (Ait.) Ait. F. flowers in sheep. J Ethnopharmacol 102:256–261
Iwalewa EO, Elujoba AA, Bankole OA (2005) In vitro spasmolytic effect of aqueous extract of *Calotropis procera* on Guinea-pig trachea smooth muscle chain. Fitoterapia 76:250–253
Jabbar S, Khan MT, Choudhuri MS, Sil BK (2004) Bioactivity studies of the individual ingredients of the Dashamularishta. Pak J Pharm Sci 17:9–17
Jain A, Katewa SS, Galav PK, Sharma P (2005) Medicinal plant diversity of Sitamata wildlife sanctuary, Rajasthan, India. J Ethanopharmacol 102:143–157
Jain A, Katewa SS, Galava P, Nag A (2008) Some therapeutic uses of biodiversity among the tribals of Rajasthan. Indian J Tradit Knowl 7:256–262
Jain SK (1991) Dictionary of Indian folk medicine and ethnobotany. Deep, New Delhi
Jang DS, Lee GY, Kim YS, Lee YM, Kim CS, Yoo JL, Kim JS (2007) Anthraquinones from the seeds of *Cassia tora* with inhibitory activity on protein glycation and aldose reductase. Biol Pharm Bull 30:2207–2210
Jimenez-Arellanes A, Meckes M, Ramirez R, Torres J, Luna-Herrera J (2003) Activity against multidrug-resistant *Mycobacterium tuberculosis* in Mexican plants used to treat respiratory diseases. Phytother Res 17:903–908
Joshi P (1995) Ethnobotany of primitive tribes in Rajasthan. Printwell, Jaipur
Joshi R, Shekhawat NS, Rathore TS (1991) Micropropagation of *Anogeissus pendula* Edgew – an arid forest tree. Indian J Exp Biol 29:615–618
Kanth VR, Diwan PV (1999) Analgesic, antiinflammatory and hypoglycaemic activities of *Sida cordifolia*. Phytother Res 13:75–77
Kar DM, Maharana L, Pattnaik S, Dash GK (2006) Studies on hypoglycaemic activity of *Solanum xanthocarpum* Schrad. & Wendl. fruit extract in rats. J Ethnopharmacol 108:251–256
Katewa SS (2009) Indigenous people and forests: perspectives of an ethanobotanical study from Rajasthan (India). In: Ramawat KG (ed) Herbal drugs: ethnomedicine to modern medicine. Springer, Berlin, pp 33–56

Katewa SS, Sharma R (1998) Ethnomedicinal observation from certain watershed areas of Rajasthan. Ethnobotany 10:46–49

Katewa SS, Chaudary BL, Jain A, Galav PK (2003) Traditional uses of plant biodiversity from Aravalli hills of Rajasthan. Indian J Tradit Knowl 2:27–39

Kavitha AV, Jagadeesan G (2006) Role of *Tribulus terrestris* (Linn.) (Zygophyllacea) against mercuric chloride induced nephrotoxicity in mice, *Mus musculus* (Linn.). J Environ Biol 27:397–400

Kayani S, Zia M, Sarwar S, Riaz-ur-Rehman, Chaudhary MF (2008) Callogenic studies of *Achyranthes aspera* leaf explant at different hormonal combinations. Pak J Biol Sci 11:950–952

Kelly RL (1995) The foraging spectrum: diversity in hunter-gatherer lifeways. Smithsonian Institution Press, Washington DC

Khalessi AM, Pack AR, Thomson WM, Tompkins GR (2004) An in vivo study of the plaque control efficacy of Persica: a commercially available herbal mouthwash containing extracts of *Salvadora persica*. Int Dent J 54:279–283

Khan B, Ahmad SF, Bani S, Kaul A, Suri KA, Satti NK, Athar M, Qazi GN (2006) Augmentation and proliferation of T lymphocytes and Th-1 cytokines by *Withania somnifera* in stressed mice. Int Immunopharmacol 6:1394–1403

Khan N, Sharma S, Alam A, Saleem M, Sultana S (2001) *Tephrosia purpurea* ameliorates N-diethylnitrosamine and potassium bromate-mediated renal oxidative stress and toxicity in Wistar rats. Pharmacol Toxicol 88:294–299

Khanna D, Sethi G, Ahn KS, Pandey MK, Kunnumakkara AB, Sung B, Aggarwal A, Aggarwal BB (2007) Natural products as a gold mine for arthritis treatment. Curr Opin Pharmacol 7:344–351

Khatoon S, Rai V, Rawat AK, Mehrotra S (2006) Comparative pharmacognostic studies of three *Phyllanthus* species. J Ethnopharmacol 104:79–86

Kim HJ, Yokozawa T, Kim HY, Tohda C, Rao TP, Juneja LR (2005) Influence of amla (*Emblica officinalis* Gaertn.) on hypercholesterolemia and lipid peroxidation in cholesterol-fed rats. J Nutr Sci Vitaminol (Tokyo) 51:413–418

Kim YM, Lee CH, Kim HG, Lee HS (2004) Anthraquinones isolated from *Cassia tora* (Leguminosae) seed show an antifungal property against phytopathogenic fungi. J Agric Food Chem 52:6096–6100

Kirira PG, Rukunga GM, Wanyonyi AW, Muregi FM, Gathirwa JW, Muthaura CN, Omar SA, Tolo F, Mungai GM, Ndiege IO (2006) Anti-plasmodial activity and toxicity of extracts of plants used in traditional malaria therapy in Meru and Kilifi Districts of Kenya. J Ethnopharmacol 106:403–407

Koko WS, Galal M, Khalid HS (2000) Fasciolicidal efficacy of *Albizia anthelmintica* and *Balanites aegyptiaca* compared with albendazole. J Ethnopharmacol 71:247–252

Krishnamurthy YL, Shashikala J (2006) Inhibition of aflatoxin B production of *Aspergillus flavus*, isolated from soybean seeds by certain natural plant products. Lett Appl Microbiol 43:469–474

Kuboyama T, Tohda C, Komatsu K (2006) Withanoside IV and its active metabolite, sominone, attenuate alpha beta (25–35)-induced neurodegeneration. Eur J Neurosci 23:1417–1426

Kulkarni SK, Dhir A (2008) *Withania somnifera*: an Indian ginseng. Prog Neuropsychopharmacol Biol Psychiatry 32:1093–1105

Kumar A, Kalonia H (2007) Protective effect of *Withania somnifera* Dunal on the behavioral and biochemical alterations in sleep-disturbed mice (grid over water suspended method). Indian J Exp Biol 45:524–528

Kumar MS, Kirubanandan S, Sripriya R, Sehgal PK (2008) Triphala promotes healing of infected full-thickness dermal wound. J Surg Res 144:94–101

Kumar S, Suri SS, Sonie KC, Ramawat KG (2003) Establishment of embryonic cultures and somatic embryogenesis in callus cultures of guggul – *Commiphora wightii* (Arnott.) Bhandari. Indian J Exp Biol 41:69–77

Kuo YH, Chen CH, Kuo LM, King ML, Wu TS, Haruna M, Lee KH (1990) Antitumor agents 112. Emarginatine B, a novel potent cytotoxic sesquiterpene pyridine alkaloid from *Maytenus emarginata*. J Nat Prod 53:422–428

Kuo YH, King ML, Chen CF, Chen HY, Chen CH, Chen K, Lee KH (1994) Two new macrolide sesquiterpene pyridine alkaloids from *Maytenus emarginata:* emarginatine G and the cytotoxic emarginatine F. J Nat Prod 57:263–269

Lee GY, Jang DS, Lee YM, Kim JM, Kim JS (2006) Naphthopyrone glucosides from the seeds of *Cassia tora* with inhibitory activity on advanced glycation end products (AGEs) formation. Arch Pharm Res 29:587–590

Li M, Qu W, Wang Y, Wan H, Tian C (2002) Hypoglycemic effect of saponin from *Tribulus terrestris*. Zhong Yao Cai 25:420–422

Lodhi S, Pawar RS, Jain AP, Singhai AK (2006) Wound healing potential of *Tephrosia purpure* (Linn.) Pers. in rats. J Ethnopharmacol 108:204–210

Machiah DK, Girish KS, Gowda TV (2006) A glycoprotein from a folk medicinal plant, *Withania somnifera*, inhibits hyaluronidase activity of snake venoms. Comp Biochem Physiol C Toxicol Pharmacol 143:158–161

Madina BR, Sharma LK, Chaturvedi P, Sangwan RS, Tuli R (2007) Purification and physico-kinetic characterization of 3beta-hydroxy specific sterol glucosyltransferase from *Withania somnifera* (L) and its stress response. Biochim Biophys Acta 1774:392–402

Malik F, Kumar A, Bhushan S, Khan S, Bhatia A, Suri KA, Qazi GN, Singh J (2007) Reactive oxygen species generation and mitochondrial dysfunction in the apoptotic cell death of human myeloid leukemia HL-60 cells by a dietary compound withaferin A with concomitant protection by N-acetyl cysteine. Apoptosis 12:2115–2133

Manu KA, Kuttan G (2008) *Boerhaavia diffusa* stimulates cell-mediated immune response by upregulating IL-2 and downregulating the pro-inflammatory cytokines and GM-CSF in B16F-10 metastatic melanoma bearing mice. J Exp Ther Oncol 7:17–29

Manu KA, Leyon PV, Kuttan G (2007) Studies on the protective effects of *Boerhaavia diffusa* L. against gamma radiation induced damage in mice. Integr Cancer Ther 6:381–388

Mares MA (1999) Encyclopedia of deserts. University of Oklahoma Press, Norman, OK

Margaret I, Reddy P, Srinivasa JK (1998) Antiinflammatory profile of *Tridax procumbens* in animal and fibroblast cell models. Phytother Res 12:285–287

Mathur M, Ramawat KG (2007) Guggulsterone production in cell suspension cultures of *Commiphora wightii* grown in shake flasks and bioreactors. Biotechnol Lett 29:979–982

Mathur N, Ramawat KG, Sonie KC (1993) In vitro propagation of *Ziziphus numularia*. Ann Arid Zones 32:219–222

Mathur N, Ramawat KG, Sonie KC (1994) Plant regeneration from seedling explants of *Ziziphus* and silver nitrate and nutrient requirement for callus morphogenesis. Gartenbau 58:255–260

Mathur N, Ramawat KG, Nandawani D (1995) Rapid in vitro multiplication of Jujube through mature stem explant. Plant Cell Tissue Organ Cult 43:75–77

Mathur S, Shekhawat GS, Batra A (2002) Micropropagation of *Salvadora persica* Linn. via cotyledonary nodes. Indian J Biotechnol 1:197–200

Matsui TA, Sowa Y, Murata H, Takagi K, Nakanishi R, Aoki S, Yoshikawa M, Kobayashi M, Sakabe T, Kubo T, Sakai T (2007) The plant alkaloid cryptolepine induces p21WAF1/CIP1 and cell cycle arrest in a human osteosarcoma cell line. Int J Oncol 31:915–922

Medeiros IA, Santos MR, Nascimento NM, Duarte JC (2006) Cardiovascular effects of *Sida cordifolia* leaves extract in rats. Fitoterapia 77:19–27

Meena PD, Kaushik P, Shukla S, Soni AK, Kumar M, Kumar A (2006) Anticancer and antimutagenic properties of *Acacia nilotica* (Linn.) on 7,12-dimethylbenz(a)anthracene-induced skin papillomagenesis in Swiss albino mice. Asian Pac J Cancer Prev 7:627–632

Middleton N, Thomas D (1997) World atlas of desertification (2nd edn). Arnold, London

Mir AI, Kumar B, Tasduq SA, Gupta DK, Bhardwaj S, Johri RK (2007) Reversal of hepatotoxin-induced pre-fibrogenic events by *Emblica officinalis* – a histological study. Indian J Exp Biol 45:626–629

Mohan L, Sharma P, Srivastava CN (2005) Evaluation of *Solanum xanthocarpum* extracts as mosquito larvicides. J Environ Biol 26:399–401

Mohan L, Sharma P, Srivastava CN (2007) Comparative efficacy of *Solanum xanthocarpum* extracts alone and in combination with a synthetic pyrethroid, cypermethrin, against malaria vector, *Anopheles stephensi*. Southeast Asian J Trop Med Public Health 38:256–260

Mohanty IR, Arya DS, Gupta SK (2008) *Withania somnifera* provides cardioprotection and attenuates ischemia-reperfusion induced apoptosis. Clin Nutr 27:635–642

Monforte MT, Trovato A, Rossitto A, Forestieri AM, D'Aquino A, Miceili N, Galati EM (2002) Anticonvulsant and sedative effects of *Salvadora persica* L. stem extracts. Phytother Res 16:395–397

Muchuweti M, Zenda G, Nadhlala AR, Kasiyamahuru A (2005) Sugars, organic acid and phenolic compounds of *Ziziphus mauritiana* fruit. Eur Food Res Technol 221:570–574

Mukherjee R, Khattar D, Jaggi M, Singh AT, Kumar M, Bala H (2006) Method for treating cancer using betulinic acid rich herbal extract. USPTO Application 20060159783

Mythilypriya R, Shanthi P, Sachdanandam P (2007) Restorative and synergistic efficacy of Kalpaamruthaa, a modified Siddha preparation, on an altered antioxidant status in adjuvant induced arthritic rat model. Chem Biol Interact 168:193–202

Nagareddy PR, Lakshmana M (2006) *Withania somnifera* improves bone calcification in calcium-deficient ovariectomized rats. J Pharm Pharmacol 58:513–519

Ndoye M, Diallo I, Gassama-Dia YK (2003) In vitro multiplication of the semi-arid forest tree, *Balanites aegyptiaca* (L.) Del. Afr J Biotechnol 2:421–424

Neychev VK, Nikolova E, Zhelev N, Mitev VI (2007) Saponins from *Tribulus terrestris* L are less toxic for normal human fibroblasts than for many cancer lines: influence on apoptosis and proliferation. Exp Biol Med 232:126–133

Okigbo RN, Igwe DI (2007) Antimicrobial effects of *Piper guineense* 'Uziza' and *Phyllantus amarus* 'Ebe-benizo' on *Candida albicans* and *Streptococcus faecalis*. Acta Microbiol Immunol Hung 54:353–366

Olaleye MT, Rocha BT (2008) Acetaminophen-induced liver damage in mice: effects of some medicinal plants on the oxidative defense system. Exp Toxicol Pathol 59:319–327

Padhy BM, Srivastava A, Kumar VL (2007) *Calotropis procera* latex affords protection against carbon tetrachloride induced hepatotoxicity in rats. J Ethnopharmacol 113:498–502

Panchabhai TS, Ambarkhane SV, Joshi AS, Samant BD, Rege NN (2008) Protective effect of *Tinospora cordifolia, Phyllanthus emblica* and their combination against antitubercular drugs induced hepatic damage: an experimental study. Phytother Res 22:646–650

Pandey R, Maurya R, Singh G, Sathiamoorthy B, Naik S (2005) Immunosuppressive properties of flavonoids isolated from *Boerhaavia diffusa* Linn. Int Immunopharmacol 5:541–553

Pareek OP (2001) Fruits of future-2 Ber. International Center for Underutilized Crops, Southampton, UK

Patil UK, Saraf S, Dixit VK (2004) Hypolipidemic activity of seeds of *Cassia tora* Linn. J Ethnopharmacol 90:249–252

Paul JP (2008) Studies on antimicrobial efficiency of *Citrullus colocynthis* (L.) Schrad: a medicinal plant. Ethnobot Leafl 12:944–947

Pavana P, Manoharan S, Renju GL, Sethupathy S (2007) Antihyperglycemic and antihyperlipidemic effects of *Tephrosia purpurea* leaf extract in streptozotocin induced diabetic rats. J Environ Biol 28:833–837

Philip BK, Muralidharan A, Natarajan B, Varadamurthy S, Venkataraman S (2008) Preliminary evaluation of anti-pyretic and anti-ulcerogenic activities of *Sida cordifolia* methanolic extract. Fitoterapia 79:229–231

Phillips OA, Mathew KT, Oriowo MA (2006) Antihypertensive and vasodilator effects of methanolic and aqueous extracts of *Tribulus terrestris* in rats. J Ethnopharmacol 104:351–355

Pinmai K, Chunlaratthanabhorn S, Ngamkitidechakul C, Soonthornchareon N, Hahnvajanawong C (2008) Synergistic growth inhibitory effects of *Phyllanthus emblica* and *Terminalia bellerica* extracts with conventional cytotoxic agents: doxorubicin and cisplatin against human hepatocellular carcinoma and lung cancer cells. World J Gastroenterol 14:1491–1497

Porchezhian E, Ansari SH (2005) Hepatoprotective activity of *Abutilon indicum* on experimental liver damage in rats. Phytomedicine 12:62–64

Purohit A, Vyas KB (2005) Hypolipidaemic efficacy of *Capparis decidua* fruit and shoot extracts in cholesterol fed rabbits. Indian J Exp Biol 43:863–866

Purohit SD, Ramawat KG, Arya HC (1979) Phenolics peroxidase and phenolase as related to gall formation in some arid zone plants. Curr Sci 48:714–716

Rahman MT, Ahmed M, Alimuzzaman M, Shilpi JA (2003) Antinociceptive activity of the aerial parts of *Solanum xanthocarpum*. Fitoterapia 74:119–121

Rahuman AA, Venkatesan P (2008) Larvicidal efficacy of five cucurbitaceous plant leaf extracts against mosquito species. Parasitol Res 103:133–139

Rahuman AA, Gopalakrishnan G, Venkatesan P, Geetha K (2008) Larvicidal activity of some Euphorbiaceae plant extracts against *Aedes aegypti* and *Culex quinquefasciatus* (Diptera: Culicidae). Parasitol Res 102:867–873

Ramani B, Reeck T, Debez A, Stelzer R, Huchzermeyer B, Schmidt A, Papenbrock J (2006) *Aster tripolium* L. and *Sesuvium portulacastrum* L.: two halophytes, two strategies to survive in saline habitats. Plant Physiol Biochem 44:395–408

Ramawat KG, Goyal S (2008) The Indian herbal drugs scenario in global perspectives. In: Ramawat KG, Merillon JM (eds) Bioactive molecules and medicinal plants, Springer, Berlin, pp 325–343

Ramawat KG, Nandwani D (1991) Regeneration in *Prosopis* species: problems, perseverance and prospects. Ann Arid Zone 30:247–258

Ramawat KG, Purohit SD, Arya HC (1979) Altered states of oxidizing enzymes and phenolic in Copdia gall. ISDT and UCDS (Jodhpur) 5:38–41

Ramawat KG, Sonie KC, Sharma MC (2004) Therapeutic potential of medicinal plants. In: Ramawat KG (ed) Biotechnology of medicinal plants: vitalizer and therapeutic. Science, Enfield, pp 1–18

Ramawat KG, Dass S, Mathur M (2009) The chemical diversity of bioactive molecules and therapeutic potential of medicinal plants. In: Ramawat KG (ed) Herbal drugs: ethanomedicine to modern medicine, Springer, Berlin, pp 7–32

Rameshree AB, Hariharan M, Unnikrishnan K (1994) Micropropagation and callus induction of *Aristolochia bracteolata* Lam. – a medicinal plant. Phytomorphology 44:247–252

Rani P, Khullar N (2004) Antimicrobial evaluation of some medicinal plants for their anti-enteric potential against multi-drug resistant *Salmonella typhi*. Phytother Res 18:670–673

Rania G, Virka GS, Avinash NB (2003) Callus induction and plantlet regeneration in *Withania somnifera*. In Vitro Cell Dev Biol Plant 39:468–474

Rao RR (2006) Conservation and bioprospection of the floristic diversity in India: prospects, constraints and urgent tasks ahead. In: Proceedings of the 29th All India Botany Conference, 9–11 October, Department of Botany, ML Sukhadia University, Udaipur, India, pp 1–5

Rasool M, Varalakshmi P (2006a) Suppressive effect of *Withania somnifera* root powder on experimental gouty arthritis: an in vivo and in vitro study. Chem Biol Interact 164:174–180

Rasool M, Varalakshmi P (2006b) Immunomodulatory role of *Withania somnifera* root powder on experimental induced inflammation: an in vivo and in vitro study. Vascul Pharmacol 44:406–410

Rasool M, Varalakshmi P (2007) Protective effect of *Withania somnifera* root powder in relation to lipid peroxidation, antioxidant status, glycoproteins and bone collagen on adjuvant-induced arthritis in rats. Fundam Clin Pharmacol 21:157–164

Rathore TS, Singh RP, Deora NS, Shekhawat NS (1992a) Cloning of *Maytenus emarginata* (Willd.) Ding Hon – a tree of the Indian Desert – through tissue culture. Plant Cell Rep 11:449–451

Rathore TS, Singh RP, Deora NS, Shekhawat NS (1992b) Clonal propagation of *Ziziphus* species through tissue culture. Sci Hortic 51:165–168

Rathore JS, Rathor V, Shekhawat NS, Singh RP, Liler G, Phulwaria M, Dagla HR (2004) Micropropagation of woody plants. In: Srivastva PS, Narula A, Srivastva S (eds) Plant biotechnology and molecular markers. Kluwer, Dordrecht, pp 195–205

Ravikanth V, Reddy NVL, Ramesh P, Rao PT, Diwan PV, Khar A, Venkateswarlu Y (2001) An immunosuppressive tryptophan-derived alkaloid from *Lepidagathis cristata*. Phytochemistry 58:1263–1266

Roat C, Ramawat KG (2009) Morphactin and 2iP markedly enhance accumulation of stilbenes in cell cultures of *Cayratia trifolia* (L.) Domin. Acta Physiol 31:411–414

Robert SD, Ismail AA, Winn T, Wolever TM (2008) Glycemic index of common Malaysian fruits. Asia Pac J Clin Nutr 17:35–39

Rogers PL, Shin HS, Wang B (1997) Biotransformation for Lephedrine production. Adv Biochem Eng Biotechnol 56:33–59

Roy PK (2008) Rapid multiplication of *Boerhaavia diffusa* L. through in vitro culture of shoot tip and nodal explants. Plant Tissue Cult Biotechnol 18:49–56

Sailaja R, Setty OH (2006) Protective effect of *Phyllanthus fraternus* against allyl alcohol-induced oxidative stress in liver mitochondria. J Ethnopharmacol 105:201–209

Saito K, Kohno M, Yoshizaki F, Niwano Y (2008) Extensive screening for edible herbal extracts with potent scavenging activity against superoxide anions. Plant Foods Hum Nutr 63:65–70

Saleem M, Ahmed Su, Alam A, Sultana S (2001) *Tephrosia purpurea* alleviates phorbol ester-induced tumor promotion response in murine skin. Pharmacol Res 43:135–144

Sandhya T, Mishra KP (2006) Cytotoxic response of breast cancer cell lines, MCF 7 and T 47 D to triphala and its modification by antioxidants. Cancer Lett 238:304–313

Sanogo R, Monforte MT, Daquino A, Rossitto A, Maur DD, Galati EM (1999) Antiulcer activity of *Salvadora persica* L.: structural modifications. Phytomedicine 6:363–366

Santos AR, De Campos RO, Miguel OG, Filho VC, Siani AC, Yunes RA, Calixto JB (2000) Antinociceptive properties of extracts of new species of plants of the genus *Phyllanthus* (Euphorbiaceae). J Ethnopharmacol 72:229–238

Satheesh MA, Pari L (2004) Antioxidant effect of *Boerhavia diffusa* L. in tissues of alloxan induced diabetic rats. Indian J Exp Biol 42:989–992

Saxena G, Singh SP, Pal R, Singh S, Pratap R, Nath C (2007) Gugulipid, an extract of *Commiphora wightii* with lipid-lowering properties, has protective effects against streptozotocin-induced memory deficits in mice. Pharmacol Biochem Behav 86:797–805

Sebastian MK, Bhandari MM (1984a) Medico-ethnobotany of Mt. Abu, Rajasthan. Indian J Ethnopharmacol 12:223–230

Sebastian MK, Bhandari MM (1984b) Medicinal plant lore of Udaipur district, Rajasthan. Bull Medico-Ethnobot Res 5:133–134

Seetharam YN, Chalageri G, Setty SR, Bheemachar (2002) Hypoglycemic activity of *Abutilon indicum* leaf extracts in rats. Fitoterapia 73:156–159

Sen J, Sharma AK (1991) Micropropagation of *Withania somnifera* from germinating seeds and shoot tips. Plant Cell Tissue Organ Cult 26:71–73

Senthil V, Ramadevi S, Venkatakrishnan V, Giridharan P, Lakshmi BS, Vishwakarma RA, Balakrishnan A (2007) Withanolide induces apoptosis in HL-60 leukemia cells via mitochondria mediated cytochrome *c* release and caspase activation. Chem Biol Interact 167:19–30

Senthilnathan P, Padmavathi R, Magesh V, Sakthisekaran D (2006) Stabilization of membrane bound enzyme profiles and lipid peroxidation by *Withania somnifera* along with paclitaxel on benzo(a)pyrene induced experimental lung cancer. Mol Cell Biochem 292:13–17

Shale TL, Stirk WA, van Staden J (2005) Variation in antibacterial and anti-inflammatory activity of different growth forms of *Malva parviflora* and evidence for synergism of the anti-inflammatory compounds. J Ethnopharmacol 96:325–330

Sharifi AM, Darabi R, Akbarloo N (2003) Study of antihypertensive mechanism of *Tribulus terrestris* in 2K1C hypertensive rats: role of tissue ACE activity. Life Sci 73:2963–2971

Sharma R, Ramawat KG (2005) Regeneration of woody plants of arid zones with special reference to *Prosopis*: problems, perseverance and prospects. In: Trivedi PC (ed) Plant biotechnology: recent advances. Panima Pub Crop, New Delhi, pp 325–349

Shetty BV, Singh V (1987–1993) Flora of Rajasthan. BSI, Calcutta, India

Shirwaikar A, Somashekar AP, Udupa AL, Udupa SL, Somashekar S (2003) Wound healing studies of *Aristolochia bracteolata* Lam. with supportive action of antioxidant enzymes. Phytomedicine 10:558–562

Shishodia S, Sethi G, Ahn KS, Aggarwal BB (2007) Guggulsterone inhibits tumor cell proliferation, induces S-phase arrest, and promotes apoptosis through activation of c-Jun N-terminal kinase, suppression of Akt pathway, and downregulation of antiapoptotic gene products. Biochem Pharmacol 74:118–130

Silva RL, Melo GB, Melo VA, Antoniolli AR, Michellone PR, Zucoloto S, Picinato MA, Franco CF, Mota Gde A, Silva Ode C (2006) Effect of the aqueous extract of *Sida cordifolia* on liver regeneration after partial hepatectomy. Acta Cir Bras 21:37–39

Singh AK (2004) Endangered economic species of Indian desert. Genet Res Crop Evolut 51:371–380

Singh AK, Sharma M, Varshney R, Agarwal SS, Bansal KC (2006) Plant regeneration from aliginate-encapsulated shoot tips of *Phyllanthus amarus* Schum and Thonn, a medicinally important plant species. In Vitro Cell Dev Biol Plant 42:109–113

Singh B, Bani S, Gupta DK, Chandan BK, Kaul A (2003) Anti-inflammatory activity of 'TAF' an active fraction from the plant *Barleria prionitis* Linn. J Ethnopharmacol 85:187–193

Singh B, Chandan BK, Prabhakar A, Taneja SC, Singh J, Qazi GN (2005) Chemistry and hepatoprotective activity of an active fraction from *Barleria prionitis* Linn. in experimental animals. Phytother Res 19:391–404

Singh I, Soyal D, Goyal PK (2006) *Emblica officinalis* (Linn.) fruit extract provides protection against radiation-induced hematological and biochemical alterations in mice. J Environ Pathol Toxicol Oncol 25:643–654

Singh OM, Subharani K, Singh NI, Devi NB, Nevidita L (2007) Isolation of steroidal glycosides from *Solanum xanthocarpum* and studies on their antifungal activities. Nat Prod Res 21:585–590

Singh P, Sharma ML, Mukherjee S (1989) Effect of indole butyric on sprouting in plant cuttings of *Commiphora wightii* (Arnott) Bhandari. Indian Drugs 26:515–516

Singh R, Singh B, Singh S, Kumar N, Kumar S, Arora S (2008) Anti-free radical activities of kaempferol isolated from *Acacia nilotica* (L.) Willd. Ex. Del. Toxicol In Vitro 22:1965–1970

Singh RK, Mittal PK, Dhiman RC (2005) Laboratory study on larvicidal properties of leaf extract of *Calotropis procera* (Family-Asclepiadaceae) against mosquito larvae. J Commun Dis 37:109–113

Sittie AA, Lemmich E, Olsen CE, Hviid L, Christensen SB (1998) Alkamides from *Phyllanthus fraternus*. Planta Med 64:192–193

Sivanesan I, Jeong BR (2007) Direct shoot regeneration from nodal explants of *Sida cordifolia* Linn. In Vitro Cell Dev Biol Plant 43:436–441

Smith MA, Vellen L, Pask J (1995) Vegetation history from archaelogical charcoals in central Australia: the late quaternary record from Puritjarra rock shelter. In: Bittman F (ed) Vegetation history and archaeobotany, Springer, Berlin, pp 171–177

Soares de Oliveira J, Pereira Bezerra D, Teixeira de Freitas CD, Delano Barreto Marinho Filho J, Odorico de Moraes M, Pessoa C, Costa-Lotufo LV, Ramos MV (2007) In vitro cytotoxicity against different human cancer cell lines of laticifer proteins of *Calotropis procera* (Ait.) R. Br. Toxicol In Vitro 21:1563–1573

Sofrata A, Lingström P, Baljoon M, Gustafsson A (2007) The effect of miswak extract on plaque pH, an in vivo study. Caries Res 41:451–454

Spelman K, Burns J, Nichols D, Winters N, Ottersberg S, Tenborg M (2006) Modulation of cytokine expression by traditional medicines: a review of herbal immunomodulators. Altern Med Rev 11:128–150

Speroni E, Cervellati R, Innocenti G, Costa S, Guerra MC, Dall' Acqua S, Govoni P (2005) Anti-inflammatory, anti-nociceptive and antioxidant activities of *Balanites aegyptiaca* (L.) Delile. J Ethnopharmacol 98:117–125

Srikumar R, Parthasarathy NJ, Shankar EM, Manikandan S, Vijayakumar R, Thangaraj R, Vijayananth K, Sheeladevi R, Rao UA (2007) Evaluation of the growth inhibitory activities of Triphala against common bacterial isolates from HIV infected patients. Phytother Res 21:476–480

Srinivasan S, Ranga RS, Burikhanov R, Han SS, Chendil D (2007) Par-4-dependent apoptosis by the dietary compound withaferin A in prostate cancer cells. Cancer Res 67:246–253

Stan SD, Hahm ER, Warin R, Singh SV (2008) Withaferin A causes FOXO3a- and Bim-dependent apoptosis and inhibits growth of human breast cancer cells in vivo. Cancer Res 68:7661–7669

Subbaraju GV, Vanisree M, Rao CV, Sivaramakrishna C, Sridhar P, Jayaprakasam B, Nair MG (2006) Ashwagandhanolide, a bioactive dimeric thiowithanolide isolated from the roots of *Withania somnifera*. J Nat Prod 69:1790–1792

Sun B, Qu W, Bai Z (2003) The inhibitory effect of saponins from *Tribulus terrestris* on Bcap-37 breast cancer cell line in vitro. Zhong Yao Cai 26:104–106

Sun B, Qu WJ, Zhang XL, Yang HJ, Zhuang XY, Zhang P (2004) Investigation on inhibitory and apoptosis-inducing effects of saponins from *Tribulus terrestris* on hepatoma cell line BEL-7402. Zhongguo Zhong Yao Za Zhi 29:681–684

Sun W, Li H, Yang SJ (2008) A triterpene saponin from *Tribulus terrestris* attenuates apoptosis in cardiocyte via activating PKC signalling transduction pathway. J Asian Nat Prod Res 10:39–48

Sutradhar RK, Rahman AM, Ahmad M, Bachar SC, Saha A, Roy TG (2007) Anti-inflammatory and analgesic alkaloid from *Sida cordifolia* linn. Pak J Pharm Sci 20:185–188

Tadeg H, Mohammed E, Asres K, Gebre-Mariam T (2005) Antimicrobial activities of some selected traditional Ethiopian medicinal plants used in the treatment of skin disorders. J Ethnopharmacol 100:168–175

Tahiliani P, Kar A (2000) *Achyranthes aspera* elevates thyroid hormone levels and decreases hepatic lipid peroxidation in male rats. J Ethnopharmacol 71:527–532

Talwar GP, Dar SA, Rai MK, Reddy KV, Mitra D, Kulkarni SV, Doncel GF, Buck CB, Schiller JT, Muralidhar S, Bala M, Agrawal SS, Bansal K, Verma JK (2008) A novel polyherbal microbicide with inhibitory effect on bacterial, fungal and viral genital pathogens. Int J Antimicrob Agents 32:180–185

Tiagi YD, Aery NC (2007) Flora of Rajasthan (South and South-East Region). Himanshu, Udaipur-New Delhi, India

Tohda C (2008) Overcoming several neurodegenerative diseases by traditional medicines: the development of therapeutic medicines and unraveling pathophysiological mechanisms. Yakugaku Zasshi 128:1159–1167

Tyagi S, Govil CM (1999) Somatic embryogenesis and micro-propagation in *Emblica officinalis* gaertn. J Indian Bot Soc 78:363–365

Upadhyay RK, Rohatgi L, Chaubey MK, Jain SC (2006) Ovipositional responses of the pulse beetle, *Bruchus chinensis* (Coleoptera: Bruchidae) to extracts and compounds of *Capparis decidua*. J Agric Food Chem 54:9747–9751

Vasudeva N, Sharma SK (2006) Post-coital antifertility activity of *Achyranthes aspera* Linn. root. J Ethnopharmacol 107:179–81

Vasudevan M, Parle M (2007a) Effect of Anwala churna (*Emblica officinalis* GAERTN.): an ayurvedic preparation on memory deficit rats. Yakugaku Zasshi 127:1701–1707

Vasudevan M, Parle M (2007b) Memory enhancing activity of Anwala churna (*Emblica officinalis* Gaertn.): an Ayurvedic preparation. Physiol Behav 91:46–54

Verma PK, Sharma A, Joshi SC, Gupta RS, Dixit VP (2005) Effect of isolated fractions of *Barleria prionitis* root methanolic extract on reproductive function of male rats: preliminary study. Fitoterapia 76:428–432

Vetrichelvan T, Jegadeesan M (2003) Effect of alcohol extract of *Achyranthes aspera* Linn. on acute and subacute inflammation. Phytother Res 17:77–79

Vinutha B, Prashanth D, Salma K, Sreeja SL, Pratiti D, Padmaja R, Radhika S, Amit A, Venkateshwarlu K, Deepak M (2007) Screening of selected Indian medicinal plants for acetylcholinesterase inhibitory activity. J Ethnopharmacol 109:359–363

Visavadiya NP, Narasimhacharya AV (2007) Hypocholesteremic and antioxidant effects of *Withania somnifera* (Dunal) in hypercholesteremic rats. Phytomedicine 14:136–142

Wang X, Bunkers GJ (2000) Potent heterologous antifungal proteins from cheeseweed (*Malva parviflora*). Biochem Biophys Res Commun 279:669–673

Wang X, Bunkers GJ, Walters MR, Thoma RS (2001) Purification and characterization of three antifungal proteins from cheeseweed (*Malva parviflora*). Biochem Biophys Res Commun 282:1224–1228

Wei FH, Xu XJ, Liu JB, Dai YH, Dussart G, Trigwell J (2002) Toxicology of a potential molluscicide derived from the plant *Solanum xanthocarpum*: a preliminary study. Ann Trop Med Parasitol 96:325–331

Widodo N, Kaur K, Shrestha BG, Takagi Y, Ishii T, Wadhwa R, Kaul SC (2007) Selective killing of cancer cells by leaf extract of Ashwagandha: identification of a tumor-inhibitory factor and the first molecular insights to its effect. Clin Cancer Res 13:2298–2306

Yadav P, Sarkar S, Bhatnagar D (1997) Action of *Capparis decidua* against alloxan-induced oxidative stress and diabetes in rat tissues. Pharmacol Res 36:221–228

Yang HJ, Qu WJ, Sun B (2005) Experimental study of saponins from *Tribulus terrestris* on renal carcinoma cell line. Zhongguo Zhong Yao Za Zhi 30:1271–1274

Yang L, Lu JW, An J, Jiang X (2006) Effect of *Tribulus terrestris* extract on melanocyte-stimulating hormone expression in mouse hair follicles. Nan Fang Yi Ke Da Xue Xue Bao 26:1777–1779

Yokozawa T, Kim HY, Kim HJ, Tanaka T, Sugino H, Okubo T, Chu DC, Juneja LR (2007) Amla (*Emblica officinalis* Gaertn.) attenuates age-related renal dysfunction by oxidative stress. J Agric Food Chem 55:7744–7752

Zafar R, Mujeeb M (2002) Rotenoids and rutin in callus culture of *Tephrosia pupurea* (L) Pers. seeds. Indian J Pharm Sci 64:217–221

Zhang JD, Cao YB, Xu Z, Sun HH, An MM, Yan L, Chen HS, Gao PH, Wang Y, Jia XM, Jiang YY (2005) In vitro and in vivo antifungal activities of the eight steroid saponins from *Tribulus terrestris* L. with potent activity against fluconazole-resistant fungal pathogens. Biol Pharm Bull 28:2211–2215

Chapter 2
Potentiality of Hydrocarbon Yielding Plants for Future Energy and Chemicals

Dipul Kalita

Abstract The development of alternative sources for energy and chemicals, particularly the use of plant biomass as a renewable source for fuel or chemical feedstock, has received much recent attention. While efforts are ongoing to find and use substitute forms of energy, the heavy dependence of most societies in the world on petroleum lends importance to the continuing supply of hydrocarbon on a self sustaining and renewable basis. This chapter attempts to review research carried out by various workers in different parts of the world on plant materials as sources of energy and chemical feedstock, and the possibilities of producing hydrocarbon and related chemical products directly or indirectly from such material. The potential of plant resources in this area might help to achieve energy autarky and reduce reliance on other sources. The chapter also introduces the exploratory work carried out at North East Institute of Science and Technology, Jorhat, and discusses some future directions that need to be considered to promote the development of petrocrops.

2.1 Introduction

The world energy scene is undergoing a period of transition. As the inevitability of the exhaustion of fossil fuels is becoming increasingly apparent, efforts are being made to find substitute sources of energy. In this context, bio-energy systems have the potential to make a significant contribution to the world's growing energy needs. Such renewable sources will be able to compete with fossil fuel resources only if special plant crops containing energy-producing hydrocarbon-like material are bred and cultivated. The Earth has vast areas of land that are unsuitable for food

D. Kalita
Cellulose, Pulp & Paper Division, North East Institute of Science and Technology (CSIR), Jorhat-785006, Assam, India
e-mail: dipkalita_2001@yahoo.co.in; kalita_d@rrljorhat.res.in

K.G. Ramawat (ed.), *Desert Plants*,
DOI 10.1007/978-3-642-02550-1_2, © Springer-Verlag Berlin Heidelberg 2010

and fodder crops, and recent experience with growing hydrocarbon-yielding plants suggests that may be possible to use these large areas for harvesting plants yielding substitutes for conventional hydrocarbons. In the last few decades, various machines have been developed that can run on liquid hydrocarbon mixtures, given appropriate chemical aggregate composition and the required physico-chemical and performance parameters. There are several compelling reasons for seriously exploring the prospects of hydrocarbon plantations. First, the prospects of increased dependence on oil imports poses a significant challenge. Second, oil prices are likely to go up substantially in the next 10–15 years. Furthermore, it is becoming increasingly evident that woody plants, which are often grown on relatively good soil, will not be able to meet the demand for fuel wood, particularly in countries like India. Petro-farming could therefore provide a welcome solution to some of these problems although no substantive claims can be made on the viability of this option at present. Various workers have conducted extensive screening programs in an attempt to identify potential bio-crude and botanochemical feed stocks.

Before discussing current efforts towards developing petroleum plantations and growing green factories for the production of hydrocarbon-like materials, it is necessary to describe briefly some of the energy sources available and their constraints, the concept of using plants for energy production in the world context, and finally the creation of a controlled experiment aimed at growing and harvesting a particular plant for its oil and hydrocarbon content. Natural gas, coal and oil (all fossilised photosynthetic products) provide a little over 95% of world energy supplies; alternative resources make only a very small contribution. The supply of fossil hydrocarbon fuels will gradually become exhausted in due course. King Hubbert, a geologist with United States Geological Survey, forecast how fuels, specifically oil and coal, will come into and go out of use (see Calvin 1978b). King Hubbert was probably correct in his guess/estimate that the use of oil might even peak before 2050. Ideally, alternative methods to increase our energy prospects would include those that use sunshine in some useful way with minimum environmental impact. This led to the idea that green plants could be used to trap sunlight and reduce carbon, particularly on the equator where plants are most productive. Through the mechanism of the photosynthetic carbon cycle, green plants capture carbon dioxide from the atmosphere and, with the aid of sunshine, separate hydrogen from water to reduce the carbon dioxide first to carbohydrate in which there is only one oxygen atom on each carbon atom. Eventually, some plants can take the carbohydrate and reduce it completely to hydrocarbons with no oxygen at all on the carbon atoms. This is essentially what petroleum is.

Though it must be admitted that information on hydrocarbon-rich plants is meager, we seem to have a great number of them. Naturally, before these plants can really be used for fuel production, substantial groundwork needs to be done. This includes screening and selection of appropriate plant species, extracting hydrocarbons, developing conversion technology and agrotechnology for different petrocrops, and studying the economics of their cultivation and conversion to make the venture viable on a commercial scale.

It would appear that, in the short term, existing processes for converting plant material to hydrocarbon will not be economic unless the plant material is available at near zero cost. Similarly, in the medium term, coal prices will have to at least double from 1975 levels before plants would become a competitive source of supply. This assessment is based on existing technology and can therefore be affected by future developments. Thus, the viability of plant-based processes could be enhanced by the development of high-yield plant systems or by the development of plants that synthesise hydrocarbons directly from carbon dioxide and water.

2.2 Screening and Processing of Biomass for Bio-Crude

The first efforts to cultivate hydrocarbon-producing plants for fuel production were made by the Italians in Ethiopia (*Euphorbia abyssinica* 1935–1936) (Frick 1938) and the French in Morocco (*Euphorbia resinifera* 1940) (Steinhell 1941). Professor Melvin Calvin (Calvin 1983b) revived the idea that hydrocarbon-producing plants could be used as future sources of oil and other chemicals. He also suggested that the energy farming concept should be given due importance. Several plant families, mainly Euphorbiaceae and Asclepiadaceae, were screened to assess their suitability as a source of low molecular weight and non-polar petroleum-like hydrocarbons. Air-dried plant materials were successively extracted with acetone and benzene, and the extracts were analysed spectroscopically for yield of rubber wax glycerides, isoprenoides and other terpenoides. Mixtures of hexane-methanol and heptane-methanol were used for extraction. Yields of up to ten barrels of oil ha^{-1} year^{-1} from each species has been claimed without any genetic improvement. Based on a biomass yield of 25 tons ha^{-1} year^{-1}, total energy yield in the form of liquid fuel from *Euphorbia lathyris* was calculated to be 48 MJ ha^{-1} year^{-1}, 26 MJ as hydrocarbon and 22 MJ as ethanol. The product of extraction of *E. lathyris* thus represented a new possibility for a future energy and material source, as was again suggested later by Nemethy and co-workers (Nemethy et al. 1979, 1981; Nemethy and Calvin 1982; Nemethy 1984). Every Euphorbia species contains latex, as an emulsion of about 30% terpenes in water. The latex hydrocarbon is present largely as a C_{30} triterpenoid, which can be cracked like oil to make high-octane gasoline. Thus, the plant *E. lathyris* was hailed as a potential ‘gasoline tree’. Some of the latex sterols of *E. lathyris* latex could also be used in the pharmaceutical industry, possibly being of even higher value in this way than as actual crude oil. Another species, *Euphorbia tirucally*, which grows prolifically in various parts of the world, was cultivated for oil but it requires more water for cultivation than *E. lathyris*.

Buchanan and his co-workers conducted an extensive program in which they screened 200 plant species covering 57 families for their suitability as renewable source of hydrocarbon, protein, carbohydrate and rubber (Buchanan et al. 1978a, 1978b, 1978c, 1979a, 1979b, 1980). Each species was rated according to its botanical characteristics and potential as a multi-use hydrocarbon-producing plant. In the botanical evaluation, plants were rated on a scale of 1 to 5, with a rating of 1 (highest value) assigned to vigorous growing perennial species that could potentially be harvested by clipping with rapid re-growth from rootstock. Species with a cumulative score of 11 or less were considered as possible candidates for development as bio-crude plants producing up to 22.4 tons ha^{-1} $year^{-1}$ dry matter.

An analytical screening program was conducted by the USDA (Bagby et al. 1981; Campbell 1983; Carr et al. 1986; Carr and Bagby 1987; Abbott et al. 1990; Seiler et al. 1991) to evaluate and identify plant species as sources of high-energy, easily extractable compounds suitable for fuel chemicals and petroleum-sparing chemical feedstock. Plant families that yielded more than one promising species were Anacardiaceae, Asclepiadaceae, Caprifoliaceae, Compositae, Eupforbiaceae and Labiaceae.

Augustus et al. (2002, 2003) screened 22 taxa found in the Western Ghats of India as potential alternative crops for the production of renewable energy oil hydrocarbon and phytochemicals. The highest hydrocarbon yields were observed in *Carissa carandas* (1.7%) and *Jatropha gossypifolia* (1.7%). The highest polyphenol fraction was observed in *Dodonaea viscosa* (17.1%), *Carissa carandas* (7.7%), *Swietenia mahagoni* (6.6%) and *Jatropha glandulifera* (6.2%). The highest oil content was observed in *Aganosma cymosa* (10.3%), *Carissa carandas* (5.8%), and *Argemone mexicana* (5.0%). *Swietenia mahagoni* yielded the highest protein content, at 8.1%. The gross heat values of 4,175.0 cal/g (17.5 MJ/kg) for *Lochnera rosea* and 4,112.0 cal/g for *Dalbergia sissoo* were the highest among the species analysed. Nuclear magnetic resonance (NMR) spectra of the hydrocarbon fractions revealed the presence of cis-polyisoprene (natural rubber) and trans-polyisoprene (gutta). Cis- and trans-polyisoprenes are potential alternative energy sources for fuel and/or industrial raw materials. The potential exists to grow these alternate crops in areas of underutilised land, subsequently stimulating industrial and economic growth.

Using cyclohexane-methanol solvent for extraction, Adams and co-workers (Adams 1982; Adams and McChesney 1983; Adams et al. 1983, 1984a, 1984b, 1986) also evaluated 80 species to establish the potential of *Asclepias speciosa* as an energy crop for production of renewable hydrocarbon fuel. The dried plant materials were extracted successively with hexane and methanol, affording 3.8 and 17.5% extracts, respectively. Chemical analyses of nonpolar extractables of the aerial parts of this plant showed that pentacyclic triterpenoids and sterols accounted for 90% of the refined hexane extract. The methanol extract consisted mainly of sugars. The residual plant material appeared to be nontoxic and equivalent to alfalfa hay in digestibility by sheep. The bio-crude content ranged from 4% to 11% on a dry weight basis. Use of hexane-extractable bio-crude of *Asclepias rotundifolia* as a source of liquid fuel was also explored. The chemical structures of the major triterpenol and their esters are shown in Fig 2.1.

Emon and Seiber (1984) conducted an analytical programme on two milkweeds, *Asclepias speciosa* and *Asclepias curassavica*. The constituents and latex extracted from both these plants showed that these milkweeds were excellent species in terms of their high calorie contents and organic composition. *A. speciosa* was found to

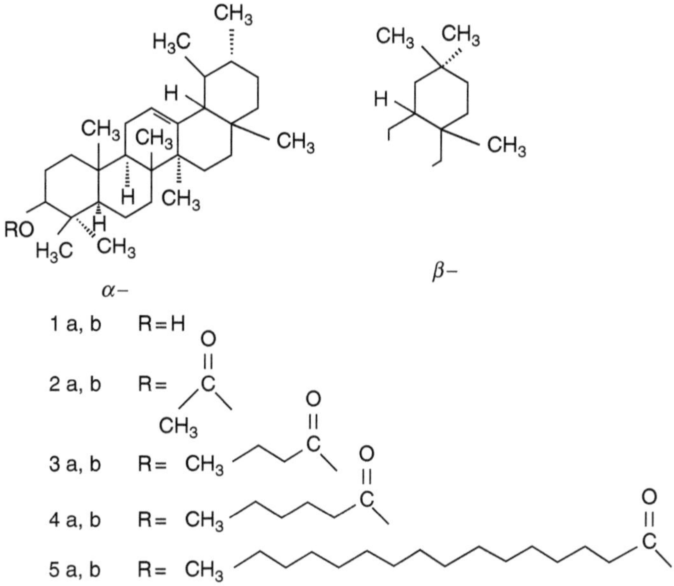

Fig. 2.1 Structure of the major triterpenols and their esters

have greater potential than *A. curassavica* for use as fuel owing to the higher calorie content of its latex. The latex of *A. speciosa* was found to contain more reduced chemicals like α and β-amyrin, amyrin acetates and cis-polyisoprene, while *A. curassavica* latex contained primarily cardiac glycosides. In fact, the physical and chemical properties of *A. speciosa* were found to be comparable to those of the *Euphorbia* spp. already considered for use as fuel and/or chemical feedstock. Due to its higher content of toxic cardiac glycosides and lower energy content, *A. curassavica* could perhaps be useful for production of speciality chemicals rather than as an energy plant. During World Wars I and II there was avid interest in cultivating various *Asclepias* species as sources of numerous strategically important plant products, particularly kapok and rubber (Woodson 1954).

During a survey of the southern United States and North Mexico, McLaughlin and Hoffman (1982) found 195 plants with potential as feedstock for bio-crude production. Cyclohexane-ethanol was used for extraction, and *Euphorbia* and *Asclepias* were found to contain high amounts of bio-crude. Bio-crude was also extracted from resinous species of the family Compositae tribe Astereae.

Roth et al. (1982, 1984, 1985) studied 508 plant species for extraction of bio-crude oil, polyphenols and proteins. Erdman and Erdman (1981) also evaluated *Calotropis procera* as a potential source of bio-crude. Dried whole plant material afforded 4.35% hexane and 16.14% methanol extracts. The hexane extract was found to be rich in hydrocarbon; the ratio of carbon to hydrogen was similar to that of crude oil, and the heat value content was comparable to that of crude oil fuel oil and gasoline. Carruthers et al. (1984) also estimated the bio-crude potential of this plant in northern Australia. Studies were also carried out on other species for their suitability as a potential sources for hydrocarbon-like materials and chemical feedstock.

In Romania, Simionescu et al. (1987a, 1987b) studied some latex-bearing plants and found that the crude latex extracted with cyclohexane (4–5%) could be separated by means of acetone into two fractions, one of which was insoluble in acetone and contained oils, fatty acids, waxes, terpenes, etc.

The fresh water alga *Botrycoccus braunii* was found to yield liquid hydrocarbons amounting to ~30% of the dried sample. Maxwell et al. (1968) reported that up to 70% of this hydrocarbon-like material consisted of a $C_{34}H_{58}$ hydrocarbon called botryococcene. These hydrocarbons were either linear compounds C_nH_{2n-2} and C_nH_{2n-4} ($n = 25, 27, 29, 31$) or branched chain compounds C_nH_{2n-10} ($n = 34, 36, 37$). The hydrocarbon oils of *B. braunii* can be recovered by solvent extraction, and would then need to be cracked, hydrogenated and reformed in order to obtain conventional transport fuels (Casadevall 1981; Wolf 1983; Metzger et al. 1982; Hillen et al. 1984).

Pittosporum resiniferum is a tall tree bearing fruits bigger than a golf ball, the orange pulp of which, when pressed, yields a sticky oil that resembles petroleum. The oil consists mainly of α-pinnene (38%) myrcene (40%) *n*-nonane (3%) and heptane (5%; Nemethy and Calvin 1982; Fernandez 1984).

Margaris and Vokou (1985) carried out some experiments in Greece to study energy-rich forest plants. Among the plant species studied were 60 species of *Euphorbia* (Euphorbiaceae) found at almost all altitudes and existing in diverse habitats from marshes to forests.

2.3 Extraction and Characterisation of Plant Extracts

Until recently, the most detailed published studies on extraction procedures were those of Buchanan et al. (1978a, 1978b, 1978c) for *Asclepias syriaca*. In these procedures, plant materials were extracted in a soxhlet apparatus, first for 48 h using a polar solvent (acetone, methanol, etc.) followed by another 48 h extraction with a non-polar solvent such as hexane or cyclohexane. The solvents used for extraction of the hydrocarbon fraction were mixtures of benzene and hexane. This fractionation procedure yields crude products. Oil fractions were examined by thin layer chromatography (TLC), and hydrocarbon fractions were examined by infrared (IR) spectroscopy to determine whether they were natural rubber waxes or mixtures. The natural rubber samples were further examined by proton nuclear magnetic resonance (PMR) and gel permeation chromatography (GPC).

Swanson et al. (1979) extracted natural rubber from different plant species by following the procedure of Buchanan (1978a, 1978b, 1978c) outlined above. Samples of the dried hydrocarbon fractions were dispersed at room temperature in tetrahydrofuran (THF). Molecular weight and molecular weight distributions were compared by GPC for rubber, Guayule and *Hevea*.

Seiler et al. (1991) extracted oil, polyphenol, hydrocarbon and protein from 28 taxa of *Helianthus* collected throughout the United States using acetone for 48 h soxhlet extraction. The acetone was then evaporated using a stream of nitrogen. The air-dried extract was partitioned between hexane and water:ethanol to obtain fractions referred to as 'oil' and 'polyphenol'. The residues were again extracted with hexane for hydrocarbons. Hydrocarbons were examined for the presence of rubber, gutta and waxes. Rubber and gutta were analysed for weight, average molecular weight (MW) and molecular weight distribution (MWD). This extraction procedure was followed by various workers in different parts of the world (Campbell 1983; Bhatia et al. 1984; Sharma and Babu 1984; Carr 1985; Carr et al. 1985, 1986, Sharma and Prasad 1986; Carr and Bagby 1987).

Roth et al. (1984) evaluated many leguminous plants at United States Northern Regional Research Center of the USDA by using soxhlet extraction, first with acetone and then with cyclohexane, from the whole plant excluding roots. The oil fractions were quantitatively analysed for classes of compound by TLC-flame ionization detection.

The major extractable components of *Asclepias linaria* and *Ilex verticillata* were extracted for 24 h in a stainless steel soxhlet extraction apparatus with 87:13 chloroform:methanol by Abbott et al. (1990) (Fig. 2.2). The residues were again extracted with water and the extracts were partitioned between hexane and methanol:water mixtures. The methanol extracts were again extracted with acetone and the residue identified as natural rubber. Initial chromatographic separation of the hexane/acetone soluble fraction was accomplished on an LC system 500 chromatograph by sequential elution with hexane, toluene, dichloromethane and methanol. Compounds were identified and confirmed by IR, HPLC, droplet counter current chromatography and TLC.

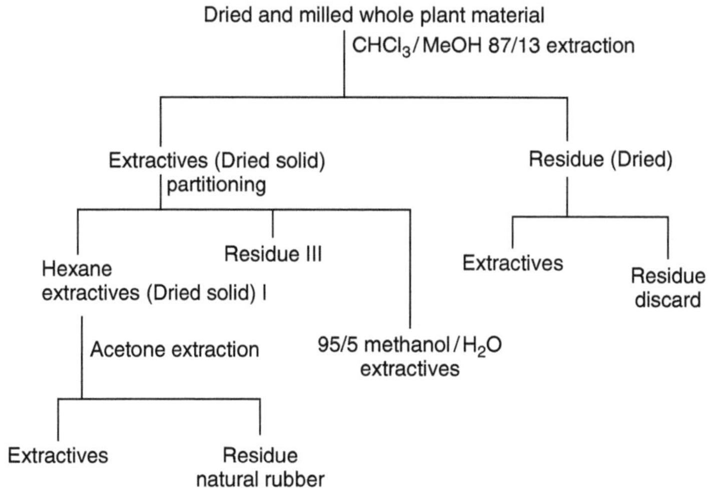

Fig. 2.2 Separation scheme for plant extracts

McLaughlin and Hoffmann (1982) used another extraction procedure from which direct estimation of the unit cost of production of biocrude and energy for each species could be evaluated. Previously, Hinman et al. (1980) had also followed this procedure, where plant materials are extracted with 300 ml cyclohexane for 12 h followed by a second extraction with 300 ml ethanol for 12 h using a soxhlet apparatus. Energy values for the two extracts and the residues were estimated from elemental analyses of each fraction from several species.

Erdman and Erdman (1981) extracted biocrude in a soxhlet extractor for 9 h with hexane. Subsequent extractions with methanol from the hexane extract residue were carried out. Gross heat value was determined by bomb calorimetry. Total carbon hydrogen and oxygen determinations were performed by pyrolysis using a Perkin Elmer model 240 analyser and a model RO 17 Leco oxygen analyser connected to an IR spectrophotometer.

McCheshney and Adams (1985) conducted experiments at the University of Mississippi to evaluate plant species as potential botanochemical sources of petroleum substitution compounds as well as for antibacterial/antifungal properties. Extracts were obtained by soxhlet extraction for 22 h with cyclohexane followed by methanol for 22 h. The methanol extracts were concentrated and dissolved in a mixture of ethyl acetate and water.

The plant species *Euphorbia lathyris* was extracted in a soxhlet apparatus with boiling heptane for 8 h, and the residue was then extracted with methanol for 8 h (Fig. 2.3; Nemethy et al. 1979). The extracts were further examined by gas chromatography (GC) with high-resolution mass spectroscopy (MS). This

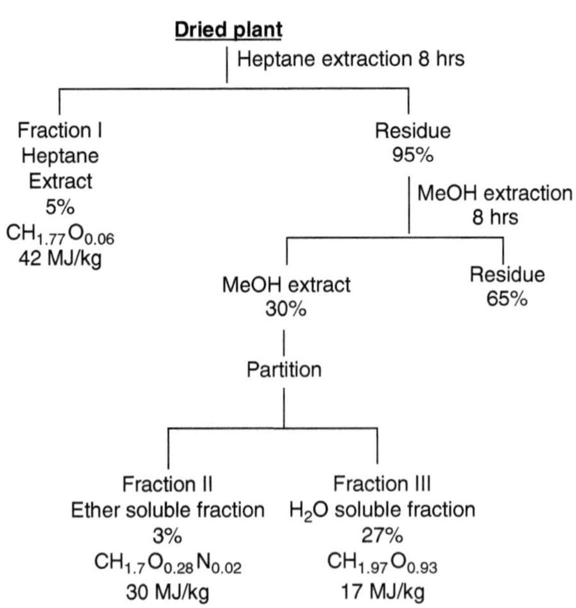

Fig. 2.3 Scheme of extraction

procedure for extraction was followed by other scientists, e.g. Ayerbe et al. (1984) in Spain, Clark et al. (1985) in the United States, and Sharma et al. (1990) in India.

In an experiment on *Euphorbia* species in Chile, Gneeco et al. (1988) extracted the biocrude with CH_2Cl_2 for 30 days at room temperature from chopped fresh plants. The plants were then oven-dried, ground and extracted again for 30 days. After removal of solvent, the material obtained was extracted with acetone. Representative fractions were characterised by quantitative analysis, IR spectroscopy, ^{13}C and ^{1}H-NMR spectroscopy and MS.

Latex collected from *Asclepias currassavica* and *Asclepias speciosa* plants was dried under vacuum at 20°C for at least 2 days, refluxed with acetone (60 ml) for 8 h, filtered and analysed by NMR, GC, GC-MS and IR spectroscopy (Emon and Seiber 1984). The acetone-insoluble residue was refluxed with methanol (40 ml) for another 8 h.

The latex of *Asclepias syriaca* was extracted with cyclohexane in a soxhlet apparatus for 10 h at 80°C (Simionescu et al. 1987a, 1987b). The extracts were cracked with the help of a suitable catalyst and the fractions analysed by GC and NMR spectroscopy.

Hammouda et al. (1984) in Egypt collected samples of latex by incision of small branches of *Euphorbia* in methanol. The methanol-preserved latex was evaporated under vacuum and the residue was exhaustively extracted with $(CH_3)_2CO$. The $(CH_3)_2CO$ extract was dissolved in 600 ml methanol:water and partitioned with *n*-hexane. The hexane extract was dissolved in hot $(CH_3)_2CO$ and left overnight. All fractions were then characterized by TLC, GLC and MS.

2.4 Processing of Bio-Crude for Hydrocarbon

Mobile Oil Corporation converted methanol into gasoline using zsm-5 zeolite as a catalyst (Weisz et al. 1979; Haag et al. 1980); later, the capability of catalytic production of high-grade fuel (gasoline) from biomass compounds such as jojoba oil, rubber latex from *Havea brassilensis*, corn and castor oils using a similar catalyst was also studied. Biomass compounds from these crops contained mostly paraffinic alcohols, esters, acids or terpenes, and their behaviour during catalytic conversion using zsm-5 zeolite as a catalyst in the presence of hydrogen was found to be similar to that of the products obtained from methanol, with the exception of their molecular dimensions, for conversion into gasoline and petroleum gases.

Vegetable oils have been hydrocracked using bi-functional catalysts at a 300–693 K/100–150 atmospheres. An 80% yield of usable fractions of gaseous hydrocarbons, gasoline and diesel was obtained (Nunes et al. 1986).

Adams et al. (1984a) and Erdman and Erdman (1981) analysed extracts from *A. speciosa* and *C. procera*, respectively, mainly for their elemental composition, and compared the data with that obtained with fossil fuel; the C and H analyses of the hexane extracts of both these plants closely resembled that of fossil fuel.

The bio-crude obtained from *Euphorbia lathyris* by Calvin et al. (1982) after extraction was subjected to catalytic cracking with special zeolite catalysts developed by Mobile Oil Corporation. The product obtained contained ethylene (10%), propylene (10%), toluene (20%), xylene (15%), C_5-C_{20} non-aromatics (21%), coke (5%), C_1-C_4 alkenes (10%), and fuel oil (10%). All these materials are useful in petro-chemical industrial processes (Calvin et al. 1983a, 1983b, 1984).

Catalytic cracking of solvent-extracted bio-crude of *Euphorbia tirucalli* was explored in the Philippines (NSTA 1982). The reaction conditions were quite drastic. High temperature and 12–15 psi pressures were employed in a catalyst to feed ratio of 1:1. In one typical run (catalyst feed ratio 1:1, temperature 650°F, reaction time 7 h) the resultant products contained non-condensable gases (69.23%), condensable gases (light oil; 21.54%), and residue (9.23%). In these experiments, the emphasis was on the production of heavy oil (Fernandez 1984). Bio-crude from *E. lathyris* and *Synadenium grantii* were also liquefied using the catalyst COM in the presence of hydrogen (Graham et al. 1991).

A comparison of cracked products obtained from *A. speciosa* (hexane extract) *E. lathyris* (acetone extract) and *Grindelia squarrosa* (methylene chloride extract) was also carried out (Largeau et al. 1981). Fluidised bed cracking of *A. speciosa* gave high yields of light gases (11%) and gasoline (58%), and low yields of diesel and heating oil. Cracking extracts of *E. lathyris* and *G. squarrosa* by the mobile group using zsm-5 zeolite as a catalyst gave good yields of middle distillates. It was observed that better liquid yields with increased middle distillates could be obtained by changing from a fluid bed to a fixed bed reactor. Cracking of *E. lathyris* and *G. squarrosa* resulted in 68% of liquid product comprising 52% gasoline and 16% middle distillates. In the case of the fixed bed reactor, the liquid yield was 78%, comprising 36% gasoline and 42% middle distillates. Cracking of *G. squarrosa* bio-crude in a fixed bed reactor resulted in 74% liquid yield, of which the yields of gasoline and middle distillates were 14% and 60%, respectively.

Hydrocracking of vegetable oils using a bi-functional catalyst (Al_2O_3-supported RH) yielded 80% of usable fractions of gaseous hydrocarbons, gasoline, and diesel (Nunes et al. 1986).

Conversion of biocrudes processed from latex of *Euphorbia royleana* and biomass of *C. gigantia* to hydrocarbon fuel preferably to middle distillate was studied. Craig and Coxworth (1984) evaluated the characteristics of plant oil extracted from *A. speciosa* using a fluidised bed pilot plant reactor and a high activity zeolite catalyst. The yields obtained from the FCC pilot were pretty much the same as commercial yields. The gasoline produced from the plant oil had a very high octane number. Analysis of the gasoline showed that the increase in octane number was due to the highly aromatic nature of the product.

Simionescu et al. (1987a, 1987b) studied the catalytic cracking of latex extracted from *Asclepias syrica*. As a result of the cracking process, the following products were obtained: gases (15–21% in weight), liquids (51–57%), and solid residue (5–8%). Further analyses showed the gases to contain C_1-C_4 saturated and unsaturated hydrocarbons. Analysis of the liquid fraction by GC and NMR revealed that it was composed of products similar to those found in autotype gasoline.

Thermogravimetric studies both on the hydrocarbon fraction and on polyisoprene rubber showed two decomposition stages. The hydrocarbon fraction showed a smaller weight loss as compared to polyisoprene rubber. At the same time, the activation energy calculated for the thermo-oxidative degradation of the polyisoprene rubber was 3.4 times greater than that of the hydrocarbon fraction. The difference lies in the higher hydrocarbon fraction, which, though a small in amount, could not be recommended as a substitute for polyisoprene rubber. However, the rough latex could be converted to fuel for diesel engines by a cracking process.

2.5 Current Research in India

Plant families like Euphorbiaceae, Asclepiadaceae, Apocynaceae, Urticaceae, Convolvulaceae, Sapotaceae, have been studied for their suitability as petrocrops by various workers in India (Bhatia et al. 1983, 1984 1989; Bhatia 1988; Kalita and Saikia 2000a, 2000b, 2001, 2004; Kalita 2008). Bio-crude potential was determined by preservation and coagulation of latex in the case of species amenable to latex tapping. In other species, the dried biomass was extracted with hexane-methanol. The bio-crude potential varied from 26% to 29%, in contrast to other species like *E. antisyphilitica* (8.46%). This study resulted in the identification of 17 potential petro-crops.

A new genus – *Capaifera* – was also evaluated as a source of fuel oil. *Capaifera lingsdorfil* and *Capaifera multijuga* are trees into which a hole can be drilled at a height of 90 cm from the ground to tap oil. The wood has a system of canals that contain oil. It was claimed that this oil could be used directly in an engine without further processing or purification. A single tree yields 20–30 l oil in 2–3 h in a single tapping, and can be tapped twice a year. Capaiba oil, as it is called, consists of 25 different compounds and each compound is a C_{15} sesquiterpene. Because of its molecular weight and volatility, the material can be used directly in diesel engines (Bhatia et al. 1983, 1984).

Aleurites moluccana was identified as a source of commercially produced lumbing oil (Pachauri and Dhawan 1987). The prospects are bright for developing a large industry for processing lumping oil. *Dipterocarpus laevis* – a species of plant comparable to Capaiba – is famous all over eastern India on account of its thin liquid balsam, which is commonly called wood oil. The properties of *D. laevis* oil are similar to those of Capaiba.

Marimuthu et al. (1989) studied 29 laticiferous taxa of different families for their suitability as alternative sources of renewable energy, rubber and other phytochemicals, and selected the most promising for large-scale cultivation. They observed that the majority of the species under investigation could be considered for large-scale cultivation as an alternative source of rubber intermediates, energy and other chemicals.

Another plant, *Pedilanthus tithymaloides*, was found to be a potential petrocrop with high biomass and hydrocarbon yields (Srivastava et al. 1985). In India this plant is cultivated as an ornamental or hedge plant, or even grown in marginal wastelands. Species such as *P. tithymaloides* var. cuculatus, *P. tithymaloides* var. verigatus, and *P. tithymaloides* (proper) have been shown to be promising varieties for development as petrocrops.

Sharma and Babu (1984) carried out a preliminary study at Dehradun, India on five latex-bearing plants. Chlorophyll, terpenes and other polar compounds could be obtained from these plants by extraction with acetone. Subsequent extraction of the plant materials with petroleum ether and benzene yielded hydrocarbon, which could be utilised as liquid fuel. *Gravellea robusta* and *Hakea saligna* contained long chain n-alkyl (C_{14}) resorcinol derivatives.

2.6 The Economics of Plant Hydrocarbon Production

There are few reports available on the economic aspects of the production and utilisation of hydrocarbon-producing plants in India. This is because we have not reached the stage of producing hydrocarbon plants on a large-scale, and conversion processes have yet to be standardised.

In the United States, Calvin (1979a, 1979b, 1979c) began his work with experimental planting of *E. lathyris* and *E. tirucalli*. His preliminary results on the chemical analysis of both whole plant extractions and latex proved that planting produced oil at a rate of 8–12% of the total dry weight. The result of a preliminary economic study of a conceptual process, including biomass operation and a processing plant that extracted the oil material, indicated a cost of US $ 3 to US $ 40 per barrel for the oil extracted. According to Calvin, the hydrocarbon yield from plants could be improved by proper selection of species and genetic manipulation, just as the Malaysians have improved the yields of rubber trees.

The Japanese had developed a plantation in Okinawa with great success. They achieved a production rate of 25–50 barrels of oil ha^{-1} $year^{-1}$. The Japanese expanded their acreage and made substantial efforts to develop a suitable extraction process for latex (Calvin 1978a, 1978b).

The economics and potential profitability of producing oil from *Pittosporium resiniferum* was also studied. Initial observations in the Philippines showed that a full bearing tree could yield as much 100 kg fruit per year. Using various oil extraction methods, a hectare of *P. resiniferum* could yield a flammable liquid at rates ranging from 2 to 28 barrels ha^{-1} (Keshri and Keshri 1994).

McLaughlin and Hoffmann (1982) carried out an economic analysis of four potential bio-crude crops. The biomass yields were 7–20 tons ha^{-1} $year^{-1}$ and the bio-crude yield 126.7 kg ha^{-1} $year^{-1}$. The cost of bio-crude per barrel for these species of plant was calculated at US $ 51–154. Keenan (1982) discussed the potential for biomass utilisation as a source of fuel petrochemicals and petroleum-sparing substances.

The direct use of photosynthetic materials for production of hydrogen fermentation to organic matter, and for thermo-chemical conversion has been proposed by many workers (Stewart et al. 1981; Weisz and Marshall 1982). Biomass offers a feasible renewable energy source, which can make a significant future contribution to the world's energy economy. The economics of processing, however, prevents the widespread use of such systems. Improvements in bioconversion efficiencies and reductions in harvesting and handling costs are needed to improve the economics of biofuels.

Little information on the cost of production of biomass and latex ha^{-1} $year^{-1}$ is available that would allow the cost of processing of bio-crude from biomass/latex to be calculated, thus allowing techno-economic feasibility studies. However, the processing costs of 1 kg bio-crude from biomass, and latex processing to liquid fuel worked out at Rupees (Rs.) 24.6 and Rs. 24.2, respectively, for a plant capacity of 25 ton/year. A continuous type reactor was the requirement for a plant capacity greater than 25 ton/year (Bhatia 1988).

The idea of using plants to create hydrocarbon-like materials as substitutes for our current energy sources has gained in importance, especially in some of the less developed areas of the world, which have a great deal of land not suitable for food production. Various efforts are being made toward this end in Japan, Thailand, Australia and Spain, and attempts are underway to improve agronomic yields, develop small-scale extraction plants, learn more about the composition of the plants, and study possible ways of modifying biosynthetic routes to produce more desirable end products (Calvin 1978a, 1978b). Therefore what is needed now is an effort on the part of the agricultural and energy community to commit itself to an energy agriculture, which would have long-term benefits for the entire world.

2.7 Research at NEIST Jorhat

The North East Institute of Science and Technology (NEIST), Jorhat, has conducted experiments to evaluate the chemical composition of some species of latex-bearing plants that grow in abundance in the forests of North Eastern India. Plant resources in the forests of North Eastern India are the richest and largest treasure house of biodiversity in this region. According to one survey, out of the total of 5,725 species endemic to India, 1,808 are endemic to this region (Nayar 1996). There are 1,500 species of trees, 337 species of climbers and climbing shrubs, 700 species of herbs, 300 species of ferns, 800 species of monocots and 350 species of grasses in the forests of this region (Baruah 1990). There are 99 species of laticiferous plants belonging to 63 genera and 25 families that occur in the plains and hilly areas of this region, and their phytogeographical status, nature of growth, frequency of occurrence and the parts used for various different purposes differ from species to species (Islam 1997). Therefore, careful preliminary chemical investigations of some abundantly available latex-bearing plant species were carried out. In the initial exploratory investigation, ten species of plant were selected and analysed for their cellulose, lignin, ash and C H N contents (Table 2.1). The plant materials were further extracted for crude protein, oil, polyphenol, hydrocarbon, unsaponifiables and free fatty acid using various

Table 2.1 Chemical composition (%) of different species of latex-bearing plants

Species	C	H	N	α-Cellulose	Lignin	Ash	Oil	Polyphenol	Hydrocarbon	Unsaponifiable matter	Fatty acid	Protein
Plumeria alba	44.89	6.72	1.26	59.56	26.42	3.76	3.56	7.89	1.36	49.3	26.8	7.87
Calotropis procera	43.49	6.65	1.01	56.42	24.43	5.62	3.07	8.42	2.04	64.3	24.3	11.26
Ficus carica	44.30	6.18	0.98	60.21	27.35	2.18	1.21	4.26	0.94	68.4	21.2	8.21
Erythrina variegata	41.21	6.01	3.83	51.72	21.64	3.41	1.01	5.26	0.29	58.9	27.4	7.62
Euphorbia nerrifolia	40.21	6.01	2.03	57.35	22.67	4.76	3.87	12.49	3.28	56.4	30.6	12.68
Allamanda catherlica	41.28	6.02	2.93	55.42	23.87	6.23	1.38	7.24	1.26	46.2	21.8	8.16
Nerium indicum	38.19	5.97	1.63	53.64	26.41	2.34	3.01	8.25	1.48	51.2	24.4	10.21
Tabernaemontana divarieta	37.81	5.83	2.03	61.24	25.51	4.56	1.36	7.42	0.86	62.1	31.5	9.26
Mimusops elengi	44.28	6.21	2.34	51.46	29.65	5.18	5.37	10.26	3.12	52.6	24.7	11.23
Euphorbia pulcherima	42.26	6.00	1.64	53.73	25.64	3.42	3.94	8.42	2.41	55.7	28.3	9.42

Table 2.2 Chemical constituents (%) of the plant biomass in different parts

Species	Family	Plant parts	Harvest moisture content	Oil	Polyphenol	Hydrocarbon
Plumeria alba	Apocynaceae	Leaf	87.5	0.21	3.86	0.26
		Stem	56.8	3.36	6.84	1.28
		Bark	89.3	4.74	7.62	1.78
		Whole plant	76.3	3.56	6.89	1.36
Calotropis procera	Asclepiadaceae	Leaf	69.1	1.68	2.58	1.06
		Stem	64.6	3.64	3.56	2.47
		Bark	76.9	3.89	3.96	2.60
		Whole plant	71.5	3.11	3.42	2.35
Euphorbia nerrifolia	Euphorbiaceae	Leaf	73.8	2.46	4.67	0.42
		Stem	62.4	3.56	9.63	2.58
		Bark	86.9	4.95	12.68	2.93
		Whole plant	78.6	3.87	11.49	2.28
Nerium indicum	Apocynaceae	Leaf	64.3	2.10	4.21	0.34
		Stem	62.4	3.71	6.23	1.36
		Bark	70.9	3.24	8.25	1.78
		Whole plant	67.2	3.24	7.54	1.45
Mimusops elengi	Sapotaceae	Leaf	65.2	1.36	1.46	1.21
		Stem	57.0	6.54	8.43	3.56
		Bark	61.4	8.21	8.91	3.92
		Whole plant	59.3	6.87	7.69	2.42

solvents. By adopting standard procedures, and after generating analytical data, five out of these ten species were finally selected for further study.

The quantitative amounts of different compounds such as hydrocarbon, oil, polyphenol, etc., in different parts of the plants as well as in whole plant extracts were determined (Table 2.2). The chemical constituents of the oil fractions extracted from these plant species were saponified and characterised for identification of compounds. The compounds thus identified were mostly sterols, fatty acids, triglycerides, and nonglyceride waxes. Hydrocarbon fractions were then characterised with the help of IR ^{1}H-NMR, ^{13}C-NMR, GC-MS and thermal analyses. The compounds were identified as polyisoprene-type rubber and other hydrocarbon-like compounds. The gross heat value and CHN of the hydrocarbon fraction were found to be comparable to that of crude oil, gasoline and lignite coal (Table 2.3). After further studies on their physico-chemical characteristics and biomass production, plant species suitable for large-scale cultivation as an alternative source for producing hydrocarbon and chemical feedstock will be suggested.

2.8 Conclusion

It is clear that any increase in our dependence on petroleum products as projected will impose unusually heavy burdens on economic development in the future. While every effort should be made to conserve stocks and utilise petroleum

Table 2.3 Characteristics of plant biomass extractives and fossil fuels

Species		Carbon (%)	Hydrogen (%)	Nitrogen (%)	Ash (%)	Gross heat (Cal/g)
Plumeria alba	Plant biomass	44.89	6.72	1.26	3.76	5,426
	Hexane extract	76.98	8.05	0.49	0.50	8,325
Calotropis procera	Plant biomass	43.49	6.65	1.01	5.62	6,145
	Hexane extract	74.13	11.34	0.37	0.65	9,837
Euphorbia nerrifolia	Plant biomass	40.21	6.01	2.03	4.76	5,132
	Hexane extract	76.30	10.88	0.30	0.82	9,218
Nerium indicum	Plant biomass	38.19	5.97	1.63	2.34	4,405
	Hexane extract	72.46	11.12	0.34	0.74	7,145
Mimusops elengi	Plant biomass	44.28	6.21	2.34	5.18	4,590
	Hexane extract	78.67	12.35	0.28	0.45	8,924
Anthracite coal[a]		79.70	2.90	–	9.60	7,156
Lignite coal[a]		40.60	6.90	–	5.90	3,889
Crude oil[b]		84.00	12.70	-	–	10,506
Gasoline[b]		84.90	14.76	–	–	11,528

[a]Bolz and Tuve 1973
[b]Ward 1978

efficiently in the Indian economy, it is essential to launch a major effort to explore and develop substitute sources of hydrocarbons. Some of the plants discussed here present interesting possibilities for the future but the viability of large-scale production will depend on the success we can achieve in research and development in this field over the next few years. The importance of sustained and large-scale research and development activity in this field requires the articulation of a well-conceived and clear strategy. Therefore, extensive research is needed on screening and selection of plant species that are well suited to specific sites, soil and climatic conditions prevalent in different parts of the country. As most of the species are wild, agro-technology for cultivation of these species needs to be standardised. Efforts may be needed to increase the biocrude potential of these species through genetic manipulation. Emphasis should also be put on developing the quality of the biocrude as well as suitable catalysts for reactions that will yield fuels of desirable quality. Therefore, an approach to maximising biomass productivity and increasing the biocrude content of quality product should be adopted, which will help alleviate the scarcity of petroleum products as well as creating a healthier environment.

The following areas of a biomass resource-based energy programme are specifically suggested for detailed investigation.

- Adoption of massive afforestation energy plantation projects in areas not used for conventional agriculture.
- Conversion of non-edible oils to diesel substitutes and large-scale trials of such a possibility.
- Bioconversion of ligno-cellulosic biomass to ethanol in an integrated system incorporating the production of biogas, bio-fertilizer and sugar for industrial applications.

- Large-scale and organised production of aquatic biomass species like algae, water hyacinth, etc., for integrated development of an energy-food-ecology system.
- Thermo-chemical and chemical conversion of biomass in small- and large-scale sectors to produce gaseous and liquid fuels and chemical feedstock.

References

Abbott TP, Patterson RE, Tjark LW, Palmer DM, Bogby MO (1990) Major extractable components in *Asclepias linaria* (Asclepiadaceae) and *Ilex verticillata* (Aquifoliaceae) – two potential hydrocarbon crops. Econ Bot 44:278–284

Adams RP (1982) Production of liquid fuels and chemical feedstocks from milkweed. In: Klass DL (ed) Energy from biomass and wastes. Institute of Gas Technology, Chicago, IL, pp 1113–1128

Adams RP, McChesney JD (1983) Phytochemicals for liquid fuel and petrochemical substitutions: extraction procedures and screening results. Econ Bot 37:207–215

Adams RP, Balandrin MF, Hogge L, Craig W, Price S (1983) Analysis of the non polar extractables of *Asclepias speciosa*. J Am Oil Chem Soc 60:1315–1318

Adams RP, Balandrin MF, Martineau JR (1984a) The showy milkweed *Asclepias speciosa*: a potential new semi arid land crop for energy and chemicals. Biomass 4:81–104

Adams RP, Baladrin MF, Martinean JR (1984b) Whole plant utilization of sunflowers. Biomass 4:87–100

Adams RP, Baladrin MF, Brown KJ, Stone GA, Gruel SM, Bagby MO (1986) Extraction of liquid fuels and chemicals from terrestrial higher plants – Part I. Yield from a survey of 614 Western United States plant taxa. Biomass 9:255–292

Augustus GDPS, Jayabalan M, Rajarathinam K, Ray AK, Seiler GJ (2002) Potential hydrocarbon producing species of Western Ghats, Tamilnadu, India. Biomass Bioenerg 23:165–169

Augustus GDPS, Jayabalan M, Seiler GJ (2003) Alternative energy sources from plants of Western Ghats (Tamilnadu, India). Biomass Bioenerg 24:437–444

Ayerbe L, Funes E, Tenorio JL, Ventas P, Mellado L (1984) *Euphorbia lathyris* as an energy crop – Part II Hydrocarbon and sugar productivity. Biomass 5:37–42

Bagby MO, Buchanan RA, Otey FH (1981) Multi use crops and botanochemical production. In: Klass DL (ed) Biomass as a non fossil fuel source. ACS Symp Ser 144:125–136

Baruah JN (1990) Natural resources of North Eastern region and their utilization. Keynote Address, Seminar on Natural Resources, Assoc Sci Soc Annual Conference pp 278–284

Bhatia VK (1988) Biomass as a non fossil fuel source – an overview. Res Ind 33:154–161

Bhatia VK, Srivastava GS, Garg VK, Gupta YK, Rawat SS, Singh S (1983) Study of laticiferous (latex bearing) plants as potential petrocrops. Fuel 62:953–955

Bhatia VK, Srivastava GS, Garg VK, Gupta YK, Rawat SS (1984) Petrocrops for fuel. Biomass 4:151–154

Bhatia VK, Mittal KG, Mehrotra RP, Mehrotra M (1989) Hydrocarbon fuels from biomass. Fuel 68:475–479

Bolz RE, Tuve GL (1973) Handbook of tables for applied engineering science. Chemical Rubber, Cleveland OH, pp 393–395

Buchanan RA, Cull IM, Otey FH, Russell CR (1978a) Hydrocarbon and rubber producing crops – evaluation of US plant species. Econ Bot 32:131–135

Buchanan RA, Otey FH, Russel CR, Cull IM (1978b) Whole plant oils, potential new industrial raw materials. J Am Oil Chem Soc 55:657–662

Buchanan RA, Cull IM, Otey FH, Russell CR (1978c) Hydrocarbon and rubber producing crops – evaluation of US plant species. Econ Bot 32:146–153

Buchanan RA, Otey FH, Russel CR (1979a) Multi use oil and hydrocarbon producing crops in adaptive systems for food material and energy production. Biores Dig 1:176–202
Buchanan RA, Swanson CL, Weisleder D, Cull IM (1979b) Gutta producing grasses. Phytochemistry 18:1069–1071
Buchanan RA, Otey FH, Bagby MO (1980) Botanochemicals. In: Wain TS, Kleiman R (eds) The resource potential in phytochemistry. Plenum, New York, pp 1–22
Calvin M (1978a) Chemistry, population, resources. Pure Appl Chem 50:407–426
Calvin M (1978b) Green factories. Chem Eng News 20:31–36
Calvin M (1979a) Petroleum plantation and synthetic chloroplast. Energy 4:851–870
Calvin M (1979b) Petroleum plantation for fuel and materials. Bioscience 29:533–538
Calvin M (1979c) Petroleum productions in solar energy: chemical conversion and storage. In: Hatula RR, Kings RB (eds) Solar energy: chemical conversion and storage, pp 221–228
Calvin M (1983a) New sources for fuel and materials. Science 291:24–26
Calvin M (1983b) Oils from plants. Lawrence Barkeley Laboratory, University Of California, pp 1–22
Calvin M (1984) Renewable fuel for the future. J Appl Biochem 6:3–18
Calvin M, Nemethy EK, Redenbaugh K, Otvos JW (1982) Plants as a direct source of fuel. Experientia 38:18–22
Campbell TA (1983) Chemical and agronomic evaluation of common milkweed. Econ Bot 37:174–180
Carr ME (1985) Plant species evaluated for new crop potential. Econ Bot 39:336–345
Carr ME, Bagby MO (1987) Tennessee plant species screened for renewable energy sources. Econ Bot 41:78–85
Carr ME, Phillips BS, Bagby MO (1985) Xerophytic species evaluated for renewable resources. Econ Bot 39:505–513
Carr ME, Bagby MO, Roth WB (1986) High oil and polyphenol producing species of the North West. J Am Oil Chem Soc 63:1460–1464
Carruthers IB, Griffiths DJ, Home V, Williams IR (1984) Hydrocarbons from *Calotropis procera* in northern Australia. Biomass 4:275–282
Casadevall E (1981) Renewable hydrocarbon production by a culture of green algae *Botryococcus braunii*, study of factors affecting the productivity. Lab Chem comm. Eur Communities (Rep) EUR 7160, pp 156–162
Clark DH, Adams RP, Lamb RC, Anderson MJ (1985) Near infrared analysis of hydrocarbon producing plant species. Biomass 8:1–11
Craig W, Coxworth EV (1984) Whole plant oils – a potential renewable fuel extender. In: Hasnain S (ed) Canadian Bioenergy R&D Seminar March, p 131
Emon JV, Seiber JN (1984) Chemical constituents and energy content of two milkweed *Asclepias curassavica* and *A. speciosa*. Econ Bot 39:47–55
Erdman MD, Erdman BA (1981) *Calotropis procera* as a source of plant hydrocarbons. Econ Bot 35:467–472
Fernandez EC (1984) The production and processing of hydrocarbon producing plants: state the art and trends in Philippines. Presented at the Regional Meeting on the production and processing of hydrocarbon-producing plants. National Science and Technology Authority, Manila May 21–25
Frick GA (1938) A new source of gasoline. Cactus Succ J 10:60
Gnecco S, Barulin J, Marticorena C, Ramirez A (1988) Chilean Euphorbiaceae species as source of fuels and raw chemicals. Biomass 15:165–173
Graham RG, Freel BA, Huffman DR, Bergougnou MA (1991) The production of liquid fuels and chemicals from biomass by Rapid Thermal Processing (RTP). In: CECDGXII, ENEA-Area Energetica (ed) Proceedings of 1st European Forum on Electricity Production from Biomass and Solid Wastes by Advanced Technologies. CEC, pp 80–86
Haag WO, Rodewald PG, Weisz PB (1980) Catalytic production of aromatics and olefins from plant materials. Symposium on alternate feedstocks for petrochemicals. American Chemists Society Meeting, Las Vegas, NV, August 24–25

Hammouda FM, Rizk AM, El Missiry MM, Radwar HM (1984) Constituents of the latex of Euphorbia royleana. Foroterpia Vol LV No 4
Hillen LW, Pollard G, Wake LV, White N (1984) Hydrocracking of the oils of *Botryococcus braunii* to transport fuels. Biotechnol Bioeng 24:193–205
Hinman CW, Hoffmann JJ, McLaughlin SP, Peoples TR (1980) Hydrocarbon production from arid land plant species. In: Proceedings of the Annual Meeting of the American Section of the International Solar Energy Society 31:110–114
Islam M (1997) Studies on certain laticiferous plants of North Eastern region, India. J Econ Tax Bot 21:1–11
Kalita D (2008) Hydrocarbon plant – new source of energy for future. Renew Sustain Energy Rev 12:455–471
Kalita D, Saikia CN (2000a) Evaluation of some latex bearing plants of N E India for energy and hydrocarbon. J Assoc Sci Soc 41:312–325
Kalita D, Saikia CN (2000b) Hydrocarbon fuels from plant sources. Chemical Weekly XLV:159–162
Kalita D, Saikia CN (2001) *Calotropis procera* and *Nerium indicum* – two potential plant sources of energy and hydrocarbon. Ind J Chem Technol 8:20–24
Kalita D, Saikia CN (2004) Chemical constituents and energy content of some latex bearing plants. Bioresour Technol 92:219–227
Keenan JD (1982) Biomass as alternative energy source. ASCE Publications, Journal of the Energy Division 108:11–22
Keshri JP, Keshri S (1994) Petroleum from plants. Invention Intelligence May:248–251
Largeau C, Casadevall E, Dif D (1981) Renewable hydrocarbon production from the alga *Botryococcus braunii*. In: Palz W, Chartier P, Hall DO (eds) Energy from biomass. 1st European Communities Conference. Applied Science, London, pp 653–658
Margaris NS, Vokou D (1985) Latex producing plant in Greece. Biomass 7:161–170
Marimuthu S, Subramanian RB, Kothari IL, Inamdar JA (1989) Laticiferous taxa as a source of energy and hydrocarbon. Econ Bot 43:255–261
Maxwell JR, Douglas AG, Eglinton G, McCormick A (1968) The Botryococcenes – hydrocarbons of novel structure from the alga *Botryococcus braunii*, Kützing. Phytochemistry 7:2157–2171
McChesney JD, Adams RP (1985) Co-evaluation of plant extracts as petrochemical substitutes and for biologically active compounds. Econ Bot 39:74–86
McLaughlin SP, Hoffmann JJ (1982) Survey of biocrude-producing plants from the Southwest. Econ Bot 36:323–339
Metzger P, Casadavall E, Coute A, Pouet Y (1982) Micro algae as a source of triglycerides. In: Strub A, Cartier P, Schleser G (eds) Energy from biomass. Applied Science, London, pp 339–343
Nayar MP (1996) Hotspots of endemic plants of India, Nepal and Bhutan, Tropical Botanic Garden and Research Institute, Thiruvananthapuram
Nemethy EK (1984) Biochemicals as an energy resource. CRC Crit Rev Plant Sci 2:117–129
Nemethy EK, Calvin M (1982) Terpenes from Pittosporaceae. Phytochemistry 21:2981–2982
Nemethy EK, Otvos JW, Calvin M (1979) Analysis of extractables from one *Euphorbia*. J Am Oil Chem Soc 55:957–960
Nemethy EK, Otvos JW, Calvin M (1981) Hydrocarbon from *Euphorbia lathyris*. Pure Appl Chem 53:1101–1108
NSTA (1982) Regional Meeting on the production and processing of hydrocarbon-producing plants. National Science and Technology Authority, Manila, Philippines
Nunes PP, Brodzki D, Bugli G, Mariadassou GD (1986) Hydrocracking of vegetable oils at high pressure and temperature. Rev Quim Ind (Rio De Janeiro) 54:8–13
Pachauri RK, Dhawan V (1987) Farming for petrol. Science Age 4 Jan:22–29
Roth WB, Cull IM, Buchanan RA, Bagby MO (1982) Whole plants as renewable energy resources checklist of 508 species analyzed for hydrocarbon, oil, polyphenol and protein. Trans Illinois State Acad Sci 75:217–231

Roth WB, Carr ME, Cull IM, Phillips BS, Bagby MO (1984) Evaluation of 107 legumes for renewable sources of energy. Econ Bot 38:358–364

Roth WB, Carr ME, Davis EA, Bagby MO (1985) *Garrya flavescens* S. Wats and *G. wrightii* Torr. – new sources of gutta-percha. Phytochemistry 24:183–194

Sharma DK, Babu CR (1984) Production of liquid hydrocarbon fuels and feedstocks from biomass – a renewable source. Fuel Sci Technol 3:49–53

Sharma DK, Prasad R (1986) Biocrude and solid fuel from laticiferous plants. Biomass 11:75–79

Sharma DK, Mbise HA, Singh SK (1990) Production of biocrude and fermentable sugar from *Croton bonplandiamum* and fermentation of hydrolyzate to get ethyl alcohol. Cellulose Chem Technol 24:193–200

Seiler GJ, Carr ME, Bagby MO (1991) Renewable resources from wild sunflowers (*Helianthus* spp, Asteraceae) Econ Bot 45:4–14

Simionescu CI, Cascaval CN, Rosu D, Rusan V (1987a) Complex and integral processing of *Asclepias syrica* L. Latex bearing plant. V. Catalytic conversion of the latex. Cellulose Chem Technol 21:77–83

Simionescu CI, Rusan V, Cascaval CN, Rosu D (1987b) Complex and integral processing of *Asclepias syrica* L. Latex bearing plant. VI. Extraction of hydrocarbons and their characterization. Cellulose Chem Technol 21:84–89

Srivastava GC, Bhatia VK, Dubey KC, Garg VK (1985) Potential of *Pedilanthus tithymaloides* as a petrocrop. Fuel 64:720–721

Steinhell P (1941) L'Euphorbe resinifere plante a caoutchouc et resine vernis. Rev Gen Caoutchouc 18:54–56

Stewart GA, Rawlins WHM, Quick GR, Begg JE Peacock WJ (1981) Oilseeds as a renewable source of diesel fuel. Search 12:107–113

Swanson CL, Buchanan RA, Otey FH (1979) Molecular weights of natural rubbers from selected temperate zone plants. J Appl Polym Sci 23:743–748

Ward CC (1978) Petroleum and other liquid fuels. In: Baumeister T (ed) Standard handbook for mechanical engineers. McGraw-Hill, New York, pp 7–14

Weisz PB, Marshall JF (1982) High grade fuel from biomass farming: potentials and constraints. Science 209:24–29

Weisz PB, Haag WO, Rodewald PG (1979) Catalytic production of high grade fuel (gasoline) from biomass compounds by shape selective catalysis. Science 206:57–58

Wolf FR (1983) *Botryococcus braunii* – an unusual hydrocarbon producing algae. Appl Biochem Biotechnol 8:249–260

Woodson RE (1954) The North American species of *Asclepias* L. Ann Mo Bot Gard 41:1–211

Chapter 3
Biology and Biotechnological Advances in *Jatropha curcas* – a Biodiesel Plant

Muppala P. Reddy and D.V.N. Sudheer Pamidimarri

Abstract Increasing global demand for energy, the impending depletion of fossil fuels, and concern over global climate change have lead to a resurgence in the development of alternative energy sources. Bio-fuels and bio-energy encompass a wide range of alternative sources of energy of biological origin, and offer excellent, environmentally friendly opportunities to address these issues. The recognition that *Jatropha* oil can yield high quality biodiesel has led to a surge of interest in *Jatropha* across the globe, more so in view of the potential for avoiding the dilemma of "food vs fuel". Hardiness, rapid growth, easy propagation, short gestation period, wide adaptation, and optimum plant size combine to make this species suitable for sustainable cultivation on wastelands. Besides biodiesel from the seed, the plant produces several useful products that also have commercial value. Large scale cultivation remains the single most important factor that will ultimately determine the success of *Jatropha* as a source of bio-fuel. The limited knowledge of the genetics of this species, low and inconsistent yields, the narrow genetic variability, and vulnerability to insects and diseases are major constraints in successful cultivation of *Jatropha* as a bio-fuel crop. Despite the optimal protein content and composition of the pressed cake, the presence of phorbol esters makes it unsuitable for consumption by livestock. A non-toxic variety with low or no phorbol ester content has been identified from Mexico, and the utility of pressed cake from this variety as livestock feed has been demonstrated successfully. In the absence of any morphological differences, identification of linked markers for toxic/non-toxic varieties will add value to the crop and facilitate further improvement. This chapter discusses current efforts towards assessing the diversity and

M.P. Reddy (✉)
Plant Stress Genomics and Technology Center, 4700 King Abdullah University of Science and Technology, Mail Box # 2053, Thuwal 23955-6900, Kingdom of Saudi Arabia
e-mail: reddymuppala@gmail.com

D.V.N.S. Pamidimarri
Central Salt and Marine Chemicals Research Institute (CSIR), G.B. Marg, Bhavnagar, Gujarat 364 021, India

K.G. Ramawat (ed.), *Desert Plants*,
DOI 10.1007/978-3-642-02550-1_3, © Springer-Verlag Berlin Heidelberg 2010

phylogeny of *Jatropha*, identification of specific markers for toxic and non-toxic varieties, and aspects of micropropagation and genetic transformation.

3.1 Introduction

Energy input is important for human social development and civilisation. Non-renewable hydrocarbons are being used as major energy sources the world over. However, their rapid depletion has brought about a situation where the whole world is facing an energy crisis, thus forcing the scientific community to search for renewable sources. In pursuit of a better life, bio-fuels and bio-energy are being used as alternatives to depleting resources, in particular to replace "petroleum products". These renewable sources of energy offer the prospect of increasing energy supplies in a self reliant way in developing countries like India, and also work to counteract the increasing levels of greenhouse gases (Fairless 2007). Plant-based fuels are among the best renewable sources, and their use can lead to a better balance of carbon dioxide and other greenhouse gases responsible for global warming.

Biodiesel, a methyl ester of fatty acids, made from edible or non-edible vegetable oils, is an appropriate alternative to petroleum-based diesel. Biodiesel derived from *Jatropha curcas* seed oil is fast emerging as an alternative to fossil fuel as it possesses the desired physiochemical characteristics, and its performance can even exceed that of conventional petro-diesel, albeit after suitable modification (Heller 1996; Openshaw 2000; Mandpe et al. 2005). *J. curcas* also has the potential to grow on eroded soils, is known for its drought endurance and is not browsed by animals.

3.2 *Jatropha curcas* – a Biodiesel Plant

Jatropha curcas belongs to the family Euphorbiaceae, is native to South America and widely distributed in South and Central America, Africa and Asia (Mandpe et al. 2005). The genus name *Jatropha* derives from the Greek word jatros (doctor) and troph'e (food). *J. curcas* is a deciduous shrub of 3 m tall and grows under wide range of arid and semi-arid climatic conditions. It can be cultivated successfully in regions with scanty to moderate rainfall and can be used to control soil erosion. Normally, five roots are formed from seedlings: one central and four peripheral. The leaves are cordate, 3–5 lobed and 10–15 cm long; stomata are hypostomatic and paracytic (Rubiaceous) type. *J. curcas* flowers twice a year, during May–June and September–November. Seeds usually mature within a month. The inflorescence is axillary paniculate polychasial cyme. The flowers are unisexual, monoecious yellowish green in glabrous or pubescent cymes at the end of the branches (Dehgan and Webster 1978). Cross pollination occurs by entomophily. Pollination is potentially influenced by both pollen depositions on stigma and nectar availability. Female flowers produce more nectar than male flowers. Fifty percent of female flowers set fruit, with a 53% fecundity rate, 32% apomixes rate and 2:3 seed:ovule ratio (Bhattacharya et al. 2005). Fruits are trilocular capsules, 1.5–3.0 cm long.

The seed contains 30–40% oil with 21% saturated fatty acids and 79% unsaturated fatty acids. Oil extracted from the seeds has been used traditionally as lamp oil and also in soap manufacture. Recognition that *Jatropha* oil can yield a high quality biodiesel has led to a surge of interest in *Jatropha* across the globe, more so in view of the potential for avoiding the "food vs fuel" dilemma (Gubitz et al. 1999; Heller 1996; Openshaw 2000; Mandpe et al. 2005; Ghosh et al. 2007). Besides biodiesel from the seed, the pressed cake contains 19.0% protein, 17.0% carbohydrates and 16.0% fibre. The deoiled cake is similar to chicken manure and can be used as an excellent organic fertiliser (Makkar and Becker 1997). Its high flash point (160–170°C) and cetane number (55–58) makes *Jatropha* diesel more eco-friendly than conventional petrochemicals (Ghosh et al. 2007; Mandpe et al. 2005).

3.2.1 *J. curcas* as Folk Medicine

Extracts of *J. curcas* are used in folk remedies for a number of physiological disorders (Duke and Wain 1991). The leaf juice is used as an external application for piles, and the leaf decoction in arthritis and venereal disease. Heated leaves are used on the breast as a lactagogue; leaf tea is used for marasmus and for jaundice (Watt and Breyer-Brandwijk 1962; Morton 1981; Perry 1980). The latex of *J. curcas* contains alkaloids such as jatrophine, jatropham, jatrophone and curcain, which are believed to have anti-cancerous properties. Latex is also used to dress sores, ulcers and inflamed tongues. The seeds are used for dropsy, gout, paralysis, and skin ailments (Watt and Breyer-Brandwijk 1962). The roots are used in decoctions as a mouthwash for bleeding gums and toothache. Antibacterial compounds have also been isolated from stem tissue of *J. curcas* (Morton 1981). In South Sudan, the seeds as well as the fruits are used as a contraceptive (List and Horhammer 1969–1979).

3.2.2 *J. curcas* Seed Cake as Fertiliser

The seed cake obtained after oil extraction contains toxic principles like curcin and phorbol esters, which makes it unsuitable for animal feed. However, potential applications as fertiliser or in biogas production have been reported (Staubmann et al. 1997; Gubitz et al. 1999; Karve 2005; Visser and Adriaans 2007). Being rich in nitrogen, the seed cake is an excellent source of plant nutrients and also helps in soil improvement (Jones and Miller 1991). Application of *Jatropha* biomass and seed cake increased yield in many crops (Sherchan et al. 1989; Kumar and Sharma 2008), and *Jatropha* seed cake could be a valuable resource in the low cost production of industrial enzymes (Mahanta et al. 2008).

3.2.3 Other Uses of *J. curcas*

Besides its medicinal value; the oil has long been used for illumination, and in the manufacture of soap and candles. Nuts collected from a non-toxic Mexican variety are roasted and consumed (Watt and Breyer-Brandwijk 1962). The latex is a strong inhibitor of watermelon mosaic virus (Tewari and Shukla 1982), and the bark is used as a fish poison (Watt and Breyer-Brandwijk 1962).

3.2.4 Toxicity of *J. curcas*

The presence of phorbol esters make the pressed cake unsuitable for livestock consumption despite the high protein content and favourable composition of other nutrients (Adolf et al. 1984; Makkar and Becker 1997). Attempts to inactivate the toxic principles through hydrothermal processing, solvent extraction, plus treatment with $NaHCO_3$ and ionising radiation did reduce the toxic principle in the meal, but these methods are time consuming and expensive (Aregheore et al. 2003; Martínez-Herrera et al. 2006). A non-toxic variety of *J. curcas* reported from Mexico and Central America was reported to contain no, or only a low amount of, toxic phorbol esters (Makkar et al. 1998a). Cultivation of non-toxic varieties would add value to the crop through utilisation of the deoiled seed cake as animal feed. However, there is no morphological difference between toxic and non-toxic varieties. Identification of linked markers for toxic/non-toxic varieties will add value to the crop and facilitate further improvement (Basha and Sujatha 2007; Sudheer et al. 2009b).

3.3 Genetic Diversity and Phylogenetics of *Jatropha* Species

Genetic diversity is one of the most valuable assets of the plant genetic resources available to mankind. An effective way of exploiting and managing available genetic resources efficiently is characterisation of germplasm. Species phylogenetics and evolution is essential not only for identification of various species but also to determine genetic relatedness within and among species. The information thus generated can be useful in breeding and molecular mapping. Assessment of diversity and phylogenetic relationships has traditionally been studied through morphological characteristics and isozyme analysis. However, such analyses have inherent disadvantages such as limited numbers of markers, and are often less effective due to their inconsistency and sensitivity to short-term environmental fluctuations (Crawford et al. 1995; Essilman et al. 1997; Francisco et al. 1996; Lesica et al. 1998; Lowrey and Crawford 1985; Soltiset al. 1992). Advances in the field of molecular biology provided many tools for studying genetic diversity at the genome level in order to investigate phylogenetic relationships among different species. DNA-based molecular analysis tools are ideal for germplasm characterisation and phylogenetic studies. Among currently available DNA fingerprinting techniques, restriction fragment length polymorphism (RFLP), amplified fragment length polymorphism (AFLP), randomly amplified polymorphic DNA (RAPD), micro-satellites markers/SSR, sequence characterised amplified regions (SCAR), sequence tagged sites (STS), and nuclear ribosomal DNA internal transcribed spacer (nrDNA-ITS) regions have been used to study genetic diversity and phylogenetics, and in generation of molecular markers for efficient use in breeding and genetic resource management.

Limited studies have been carried out on the genetic diversity and phylogenetics of genus *Jatropha*. Puangpaka and Thaya (2003) studied the karyology of five *Jatropha* species by staining chromosomes of the microsporocyte, and reported that, in most species, the chromosomes paired as bivalents at first metaphase and separated to 11:11 at first anaphase. The bivalent length ranged from 1 to 3.67 μm and most species had chromosome numbers of $2n = 22$. This study reported that *J. curcas* and *J. multifida* were chromosomally similar. *J. integerrima*'s 'Red' and 'Pink' flower had the same meiotic configuration of six ring II + five rod II, while the meiotic configuration of *J. podagrica* was eight ring II + three rod II. The karyology of *J. gossypifolia* was determined from first anaphase cells and the chromosomes separated to 11:11. Thus, according to Puangpaka and Thaya (2003), *J. curcas*, *J. multifida*, and *J. gossypifolia* appear to be related closely to each other, based on their meiotic configuration and morphological similarity. Carvalho et al. (2008) studied the genome size, base composition and karyotype of *J. curcas*. The results showed that the 2C value of the *J. curcas* genome was 0.85 pg, with an average GC base composition of 38.7%. The karyotype of *J. curcas* is made up of 22 relatively small metacentric and submetacentric chromosomes whose size ranges from 1.71 to 1.24 μm. Based on cytological and peroxidase isozyme studies, Prabakaran and Sujatha (1999) reported *J. tanjorensis* as a natural interspecific hybrid of *J. curcas* and *J. gossypifolia*.

Genetic diversity and phylogenetic analysis using RAPD, AFLP and nrDNA ITS analyses suggested interspecific genetic divergence of 97.74% (RAPD), 97.25% (AFLP) and 0.419 by nrDNA ITS sequence (Sudheer et al. 2009a, 2009c). Phylogenetic trees constructed using RAPD, AFLP and nrDNA ITS sequence data supported the highest genetic similarity between *J. curcas* and *J. integerrima*. The geographical distribution of *J. glandulifera* is wider, and it has morphologically distinct features that separate major and minor clades. The claim that *J. tanjorensis* is a spontaneous hybrid between *J. curcas* and *J. gossypifolia* (Prabakaran and Sujatha 1999) could not be supported by molecular data (Sudheer et al. 2009a, 2009c). AFLP fingerprint profiles and dendrograms showing the phylogenetic relationships between *Jatropha* species based on nrDNA ITS data are shown in Figs. 3.1 and 3.2. RAPD analysis revealed genetic diversity among different species of *Jatropha* of 80.2%, and confirmed the distinct genetic background of *J. glandulifera*. Species diagnostic markers reported using the RAPD technique (Fig. 3.3) could be utilised for identifying *Jatropha* species from any

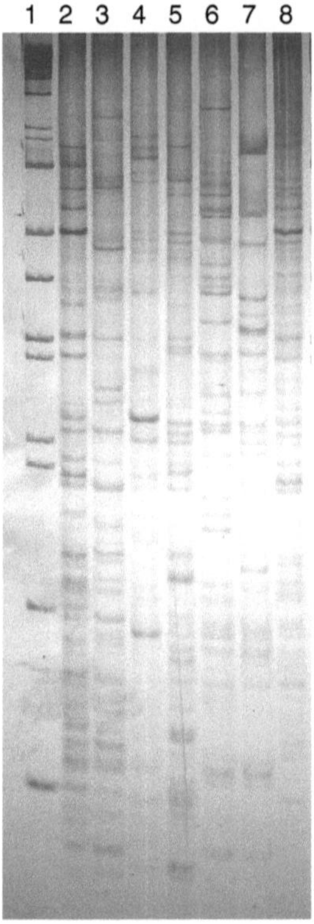

Fig. 3.1 Amplified fragment length polymorphism (AFLP) profiles of different species of Jatropha with selective amplification with primer set E ACA/M CAT, Lanes: *1* 1 kb + 100 bp DNA ladder mix, *2 J. curcas*, *3 J. tanjorensis*, *4 J. glandulifera*, *5 J. gossypifolia*, *6 J. multifida*, *7 J. podagrica*, *8 J. integerrima*

mixed population, and for characterisation of intraspecific hybrids (Ganesh Ram et al. 2008).

3.3.1 Intraspecific Genetic Diversity in J. curcas

Systematic provenance trials on performance evaluation in *J. curcas* are limited. Nicaraguan and Cape Verde provenances were tested in two different sites in Northern Nicaragua. The Nicaraguan genotype has fewer branches, larger and pale leaves with bigger seeds, whereas Cape Verde provenance produced a high seed yield. In the Nicaraguan germplasm, a male sterile plant that produces more fruit than hermaphrodite types was observed. Heller (1996) assessed the performance of 13 provenances in multi-location field trials in Senegal and Cape Verde and observed significant differences in vegetative growth.

Genetic similarity studies of toxic and non-toxic varieties by RAPD indicated 96.3% similarity (Basha and Sujatha 2007). In another study, Sudheer et al. (2009b) reported 84.91% and 83.59% similarity by RAPD and AFLP, respectively. Inter- and intra-population studies using RAPD and inter simple sequence repeat (ISSR)

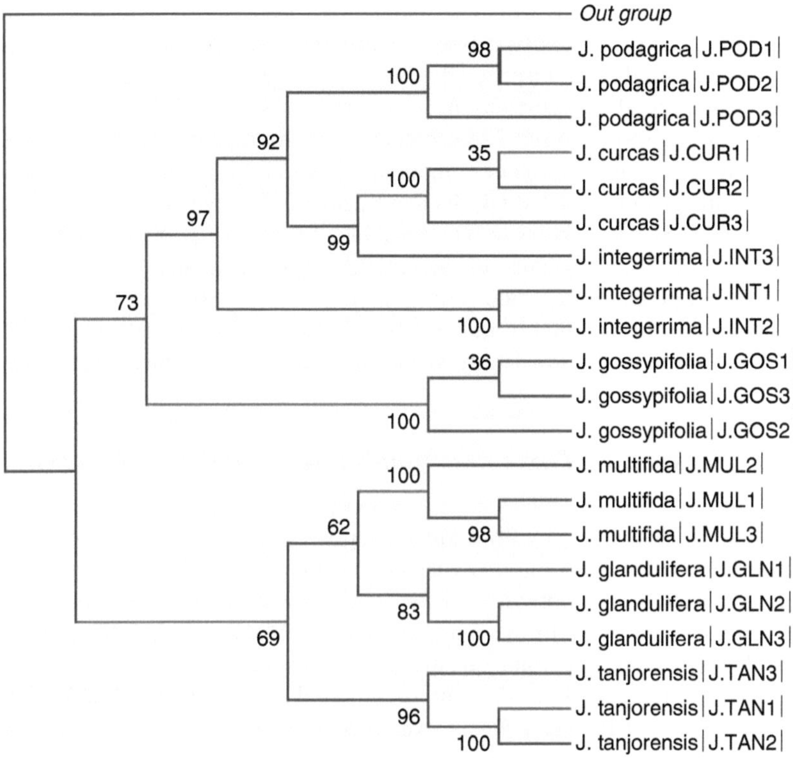

Fig. 3.2 Phylogenetic tree of genus *Jatropha* generated using nrDNA ITS sequence (Sudheer et al. 2009c)

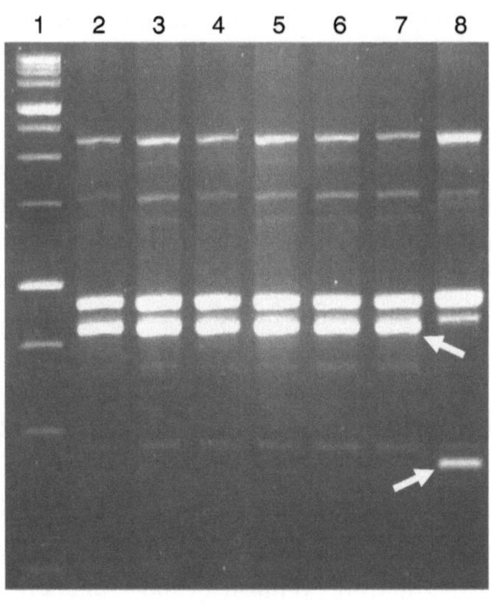

Fig. 3.3 RAPD profile with primer OPQ15 showing toxic and non-toxic specific diagnostic markers of *J. curcas*. Lanes: *1* 1 kb marker; *2–7* toxic varieties of *J. curcas*, *8* non-toxic variety of *J. curcas* (Sudheer et al. 2009b). *Arrows* Amplification products specific for toxic (lanes 1–7) and non-toxic (lane 8) varieties of *J. curcas*

analyses in 42 germplasms of *J. curcas* collected from different regions in India along with a non toxic genotype from Mexico showed 42% and 37.4% polymorphism by RAPD and ISSR, respectively (Basha and Sujahta 2007). Ganesh Ram et al. (2008) characterised 5 accessions from one region with 18 RAPD primers, and Ranade et al. (2008) analysed 22 accessions from six regions using seven RAPD and four directed amplification of minisatellite DNA (DAMD) primers. Population-specific markers were isolated (Basha and Sujahta 2007), which can be utilised to develop robust SCAR markers for identification of particular populations by single-step PCR. Intraspecific diversity analysis by RAPD and AFLP of *J. curcas* collected from different geographical regions of India indicated the existence of low genetic diversity (Sudheer et al. 2009d), and microsatellite analysis within a local population of *J. curcas* showed very low Hardy-Weinberg equilibrium (Sudheer et al. 2009d).

3.3.2 Markers for Toxic and Non-Toxic Varieties of J. curcas

A non-toxic *J. curcas* variety from Mexico has been reported, and its seeds can be used for human consumption after roasting (Makkar et al. 1998b). Cultivation of non-toxic varieties could provide oil for biodiesel and de-oiled cake as livestock feed, and thus add value to the crop (Becker and Makkar 1998). No significant morphological, qualitative and quantitative differences were observed between toxic and non-toxic varieties except for phorbol ester content (Makkar et al. 1998b; Makkar and Becker 1997). Development of markers that can help distinguish non-toxic from toxic varieties will not only add value to the product, but will help in the selective cultivation of non-toxic varieties.

Molecular markers specific to toxic and non-toxic varieties were identified using selective primers in a single PCR reaction using RAPD and AFLP techniques (Sudheer et al. 2009b). The primers IDT E-18 and OPL14, and the AFLP selective primer combination E-ACC/M-CAC resulted in polymorphic markers for both toxic and non-toxic varieties. Novel microsatellite markers were also tested for their polymorphism in toxic and non-toxic varieties, and 7 out of 12 markers succeeded in showing polymorphism (Sudheer et al. 2009b). The isolated SCAR markers confirmed their specificity and reproducibility in discriminating toxic and non-toxic genotypes (Basha and Sujahta 2007). The specific markers generated will be useful in distinguishing non-toxic from toxic varieties of *J. curcas* and can be utilised further in marker assisted selection (MAS), quantitative trait loci (QTL) analysis and further molecular breeding studies.

3.4 Tissue Culture and Genetic Transformation

During the last few decades, studies on tissue culture and micropropagation have increased tremendously due to their wide utility in large-scale multiplication and in genetic transformation. Winton (1968) regenerated the first complete plant from tissue culture of a tree species in 1968; since then, protocols for micropropagation of many plant species have been established and exploited commercially. Plant tissue culture and genetic transformation methods offer an important option for effective multiplication and improvement of plants within a limited time frame. To meet the large-scale demand and ensure a constant supply of quality planting material required the development of mass multiplication techniques. The initial steps in these approaches, i.e. micropropagation, plant regeneration and transformation, in improving *Jatropha* spp. have been discussed (Sujatha et al. 2008).

3.4.1 In Vitro Micropropagation of Jatropha

Several efforts have been made to establish protocols for micropropagation and regeneration from different explants of *Jatropha*. Induction of callus has been achieved from all parts of the plant, i.e. hypocotyl, cotyledon, leaf, petiole and stem (Sujatha et al. 2008; Rajore and Batra 2007; Kumar et al. 2008). Morphogenesis and regeneration (Sujatha and Mukta 1996; Shrivastava and Banerjee 2008), somatic embryogenesis (Sardana et al. 2000; Jha et al. 2007), regeneration from epicotyl callus (Qin et al. 2004; Rajore and Batra 2007), multiple shoot proliferation from shoot tips (Rajore and Batra 2005; Datta et al. 2007), and shoot bud proliferation from axillary nodes and leaf sections of non-toxic *J. curcas* have been reported (Sujatha et al. 2005). Despite the many regeneration systems developed from different explants of *J. curcas*, the presence of intermediary callus or callus-mediated regeneration should be avoided for the production of true-to-type plants. Recently, direct regeneration from leaf explants has been reported (Deore and Johnson 2008; Kumar 2008). According to Kumar (2008), regeneration efficiency varies significantly according to the source and type of explant. Regeneration in *J. curcas* is also reported to be highly genotype dependent (da Camara Machado et al. 1997). The comprehensive efforts towards developing suitable micropropagation protocols in *Jatropha* species, together with morphogenesis responses, are provided in Table 3.1 for ready reference.

3.4.2 Genetic Transformation Studies

Recent advances in genetic transformation and the availability of characterised genes with variety of advantages have made it possible to transfer chimeric gene/genes of academic/agronomic importance to the genome of recipient species to produce transgenic progeny with desired characteristics. This technology may help bypass some of the limitations of classical breeding programmes and reduce the time required to produce improved varieties. *Agrobacterium* (Ag)-mediated

Table 3.1 In vitro responses in *Jatropha* species. *pde* Proliferation and differentiation from endosperm, *ci* callus induction, *asr* adventitious shoot regeneration, *se* somatic embryogenesis, *dsr* direct shoot regeneration, *msbi* multiple shoot bud induction, *msp* multiple shoot proliferation

Jatropha species	Response	Reference
J. panduraefolia	pde	Johri and Bhojwani 1965
J. panduraefolia	pde	Srivastava 1971
J. panduraefolia	pde	Srivastava and Johri 1974
J. integerrima	asr	Sujatha and Dhingra 1993
J. curcas	dsr, cmr	Sujatha and Mukta 1996
J. curcas	dsr	Sardana et al. 1998
J. tanjorensis	asr	Prabakaran and Sujatha 1999
J. integerrima	asr	Sujatha and Reddy 2000
J. curcas	se	Sardana et al. 2000
J. curcas	msp	Rajore et al. 2002
J. curcas	Cl	Lin et al. 2002
J. curcas x *J. integerrima hybrids*	asr	Sujatha and Prabakaran 2003
J. curcas	Cl	Lu et al. 2003
J. curcas	Cl	Weida et al. 2003
J. curcas	dsr, cmr	Wei et al. 2004
J. curcas	msbi, asr	Sujatha et al. 2005
J. curcas	msp	Rajore and Batra 2005
J. curcas	cmr	Qin et al. 2006
J. curcas	cmr	Sharma et al. 2006
J. curcas	se	Jha et al. 2007
J. curcas	msbi	Datta et al. 2007
J. curcas	msp, se	Kalimuthu et al. 2007
J. curcas	cmr	Rajore and Batra 2007
J. curcas	dsr	Deore and Johnson 2008
J. curcas	asr	Li et al. 2008
J. curcas	msbi	Shrivastava and Banerjee 2008
J. curcas	msp, cmr	Thepsamran et al. 2008
J. curcas	dsr	Kumar 2008

transformation is the main method used for developing transgenic plants. Regeneration of plants from single cells is a prerequisite for Ag-mediated gene transfer in order to achieve homogenetically transformed plants. The choice of explant having competence for transformation and regeneration is also a crucial factor, as is an efficient micropropagation protocol for genetic transformation.

Li et al. (2006) were the first to report genetic transformation from callus cultures of *J. curcas* through Ag-mediated transformation. They further reported that *Agrobacterium* strain, explant type, marker gene selection, and bacterial incubation time all play an important role in transformation efficiency. The overall transformation efficiency was 13% (Li et al. 2006, 2008), and further studies will be needed in order to improve this figure.

3.5 Conclusions and Prospects

Jatropha is seen as a very promising option for the production of biofuel from degraded areas, generating rural employment, increasing environmental quality and providing primary energy carriers to energy-deficient areas. However, the adoption and implementation of the concept has advanced comparatively slowly so far. The major constraints limiting large-scale profitable cultivation are low and inconstant yields due to the non-availability of quality germplasm and proper agronomic practices. Varietal improvement in *Jatropha* should concentrate on enhancing and stabilising *Jatropha* productivity in various production systems, and on improving the quality of oil and seed meal for diversified utilisation.

Genetic diversity in *J. curcas* was found to be limited and there is an immediate need for genetic enhancement of *Jatropha*, which could attract more studies. Selection for broad genetic variance is positively associated with a reduction in the yield potential of the crop. Identification of genotypes capable of ensuring both profitable yield and wide genetic variability will be a challenging task requiring a complete set of information in order to understand how a given phenotype is constituted at the molecular, biochemical, reproductive and agronomic level. Therefore, rapid identification of molecular and metabolic markers that can define a required phenotype is important. A multidisciplinary approach based on molecular genetics, functional genomics, plant reproductive biology, biochemistry and agronomy will provide the accurate information required to identify genotypes for the constitution of new, improved seed populations.

Markers capable of discriminating between toxic and non-toxic accessions will provide an extremely useful tool for selection and assisted breeding towards the production of *J. curcas* varieties with low or null phorbol esters. Reducing phorbol esters through a transgenic approach would be a protracted balancing act. Molecular genetic tools may allow genetic improvement; however, one needs to be cautious regarding the stability of integration and expression of foreign gene/genes when taking a transgenic approach in long-lived plants like *Jatropha*.

References

Adolf W, Opferkuch HJ, Hecker E (1984) Irritant phorbol derivatives from four *Jatropha* species. Phytochemistry 23:129–132

Aregheore EM, Becker K, Makkar HPS (2003) Detoxification of a toxic variety of *Jatropha curcas* using heat and chemical treatments, and preliminary nutritional evaluation with rats. S Pac J Nat Sci 21:50–56

Basha SD, Sujatha M (2007) Inter- and intra-population variability of *J. curcas* (L.) characterized by RAPD and ISSR markers and development of population-specific SCAR markers. Euphytica 56:375–386

Becker K, Makkar HPS (1998) Toxic effects of phorbolesters in carp (*Cyprinus carpio* L.). Vet Hum Toxicol 40:82–86

Bhattacharya A, Datta K, Datta SK (2005) Floral biology, floral resource constrains and pollination limitation in *Jatrapha curcas* L. Pak J Biol Sci 8:456–460

Carvalho CR, Clarindo WR, Praca MM, Araujo FS, Carels N (2008) Genome size, base composition and karyotype of *Jatropha curcas* L., an important biofuel plant. Plant Sci 174:613–617

Crawford AM, Dodds KG, Ede AJ, Pierson CA, Montgomery GW, Garmonsway HG, Beattie AE, Davies K, Maddox JF, Kappes SW, Stone RT, Nguyen TC, Penty JM, Lord EA, Broom JE, Buitkamp J, Schwaiger W, Epplen JT, Matthew P, Matthews ME, Hulme DJ, Beh KJ, Mc-Graw RA, Beattie CW (1995) An autosomal genetic linkage map of the sheep genome. Genetics 140:703–724

Da Camara Machado A, Frick NS, Kremen R, Katinger H, da Camara Machado LM (1997) Biotechnological approaches to the improvement of *Jatropha curcas*. In: Proceedings of the first international symposium on biofuel and industrial products from *Jatropha curcas* and other tropical oil seed plants. 23–27 February 1997, Managua, Nicaragua, Abstract 15

Datta MM, Mukherjee P, Ghosh B, Jha TB (2007) In vitro clonal propagation of biodiesel plant (*Jatropha curcas* L.). Curr Sci 93:1438–1442

Dehgan B, Webster GL (1978) Three new species of Jatropha (Euphorbiaceae) from Western Mexico. Madron 25:30–39

Deore AC, Johnson TS (2008) High-frequency plant regeneration from leaf-disc cultures of *Jatropha curcas* L.: an important biodiesel. Plant Biotech Rep 2:7–11, doi: 10.1007/s11816-008-0042-y

Duke JA, Wain KK (1991) Medicinal plants of the world. USDA Computer listing of folk medicine of the world. http://www.ars-grin.gov/duke/

Essilman EJ, Crawford DJ, Brauner S, Stuessy TF, Anderson GJ, Silva OM (1997) RAPD marker diversity within and divergence among species of *Dendroseris* (Asteraceae: Lactuceae). Am J Bot 4:591–596

Fairless D (2007) Biofuel: the little shrub that could – maybe. Nature 449:652–655

Francisco OJ, Crawford DJ, Santos-Guerra A, Cravalho JA (1996) Isozyme differentiation in the endemic genus *Argyranthemum* (Asteraceae, Anthemideae) in the Macaronesian islands. Plant Syst Evol 202:137–152

Ganesh Ram S, Parthiban KT, Senthil KR, Thiruvengadam V, Paramathma M (2008) Genetic diversity among *Jatropha* species as reveled by RAPD markers. Genet Resour Crop Evol 55:803–809

Ghosh A, Chaudhary DR, Reddy MP, Rao SN, Chikara J, Pandya JB, Patolia JS, Gandhi MR, Adimurthy S, Vaghela N, Mishra S, Rathod MR, Prakash AR, Shethia BD, Upadhyay SC, Balakrishna V, Prakash CHR, Ghosh PK (2007) Prospects for *Jatropha* methyl ester (biodiesel) in India. Int J Environ Stud 64:659–674

Gubitz GM, Mittelbach M, Trabi M (1999) Exploitation of the tropical oil seed plant *Jatropha curcas* L. Bioresour Technol 67:73–82

Heller J (1996) Physic nut. *Jatropha curcas* L. Promoting the conservation and use of underutilized and neglected crops. International Plant Genetic Resources Institute, Rome, Italy, http:// www.ipgri.cgiar.org/publications/pdf/161.pdf

Jha TB, Mukherjee P, Datta MM (2007) Somatic embryogenesis in *Jatropha curcas* Linn., an important biofuel plant. Plant Biotechnol Rep 1:135–140

Johri BM, Bhojwani SS (1965) Growth responses of mature endosperm in cultures. Nature 208:1345–1347

Jones N, Miller JH (1991) *Jatropha curcas* a multipurpose species for problematic sites. Land Res Series 1:1–12

Kalimuthu K, Paulsamy S, Kumar RS, Sathya M (2007) In vitro propagation of the biodiesel plant *Jatropha curcas* L. Plant Tissue Cult Biotechnol 17:137–147

Karve AD (2005) Compact biogas plant-compact, low-cost digester for biogas from waste starch. Mimeo, http://www.bioenergylists.org/en/compactbiogas

Kumar A, Sharma S (2008) An evaluation of multipurpose oil seed crop for industrial uses (*Jatropha curcas* L.): a review. Ind Crops Prod 28:1–10

Kumar N (2008) Studies on regeneration and genetic transformation in *J.curcas*. PhD thesis, Bhavngar University, Bhavngar, India

Lesica P, Leary RF, Allendort FR, Bilderbecl DE (1998) Lack of genetic diversity within and among populations of an endangered plant, *Hawellia aquatilis*. Conserv Biol 2:275–282

Li M, Li H, Jiang H, Pan X, Wu G (2008) Establishment of an *Agrobacterium*-mediated cotyledon disc transformation method for *Jatropha curcas*. Plant Cell Tissue Organ Cult 92:173–181

Li MR, Li MQ, Wu GJ (2006) Study on factors influencing Agrobacterium-mediated transformation of *Jatropha curcas*. Fen Zi Xi Bao Sheng Wu Xue Bao 39:83–89

Lin J, Tang L, Chen F (2002) Tissue culture and plantlet regeneration of *Jatropha curcas*. Plant Physiol Commun 38:252

List PH, Horhammer L (1969–1979) Hager's handbuch der pharmazeutischen praxis, vols 2–6, Springer, Berlin

Lowrey TK, Crawford DJ (1985) Allozyme divergence and evolution in *Tetramolopium* (Compositae: Astereae) on the Hawaiian Islands. Syst Bot 10:64–72

Lu WD, Wei Q, Tang L, Yan F, Chen F (2003) Induction of callus from *Jatropha curcas* and rapid propagation. Chin J Appl Environ Biol 9:127–130

Mahanta N, Gupta A, Khare SK (2008) Production of protease and lipase by solvent-tolerant *Pseudomonas aeruginosa* PseA in solid-state fermentation using *Jatropha curcas* seed cake as substrate. Bioresour Technol 99:1729–1735

Makkar HPS, Becker K (1997) Potential of Jatropha seed cake as protein supplement in livestock feed and constraints to its utilization. In: Proceedings of the first international symposium on biofuel and industrial products from *Jatropha curcas* and other tropical oil seed plants. 23–27 February 1997, Managua, Nicaragua, pp 23–27

Makkar HPS, Aderibigbe AO, Becker K (1998a) Comparative evaluation of non-toxic and toxic varieties of *Jatropha curcas* for chemical composition, digestibility, protein degradability and toxic factors. Food Chem 62:207–215

Makkar HPS, Becker K, Schmook B (1998b) Edible provenances of *Jatropha curcas* from Quintana Roo state of Mexico and effect of roasting on antinutrient and toxic factors in seeds. Plant Foods Human Nutr 52:31–36

Mandpe S, Kadlaskar S, Degen W, Keppeler S (2005) On road testing of advanced common rail diesel vehicles with biodiesel from the *Jatropha curcas* plant. Soc Automot Eng 26: 356–364

Martínez-Herrera J, Sibdhiraju S, Francis G, Davila-Ortiz G, Becker K (2006) Chemical composition, toxic/antimetabolic constituents and effects of different treatments on their levels, in four provenances of *Jatropha curcas* L. from Mexico. Food Chem 96:80–89

Morton JF (1981) Atlas of medicinal plants of middle America. Development study of *Jatropha curcas* (Sabu Dum) oil as a substitute for diesel engine oil in Thailand. J Agric Assoc China 120:1–8

Openshaw (2000) A review of *Jatropha curcas:* an oil plant of unfulfilled promise. Biomass Bioenergy 9:1–15

Perry LM (1980) Medicinal plants of East and Southeast Asia. MIT Press, Cambridge, MA

Prabakaran AJ, Sujatha M (1999) *Jatropha tanjorensis* Ellis & Saroja: a natural interspecific hybrid occurring in Tamil Nadu, India. Genet Resour Crop Evol 46:213–218

Puangpaka SS, Thaya JJ (2003) Karyology of *Jatropha* (Euphorbiaceae) in Thailand. Thai For Bull (Bot) 31:105–112

Qin H, Song S-Q, Long C-L, Cheng H-Y (2006) Tissue culture and plant regeneration of *Jatropha curcas* (Euphorbiaceae). Acta Bot Yunnan 28:649–652

Qin W, Wei-Da L, Yi L, Shu-Lin P, Ying XU, Lin T, Fang C (2004) Plant regeneration from epicotyl explants of *Jatropha curcas*. J Plant Physiol Mol Biol 30:475–478

Rajore S, Batra A (2005) Efficient plant regeneration via shoot tip explant in *Jatropha curcas*. J Plant Biochem Biotechnol 14:73–75

Rajore S, Batra A (2007) An alternative source for regenerable organogenic callus induction in *Jatropha curcas* L. Indian J Biotechnol 6:545–548

Rajore S, Sardana J, Batra A (2002) In vitro cloning of *Jatropha curcas* L. J Plant Biol 29:195–198

Ranade SA, Srivastava AP, Rana TS, Srivastava J, Tuli R (2008) Easy assessment of diversity in *Jatropha curcas* L. plants using two single-primer amplification reaction (SPAR) methods. Biomass Bioenergy 32:533–540

Sardana J, Batra A, Sharma R (1998) In vitro plantlet formation and micropropagation of *Jatropha curcas* (L.). Adv Plant Sci Res India 11:167–169

Sardana J, Batra A, Ali DJ (2000) An expeditious method for regeneration of somatic embryos in *Jatropha curcas* L. Phytomorph 50:239–242

Sharma A, Kansal N, Shekhawat GS (2006) In vitro culture and plant regeneration of economically potent plant species *Jatropha curcas*. Biochem Cell Arch 6:323–327

Sherchan DP, Thapa YB, Khadka RJ, Tiwari TP (1989) Effect of green manure on rice production. PAC Occasional Paper 2, Pakhribas Agricultural Centre. Dhankuta, Koshi Zone, 1:12

Shrivastava S, Banerjee M (2008) In vitro clonal propagation of physic nut (*Jatropha curcas* L); influence of additives. Int J Integr Biol 3:73–79

Soltis PS, Soltis DE, Tucker TL, Lang A (1992) Allozyme variability is absent in the narrow endemic *Bensoniella oregona* (Saifragacear). Conserv Biol 6:131–134

Srivastava PS (1971) In vitro induction of triploid roots and shoots from mature endosperm of *Jatropha panduraefolia*. Z Pflanzenphysiol 66:93–96

Srivastava PS, Johri BM (1974) Morphogenesis in mature endosperm cultures of *Jatropha panduraefolia*. Beitr Biol Pflanz 50:255–268

Staubmann R, Foidl G, Foidl N, Gubitz GM, Lafferty RM, Arbizu VM, Steiner W (1997) Biogass production from *Jatropha curcas* cake. Appl Biochem Biotechnol 63:457–467

Sudheer DVNP, Pandya N, Reddy MP, Krishnan TR (2009a) Comparative study of interspecific genetic divergence and phylogenic analysis of genus *Jatropha* by RAPD and AFLP. Mol Biol Rep 36:901–907, doi: 10.1007/s11033-008-9261–0

Sudheer PDVN, Singh S, Shaik MG, Patel J, Reddy MP (2009b) Molecular characterization and identification of markers for toxic and non–toxic varieties of *Jatropha curcas* L. using RAPD, AFLP and SSR markers. Mol Biol Rep 36:1357–1364, doi 10.1007/s11033-008-9320-6

Sudheer PDVN, Balaji C, Reddy MP (2009c) Genetic diversity and phylogenetic analysis of genus *Jatropha* based on nuclear ribosomal DNA ITS sequence. Mol Biol Rep (in press) doi 10.1007/s11033-008-9401-6

Sudheer PDVN, Sinha R, Kothari P, Reddy MP (2009d) Isolation of novel microsatellites from *Jatropha curcas* L. and their cross species amplification. Mol Ecol Resour 9:431–433

Sujatha M, Dhingra M (1993) Rapid plant regeneration from various explants of *Jatropha* integerrima. Plant Cell Tissue Organ Cult 35:293–296

Sujatha M, Mukta N (1996) Morphogenesis and plant regeneration from tissue cultures of *J. curcas* L. Plant Cell Tissue Organ Cult 44:135–141

Sujatha M, Prabakaran AJ (2003) New ornamental *Jatropha* hybrids interspecific hybridization. Genet Resour Crop Evol 50:75–82

Sujatha M, Reddy TP (2000) Morphogenic responses of *Jatropha integerrina* explants to cytokinins. Biologia 55:99–104

Sujatha M, Makkar HPS, Becker K (2005) Shoot bud proliferation from axillary nodes and leaf sections of non-toxic *Jatropha curcas* L. Plant Growth Regul 47:83–90

Sujatha M, Reddy TP, Mahasi MJ (2008) Role of biotechnological interventions in the improvement of castor (*Ricinus communis* L.) and *Jatropha curcas* L. Biotechnol Adv 26:424–435

Tewari JP, Shukla IK (1982) Inhibition of infectivity of two strains of watermelon mosaic virus by latex of some angiosperms. Geobios 9:124–126

Thepsamran N, Thepsithar C, Thongpukdee A (2008) In vitro induction of shoots and roots from *Jatropha curcas* L. explants. J Hortic Sci Biotechnol 83:106–112

Visser J, Adriaans T (2007) Anaerobic digestion of *Jatropha curcas* press cake. Report produced for FACT, Ingenia, Eindhoven

Watt JM, Breyer-Brandwijk MG (1962) The medicinal and poisonous plants of Southern and Eastern Africa, 2nd edn. Livingstone, Edinburgh

Wei Q, Lu W-D, Liao Y, Pan S-L, Xu Y, Tang L, Chen F (2004) Plant regeneration from epicotyl explant of *Jatropha curcas*. J Plant Physiol Mol Biol 30:475–478
Weida L, Qim W, Lin Tang, Fang Y, Fang C (2003) Induction of callus from *Jatropha curcas* and its rapid propagation. Ying Yong Yu Huan Jing Sheng Wu Xue Bao 9:127–130
Winton LL (1968) Plantlet formation from aspen tissue culture. Science 160:1234–1235

Chapter 4
Biology of Annual Plants in Arid and Semi-Arid Desert Regions of China

Xuehua Li

Abstract Annual species are a major component of desert vegetation, with their unique traits playing a key role in vegetation restoration in arid and semi-arid deserts. This chapter presents a systematic discussion of current research status and expectations of future studies with respect to annual species in the deserts of China. To date, studies on annual species have been concerned mainly with vegetation distribution pattern, seed germination, soil seed banks, competitive mechanisms in annual species population, and so on. Due to the important ecological functions of annual species in desert floral systems, further studies on annual species, especially on seed adaptability and plant coexistence mechanisms, would benefit ecosystem conservation and restoration efforts, rational utilisation of resources, and sustainable development of arid and semi-arid deserts.

4.1 Introduction

Arid and semi-arid deserts worldwide are often characterised by abundant annual plants. Ludwig et al. (1988) reviewed the distribution of annual plants in North American deserts, and concluded that the winter-annual flora was richest in the Californian deserts and became relatively poorer towards the east, whereas summer-annuals were richest in the deserts of western Texas with a progressive decline in richness towards eastern California. Boulos and Al-Dosari (1994) reported that about 70% of the plant species known in Kuwait are annuals, with the proportion being substantially higher in desert areas. Areas with Mediterranean climates, such as the Mediterranean Basin, are rich in annuals (Goldblatt 1978).

X. Li
Institute of Applied Ecology, Chinese Academy of Sciences, Shenyang 110016, People's Republic of China
e-mail: lixuehua@iae.ac.cn

K.G. Ramawat (ed.), *Desert Plants*,
DOI 10.1007/978-3-642-02550-1_4, © Springer-Verlag Berlin Heidelberg 2010

In the arid and semi-arid deserts of China, annual plants also play a very important role in the vegetative flora. Liang et al. (2003) found that annual plants are constant synusiae, and their species and flora diversity are very high, accounting for about 16.7% of endemic species in the Alashan Desert. In Horqin Sandy Land, annual plants form an important part of the vegetation, accounting for about 25.2% of total plant species (The Inner Mongolia and Ninxia Survey Team of CAS 1985). Ephemerals, i.e. special annuals, fulfil an important function in sand fixation, and account for about 37.1% of the floristic composition of the Gurbantunggut Desert in northwest China (Zhang and Chen 2002).

Many studies on different aspects of annual plants, including micro-distribution characteristics, seed germination, seedling emergence and survival, soil seed bank pattern and dynamics, population regulation, and environmental factors and survival strategies, etc., have been carried out in arid and semi-arid regions all over the world (Gutierrez and Whitford 1987; Wang et al. 1998; Gutterman 2000; Xu and Li 2002; Long and Li 2003; Li et al. 2006). Guo (1998) reasoned that annual plants are good candidates for answering many key bio-ecological questions because they: (1) compose a large fraction of the diverse flora in arid and semi-arid deserts, (2) have a short life history – completing their life cycle in a relatively short and favourable lifespan, and (3) have highly predictable phenologies.

Most studies on annual plants have been focussed mainly on North American deserts, such as the Great Basin, and the Sonoran, Mojave, and Chihuahua Deserts, and the Negev Desert in Israel, and studies of other regions are few and unsystematic. In the arid and semi-arid deserts of China, although some the features of annual plants listed above have been investigated, the information available on annual plants in this region is very limited. Owing to the importance of annual plants in floral composition, and their unique traits as optimal experimental subjects for ecological studies, we aim to summarise the general biology of annual plants in the arid and semi-arid deserts of China. Our goal is to summarise current research in this area, and to ascertain the key aspects required of future studies on annual plants in the arid and semi-arid regions of China.

4.2 Species Diversity and Distribution Characteristics

Many years ago, Chinese researchers realised the important role played by annual plants in the rational utilisation of natural resources and regional sustainable development in arid and semi-arid deserts. In the 1960s, a comprehensive surveying team for the Inner Mongolia and Ningxia autonomous regions organised by the Chinese Academy of Sciences (The Inner Mongolia and Ninxia Survey Team of CAS 1985) summarised the specific diversity and ecological function of annual synusiae in deserts of Inner Mongolia and adjacent regions, and first suggested the concept of summer-rain-annuals. In the 1980s, Mao and Zhang (1994) surveyed the species types and distribution of ephemeral flora in the deserts of northern

Xingjiang Autonomous region. To date, no systematic information on species diversity and distribution of annual plants is available for other deserts of China.

4.2.1 Annual Plants in the Alashan Desert of Inner Mongolia

Based on the results of the comprehensive surveying team in Inner Mongolia, Liang et al. (2003) made further studies on annual plants in the Alashan Desert. There were 61 annual species in this region, belonging to 12 families and 35 genera, which could be divided into 4 annual species types and 12 areal types (Table 4.1).

In the Alashan Desert, annual plants synusiae are composed mainly of annual *Salsola* including 32 Chenopods, annual short grass including 9 Gramineous species, and annual *Artemisia* including 6 Composite species, which together account for 77% of the total annual plants. Annual weeds included 14 species from 9 different families, such as Boraginaceae, Cruciferae, Caryophyllaceae, Leguminosae, etc. Consistent with the floristic diversity, 26 annual plants belong to temperate zone types and 30 annual plants belong to typical Asia desert types. In general, annual plants are continuous synusiae with important ecological functions and showing highly adaptability to desert ecosystems, as reflected by the predominance of Chenopods and Gramineous species and the presence of both temperate zone and desert types in annual synusiae.

4.2.2 Ephemeral Plants in the Gurbanturggut Desert of Xingjiang

Ephemeral plants usually include ephemerals and ephemeroid. The present discussion focusses mainly on ephemerals, i.e. special annual plants. Ephemerals are distributed mainly in Central Asia, the Dzungarian Basin, the Mediterranean seashore, West Asia and North African. In China, ephemerals are found only in northern Xingjiang, with their eastern boundary defined by the eastern edge of the Dzungarian Basin.

Table 4.1 Species diversity and distribution of annual plants in Alashan Desert

Species types	Species number	Family	Habitats
Annual *Salsola*	32	Chenopodiaceae	Sand, gritty, gravel, and saline desert
Annual short grass	9	Gramineae	Sand, gritty, gravel, saline, and clayey desert
Annual *Artemisia*	6	Composiae	Sand, gritty, gravel, rocky and clayey desert
Annual weeds	14	Belong to nine families	Sand, gritty, gravel, rocky, and clayey desert

Due to the presence of ephemeral plants in flora groups in the deserts of China, and their importance to floristic and phytogeographical research, several studies of ephemerals in the Gurbanturggut Desert of Xingjiang have been conducted. Based on Mao and Zhang (1994), there are about 205 ephemeral species, of which 94% live only in Xingjiang province of China; however, this still leaves over 42 species belong to 35 genera distributed throughout the rest of China. Wang et al. (2003) conducted a further study on the distribution of ephemerals on the longitudinal dune surface in the southern area of the Gurbanturggut Desert. A total of 45 species were recorded at the study site, of which 29 were ephemerals. Annual ephemeral plants formed special spring synusiae communities, of which *Alyssum linifolium* and *Erodium oxytthynehum* were the dominant species. Most ephemerals usually germinate from late March to early April, and the end of their life-cycle usually occurs sometime between the last 10 days of June and the first 10 days of July, depending on climatic conditions. In northern Xingjiang, serious wind and sand disturbance usually occurs from March to June; here, average coverage of shrubs and perennials falls short of 10% on most of the sand dune surface, but average coverage of ephemerals on the interdune corridor and the two plinths reached 13.9% in March, 40.2% in April, and 14.1% in June 2002. It was concluded that ephemerals have a very important eco-function in stabilisation of the Gurbantunggut Desert.

4.3 Seed Germination Traits and Strategies

Seed germination behavior begins the life cycle of annual species and plays a key role in the persistence and dynamics of annual desert species (Tevis 1958; Mott 1974). Most germination studies on desert annuals have focussed on the proportion of seeds that could germinate in a growing season, and the effects of specific environmental factors on seed germination (Mott 1974; Grime et al. 1981; Baskin et al. 1993; Gutterman 1994). In China, there few sporadic studies available focus mainly on germination of shrubs, perennials and annuals, but germination data of plants, especially for annual plants in arid and semi-arid desert, is severely lacking. Therefore, further studies on the germination strategies and patterns of annual species will provide valuable information towards understanding species adaptability to changing environments.

4.3.1 Seed Germination Traits and Process

Fresh and mature seeds of 22 annual species were collected directly from plants in August to early November in 2003; a portion of seeds from each species was stored under chilled (–10 to 0.5°C) or dry (15–20°C) conditions. In December 2003, the capacity of freshly collected seeds for immediate germination was evaluated. At the beginning of the following growing season (early April), sub-samples of the stored seeds were removed from cold or dry storage, and their germination tested.

Table 4.2 Seed germination percentage of 22 annual species in Horqin Sandy Land

Species	Freshly collected	Chilling for 150 days	Dry storage for 150 days	Dry storage for 365 days
Artemisia sieversiana	99.6±0.42 a	97.6±1.17 a	92.4±1.17 b	97.9±0.90 a
Artemisia scoparia	54.8±7.58 a	58.4±2.56 a	68.0±1.41 a	82.3±1.68 b
Chloris virgata	98.2±0.44 a	100±0 a	90.0±3.52 b	99.6±0.41 a
Enneapogon brachystachyus	92.0±1.71 a	94.6±0.84 ab	95.5±0.98 b	99.5±0.50 c
Eragrostis pilosa	24.8±4.50 a	20.8±2.24 a	14.8±2.94 a	61.2±4.08 b
Setaria viridis	0	68.2±2.44 a	1.84±0.16 b	0
Setaria lutescens	44.0±3.23 a	98.5±0.63 b	89.3±4.18 c	90.2±0.97 c
Digitaria cilliaris	70.6±3.52 a	82.8±3.33 b	100±0 c	99.6±0.41 c
Echinlchloa crusgallis	8.48±1.82 a	100±0 b	99.2±0.80 b	100±0 b
Aristida adscensionis	90.2±3.12 a	89.9±2.06 a	91.9±5.51 a	90.6±2.87 a
Agriophyllum squarrosum	91.6±1.67 a	42.1±10.9 b	39.2±6.02 b	51.6±4.53 c
Corispermum candelabrum	3.60±1.13 a	6.16±1.92 a	5.13±1.49 a	9.60±3.21 a
Bassia dasyphylla	53.7±3.83 a	100±0 b	100±0 b	0
Kochia sieversiana	26.0±4.47 a	100±0 b	92.9±2.04 b	96.4±0.4 b
Chenopodium acuminatum	76.8±2.95 a	88.4±3.25 b	82.7±4.01 ab	26.0±2.76 c
Chenopodium glaucum	38.8±2.64 a	63.9±2.28 b	41.5±2.75 a	48.4±0.98 c
Chenopodium aristatum	55.6±3.06 a	92.8±1.74 b	79.6±2.93 c	60.3±2.88 a
Salsola ruhtenica	93.9±1.29 a	90.3±1.33 a	92.8±1.66 a	90.4±2.67 a
Leonurus sibiricus	70.2±4.68 a	80.5±1.17 b	94.5±0.98 c	80.8±6.40 a
Potulaca oleracea	28.4±2.79 a	80.5±2.31 b	92.7±3.06 c	98.0±0.91 c
Lappula myosotis	6.98±1.95 a	44.0±1.33 b	71.6±2.64 c	96.0±0.89 c
Euphorbia humifusa	89.1±1.88 a	48.0±5.22 b	24.8±2.06 c	0

All germination experiments were conducted in the same germination chamber, under visible radiation at a flux of 9,000 lux, over a 15-h day at 28°C, and dark for 9 h at 16°C, approximating the average daily maximum and minimum temperature at a 2–5 cm soil depth during the growing season in Horqin Sandy Land.

The 22 annual species differed considerably with respect to the capacity of freshly collected seeds to germinate (Table 4.2). The germination percentage of *Artemisia sieversiana*, *Chloris virgata*, *Enneapogon brachystachyus*, *Aristida adscensionis*, *Agriophyllum squarrosun*, *Salsola ruhtenica* and *Euphorbia humifusa* was near or above 90%, i.e., a high germination capacity. *Digitaria ciliaris*, *Chenopodium acuminatum* and *Leonurus sibiricus* exhibited germination percentages between 70% and 80%, and the remaining 14 species failed to achieve 60% germination, which meant that the seeds of these species exhibited some dormancy. Indeed, in these latter 14 species, seed dormancy was perhaps even stronger than that of other species.

According to Grime et al. (1981), dry storage and cold are the main treatments facilitating germination. After cold- or dry-storage for 150 days, the germination percentages of nine species with lower germination rates (below 60%) increased significantly to exceed 80%, meaning that these seeds can lose their dormancy gradually during chilling or dry storage. This breaking of dormancy, usually termed 'after-ripening', is a well-known phenomenon in seeds of many species, such as *Avena fatua* L. (Simpson 1965), and some desert plants, e.g., *Plantago coronopus* L. subsp. *commutata* (Guss.) Pilger (Plantaginaceae; Gutterman et al. 1998). Cold and dry storage did not affect the germination capacity of five species, including *A. sieversiana*, *C. virgata*, *E. brachystachyus*, *A. adscensionis*, and *S. ruhtenica*, but significantly reduced germination of *A. squarrosun* and *E. humifusa* to below 50%, which perhaps meant that seeds of these two annual species are dormant or lose vigor during the growing season. After cold and dry storage, five species, including *A. scoparia*, *E. pilosa*, *S. viridis*, *C. candelabrum*, *C. glaucum*, and *L. myosotis*, still showed strong dormancy in the growing season.

Seed vigor or longevity is an important factor in the formation of a persistent soil seed bank, so we examined this feature of 22 annual species stored under dry storage conditions for 365 days in the laboratory. Seeds of 11 annual species maintained high viability and germination capacity. Longer dry storage increased germination percentages of four species (*A. scoparia*, *C. glaucum*, *E. pilosa*, *L. myosotis*), but could not entirely remove seed dormancy of *C. glaucum* and *E. pilosa*. Seed germination of *A. squarrosun*, *C. acuminatum* and *C. aristatum* decreased but seed vigor was high, which suggested that secondary dormancy was induced by longer dry storage. Seeds of *B. dasyphylla* and *E. humifusa* are short-lived, with a longevity of less than 1 year. This may be attributed to the injurious effects of higher temperatures during the longer dry storage.

Regulation of the seed germination process is one form of adaptation to environmental uncertainty in desert annuals. In our study, 22 annual species showed only two different germination processes. For *S. glauca*, *A. squarrosun*, *K. sieversiana*, *C. glaucum* and *S. ruhtenica*, the germination curve of different experiments had two or more peaks, i.e., was multi-modal, as represented by *A. squarrosun*, *K. sieversiana* and *C. glaucum* (Fig. 4.1A). For the other 17 species, the germination curve was unimodal, i.e., had one evident peak, as represented by *C. virgata*, *P. aleracea* and *E. pilosa* (Fig. 4.1B).

4.3.2 Seed Germination Strategies and Adaptability to Different Environments

In Horqin Sandy Land, the life cycle of annual species usually takes place from April to October. For the 22 annual species considered here, seeds of 14 species showed a high germination capacity in the following growing season, and seeds of 6 of these species showed an obvious after-ripening mechanism. After-ripening is a very important survival mechanism for these annual species as dormancy prevents germination in the winter season, and the disappearance of dormancy in the growing season renders the seeds ready to germinate when the first rain falls. Seeds of eight species showed obvious dormancy, with only a fraction of the plant's seeds being able to germinate during the growing season. Cohen (1966) and Levins (1969) predicted that innate dormancy could be a viable evolutionary strategy that can help plants cope with environmental variability.

Based on the characteristics of germination percentages and processes in our 150-day chilling and dry storage experiments, we concluded that 22 annual species

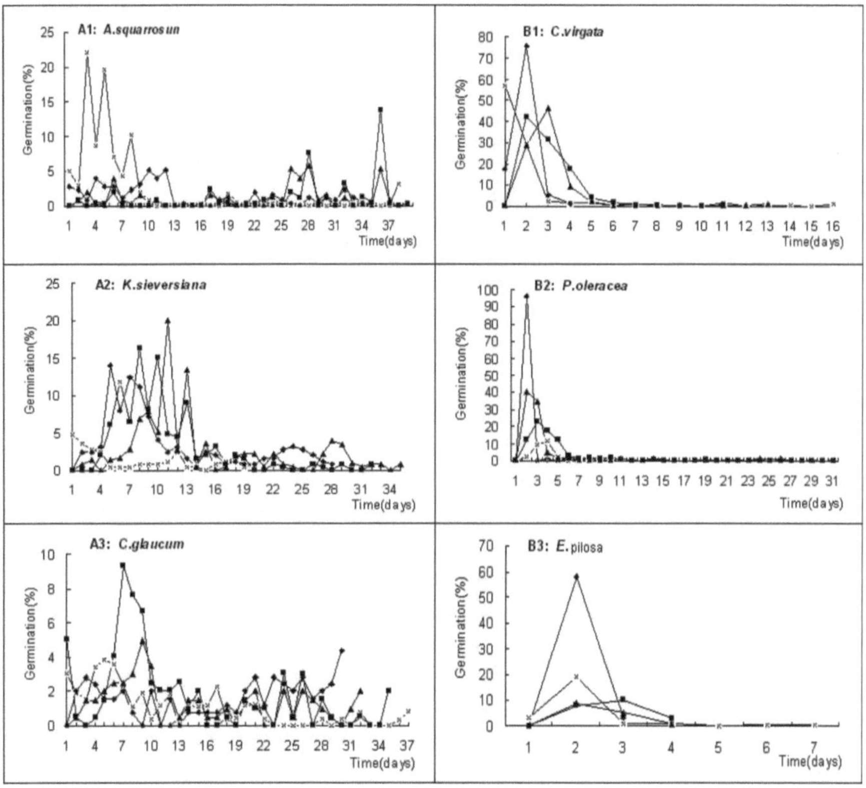

Fig. 4.1 Seed germination process with multi-modal (A1–A3) and unimodal germination (B1–B3) curves of six representative annual species in Horqin Sandy Land. × Freshly collected seeds, ■ seeds subjected to chilling for 150 days, ▲ seeds subjected to dry storage for 150 days, ♦ Seeds subjected to dry storage for 365 days

produced four germination patterns and showed two germination strategies during the growing season (Fig. 4.2). Eleven species (*C. virgata*, *E. brachystachyus*, *D. ciliaris*, *S. lutescens*, *E. crusgallis*, *A. sieversiana*, *C. acuminatum*, *B. dasyphylla*, *E. humifusa*, *L. sibiricus*, *P. oleracea*) had high germination percentages and exhibited a short germination duration and unimodal curve mode. Three species (*S. lutescens*, *K. sieversiana*, *S. ruhtenica*) had a high germination percentage, but exhibited longer duration and a multi-modal curve mode, indicating a within-season delay in germination. Eight species had a strong seed dormancy mechanism and lower germination percentages in the growing season, but five species (*E. pilosa*, *S. viridis*, *C. candelabrum*, *L. myosotis*, *E. humifusa*) exhibited short germination duration and a unimodal germination curve. The remaining three species (*A. scoparia*, *A. squarrosun*, *C. glaucum*) exhibited longer germination duration and a multi-modal curve. According to Bowers (1996), if species are divided into risk-taking and risk-avoiding species, then the 11 annual species with high germination percentages, short germination duration and a unimodal curve, showed a

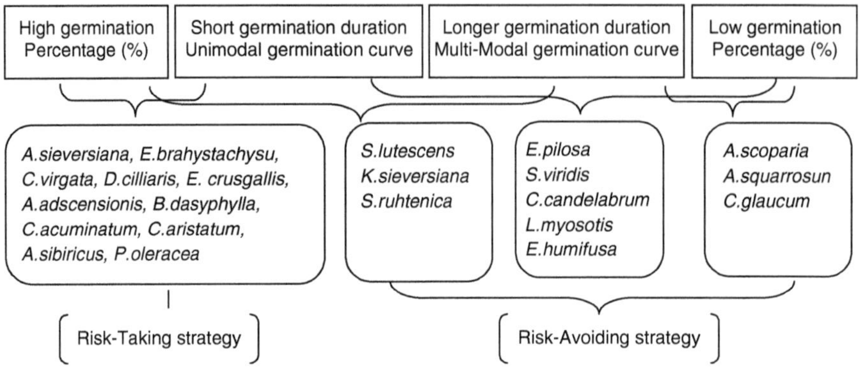

Fig. 4.2 Seed germination strategies of 22 annual species in Horqin Sandy Land

risk-taking strategy by germinating early in the growing season, and other 11 species showed a risk-avoiding strategy by decreasing the amount of seeds germinating, prolonging germination duration, or both.

Seed germination behaviour plays a key role in the persistence and dynamics of annual desert plants, and differences in this trait will obviously be related closely to distribution and other species traits (Bungard et al. 1997). For most annual species with a risk-taking strategy, such as *A. sieversiana*, *C. virgata*, *E. brachystachyus* and so on, a rapid germination strategy favours increasing the competitive advantage in their microhabitats. For species with low germination percentages, only a fraction of seeds germinate in favourable environmental conditions. Delays in germination within or beyond the growing season could help create an age structure within the belowground population of seeds for some annual species (Kalisz 1991). On the other hand, these annual species usually produce a large number of small seeds that tend to persist for longer in the soil (Yan et al. 2005), thus favouring the existence of persistent soil seed banks still further (Thompson et al. 1993). Although germination delays could reduce the competitive advantage gained by early emergence and establishment, in the case of *E. pilosa*, *A. scoparia* and *C. glaucum*, their tiny seed mass could compensate quantitatively for the shortage of germinated seeds, which in turn will increase the competitive advantage in many microhabitats. Thus, these annual species have a relatively broad range distribution and strong resistance to disturbance in semi-arid regions (Wang et al. 1959, 2004; Qin 2004).

4.4 Soil Seed Bank and its Relationship to Vegetation

4.4.1 Study Status of Soil Seed Banks in Deserts

The soil seed bank means all living seeds that exist both in litter and in soil (Simpson 1989). Soil seed banks play an important role in deserts and allow a great variety of species, mainly annuals, to survival in arid and semi-arid regions. Soil seed banks have long been the focus of research in many countries. In general, hotspots of soil seed bank research can be classified into several fields (Yu and Jiang 2003): (1) methodology (Bigwood and Inouye 1988; Grime 1989; Peng et al. 2000); (2) classification (Thompson and Grime 1979; Nakagoshi 1985; Garwood 1989; Hodgson and Grime 1990); (3) temporal and spatial patterns (Henderson et al. 1988; Kemp 1989; Levassor et al. 1990; Guo 1998); (4) dynamic modelling (Simpson 1989; Thomson 1995; Leishman et al. 2000); and (5) relationship of soil seed banks with aboveground vegetation (Russi et al. 1992; Holzapfel et al. 1992; Maranon 1998). In China, studies on soil seed banks began later than in other countries, and focussed mainly on soil seed banks patterns in the vegetation restoration and succession processes of forest and typical grassland. In the past 10 years, studies on the seed banks in the deserts of China have been expanded to include characteristics of soil seed banks of different desertification processes, and the dynamic mechanisms of soil seed banks affected by different disturbances such as grazing, enclosure, etc.

4.4.2 *Proportion of Annual Species in Soil Seed Banks*

The proportion of annual species in soil seed banks differs in different vegetation types and microhabitats. Tong et al. (2008) found that the proportion of annual species in soil seed banks was about 50% or lower, and seeds of several annuals, such as *Chenopodium album*, and *Salsola collina*, dominate in terms of soil seed density at different grazing exclusion restoring succession stages in degraded steppe in Xiligole dominated by two perennial species (*Artemisia frigida*+*Cleistogenes squarrosa*). In Ertok Front Banner of west Inner Mongolia, annual herbs represented around 90% of the life forms in the soil seed bank at different grazing intensities in steppe desert (Deying et al. 2008).

In Horqin Sandy Land, the characteristics of soil seed banks of sand dune successions were investigated. Annuals dominated in terms of both species and soil seed density in sand dune succession processes, accounting for ~90% of seeds (Fig. 4.3). Particularly in semi-fixed and fixed sand dunes, soil seed density was

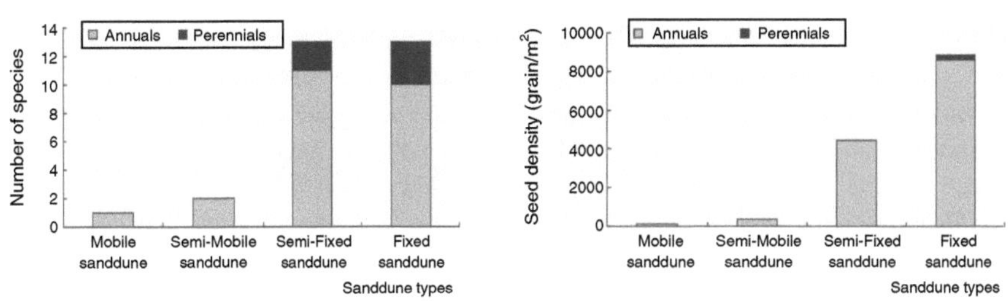

Fig. 4.3 Number of species and seed density in soil seed bank of different sand succession stages in Horqin Sandy Land

about 4,402 and 8,880 grains/m^2, respectively, and was dominated by three annual species (*E. pilosa*, *C. acuminatum*, *S. viridis*). In meadow grassland under grazing and harvesting (Jiang et al. 2004), the soil seed bank was composed mainly of some dwarf and short-life annuals, with their seeds accounting for 81.66% and 68.08% of the soil seed bank, respectively. Four annual species (*C. virgata*, *C. glaucum*, *D. cilliaris*, *S. viridis*) accounted for a high proportion of the soil seed bank of grazing meadow grassland, at 38.55%, 15.42%, 14.95% and 9.83%, respectively, and the proportion of *S. viridis* was the highest (52.7%) in harvesting meadow grassland. Other studies found similar results (Zhao and Bai 2001; Zhao et al. 2006).

4.4.3 Temporal and Spatial Patterns of Soil Seed Banks

The temporal and spatial patterns of soil seed banks have proved very important in helping us understand the vegetation processes of deserts, such as conservation, restoration and management. In Horqin Sandy Land, soil seed density at a depth of 0–50 mm were studied at the beginning and end of the growing season of four sand dune succession processes (Fig. 4.4). Only seeds of *A. squarrosum* were found in the soil of mobile sand dunes, with seeds of both *A. squarrosum* and *C. candelabrum* being present in the soil of semi-mobile sand dunes. For mobile and semi-mobile sand dunes, soil seed density did not differ between different sand depths (0–20 mm, 20–50 mm) or at different times (March, September). One reason for this is the severe shortage of species and seed density, and the second is that mobile dunes are subject to more severe drought and sand disturbance, such as sand erosion or sand burial. The temporal and spatial patterns of soil seed density of semi-fixed and fixed sand dunes differed significantly from that of mobile sand dunes.

Seed density in the soil surface layer (0–20 mm) was highest, and decreased significantly in underground soil (20–50 mm). This "V" spatial distribution pattern of seeds in soil is a good adaptive strategy for annual species because a surface location favours seed germination in the event of rain. The temporal pattern of soil seed banks of semi-fixed and fixed sand dunes follows the spatial pattern, i.e. seed density in March was significantly higher than that in September. We concluded that the soil seed banks of semi-fixed and fixed sand dunes are characterised by the dominance of immediately germinable seed banks (IGSB) in March, and non-immediately germinable seed banks (NIGSB) in September, of which the type was similar to type IIIa described by Thompson and Grime (1979).

Since *A. squarrosum* is the only species whose seeds are found in soil seed banks in mobile sand dunes, its spatial and temporal patterns have attracted much attention. Liu et al. (2007) investigated the distribution of soil seed banks in a deep sand profile along a transect of a mobile sand dune. Based on the temporal patterns of soil seed banks, seed persistence of *A. squarrosum* was concluded, which could ensure the sustainable supply of viable seeds, thus contributing to seedling recruitment in disturbed environments (Cavieres and Arroyo 2001; Urban 2005). In terms of spatial distribution, seeds of *A. squarrosum* were stored in the middle and upper

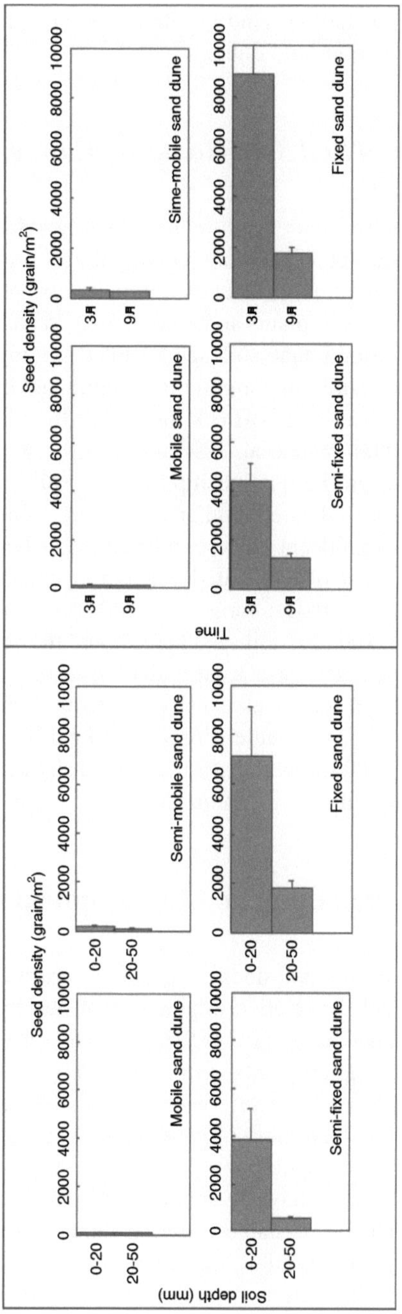

Fig. 4.4 Temporal and spatial patterns of soil seed banks of different sand succession stages in Horqin Sandy Land

windward slopes, and tended to be vertically concentrated at depths of 20–70 cm. These results validated two hypotheses: (1) the seed bank structure is related to position of the dune; (2) a shallow sand profile is not enough to elucidate the soil seed bank pattern of the active sand dune.

4.4.4 Relationship of Soil Seed Banks to Aboveground Vegetation

Most studies have shown two main results concerning the relationship between soil seed banks and aboveground vegetation: comparability and not, depending on community differences. In some grasslands dominated by perennials, there is no correlation between densities of plants and seed density in soil. But in recent years, some researchers have found that soil seed banks are highly consistent with aboveground vegetation, mainly in communities dominated by annual species and suffering unpredictable disturbances (Leck and Gravelline 1979; Levassor et al. 1990; Holzapfel et al. 1992; Peco et al. 1998).

In China, most studies on the relationship between soil seed banks and aboveground vegetation in deserts showed similar results, i.e. comparability. Zhao and Bai (2001) found that a significant correspondence exists between the number of soil seeds and the frequency of associated vegetation in sandy grassland after 7 years enclosure ($r = 0.47$, $P < 0.01$). Jiang et al. (2004) concluded that there was no significant correlation between the soil seed density and the vegetation of grassland under grazing, but a significant correlation under harvesting ($r = 0.76$, $P < 0.01$). Zhao et al. (2006) found that seed bank density was strongly related to the standing vegetation density in both grazed and enclosed grassland ($r_{grazed} = 0.88$, $r_{enclosed} = 0.76$, $P < 0.01$). Li et al. (2007) found that soil seed density in sand dune succession processes was correlated with species frequency ($r = 0.57$, $P < 0.01$).

4.5 Population Competition and Environmental Effects

Plant competitive mechanisms and density regulation are the important questions in population ecology, and the relative importance of density-dependence (Grime 1973) and non-density-dependence (Grubb 1985; Tilman 1988; Coomes et al. 2002) in population regulation of desert plants is still not widely recognised by ecologists.

In arid and semi-arid deserts, the responses of annual species to environmental variation are more sensitive than those of perennials. In the Shapotou region of the Tengger desert in China, the population dynamics and the competitive aspects of the annual species *E. poaeoides* have been studied (Zhang et al. 2001; Wang et al. 2001; Xu and Li 2002). The results can be summarised as follows: (1) the key factor restricting population size was rainfall, i.e. when the environmental fitness (water stress) is small, precipitation restricts population size, but density-dependent competition regulates population size if environmental fitness is large; (2) for individual plant growth, average individual weight decreases as population density increases due to the effect of intra-specific competition; (3) the survival curve of *E. poaeoides* shows a Deevey β type, which means that this species is a r-strategist (adopts a risk-taking strategy).

In addition, the population dynamics of *E. poaeoides* are closely related to seed germination behaviour in soil, the thickness of the soil crust and nutrient content. Long and Li (2003) conducted further studies on the effects of soil crusts on seedling survival and growth. *E. poaeoides* seedling survival was high in moss crust but low in sand, and survival was correlated positively with the water content of the upper soil. Another annual species, *B. dasyphylla*, showed a similar response. Zhao et al. (2008) studied how human activities and climate changes affect plants diversity in Horqin Sandy Grassland, and found that the diversity of annuals increased significantly under different grazing intensities over 14 years, but that there was no significant correlation with climate change.

4.6 Effects of Animals on Annual Species

Desert animals such as rodents, ants and birds, affect the composition and distribution of annual populations by selecting, fetching and conveying seeds (Beatley 1974; Gutierrez and Whitford 1987). In the Shapotou region of the Tengger desert, Xin and He (2003) found that a particular kind of ants, *Messor aciculatus* (F. Smith) harvested mainly seeds of five annual species: *S. viridis*, *S. collina*, *B. dasyphylla*, *E. humifusa*, and *S. ruthenica*. Especially for *S. viridis* and *S. collina*, the distribution of seeds in the nests of this ant altered the seed density in soil, and further affected the vegetation distribution of these two annual species in desert grassland. This study also found that seed germination was significantly decreased by storage in ant nests, especially for *S. collina*, *B. dasyphylla*, and *S. ruthenica*.

In the Junggar Desert, in the Xingjiang Autonomous Region of China, Yang et al. (2006) studied how the burrows of the great gerbil (*Rhombomys opimus*) affect small-scale plant communities. They found that disturbed soil on the bare ground on burrow sites promotes germination and growth of eight annual herbs, with an especially significant increase in the density and frequency of occurrence of five annual herbs growing at burrow sites: *Halogeton glomeratus*, *Girgensohnia oppositiflora*, *Plantago minuta*, *Lappula semiglabra*, and *Astragalus* sp., possibly facilitated by the burrowing activities of the great gerbil.

4.7 Conclusion

In the arid and semi-arid deserts of China, studies on many aspects of annual species have long been carried out, but to date such studies have not focussed on annual species. In recent years, many researchers have realised the important biological function of annual species in desert flora, and have begun to study some of the many aspects discussed above. Indeed, it is necessary to further understand the vegetation distribution patterns of annual species in deserts, the mechanisms used for seed adaptation to variation in habitat, and especially the co-existence mechanisms that exist between annual species and other species within desert environments. Such studies would facilitate predictions of environmental variation, and aid in restoring the vegetation and conserving biological diversity in Chinese arid and semi-arid deserts.

Acknowledgements The author thanks Prof. K.G. Ramawat for providing the chance to participate in this book, and also thanks many authors in China for permission to use their data and the results of published articles.

References

Baskin CC, Chesson PL, Baskin JM (1993) Annual seed dormancy cycles in two desert winter annuals. J Ecol 81:551–556

Beatley JC (1974) Phenological events and their environmental triggers in Mojave Desert ecosystems. Ecology 55:551–556

Bigwood DW, Inouye DW (1988) Spatial pattern analysis of seed banks: animal proved method and optimized sampling. Ecology 69:639–650

Boulos L, Al-Dosari M (1994) Checklist of the flora of Kuwait. J University Kuwait (Science) 21:203–218

Bowers JE (1996) Seedling emergence on Sonoran Desert dunes. J Arid Environ 33:63–72

Bungard RA, Daly GT, McNeil DL, Jones AV (1997) *Clematis vitalba* in a New Zealand native forest remnant: does seed germination explain distribution? N Z J Bot 35:525–534

Cavieres LA, Arroyo TK (2001) Persistent soil seed bank in *Phacelia secunda* (Hydrophyllaceae): experimental detection of variation along an altitudinal gradient in the Andes of central Chile (33°S). J Ecol 34:1293–1310

Cohen D (1966) Optimizing reproduction in a randomly varying environment. J Theor Biol 12:119–129

Coomes DA, Rees M, Turnbull L, Ratcliffe S (2002) On the mechanisms of coexistence among annual plant species, using neighborhood technique and simulation models. Plant Ecol 163:23–38

Deying, Aotegen, Burenjiya (2008) Effects of different grazing intensities on species diversity and life form of soil seed bank in steppe desert (in Chinese). Arid Zone Res 25:637–641

Garwood NC (1989) Tropical soil seed banks: a review. In: Leck MA (ed) Ecology of soil seed bank. Academic, San Diego, pp 149–209

Goldblatt P (1978) Analysis of the flora of southern Africa: its characteristics, relationships and origins. Ann Mo Bot Gard 65:369–436

Grime JP (1973) Control of species density in herbaceous vegetation. J Environ Manage 1: 152–167

Grime JP (1989) Seed bank in ecological perspective. In: Leck MA, Parker WT, Simpson RL (eds) Ecology of soil seed bank. Academic, San Diego, pp XVI–XXI

Grime JP, Mason G, Curtis AV, Rodman J, Band SR, Mowforth MAG, Neal AM, Shaw S (1981) A comparative study of germination characteristics in local flora. J Ecol 69:1017–1059

Grubb PJ (1985) Plant populations and vegetation in relation to habitats disturbance and competition problems of generalizations. In: White J (ed) The population structure of vegetation. Junk, Dordrecht

Guo Q (1998) Microhabitat differentiation in Chihuahua Desert plant communities. Plant Ecol 139:71–80

Gutierrez JR, Whitford WG (1987) Chihuahua Desert annuals: importance of water and nitrogen. Ecology 68:2032–2045

Gutterman Y (1994) Strategies of seed dispersal and germination in plants inhabiting deserts. Bot Rev 60:373–425

Gutterman Y (2000) Environmental factors and survival strategies of annual plant species in the Negev Desert, Israel. Plant Species Biol 15:113–125

Gutterman Y, Shem-Tov S, Gozlan S (1998) The effect of post maturation temperatures and duration on seed germinability of *Plantago coronopus* occurring in natural populations in the Negev Desert highlands. J Arid Environ 38:451–463

Henderson CB, Petersen KE, Redak RA (1988) Spatial and temporal patterns in the seed bank and vegetation of desert grassland community. J Ecol 76:717–728

Hodgson JG, Grime JP (1990) The role of dispersal mechanisms strategies and seed banks in the vegetation dynamics of the British landscape. In: Bunce RGH, Howard DC (eds) Species dispersal in agriculture habitats. Belhaven, London, pp 65–81

Holzapfel C, Schmidt W, Shmida A (1992) Effects of human-caused disturbances on the flora along a Mediterranean desert gradient. Flora 186:261–270

Jiang DM, Li RP, Liu ZM, Yan QL (2004) Soil seed bank in Horqin meadow grassland under grazing and harvesting (in Chinese). Chin J Appl Ecol 15:1860–1864

Kalisz S (1991) Experimental determination of seed bank age structure in the winter annual *Collinsia verna*. Ecology 72:575–585

Kemp PR (1989) Ecology of soil seed banks: In: Leck MA, Parker VT, Simpson RL (eds) Seed banks and vegetation processes in deserts. Academic, San Diego, pp 257–282

Leck MA, Gravelline KJ (1979) The seed bank of a freshwater tidal marsh. Am J Bot 66: 1006–1015

Leishman MR, Masters GJ, Clarke IP, Brown VK (2000) Seed bank dynamics: the role of fungal pathogens and climate change. Funct Ecol 14:293–344

Levassor C, Ortega M, Peco B (1990) Seed bank dynamics of Mediterranean pastures subjected to mechanical disturbance. J Veg Sci 1:339–344

Levins R (1969) Dormancy as an adaptive strategy. In: Woolhouse HW (ed) Dormancy and survival, symposia of the society for experimental biology. Academic, New York, p 23

Li XH, Li XL, Jiang DM, Liu ZM (2006) Germination strategy and ecological adapability of *Eragrostis pilosa* (in Chinese). Chin J Appl Ecol 17:607–610

Li XH, Han SJ, Zong WJ, Li XL, Jiang DM (2007) Characteristics of soil seed banks of sand dune succession process in Horqin Sandy Land (in Chinese). J Beijing Forestry University 29(2):66–69

Liang CZ, Liu ZL, Zhu ZY, Wang W (2003) Specific diversity and distribution characteristics of annual synusia in Alashan Desert (in Chinese). Chin J Appl Ecol 14:897–903

Liu ZM, Yan QL, Liu B, Ma JL, Luo YM (2007) Persistent soil seed bank in *Agriophyllum squarrosum* (Chenopodiaceae) in a deep sand profile: variation along a transect of an active sand dune. J Arid Environ 71:236–242

Long LQ, Li XR (2003) Effects of soil microbiotic crusts on seedling survival and seedling growth of two annual plants (in Chinese). J Desert Res 23:656–660

Ludwig JA, Cunningham GL, Whitson PD (1988) Distribution of annual plants in North American deserts. J Arid Environ 15:221–227

Mao ZM, Zhang DM (1994) The conspectus of ephemeral flora in Northern Xinjiang (in Chinese). Arid Zone Res 11:1–26

Maranon T (1998) Soil seed bank and community dynamics in an annual-dominated Mediterranean salt-marsh. J Veg Sci 9:371–378

Mott JJ (1974) Factors affecting seed germination in three annual species from an arid region of western Australia. J Ecol 62:699–709

Nakagoshi N (1985) Buried viable seeds in temperate forests. In: White J (ed) The population structure of vegetation. Junk, Dordrecht, pp 551–570

Peco B, Ortego M, Levassor C (1998) Similarity between seed bank and vegetation in Mediterranean grassland: a predictive modal. J Veg Sci 9:815–828
Peng J, Li XG, Fu YC, Liu YC (2000) Seed rain and seed bank of constructive species in evergreen broadleaved forest at Chongqing Simian Mountain (in Chinese). Chin J Appl Ecol 11: 22–24
Qin ZS (2004) Eragrostis Wolf (in Chinese). In: Qin ZS (ed) Flora plantarum herbacearum chinae boreali-orientalis, vol 10. Science Press, Beijing, p 89
Russi L, Cocks PS, Roberts EH (1992) Seed bank dynamics in a Mediterranean grassland. J Appl Ecol 29:763–771
Simpson RL (1989) Ecology of soil seed bank. Academic, San Diego, pp 149–209
Simpson GM (1965) Dormancy studies in seed of *Avena fatua* L. The role of gibberellin in embryo dormancy. Can J Bot 43:793–816
Tevis L Jr (1958) A population of desert ephemerals germinated by less than one inch of rain. Ecology 39:681–688
The Inner Mongolia and Ninxia Survey Team of CAS (1985) Vegetation in Inner Mongolia (in Chinese). Science Press, Beijing
Thompson K, Grime JP (1979) Seasonal variation in the seed bank of herbaceous species in ten contrasting habitats. J Ecol 67:893–921
Thompson K, Band SR, Hodgson JC (1993) Seed size and shape predict persistence in soil. Funct Ecol 7:236–241
Thomson DL (1995) The seasons, global temperature, and precession. Science 269:59–68
Tilman D (1988) Plant strategies and the dynamics and structure of plant communities. Princeton University Press, Princeton, pp 67–69
Tong C, Feng X, Zhang YM, Zhong YK (2008) Soil seed banks in different grazing exclusion restoring succession stages in the Xiligole degraded steppe (in Chinese). Acta Ecol Sin 28 (5):1991–2002
Urban KE (2005) Plant species dynamics during restoration of health ponds in northwestern Germany. Phytocoenologia 35:511–532
Wang V, Fu PY, Li CY (1959) Chenopodiaceae (in Chinese). In: Liou NT (ed) Flora plantarum herbacearum chinae boreali-orientalis, vol 2, Science Press, Beijing, p 226
Wang XQ, Jiang J, Lie JQ, Zhang WM, Qian YB (2003) The distribution of Ephemeral vegetation on the Longitudinal dune surface and its stabilization significance in the Gurbantunggut Desert (in Chinese). Acta Geogr Sin 58:598–605
Wang XQ, Jiang J, Lie JQ, Zhao CJ (2004) Relationship between ephemeral plants distribution and soil moisture on longitudinal dune surface in Gurbantonggut Desert (in Chinese). Chin J Appl Ecol 15:556–560
Wang XZ, Zhang JG, Shi SY, Wang G (2001) Competitive regulation of annual plant *Eragrostis poaeoides* in planted vegetation for fixing sand dunes (in Chinese). J Desert Res 31:240–243
Wang ZL, Wang G, Liu XM (1998) Germination strategy of the temperate sandy desert annual Chenopod *Agriophyllum squarrosum*. J Arid Environ 40:69–76
Xin M, He DH (2003) Storage patterns of seed in natural nest of *Messor Actculatus* (F.smith) in the protected desert (in Chinese). J Ningxia Agricultural College 24:9–13
Xu CL, Li ZZ (2002) Population regulation and dynamical simulation of annual plant *Eragrostis poaeoides* in desert region (in Chinese). Acta Bot Boreal Occident Sin 22:1415–1420
Yan QL, Liu ZM, Li RP, Luo YM, Wang HM (2005) Relationship of seed production, seed morphology and life form of plant species (in Chinese). Acta Pratacultur Sin 14:21–28
Yang WK, Qiao JF, Gao XY, Jiang HP (2006) Impact of burrows of the great gerbil on small-scale plant community in Junggar Desert, China (in Chinese). Arid Land Geogr 29:219–224
Yu SL, Jiang GM (2003) The research development of soil seed bank and several hot topics (in Chinese). Acta Phytoecol Sin 27:552–560
Zhang JG, Li XR, Wang XP, Wang G, Li JG, Wang ZL (2001) Population dynamics of annual plant *Eragrostis poaeoides* in fixed and dune in Shapotou area (in Chinese). J Desert Res 31:232–235

Zhang LY, Chen CD (2002) On the general characteristics of plant diversity of Gurbantunggut Sandy Desert (in Chinese). Acta Ecol Sin 22:1923–1932
Zhao HL, Okuro T, Li YL, Zuo XA, Huang G, Zhou RL (2008) Effects of human activities and climate changes on plant diversity in Horqin sandy grassland, Inner Mongolia (in Chinese). Acta Pratacultur Sin 17:1–8
Zhao LY, Li ZH, Zhao JH, Zhao HL, Zhao XY (2006) Comparison on the difference in soil seed bank between grazed and enclosed grasslands in Horqin Sandy Land (in Chinese). J Plant Ecol (formerly Acta Phytoecol Sin) 30:617–623
Zhao WZ, Bai SM (2001) Characteristics of seed bank at fenced grassland in Horqin Sandy Land (in Chinese). J Desert Res 21:204–208

Chapter 5
Soil Biology in Traditional Agroforestry Systems of the Indian Desert

S. Sundaramoorthy, Santosh Kumar Mehar, and Manohar Singh Suthar

Abstract Traditional arid and semi-arid agroforestry systems serve as models for the 'protective–productive' rehabilitation strategies of agrarians. The nature of the organic materials present and their decomposition governs nutrient availability in soil systems. Indian desert vegetation has various non-fodder plants in and around farms that can well be used, along with diverse microbes, for improving soil biological fertility. The higher percentage loss of non-fodder plant litter, and the comparatively high phosphorus mineralisation offer the possibility of improving the biological characteristics of arid soils. *Prosopis cineraria*-based systems govern soil microbes according to their soil organic matter content. Enzyme activities and their dynamics with respect to litter amendment have been comparatively well-studied for desert soils. Changes in microflora due to addition of *Acacia leucophloea* (AL) and *Prosopis juliflora* (PJ) litter in wetting–drying cycles in microcosm experiments revealed that a higher proportion of litter enhanced population levels of cellulolytic fungi, with a further rise following rewetting. However, populations of lipolytic fungi, proteolytic bacteria and actinomycetes were higher during the initial periods of decomposition. The results indicate a strong relationship between lipolytic fungi and sugar fungi, lignolytic fungi, actinomycetes, and proteolytic bacteria. Bacteria and actinomycetes play a synergistic role during decomposition of desert soils.

S. Sundaramoorthy (✉) and M.S. Suthar
Department of Botany, JNV University, Jodhpur 342001, Rajasthan, India
e-mail: jnvusundar@rediffmail.com

S.K. Mehar
Department of Biological and Earth Sciences, Shri Venkateshwara University, Tirupathi, Andra Pradesh, India

K.G. Ramawat (ed.), *Desert Plants*,
DOI 10.1007/978-3-642-02550-1_5,

5.1 Introduction

Agroforestry is neither an invention nor a new concept. The practice of combining tree species with crops had been practiced in some form or other in many countries worldwide. However, agroforestry as an applied science is of recent origin. With the increase in population of both cattle and human beings in geometrical progression, and land area remaining finite, there is no alternative but to grow food, fodder, feed and fibre in an integrated manner on the same unit of land. Consequently, agroforestry is progressively emerging as a new discipline of human knowledge. Rao and MacDicken (1991) defined agroforestry as "a land use that involves deliberate retention, introduction or mixture of trees or other woody perennials in crop/animal production fields to benefit from the resultant ecological and economic interaction".

Agroforestry systems have been practiced in different parts of India since time immemorial. A fairly large number of agroforestry systems exist throughout the country. All these systems have evolved independently under diverse climatic, edaphic, floristic and socio-economic conditions, and, accordingly, components of such systems vary from one place to another. Awareness of traditional agroforestry systems reveals the importance of developing a greater knowledge of the past and the present activities of rural populations pertaining to agroforestry. On the basis of climatic classification, India has been divided into eight different regions: (1) Upper Himalayan region, (2) Western Himalayas, (3) Eastern Himalayas, (4) Indus plains, (5) Upper Gangatic plain, (6) Lower Gangatic plain, (7) Deccan region, and (8) Malabar region (Harsh et al. 1989).

5.2 Agroforestry in Western Rajasthan

More than 60% of the total arid area in India is present in Rajasthan. Therefore, agroforestry in this region has its own distinctive features. The vegetation of arid regions is very sparse and consists of scattered thorny trees, shrubs and grasses, classified as tropical thorn forest. Cultivation of crops has always been a risky endeavour in arid lands. In this part of the country, cultivation is in many ways a battle against all odds, as the adverse climatic conditions attempt to destroy the crop at each step of its life cycle, i.e. from seed germination to seed ripening. The uncertainty of rainfall and its erratic distribution, high temperature, intense solar radiation, unavailability of irrigation (or unsuitable ground water even if available) are some of the many factors that have framed agroforestry practices in arid regions. Most desert farmers raise livestock as a subsidiary occupation, and allow trees and shrubs to grow on cultivated tracts in order to cover the risk and uncertainty of crop production.

In different parts of the region, various types of tree–crop combinations are practiced, giving rise to different agroforestry systems. *Acacia albida* and *Holoptelia integerifolia* with cluster bean and moong bean, *Prosopis cineraria* with various crops, *Acacia nilotica* sub. *indica*, *A. nilotica* sub. *cupressiformis* with cicer, clusterbean, sorghum, and moong bean are some of the tree–crop combinations practiced in the form of agrisilviculture systems. *Azadirachta indica* and *Albizia lebbeck* are trees that are used widely in the silvi-pastoral system of the region, along with *Dichanthium annulatum*, *Cenchrus ciliaris*, *Cenchrus setigerus* and *Panicum antidotale* (Faroda and Singh 1998). Grass distribution also varies with rainfall. In rainfall less than 150 mm *Lasiurus sindicus* is the dominating grass; as rainfall increases to 250–300 mm, *C. ciliaris* and *C. setigerus* are the main grass species. However, in saline soils, *Sporobolus helvolus* and *Desmotachys binnata* are common. In lean periods, the leaves of these tree species are fed to animals.

Harsh et al. (1992) described the distribution and structural components of traditional agroforestry systems of the Thar Desert as part of their reconnaissance of the administrative districts of western Rajasthan. Only the prominent components (both woody and herbaceous) of the system were listed. The growing of trees (*P. juliflora*, *Acacia tortilis*, *Acacia senegal*, *Euphorbia caducifolia*, etc.) on bunds as windbreaks is also very common in all agroforestry zones in Rajasthan. Management of arid land by practicing various agroforestry systems is meant not only for the production of fuel, fodder, fertiliser, fibre and fruits but also for the cultivation and rehabilitation of land resources in the region (Faroda and Singh 1998).

5.2.1 Desert Soils

Sand dunes and sandy plains with a very weak manifestation of pedogenesis are the dominant features of arid Rajasthan, but degraded plains and buried pediments cover significant areas of the southeastern and central parts of this region. The source-rock-dependent variation in parent material, and the long periods of time required for arid to semi-arid geomorphic processes during a complex landscape evolutionary history have resulted in a variety of fairly well developed soils (Dhir et al. 1997). The soils of arid regions have been variously categorised and classified. The soils of the central arid zone are covered mostly by a thick mantle of aeolian sand of late Pleistocene age, with soil profile development by accumulation of lime. Soils here appear as zonal formations and were described as desert soils by Raychaudhuri et al. (1963), and as light brown sandy soil by Dhir (1977). The soils of arid regions of Rajasthan have been identified and mapped through soil surveys by the Central Arid Zone Research Institute (CAZRI) and the State Government Departments during the past 35 years or more. As a result, various classifications have been proposed by workers like Roy and Sen (1968), Mathur et al. (1972), Dhir and Mann (1978) and Lodha et al. (1982). Dhir et al. (1997) classified arid soils of Rajasthan at the family level category of soil taxonomy of the United States Department of Agriculture (USDA; Soil Survey Staff 1992).

5.3 Soil Fertility

Tree-based agricultural systems are considered more efficient in nutrient cycling than many herbaceous systems because the root systems of trees are more extensive and deeper than those of herbaceous species (Nair 1995). Soil fertility is maintained through production and decomposition of tree roots, agricultural crops and litter fall, which in turn increases organic matter and the biological activity of the soil (Szott et al. 1991), enhancing soil nutrient status. Thus, agroforestry systems promote closed nutrient cycling by taking up soil nutrients through tree roots and recycling them as litter, including root residues, and helping to synchronise nutrient release with crop requirements by controlling the quality, timing and manner of addition of plant residues (Young 1991). Agroforestry has important effects on soil physical and chemical characteristics, the rate of soil nutrient cycling, and the role of soil in long-term storage of C and N. These processes are regulated by various parameters such as the physical and chemical properties of litter, soil properties, climate, and decomposer communities consisting of microorganisms and soil invertebrates (Upadhyay and Singh 1989). Soil patches found beneath tree canopies are important local and regional nutrient reserves that influence community structure and ecosystem function (Rhoades 1997). The effect of individual plants on soil heterogeneity influences the functioning of several ecosystems (Hook et al. 1991; Mazzarino et al. 1991; Schlesinger et al. 1996; Schlesinger and Pilmanis 1998). Zinke (1962) pointed out that individual trees have an influence proportional to their crown area projected onto the surface; however, the concentration and availability of nutrients under tree crown varies with depth of soil and distance from tree bole (Mazzarino et al. 1991).

Studies in different parts of the world under various ecosystems concur with the conclusion that soils under tree canopy are better off in terms of soil nutrients and structure. For example, in the Mediterranean area of California, Jackson et al. (1990) reported that soils under deciduous blue oak canopies had higher N turnover and inorganic N availability than the surrounding open grassland soils. Similar results have been reported for semi-arid, mesic and humid savannahs (Belsky et al. 1989, 1993; Ischei and Mroghalu 1992; Mordelet et al. 1993). In semi-arid and arid regions, isolated trees are found within a grass matrix. The net effect on grass production can be negative, neutral or positive, and can change with tree age or size and density (Scholes and Archer 1997). Similarly, studies in the Dehesa ecosystem have shown that soils developing under tree canopies are richer in organic matter, nutrients and water availability (Joffre et al. 1988; Joffre and Rambal 1993).

Plant production in organic farming depends mainly on nutrient release as a function of mineralisation processes in soils. An active soil microflora and a considerable pool of accessible nutrients, therefore, is an important priority in organic farming. Fertilising the soil rather than the plant in order to assure sufficient nutrient mineralisation to meet his economic needs is the organic farmers' goal (Fließbach and Mader 2000). In general, release of potentially mineralisable nutrients, decomposition, and the biophysical manipulation of soil structure are functions of the soil microbial biomass and its activity. Substrates for microbial biomass and its activity depend upon plant production and other organic inputs (Franzluebbers et al. 2000).

The use of ecological principles, viz. productivity, decomposition and nutrient cycling, could be useful in increasing the sustainability of high production agriculture and reducing environmental problems (Matson et al. 1997). The development of sustainable agricultural systems will require new techniques that help to minimise fertiliser application rates while maintaining adequate crop yields. The application of biological resources to exploit nutrients present in the soil may hold promise for the future (Jeffries and Barea 1994). Many farmers in Kenya's semi-arid lands cannot afford to purchase inorganic fertilisers to improve their crop yields. Thus, they rely on traditional agronomic practices such as addition of crop residues, animal manures or intercrops of cereals and legumes (Muniafu and Kinyamario 2007).

Organic inputs are commonly used in the maintenance of soil productivity (Bekunda and Woomer 1996), but use is constrained by inadequate supply and labour requirements. Often, naturally occurring plants are present in sufficient abundance around farmers' fields, along paths, in hedges, and in fields under fallow to act as significant resources of nutrient supply to nearby cultivated areas (Lauriks et al. 1999). Man has recognised the importance of organic manures for maintaining or improving the fertility of soils since farming began (Allison 1973). Organic amendments, such as green manure crops or animal manures, influence soil biota both immediately, through increased food supply, and indirectly, by changes in soil chemical and physical variables (Kautz et al. 2006). In agriculture systems, organic materials enter the soil system via green manure, litter fall, root detritus and exudates that are subsequently decomposed by heterotrophic microorganisms to obtain C and nutrients for growth and maintenance (Teklay et al. 2007).

In many developing countries, organic materials, which include farmyard manure, sewage sludge, and town refuse, are the major sources of plant nutrients, and play an important role in maintaining aggregate stability, water-holding capacity and soil structure (Brown et al. 2000; Nkongolo et al. 2000), and these soil properties play an important role in increasing soil microbial biomass (Kanazawa et al. 1988; Ibrahim and Shindo 1999). It is a well-established fact that organic manure generally has beneficial effects on soil microbial biomass and microbial activity (Fraser et al. 1988; Kanazawa et al. 1988). The highest microbial biomass C and N contents have been observed under the integrated use of organic manure (Santhy et al. 1999). The importance of the quality of organic additions to N dynamics in natural and managed ecosystems has long been recognised (Swift et al. 1979; Melillo et al. 1982). The quality or fertility of soils is often discussed in relation to the sustainability of agricultural practices. Soil quality has been defined as 'the continued capacity of soil to function as a vital living system, within ecosystem and land use boundaries, to sustain biological productivity, promote the quality of air and water environments, and maintain plant, animal and human health' (Doran and Parkin 1994; Doran 2002). Soil fertility is an integral part of soil quality, focussing more on the productivity of the soil, and has been defined by

Persson and Otabbong (1994) as 'the sustainable capacity of a soil to produce good yields of high quality on the basis of chemical, physical and biological quality factors'.

5.4 Nature of the Organic Matter

Organic resources entering the soil can be categorised by their chemical composition and are composed of a complex mixture of compounds, including specialised polymers associated with the cell walls of plants (e.g. cellulose, hemicellulose, lignin) and fungi (e.g. chitin, tannin, melanin), as well as universal biomolecules such as fats, nucleic acids, proteins, other polysaccharides, and the monomeric constituents of these polymers: sugars, amino acids, nucleotides and nucleosides, fatty acids and other aliphatics, and aromatics (Moore et al. 2004). The decomposition of plant materials in soil is affected by biotic and abiotic factors, one of the most important of which is their biochemical quality, i.e. litter-quality (Swift et al. 1979). The initial concentrations of soluble C and N, lignin and polyphenols are generally recognised as the main litter-quality variables controlling rates of decomposition (Palm 1995; Heal et al. 1997). However, the importance of any one (or any combination) of these constituents may vary depending on the process, timeframe, or type of plant material involved (Palm 1995; Palm and Rowland 1997).

Decomposition and nutrient release rates are determined by the resource quality of the organic material, environmental conditions and the decomposer organisms present (Swift et al. 1979). Resource quality includes N concentration, C-N ratio, lignin concentration, lignin-N ratio, soluble polyphenols and polyphenol-N ratios (Melillo et al. 1982; Palm and Sanchez 1991; Constantinides and Fownes 1994; Mtambanengwe and Kirchmann 1995).

Plant material that is low in lignin and other polyphenols and high in nitrogen and soluble carbohydrates generally decomposes relatively quickly (Tian et al. 1995). Initial lignin concentration or the lignin/nitrogen ratio has proved to be an effective index of decomposition rates and nitrogen release from litter (Melillo et al. 1982). Some polyphenols, such as condensed tannins, could inhibit exoenzyme activity and have been found to reduce decomposition rates (Palm and Sanchez 1990; Tian et al. 1995). Phenols are the decomposition products of lignin that have been metabolised by fungi (Nord 1964). Microorganisms oxidise phenols and polyphenols by splitting their benzene rings to form carboxylic acids such as succinic and acetic acid (Dagley 1967). These acids can then enter the tri-carboxylic acid cycle and supply energy to microorganisms as they are oxidised through a series of carboxylic acids. The high wax, saponin and phenolic content characteristic of the leaves of woody species in arid and semi-arid environments exerts a significant influence over the mineralisation process (Whitford 2002). Particular to some arid environments is the photo-oxidation of lignin and complex molecules, which partly explains the relatively faster decomposition of organic matter reported in some arid ecosystems (Schaefer et al. 1985).

Quality of plant litter with respect to decomposition can be defined as the relative simplicity of its mineralisation by decomposing organisms (Paustian et al. 1997). Plant litter quality involves intrinsic characteristics of plant material that affect its assimilation by decomposers. These characteristics are both chemical and physical in nature. Most experimental studies on plant litter quality defined chemical quality indices using various ratios of C, N, lignin and polyphenols (Vanlauwe et al. 1997). From various studies, at least two general paradigms have emerged. First, the C:N ratio is accepted as a general index of quality (Seneviratne 2000): mineralisation rates tend to decrease with increasing C:N ratio. Second, lignin contributes to the recalcitrance of plant litter. Changes in substrate quality can affect the relative abundance of bacteria and fungi; whereas materials of high C:N ratios favour colonisation by fungi, more labile materials with low C: N ratios favour bacteria. With these changes come shifts in the abundances of the consumers of bacteria and fungi (Moore et al. 2004). Soil microfauna respond indirectly to litter quality via the microflora they consume, whereas meso- and macro-fauna ingest litter directly to a greater extent and have therefore been suggested to be more strongly influenced by differences in litter quality (Wardle 2002).

Several litter quality parameters have been proposed in an effort to develop a quantitative prediction of the time course of nutrient release and decomposition. For agricultural crops, incorporation into soil of plant materials with a C:N ratio $<$ 20 may result in net N mineralisation, and materials with C:N $>$ 20 might tend to cause net immobilisation (Wagner and Wolf 1999). The higher the initial lignin content, the greater the amount of N immobilised per unit carbon respired (Melillo et al. 1982). The relative importance of litter quality indices is influenced by soil properties, and may change as decomposition progresses (Berg and McClaugherty 2003). During decomposition, the C:N ratio undergoes a progressive change. As decomposition proceeds, CO_2 is liberated, but nitrogenous materials tend to accumulate in the organisms carrying out the decomposition and also in microbial metabolites. This progressively reduces the C:N ratio (Tarafdar et al. 2001a). Lignin and polyphenol contents above 15 and 4%, respectively, may retard litter decomposition and N mineralisation (Schroth 2003). Lignin is considered an important component determining the rate of decomposition (Palm and Rowland 1997), because it physically protects cellulose and other carbohydrates and also inhibits the synthesis and activity of cellulolytic enzymes. Soluble polyphenols react with N compounds in the soil to form more recalcitrant complexes that may retard N mineralisation (Schroth 2003).

Litter quality (carbon, nitrogen, phosphorus, cellulose, hemi-cellulose, lignin, polyphenols and the relative proportion of nitrogen with carbon, lignin, polyphenols and lignin+ polyphenols), and evaluation of non-fodder plants (*Tephrosia purpurea*, *Aerva persica*, *Clerodendrum phlomidis*, *Calotropis procera*, *Colophospermum mopane*, *Salvadora persica* and *S. oleoides* from different localities) of the Indian desert revealed non-significant variation in nitrogen content between herb and shrub species. However, lignin:N and (lignin + polyphenols):N ratios showed highly significant variations between these species because of the differences in the total lignin and polyphenols (Table 5.1).

Table 5.1 Relative proportion of C, N, lignin and polyphenols in litter of some herbs, shrubs and tree species collected from different localities of arid regions

Plant	Site	C:N	Lignin:N	Polyphenols: N	(Lignin + polyphenols):N
Tephrosia purpurea	Barmer	16.822	3.669	0.59	4.26
	Jalore	16.882	3.869	0.572	4.442
	Jodhpur	17.361	2.996	0.699	3.696
Aerva persica	Barmer	20.649	3.286	1.136	4.422
	Jalore	18.223	3.115	0.933	4.049
	Jodhpur	17.142	2.825	0.995	3.82
Clerodendrum phlomidis	Barmer	14.331	5.523	0.431	5.954
	Jalore	15.002	6.15	0.453	6.604
	Jodhpur	14.586	5.206	0.531	5.738
Calotropis procera	Barmer	15.415	8.36	0.436	8.796
	Jalore	15.044	8.609	0.418	9.028
	Jodhpur	18.529	12.02	0.578	12.6
Salvadora persica	Barmer	21.709	3.669	0.685	4.355
	Jalore	19.458	4.368	0.661	5.03
	Jodhpur	14.675	2.276	0.631	2.908
Salvadora oleoides	Barmer	19.557	2.805	0.447	3.253
	Jalore	20.519	3.414	0.476	3.891
	Jodhpur	15.15	2.246	0.444	2.691
Colophospermum mopane	Jodhpur	20.37	3.936	1.954	5.89

(Unpublished data; DST Project)

5.5 Litter Decomposition

Decomposition of plant residues is the microbially mediated progressive breakdown of organic material to C (biomass and CO_2) and other nutrients (Kumar and Goh 2000). Litter decomposition is an important component of global nutrient cycles and, as climate exerts strong controls over rates of litter decomposition, climate change may alter such cycles. The relative importance of climate and litter quality in controlling rates of decomposition varies across ecosystem types (Shaw and Harte 2001).

Litter decomposition exhibits two phases: an initial phase with rapid degradation of carbohydrates and amino acids, and a second, slower phase corresponding to the decomposition of cell walls and structural components (Sorensen 1975). Litter decomposition and concomitant nutrient release is regulated by the chemical composition of the litter (Oglesby and Fownes 1992) as well as abiotic factors, mesofauna and microbial action (Li et al. 2001). Although the C/N ratio or N-related indices of residues have often been found to be the major factors determining decomposition processes (Jarvis et al. 1996; Kemp et al. 2003), they are not the only determinants. Other factors such as the lignin, cellulose, polyphenolic and tannin content of the litter also affect nutrient release dynamics during decomposition (Palm and Sanchez 1991; Mafongoya and Nair 1997).

During litter decomposition, easily decomposable polysaccharides are dissipated faster (Melillo et al. 1982), whereas other chemical compounds, such as polyphenols, may interfere with enzyme functions or decomposer metabolism, resulting in a reduction in the decomposition rate of other litter constituents (Palm and Sanchez 1991). A decrease in quality or decomposability of plant substrate as decomposition progresses is a general concept. The decomposing material becomes enriched in recalcitrant chemical compounds, due to direct chemical changes in the substrate itself, and the succession of microorganisms able to assimilate the substrate (Berg and Staaf 1981). Substrate chemistry can influence decay rates in various ways. Shifts in microbial population or changes in decomposition rates can be linked tightly with modification of litter chemistry. Higher nitrogen content speeds the decomposition process, whereas high content of lignin and polyphenolics may delay and inhibit the process (Palm and Sanchez 1991; Vanlauwe et al. 1996; Bernhart-Reversat 1999).

Plant residue size is a factor that could influence decomposition of residues added to soil, besides a number of other factors such as soil temperature, soil type, nutrient and water availability in soil, chemical nature and amount of residues and soil–residue contact (Swift et al. 1979). With a reduction in particle size, the ratio of surface area of plant residue to mass of soil would increase (Angers and Recous 1997); thus, finer-sized plant residues could be expected to decompose faster than coarse ones.

The litter bag technique as introduced by Bocock and Gilbert (1957) is the most widely used method for examining litter decomposition rates when comparing species, sites and experimental manipulations. This technique assumes that litter-bags reflect the characteristic trends of unconfined decomposing litter (Singh and Gupta 1977; Wieder and Lang 1982; Gill and Burke 2002; Kemp et al. 2003; Moretto and Distel 2003; Albers et al. 2004; Tekley and Malmer 2004; Bragazza et al. 2007). Measurement of weight loss, often using litterbags, is the method used most extensively to assess rates of litter decomposition, especially in long-term decomposition studies in the field (Berg et al. 1984; McClaugherty and Berg 1987; Coûteaux et al. 1995).

The daily decay constants of leaf and scale-leaf of *Bambusa balcooa* and *B. pallida* did not differ significantly, and both species showed positive correlations with incubation period and mass loss rates during decomposition (Arunachalam et al. 2005). The decomposition of spruce needles and beech leaves in beech, spruce and mixed (beech/spruce) forest revealed that the accumulation of litter materials in spruce forests is not due to the recalcitrance to decay of spruce needles but rather to adverse environmental conditions such as the high polyphenol contents in the litter layer of spruce stands, which retards decomposition processes; spruce needles appear to be more sensitive to this retardation than beech leaves (Albers et al. 2004). Climatic factors such as precipitation and drought may cause contrasting results; relatively large changes in precipitation produce comparatively small changes in rates of decay of *Larrea tridentate* and *Prosopis glandulosa* leaf and root litter, and the effect of drought on decomposition depends on the whether the litter consists of leaves or roots, and on the particular chemical fraction undergoing decay (Kemp et al. 2003).

Agrarians in the Thar Desert have been cultivating crops since historical times by growing them with trees (Mann and Muthana 1984; Nagarajan and Sundaramoorthy 2000a, 2000b) and allowing many weeds to grow alongside the crop for considerable periods (Sen 1982; Sundaramoorthy 1987). The bravery of such agrarians in the face of hostile environmental conditions, viz., low and erratic rainfall, intense solar radiation, high seasonal and diurnal variations in temperature, high wind velocity, etc., has resulted in the availability of fodder and non-fodder organic sources in the desert that can be used to improve soil fertility. Incorporation of tree leaves enhances soil respiration and dehydrogenase activity (Tarafdar and Rao 1990, 1992), whereas crop residues and farmyard manure (FYM) generally enhances soil fertility status (N and P availability, organic matter, enzyme activity) by 10–20% as well as the yield of the pearl millet crop (Aggarwal et al. 1997). Our studies have indicated that incorporation of tree litter (highly palatable) in soils, even at low soil water content, enhances soil microflora (fungi, actinomycetes, bacteria and nitrifying bacteria), and microbial biomass – C, N and P and available forms of N and P (Purohit et al. 2002; Mehar et al. 2002). Aggarwal and Venkateshwarlu (1989) suggested supplementation of chemical fertilisers with bulky organic manures. Singh et al. (1981) observed that, under the arid conditions of Jodhpur, continuous application of sheep manure in general gave substantially higher yields than application of inorganic fertiliser (urea) alone. Application of urea may also cause environmental pollution due to the high potential for volatilisation losses (Aggarwal et al. 1987). Rao and Singh (1993) showed that substitution of 50% of the fertiliser requirement by FYM resulted in almost equal levels of yield compared to 100% chemical fertiliser. Aggarwal and Praveen-Kumar (1996) showed that FYM application increases the utilisation efficiency of fertiliser N. Incorporation of crop residues into soil has been reported to increase the organic matter content of soil (Aggarwal et al. 1997). Hegde et al. (1982) reported higher organic C, available P and exchangeable K content in arid soils after 5 years of continuous incorporation of maize residue into the soil. No doubt, these studies on arid soils have provided information to science, and opened up the possibility of organic source incorporation for protective production in arid regions, but as yet such methods cannot be practiced by our agrarians as demand for fodder is their highest priority. Hence, our recent research endeavours have focussed on the utilisation of non-fodder plants that occur naturally but are not equally distributed throughout arid regions. *Aerva persica* is dominant in zones deficient in rainfall in the extensive silvipastoral systems of western Rajasthan. *A. persica* is a co-dominant species along with *Panicum turgidum* and *Haloxylon salicornicum*, and has a density of 250 individuals ha^{-1} (Shankar and Kumar 1987). *Tephrosia purpurea* – a common legume that occurs throughout the desert region – forms pure stands in many farm fields. *Calotropis procera* occurs abundantly at 200–250 mm rainfall zones in the sandy undulating plains of the Barmer area, where its density ranges from 200–400 individuals ha^{-1} (Saxena 1997). *Salvadora persica* and *S. oleoides* dominate in the heavy soils of the alluvial plains of the Pali-Jalore belt (Saxena 1997; Tewari et al. 1999). *Clerodendrum phlomidis* occurs abundantly in the sandy undulating plains of the western part of the arid region. *Colophospermum mopane* is a tree species that has been introduced into the silvipastoral system. Such

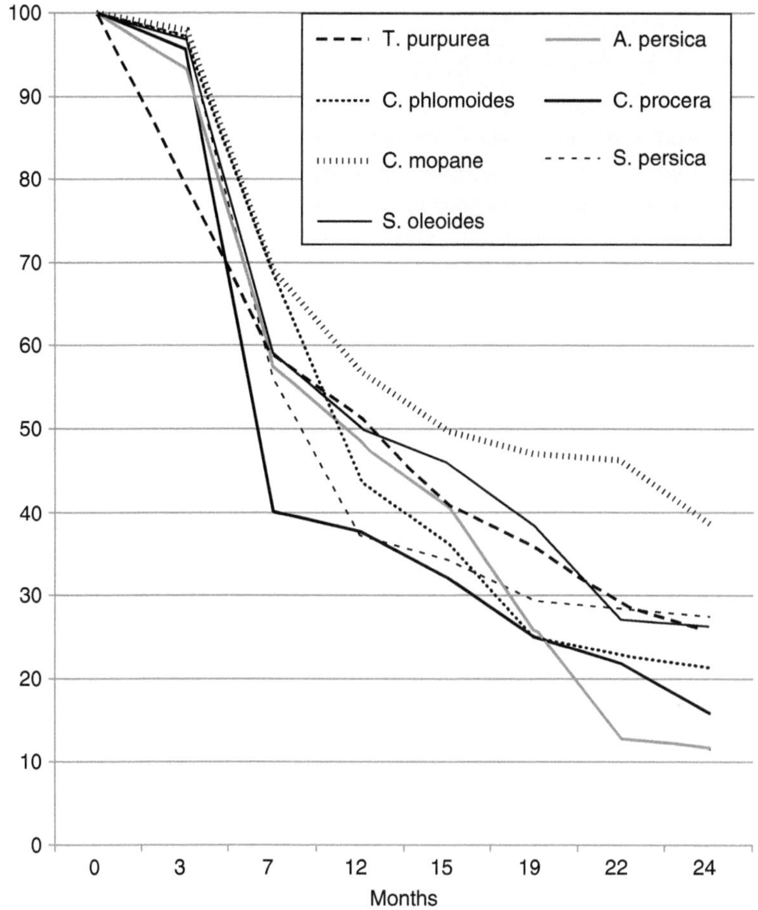

Fig. 5.1 Percent litter remaining after 2 years for some non-fodder plants of the Indian desert

species provide leaf/stem litter; root mass and root activity that contribute to the development of soil organic matter and biological activities in desert soils. The percent litter loss after 2 years of field incubation (litter bags buried at 10 cm depth) ranged from 62% to 88% (Fig. 5.1). P mineralisation/loss from these litters was comparatively more (95–98%) than N mineralisation (59–90%), which was comparable to C mineralisation (62–87%).

5.6 Microorganisms

Soil organisms consist of microflora (bacteria and fungi) and soil fauna (protozoa and invertebrate groups such as nematodes, mites, and earthworms). Such organisms influence the availability of nutrients for crop production via a range of activities such as decomposition of crop residues, immobilisation of nutrients, mineralisation, biological nitrogen fixation, and bioturbation. The soil fauna is crucial for the initial comminution and mixing of residues into the soil, whilst the microflora has a greater suite of enzymes for chemical breakdown of organic material (Paul and Clark 1996). The soil food web is composed of a diverse community of organisms, which can be grouped according to body width: microorganisms, i.e. bacteria (<2 μm) and fungi (2–100 μm); microfauna, i.e. nematodes, protozoa and rotifers (<100 μm); mesofauna, e.g. mites, springtails and enchytraeids (0.1–2 mm); and macrofauna, e.g. earthworms and millipedes (>2 mm; Swift et al. 1979). Soil organisms are responsible for decomposition of the organic resources that enter the soil, and thereby for the cycling of nutrients bound in these resources.

Numerically, fungi are the most abundant soil organisms in terrestrial ecosystems and are the primary decomposer of organic residue in soil, thus regulating the ecosystem structure and function (Parkinson and Coleman 1991; Lodge and Cantrell 1995; Miller and Lodge 1997). Fungi are key to the recycling of nutrients from aboveground and belowground plant litter (Swift 1976; Rayner and Boddy 1988) and through decomposition of organic matter, and serve alternatively as a source and sink of labile nutrients that are necessary for plant growth (Swift et al. 1979; Marumoto et al. 1982; Lodge 1985; Young and Insam 1991). Among soil microorganisms, microfungi are considered to be one of the most important microbial decomposers, dominating the microbial biomass and activity of many types of soil (Kjoller and Struwe 1982; Behera et al. 1990; Basu and Behera 1993). Extracellular enzymes degrade more complex carbohydrates, such as cellulose and non-cellulosic polysaccharides, and the lignin that remains at later decomposition stages; fungi are in general better adapted to utilise these resources (Miller 1993; Lavelle and Spain 2001).

The number of filamentous fungi in soil varies directly with content of utilisable organic matter. According to Richle et al. (1975), soil microorganisms contribute appreciably to total soil respiration. Our studies of *Prosopis cineraria*-based agroforestry systems revealed that fungal population and biomass are directly proportional at 99% significance to organic matter ($r = 0.73$; 0.78, respectively). It was commonly assumed that high temperature and soil drying during the summer months brings down the microbial population to very low levels in many desert soils (Sasson 1972). Hot months are associated with high soil temperature and very low moisture, causing desiccation to fungal filaments (Satpathy et al. 1982), and spores might be in resting form resulting in low biovolume. An evaluation of factors (soil and climatic) that might be responsible for changes in fungal biomass using a stepwise regression model considering fungal biomass (dependent variable), the preceding month's rainfall and temperature, and the current status of pH, electrical conductivity (EC), organic carbon and soil nitrogen (independent variables), revealed that organic carbon (X) in the case of undercanopy, and rainfall (X) in the preceding month in open areas, are the critical factors affecting fungal biomass (Y). The final equations:

$$Y = -1.03392 + 0.1870777\,X \qquad (5.1)$$

$$Y = 16.49569 + 0.2185182\,X \qquad (5.2)$$

explain 76.48% and 73.37% of variation, with 7.42 and 6.40 standard deviation of residuals for undercanopy and open areas, respectively (Purohit et al. 2002).

Fungi are capable of translocating N relatively long distances, and their occurrence and characteristics could dramatically affect the amount and distances that N is transferred in natural litter layers (Hart and Firestone 1991; Frey et al. 2003). Fungi also have different decomposition characters from bacteria (Balser 2005) and thus may affect both the effective source and sink strength of different litters as well as regulating the transfer rate of N (Schimel and Hattenschwiler 2007). Recent studies (Gehlot et al. 2008) on legume symbioses with *Rhizophia* in the stressed Indian Thar Desert ecosystem revealed the presence of nodulating symbionts of alien (*Mimosa pudica*) as well as native (*Mimosa hamata*) Mimosa on the basis of 16s rRNA gene sequencing and immunodetection, and confirmed the presence of beta-rhizobia (*Burkholderia phymatum* and *Cupravidus taiwanensis*) in *M pudica* isolates and *Sinorhizobium* sp. in *M hamata.* Diverse types of rhizobia (*Rhizobium*, *Bradyrhizobium*, *Sinorhizobium*, *Burkholderia*, *Ochrobactrum* and *Methylobacterium*) have been isolated from root nodules of native legumes.

Microorganisms are actively engaged in P transformation in soil and transport to plants. The rhizosphere is the most active site for P transformation, which differs considerably from bulk soils (Helal and Sauerbeck 1984). The rhizosphere effect concerns changes in phosphate availability from organic and inorganic sources. The release of phosphate from organic sources is catalysed by phosphatases and phytases, which have been shown to be present in rhizosphere components: soil (Tarafdar and Jungk 1987), microbes (Tarafdar and Chhonkar 1979) and plant roots (Tarafdar 1989). Although organic P (Po) is present in soil solution at higher concentration than inorganic phosphate (Seeling and Jungk 1996), direct uptake of Po compounds by plants is considered unlikely. Rather, plant roots acquire P as inorganic phosphate (Pi) from the soil solution (Barber 1984). Thus, to contribute to plant P nutrition, soil Po must be catalysed through hydrolysis of C–O–P bonds by phosphatases or phytase, which are therefore important in the P nutrition of plants (Beck et al. 1989). The relative contribution of acid phosphatase to soil Po hydrolysis from plant and fungal sources has been demonstrated (Tarafdar et al. 2001b), and the results suggested that fungal acid phosphatase is more efficient than that from plant sources. A strong linear relationship between intra versus extracellular fungal acid phosphatase ($R^2 = 0.94$), alkaline phosphatase ($R^2 = 0.96$) and phytase ($R^2 = 0.97$) was observed (Tarafdar et al. 2002). Three-fourths of acid phosphatase are generally present inside fungal cells and only 25% is released extracellularly in culture media. Phytase showed the reverse trend, with 39 times higher extracellular phytase activity noted than was present inside fungal cells.

5.7 Soil Enzymes

Enzymes in soils are similar to enzymes in other systems, in that their reaction rates are markedly dependent upon pH, ionic strength, temperature, and the presence or absence of inhibitors (Burns 1978; Dick and Tabatabai 1978). Extracellular enzymes may play an ecological role for some microbial community members by hydrolysing substrates that are too large or insoluble for direct absorption by microbial cells. Soil enzyme activities can be sensitive and early indicators of both natural and anthropogenic disturbances (Dick 1997; Giller et al. 1998). Enzyme activities hold potential as a soil quality indicator because their assays are often relatively simple procedures, they are integrative in nature and are sensitive to land management (Dick 1997; Bandick and Dick 1999; Nadiaye et al. 2000). Kourtev et al. (2002) suggested that soil enzyme expression is a labial property of the microbial community. Soil enzyme activities are responsible for changes in soil management that increase soil productivity. Such activities provide knowledge of the rate and extent of the recovery of fertility (Caravaca et al. 2002). The soil enzymes most often studied are oxidoreductase, transferase and hydrolases, because of their role in the decomposition of various organic compounds; thus, they are important in nutrient cycling and formation of soil organic matter. Enzymes involved in the C cycle (glucosidases, xylanase, amylase and cellulase), the N cycle (amidase, urease and protease), and the P cycle (acid- and alkaline phosphatase) are also involved. In general, hydrolytic enzymes are good choices as soil quality indexes because it is likely that organic-residue-decomposing organisms are the major contributors to soil enzyme activity (Speir and Ross 1976; Speir 1977). Our estimates, and comparison of extracellular enzymes, viz., β-glucosidase, urease, acid and alkaline phosphatase from culture isolates of 28 soil fungi revealed that *Alternaria alternata* had the highest β-glucosidase and urease activity, whereas *Aspergillus niger* had the highest alkaline phosphatase activity. *Cunninghamella* sp. showed highest acid phosphatase and lowest urease activity. The urease activity pattern separates the fungi in two groups of similar activity (higher activity group, i.e. >500 μg urea hydrolysed h^{-1} ml^{-1}: *Alternaria alternate*, *A. brassicola*, *Aspergillus flavus*, *A. nidulans*, *A. terreus*, *A. flaviceps*, *A. fumigatus*, *A. niger*, *A. parasiticus*, *Curvularia lunata*, *Paciliomycetes lilacicum*, *Penicillium citrinum*, *Trichoderma kunghii*, *T. virens*, *T. viridi*, *T. harzianum*; and lower activity group i.e. <500 μg urea hydrolysed h^{-1} ml^{-1}:, *Alternaria solani*, *Fusarium solani*, *F. moniliformis*, *F. accessi*, *F. oxisporum*, *Mucor racemosus*, *Rhizoctonia solani*, and *Trichothecium roseam*).

5.8 Microcosm Experiment – Drying and Wetting

Drying and wetting of soil and the associated nutrient dynamics is very important, and is the major determinant of mineralisation and immobilisation of nutrients in deserts. Most studies on microbial succession during decomposition following wetting of dried desert soil come from Mediterranean climates, which are characterised by major rainfall spells during winters. Under such conditions, the decomposition process commences shortly after litter fall (Schlesinger 1985; Gallardo and Merino 1993). In contrast, there is little information on regions with monsoon climates, having more or less dry winters (Morris et al. 1982; Whitford et al. 1986), and there is a paucity of such information for the extensive Thar Desert. In microcosm experiments, litter in the form of mature leaves and small twigs of *Acacia leucophloea* (AL) and *Prosopis juliflora* (PJ) was mixed with 100 g soil so as to achieve three fertility levels equivalent to 40, 80 and 160 kg N ha^{-1}. After litter addition, soil was wetted to 80% water holding capacity (WHC) for the experimental set up. Of the two sets, one was subjected to continuous drying and sampled at three intervals, i.e. after 15, 30 and 45 days of continuous drying (15CD, 30CD and 45CD, respectively). The other set was allowed to dry for 15 days after which it was rewetted to its original moisture content, i.e. 80%, and estimated after 30 and 45 days of drying (30DRW and 45DRW, respectively).

The results revealed that fungal populations were higher in treatments with a higher level of litter addition in both species; at 15CD and 30CD, the density of both species in the litter-amended soil was comparable; however, during the later period (45CD and 30, 45DRW), fungal density was higher in AL-amended than in PJ-amended soil (Fig. 5.2). The density of cellulolytic fungi fluctuated with the availability of more cellulose, as hard-to-degrade substances pave the way for the action of these fungi as decomposition progresses. With the passage of time, the density of cellulolytic fungi increased in AL litter-amended soil (Fig. 5.3). The more the litter in the treatment, the greater the density of cellulolytic fungi it supported, and there was a further rise in fungal density with rewetting of the soil. The density of both lignolytic and lipolytic fungi was higher during the initial phase of decomposition, and their density decreased during incubation up to 45CD. After rewetting, a small rise in numbers was noted in almost all cases; after this rise, values again decreased during 45DRW (Figs. 5.4 and 5.5). Proteolytic bacteria were in greatest numbers during the initiation of decomposition as revealed by population counts after 15CD, and numbers continued to reduce as the process of decomposition progressed, rewetting of soil caused only a marginal increase in the proteolytic bacterial count (Fig. 5.6). The population of actinomycetes was higher during the initial periods, i.e. 15CD and 30CD, but declined at 45CD. Rewetting of soil caused a large increase in the actinomycetes population, and it remained high at 30DRW and 45DRW (Fig. 5.7). The bacterial population was affected greatly by the reduction in the moisture content, being higher at 15CD and 30CD. The lowest bacterial density was observed at 45CD. Upon addition of water, bacterial density increased and it was noted to be particularly high at 30DRW and 45DRW (Fig. 5.8). Generally, the initial decomposition of a plant material is performed by so-called "opportunistic bacteria" as well as sugar fungi specialised in exploiting readily metabolisable non-polymeric resources such as simple sugars, amino acids, starch and, to some extent, also pectic substances. These organisms die as their nutrient resource reduces, and decomposition is taken over by other species of bacteria and fungi with a slower growth rate and the capacity to decompose more refractory

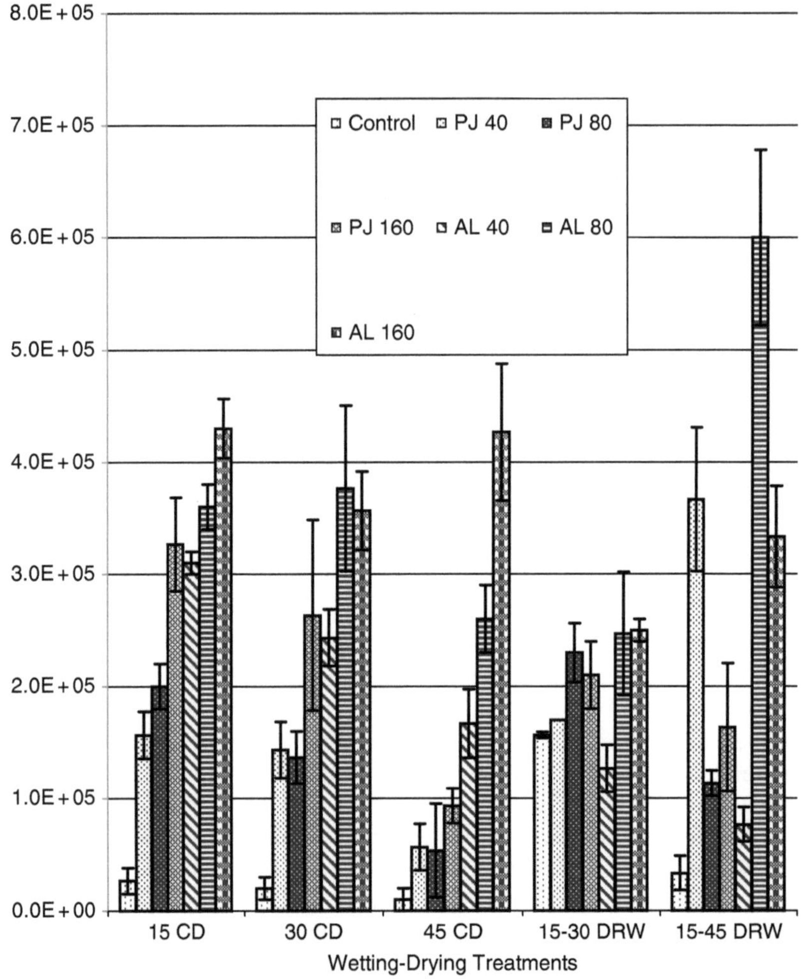

Fig. 5.2 Fungal density in litter-amended soils under wetting–drying conditions

structural carbohydrates, which they break down enzymatically to smaller molecules for their use.

Two important attributes of the decomposers are (1) to be able to penetrate the protective surface of a substrate, and (2) to invade it at the cellular and molecular level (Heal et al. 1997). Many fungi accomplish this by producing specific hyphae that mechanically penetrate the cuticularised surface of plants. This penetration is followed by exploitation of available substrates found within and between the cells that can be enzymatically attacked. An alternative course of action used by both bacteria and fungi is that of enzymatic attack on cutin and cell wall polymers by extracellular enzymes (Hammel 1997). The water-soluble parts inside the plant cell may leach out of the cell if they come into contact with soil water, and are

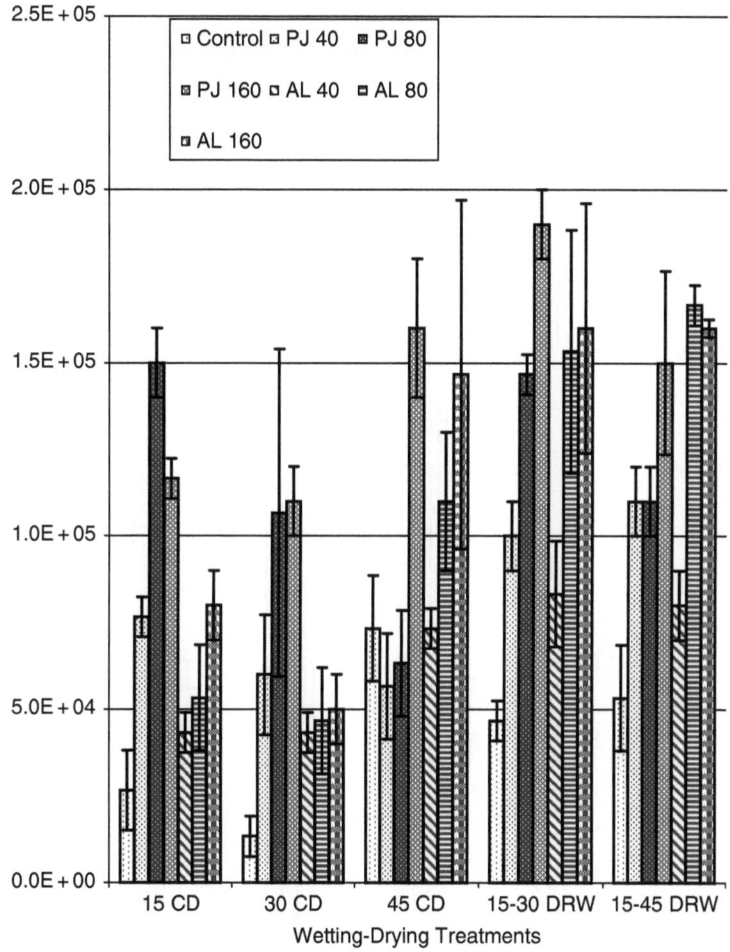

Fig. 5.3 Cellulolytic fungal density in litter-amended soils under wetting–drying conditions

decomposed (Collins et al. 1990). Thus, during decay, changes in the microbial community occur according to the catabolic capabilities of the organisms involved.

Plant proteins are decomposed by a variety of organisms (Kögel-Knabner 2002). Generally, being non-structural, proteins outside the cell wall are degraded more easily than those that are structural and embedded in the cell wall. Approximately 5% of the cell wall is made of proteins such as glycoproteins and extensins linked to cellulose (Hatfield et al. 1999). Thus, it is presumed that, as the outer parts of cell wall are degraded and proteins are made available to the decomposing community during decomposition, there is rapid decomposition of proteins. These factors working in the background were responsible for the strong relationships observed in our study between proteolytic bacteria and cellulolytic fungi, lipolytic fungi, bacteria and actinomycetes. Dilly et al. (2001) had previously reported a strong

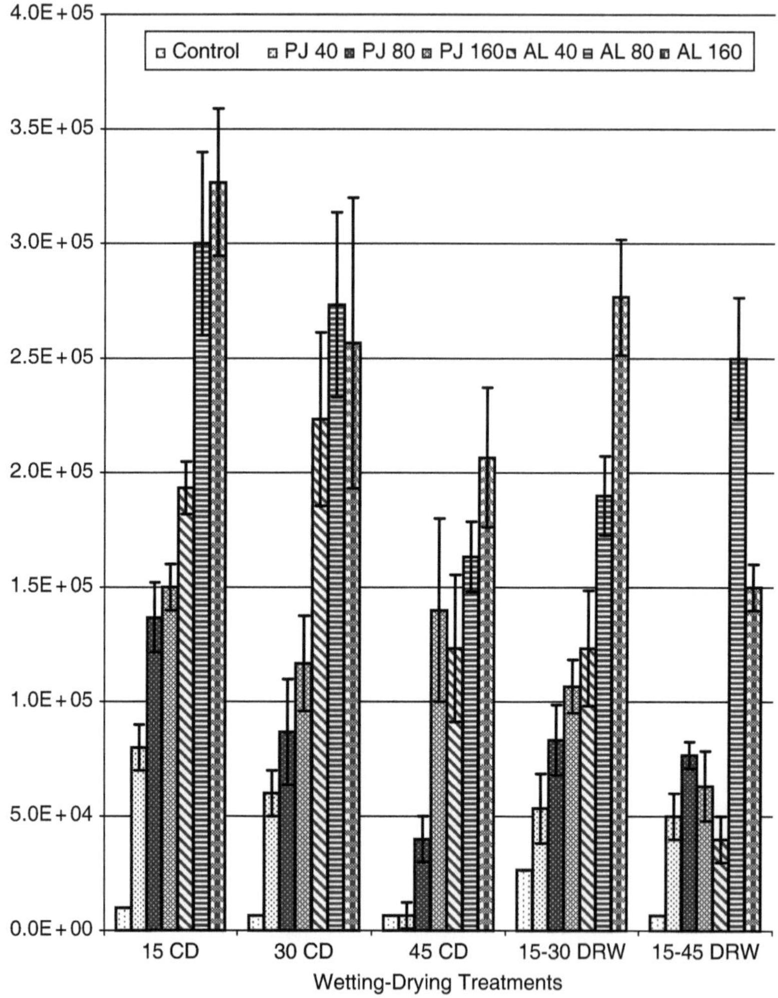

Fig. 5.4 Lignolytic fungal density in litter-amended soils under wetting–drying conditions

correlation between proteolytic and lipolytic degradation potential and ascribed it to bacterial mineralisation following fungal degradation of polymers. Moreover, these bacteria were related strongly to the soil ammoniacal N content. We reasoned this to be due to their involvement in N cycling, whereby they convert proteins into the amino acids that are subsequently used by other microbial groups for ammonification.

External enzymes, produced mainly by fungi, are required to break down cellulose polymers. The population of cellulolytic fungi revealed a more or less increasing trend during the progress of decomposition, showing that, as the decomposition process progresses and deeper tissues are attacked, more and more cellulosic structures are made available. Correlation analysis revealed that cellulolytic

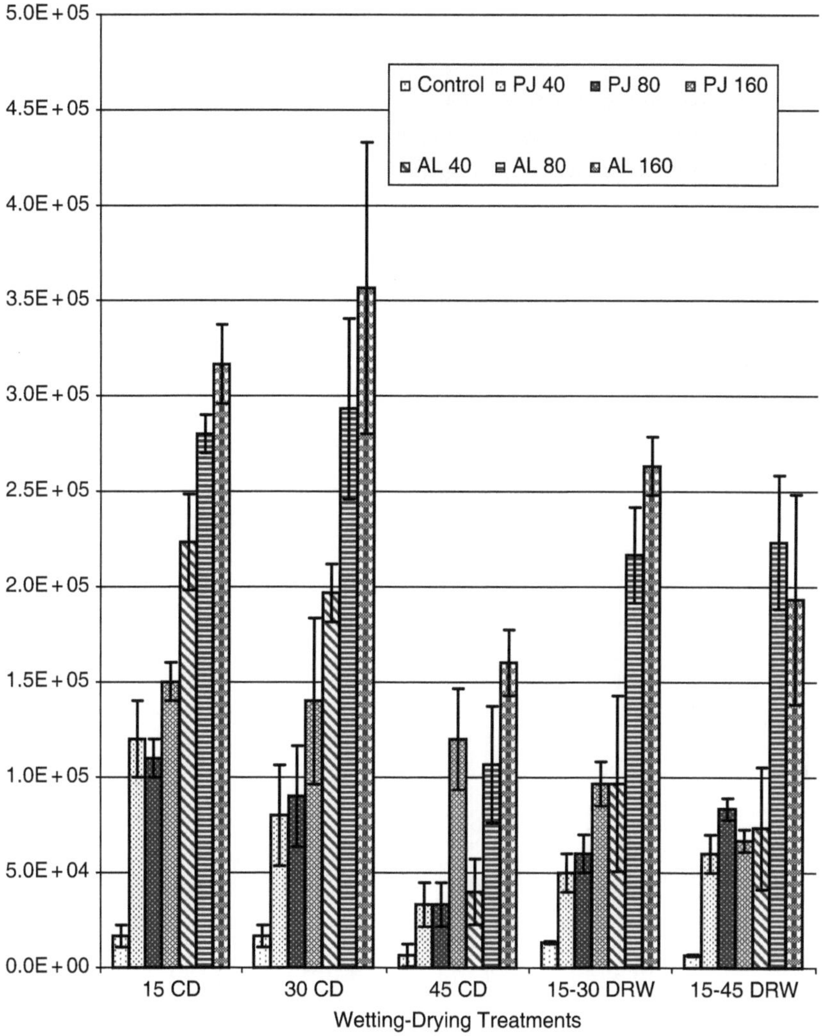

Fig. 5.5 Lipolytic fungal density in litter-amended soils under wetting–drying conditions

fungi were related strongly to the nitrifying bacterial population as well as to other bacterial populations. As more and more refractory substances begin to accumulate as decomposition progresses and the C/N ratio of soil changes, the degrading community becomes N limited; hence, the nitrifying potential, due to its involvement in N supply, can feed into other degrading pathways during the later stages of decomposition, and the same sort of scenario could also explain the correlation between lignolytic fungi and nitrifying bacteria. Additionally, cellulolytic fungi also correlated strongly with actinomycetes due to the synergistic approach of these two types of organisms in the decomposition process.

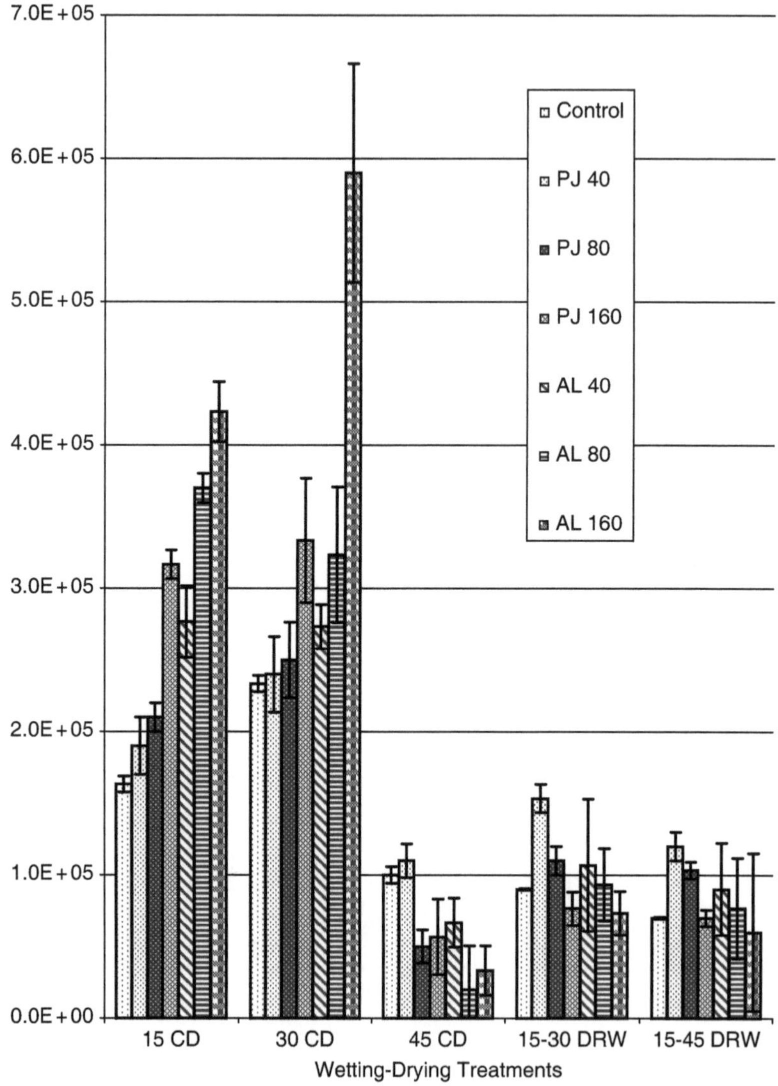

Fig. 5.6 Proteolytic bacterial density in litter-amended soils under wetting–drying conditions

Lipids play a significant role in microbial growth and the development of high microbial densities. Extractable lipids are noted throughout decomposition (Dilly et al. 1997) but their content decreases gradually with decomposition (Bridson 1985; Cortez et al. 1996), as does the population of lipolytic decomposers (Dilly et al. 2001). Readily available lipids are mineralised at the start of decomposition by zymogenous fungi (Rosenbrock et al. 1995). In contrast, more refractory litter-derived lipids that resist the early zymogenous attack are mineralised along with the microbially produced lipids during the later stages of decomposition. Earlier reports

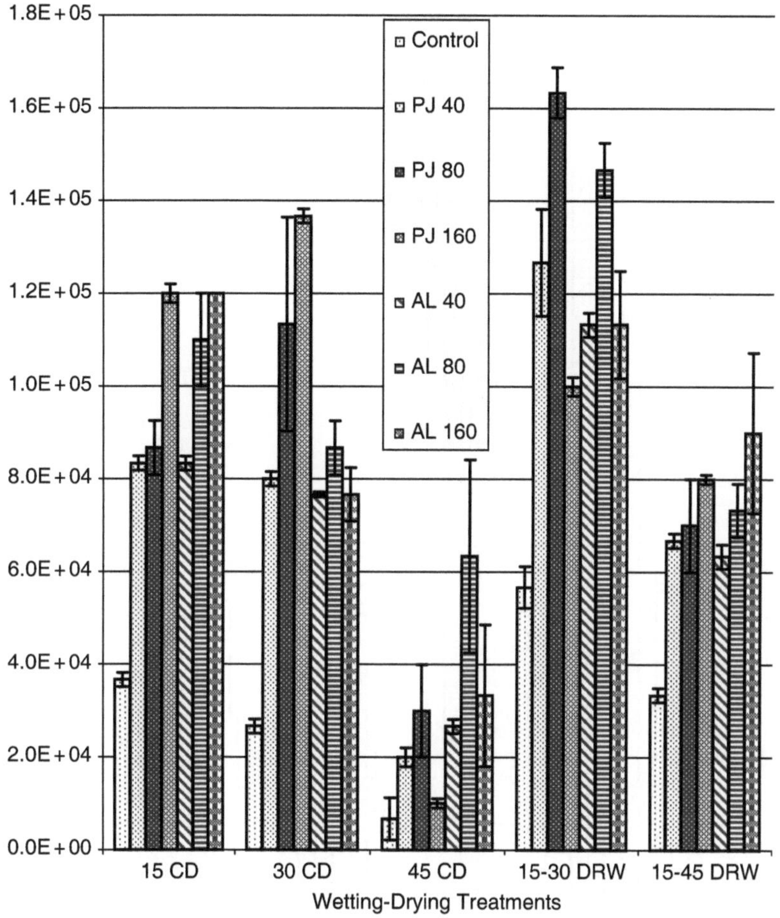

Fig. 5.7 Actinomycete density in litter-amended soils under wetting–drying conditions

(Zvyaginstev 1994) have proposed that such zymogenous fungi were probably present on leaf surfaces on the tree before litter fall. Our results support many of these earlier studies in the sense that we also noted a higher lipolytic fungal population during the initial stages of decomposition, which decreased as decomposition progressed. Our results indicate the strong relatedness of lipolytic fungi with sugar fungi, lignolytic fungi, actinomycetes, proteolytic bacteria and general bacteria.

Bacteria may be associated with other microorganisms and contribute to the cell wall degradation process or they can be antagonistic and inhibit the growth of other microbes. Pellinen et al. (1984, 1987) and Daniel et al. (1987) have described the lignin degradation potential of bacteria. Streptomyces species form an important group with antagonistic properties against other wood-invading microorganisms, while some bacteria have been detected in connection with other fungi.

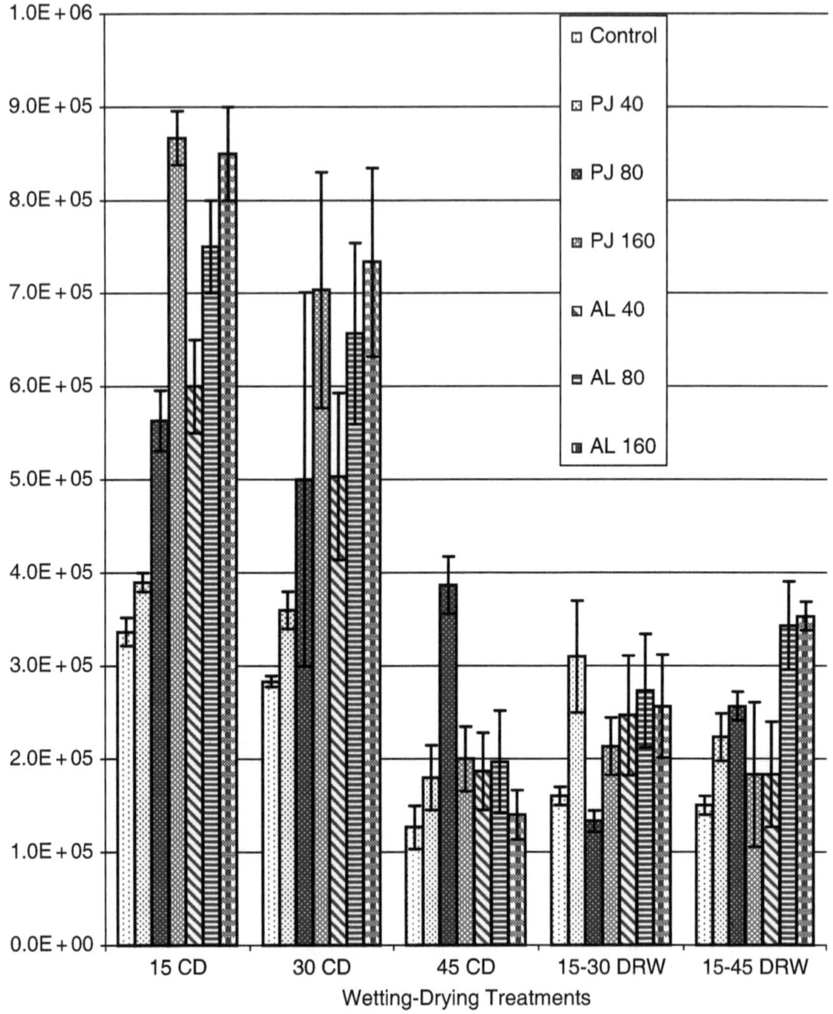

Fig. 5.8 Bacterial density in litter-amended soils under wetting–drying conditions

Such bacteria provide substances that promote hyphal growth and may also diminish the catabolic inhibition of cellulose degradation caused by cellobiose and glucose by utilising these sugars (Erikson et al. 1990).

Actinomycetes have varied capabilities and can act as useful tools for decomposition. We found a strong correlation of bacteria and actinomycetes with many of the microbial groups studied, suggesting a synergistic role in desert soils during decomposition of litter. Changes in substrate composition and increases in moisture were found to be the reasons for the large increase in the actinomycete population after rewetting of soil. Upon rapid rewetting, the cells of some microorganisms burst due to the influx of water into the cells, while others that survive by release of intracellular solutes to maintain proper cell turgor pressure rapidly mineralise the compounds released by the dead microorganisms (Halverson et al. 2000). Our observations of the population counts of actinomycetes is further confirmed by their strong correlation with soil moisture. A significant fact that was also revealed by our study of bacteria/fungi and proteolytic bacteria/fungi ratios is that these ratios were higher during continuous drying in the control treatment, or in treatments with lower levels of litter addition. We believe that the comparatively simple organic substrates in these treatments could have favoured higher bacterial population due to their limited degrading potential as compared to fungi. When the soil was rewetted and further subjected to drying, conditions were found to be similar in more of the treatments. This might be due to the fact that, upon hydration, death of microbial biomass could release easily degradable organic substrates that could be utilised by bacteria. Alternatively, the substrate composition could have changed by this period, and/or bacteria could be utilising the degradation intermediates produced by fungi.

Acknowledgements The commendable academic assistance of Ph.D. scholars, viz., Dr. M. Nagarajan, Dr. Usha Purohit, Dr. Amar Chand Chouhan and Ms. Harshadha Purohit – all members of project teams working in the soil biology of traditional agroforestry – furthered our understanding of thirsty and hungry arid soils and their biological status. Mr. Pradeep Kumar Panwar and Ms. Deepti Sharma took considerable pains in arranging the references of this chapter and we record our gratitude. I (S. Sundaramoorthy) have freely used some of the research findings from R&D projects sanctioned to me by the Indian Council of Agricultural Research, Department of Science and Technology, and University Grants Commission; accordingly, the financial assistance received is gratefully acknowledged.

References

Aggarwal RK, Praveen-Kumar (1996) Integrated use of farm yard manure and fertilizer nitrogen for sustained yield of pearl millet (*Pennisetum glaucum*) in a desert sandy soil. Ann Arid Zone 35:29–35

Aggarwal RK, Venkateshwarlu J (1989) Long-term effect of manure and fertilisers on important cropping systems of arid region. Fertilizer News 34:67–70

Aggarwal RK, Raina P, Praveen-Kumar (1987) Ammonia volatilization losses from urea and their possible management for increasing nitrogen use efficiency in an arid region. J Arid Environ 13:163–168

Aggarwal RK, Praveen-Kumar, Power JF (1997) Use of crop residue and manure to conserve water and enhance nutrient availability and pearl millet yield in an arid tropical region. Soil Tillage Res 41:43–51

Albers D, Migge S, Schaefer M, Scheu S (2004) Decomposition of beech leaves (*Fagus sylvatica*) and spruce needles (*Picea abies*) in pure and mixed stands of beech and spruce. Soil Biol Biochem 36:155–164

Allison FE (1973) Soil organic matter and its role in crop production. Elsevier, New York

Angers DA, Recous S (1997) Decomposition of wheat straw and rye residues as affected by particle size. Plant Soil 189:197–203

Arunachalam A, Upadhyaya K, Arunachalam K, Pandey HN (2005) Litter decomposition and nutrient mineralization dynamics in two bamboo species growing in a 9-year-old "jhum" fallow. J Trop For Sci 17:33–44

Balser TC (2005) Humification. In: Hillel D, Rosenzweig C, Powlson D, Scow K, Sparks D (eds) Encyclopedia of soils in the environment, 2nd edn. Elsevier, Oxford, pp 195–207
Bandick A, Dick RP (1999) Field management effects on soil enzyme activities. Soil Biol Biochem 31:1471–1479
Barber SA (1984) Soil nutrient bioavailability: a mechanistic approach. Wiley, New York
Basu S, Behera N (1993) Effects of tropical forest conservation on soil microbial biomass and activity. Biol Fertil Soil 16:302–304
Beck E, Fußeder A, Kraus M (1989) The maize root system in situ: evaluation of structure and capability of utilization of phytase and inorganic soil phosphates. J Plant Nutr Soil Sci 152:159–167
Behera N, Joshi SK, Pati DP (1990) Root contribution to total soil metabolism in a tropical forest soil from Orissa, India. For Ecol Manage 36:125–134
Bekunda MA, Woomer PL (1996) Organic resource management in banana based cropping systems of the Lake Victoria basin. Agric Ecosys Environ 59:171–180
Belsky AJ, Amundson RG, Duxbury JM, Riha SJ, Mwonga SW (1989) The effects of trees on their physical, chemical and biological environments in a semiarid savannah in Kenya. J Appl Ecol 26:1005–1024
Belsky AJ, Mwonga SM, Amundson RG, Duxbury JM, Ali AR (1993) Comparative effects of isolated trees on their undercanopy environments in high-low-rainfall savannahs. J Appl Ecol 30:143–155
Berg B, McClaugherty C (2003) Plant litter: decomposition, humus formation, carbon sequestration, Springer, Berlin
Berg B, Staaf H (1981) Leaching accumulation and release of nitrogen in decomposing forest litter. Ecol Bull 33:163–178
Berg B, Ekbohm G, McClaugherty C (1984) Lignin and hemicelluloses relations during long-term decomposition of some forest litters. Can J Bot 62:2540–2550
Bernhart-Reversat F (1999) Changes in relationships between initial litter quality and CO_2 release during early decomposition of tropical leaf litters. Eur J Soil Biol 34:117–122
Bocock KL, Gilbert OJW (1957) The disappearance of litter under different woodland conditions. Plant Soil 9:179–185
Bragazza L, Siffi C, Iacumin P, Gerdol R (2007) Mass loss and nutrient release during litter decay in peat land: the role of microbial adaptability to litter chemistry. Soil Biol Biochem 39:1428–1436
Bridson JN (1985) Lipid fraction in forest litter: early stages of decomposition. Soil Biol Biochem 17:285–290
Brown SMA, Cook HF, Lee HC (2000) Topsoil characteristics from a paired farm survey of organic versus conventional farming in southern England. Biol Agric Hortic 18:37–54
Burns RG (1978) Enzyme activity in soil: some theoretical and practical considerations. In: Burns RG (ed) Soil enzymes. Academic, New York, pp 295–340
Caravaca F, Barea JM, Figueroa D, Roldan A (2002) Assessing the effectiveness of mycorrhizal inoculation and soil compost addition for reafforestation with *Olea europaeea* subsp. *sylvestris* through changes in soil biological and physical parameters. Appl Soil Ecol 20:107–118
Collins HP, Elliott LF, Rickman RW, Bezdicek DF, Papendick RI (1990) Decomposition and interactions among wheat residue components. Soil Sci Soc Am J 54:780–785
Constantinides M, Fownes JH (1994) Nitrogen mineralization from leaves and litter of tropical plants: relationship to nitrogen, lignin and soluble polyphenols concentrations. Soil Biol Biochem 26:49–55
Cortez J, Demard JM, Bottner P, Monrozier LJ (1996) Decomposition of Mediterranean leaf litters: a microcosm experiment investigating relationships between decomposition rates and litter quality. Soil Biol Biochem 28:443–452
Coûteaux MM, Bottner P, Berg B (1995) Litter decomposition, climate and litter quality. Trends Ecol Evol 10:63–66
Dagley S (1967) The microbial metabolism of phenolics. In: McLaren AD, Peterson GH (eds) Soil biochemistry. Dekker, New York, pp 287–317

Daniel GF, Nilsson T, Singh AP (1987) Degradation of lignocellulosics by unique tunnel-forming bacteria. Can J Microbiol 33:943–948

Dhir RP (1977) Western Rajasthan soils: their characteristics and properties. In: Jaiswal PL (ed) Desertification and its Control. ICAR, New Dehli, pp 102–115

Dhir RP, Mann HS (1978) Arid soils, their characteristics and management. National Symposium on Land and Water Management in Indus Basin, ICAR/PAU, Ludhiana, pp 422–437

Dhir RP, Joshi DC, Singh N, Sharma BK, Kolarkar AS (1997) Taxonomy and distribution of arid soils of Rajasthan. Ann Arid Zone 36:327–333

Dick RP (1997) Enzyme activities as integrative indicators of soil health. In: Parkhurst CE, Doube BM, Gupta VVSR (ed) Biological indicators of soil health, CAB International, Oxon, pp 121–156

Dick WA, Tabatabai MA (1978) Inorganic pyrophosphatase activity of soils. Soil Biol Biochem 10:59–65

Dilly O, Wacherndorf C, Irmler U, Blume HP, Munch JC (1997) Changes of abiotic and biotic parameters in the course of decomposition of leaf litter in a black alder [(*Alnus glutinosa*) Gaertn. (L.)] forest of a moronic landscape in Northern Germany. EcoSys 6:31– 42

Dilly O, Bartsch S, Rosenbrock P, Buscot F, Munch JC (2001) Shifts in physiological capabilities of the microflora during the decomposition of leaf litter in a black alder [*Alnus glutinosa* (Gaertn.) L.] forest. Soil Biol Biochem 33:921–930

Doran JW (2002) Soil health and global sustainability: translating science into practice. Agric Ecosys Environ 88:119–127

Doran JW, Parkin TB (1994) Defining and assessing soil quality. In: Doran JW, Coleman DC, Bexdicek DF, Stewart BA (eds) Defining soil quality for a sustainable environment, SSSA Special Publication No. 35, Soil Sci Soc Am Am Soc Agro, Madison, WI, pp 3–21

Erikson KE, Blanchette RA, Ander P (1990) Microbial and enzymatic degradation of wood and wood components. Springer Series in Wood Sciences, Springer, Berlin

Faroda AS, Singh M (1998) Fifty years of arid zone research in India. CAZRI, Jodhpur, India

Fließbach A, Mader P (2000) Microbial biomass and size-density fractions differ between soils of organic and conventional agricultural systems. Soil Biol Biochem 32:757–768

Franzluebbers AJ, Haney RL, Honeycutt CW, Schomberg HH, Hons FM (2000) Flush of carbon dioxide following rewetting of dried soil relates to active organic pools. Soil Sci Soc Am J 64:613–623

Fraser DG, Doran JW, Sahs W, Lesoing GW (1988) Soil microbial population and activities under conventional and organic management. J Environ Qual 17:585–590

Frey SD, Six J, Elliot ET (2003) Reciprocal transfer of carbon and nitrogen by decomposer fungi at the soil-litter interface. Soil Biol Biochem 35:1001–1004

Gallardo A, Merino J (1993) Leaf decomposition in two Mediterranean ecosystems of Southwest Spain: influence of substrate quality. Ecol 74:152–161

Gehlot HS, Panwar D, Parihar R, Khichar N, Tak N, Shekhawat NS (2008) Nodulating legumes and associated rhizobia found in arid regions of the Thar Desert. In: Parihar P, Parihar L (eds) Advances in applied microbiology. Agrobios India, Jodhpur, pp 27–36

Gill RA, Burke IC (2002) Influence of soil depth on the decomposition of *Bouteloua gracilis* roots in the shortgrass steppe. Plant Soil 241:233–242

Giller KE, Witter E, McGrath SP (1998) Toxicology to heavy metals to microorganisms and microbial processes in agricultural soils: a review. Soil Biol Biochem 30:1389–1414

Halverson LJ, Jones TM, Firestone MK (2000) Release of intracellular solutes by four soil bacteria exposed to dilution stress. Soil Sci Soc Am J 64:1630–1637

Hammel KE (1997) Fungal degradation of lignin. In: Cadish G, Giller KE (ed) Driven by nature: plant litter quality and decomposition. CAB International, Cambridge, pp 33–44

Harsh LN, Tewari JC, Venkateshwarlu J (1989) Agroforestry systems in India. In: National Seminar on Agroforestry Systems in India. CRIDA, Hyderabad

Harsh LN, Tewari JC, Burman U, Sharma SK (1992) Agroforestry in arid regions. Indian Farming 45:32–37

Hart SC, Firestone MK (1991) Forest floor mineral soil interaction in the internal nitrogen cycle of an old growth forest. Biogeochemistry 12:103–127

Hatfield RD, Ralph J, Grabber JH (1999) Cell wall structural foundations: molecular basis for improving forage digestibilities. Crop Sci 39:27–37

Heal OW, Anderson JM, Swift MJ (1997) Plant litter quality and decomposition: an historical overview. In: Cadisch G, Giller GE (eds) Driven by nature: plant litter quality and decomposition. CAB International, UK, pp 3–30

Hegde BR, Havanagi GV, Reddy M, Venugopal N, Viswanath AP, Satyanarayanam T (1982) Studies on incorporation of maize residue on soil properties and yield of maize under rainfed conditions. Indian J Agron 27:254–258

Helal HM, Sauerbeck DR (1984) Influence of plant roots on C and P metobolism in soil. Plant Soil 76:175–182

Hook PB, Burke IC, Lauenroth WK (1991) Heterogeneity of soil and plant N and C associated with individual plants and openings in North American shortgrass steppe. Plant Soil 138: 247–256

Ibrahim SM, Shindo H (1999) Relationships between aggregation and hyphal length and microbial biomass-C in soil amended with rice straw and *Azolla*. Pedologist 43:82–87

Ischei AO, Mroghalu JI (1992) The effects of tree canopy covers on soil fertility in a Nigerian savannah. J Trop Ecol 8:329–338

Jackson LE, Strauss RB, Firestone MK, Bartolome JW (1990) Influence of tree canopies on grassland productivity and nitrogen dynamics in deciduous oak savannah. Agric Ecosyst Environ 32:89–105

Jarvis CS, Stockdale EA, Shepherd MA, Powlson DS (1996) Nitrogen mineralization in temperate agricultural soils: processes and measurement. Adv Agron 57:187–235

Jeffries P, Barea JM (1994) Biogeochemical cycling and arbuscular mycorrhizas in the sustainability of plant–soil systems. In: Gianinazzi S, Schuepp H (eds) Impact of arbuscular mycorrhizas on sustainable agriculture and natural ecosystems, Birkhauser, Basel, pp 101–116

Joffre R, Rambal S (1993) How tree cover influences the water balance of Mediterranean rangelands. Ecol 74:570–582

Joffre R, Vacher J, De los Llanos C, Long G (1988) The dehesa: an agrosilvopastoral system of the Mediterranean region with special reference to the Sierra Morena area of Spain. Agrofor Syst 6:71–96

Kanazawa S, Asakawa S, Takai Y (1988) Effect of fertilizer and manures application on microbial numbers, biomass, and enzyme activities in volcanic soils. Soil Sci Plant Nutr 34:429–439

Kautz T, López-Fando C, Ellmer F (2006) Abundance and biodiversity of soil micro arthropods as influenced by different types of organic manure in a long-term field experiment in Central Spain. Appl Soil Ecol 33:278–285

Kemp PR, Reynolds JF, Virginia RA, Whitford WG (2003) Decomposition of leaf and root litter of Chihuahuan desert shrubs: effects of three years of summer drought. J Arid Environ 53: 21–39

Kjoller A, Struwe S (1982) Microfungi in ecosystems: fungal occurrence and activity in litter and soil. Oikos 39:389–422

Kögel-Knabner I (2002) The macromolecular organic composition of plant and microbial residues as inputs to soil organic matter. Soil Biol Biochem 34:139–162

Kourtev PS, Ehrenfeld JG, Huang WZ (2002) Enzyme activities during litter decomposition of two exotic and two native plant species in hard wood forest of New Jersey. Soil Biol Biochem 34:1207–1218

Kumar K, Goh KM (2000) Crop residues and management practices: effects on soil quality, soil nitrogen dynamics, crop yield and nitrogen recovery. Adv Agron 68:197–219

Lauriks R, Dewulf R, Carter SE, Niang A (1999) A methodology for the description of border hedges and the analysis of variables influencing their distribution: a case study in western Kenya. Agrofor Syst 44:69–86

Lavelle P, Spain AV (2001) Soil ecology. Kluwer, Dordrecht

Li Z, Peng SL, Rae DJ, Zhou GY (2001) Litter decomposition and nitrogen mineralization of soils in subtropical plantation forests of southern China, with special attention to comparison between legumes and non-legumes. Plant Soil 229:105–116

Lodge DJ (1985) Preliminary estimates of fungal biomass and nutrient stores in the litter and soil of a tropical rain forest. Agron Abstr Soil Sci Soc Am, Chicago, IL, pp 158–159

Lodge DJ, Cantrell S (1995) Fungal communities in wet tropical forests: variation in time and space. Can J Bot 73:391–398

Lodha BK, Joshi DC, Jain SV (1982) Physiology and soil association in Rajasthan. J Indian Soc Soil Sci 30:326–333

Mafongoya PL, Nair PKR (1997) Multipurpose tree prunings as a source of nitrogen to maize (*Zea mays* L.) under semiarid conditions in Zimbabwe. I. Nitrogen recovery rates in relation to pruning quality and method of application. Agrofor Syst 35:31–46

Mann HS, Muthana KD (1984) Arid zone forestry. CAZRI Monogr. 23, CAZRI, Jodhpur

Marumoto T, Anderson JPF, Domasch KH (1982) Mineralization of nutrients from microbial biomass. Soil Biol Biochem 14:469–475

Mathur CM, Ganu SN, Moghe VB, Jain SV (1972) Soils of Rajasthan: survey and classification in retrospect and prospect. Soil Survey Organisation, Department of Agriculture, Rajasthan

Matson PA, Parton WJ, Power AG, Swift MJ (1997) Agricultural intensification and ecosystem properties. Science 277:504–509

Mazzarino MJ, Oliva J, Nunez ANG, Buffa E (1991) Nitrogen mineralisation and soil fertility in the dry Chaco ecosystem (Argentina). Soil Sci Soc Am J 55:515–522

McClaugherty C, Berg B (1987) Cellulose, lignin and nitrogen concentrations as rate regulating factors in late stages of forest litter decomposition. Pedobiologia 30:101–112

Mehar SK, Purohit U, Sundaramoorthy S (2002) Microbial biomass dynamics as affected by moisture and added litter content. Indian J Environ Sci 6:27–34

Melillo JM, Aber JD, Muratore JF (1982) Nitrogen and lignin control of hardwood leaf litter decomposition dynamics. Ecol 63:621–626

Miller F (1993) Composting as a process based on the control of ecologically selective factors. In: Blaine MF Jr (ed) Soil microbial ecology applications in agricultural and environmental management, Dekker, New York, pp 515–544

Miller RM, Lodge DJ (1997) Fungal responses to disturbance: agriculture and forestry. In: Wicklow DT, Söderström BE (eds) The Mycota, vol IV. Environmental and microbial relationships, 2nd edn. Springer, Berlin, pp 65–84

Moore JC, Berlow EL, Coleman DC, de Ruiter PC, Dong Q, Hastings A, Johnson NC, McCann KS, Melville K, Morin PJ, Nadelhoffer K, Rosemond AD, Post DM, Sabo JL, Scow KM, Vanni MJ, Wall DH (2004) Detritus, trophic dynamics and biodiversity. Ecol Lett 7:584–600

Mordelet P, Abbadie L, Menaut JC (1993) Effects of tree clumps on soil characteristics in a humid savannah of West Africa (Lamto, Cote d'Ivoire). Plant Soil 153:103–111

Moretto AS, Distel RA (2003) Decomposition of and nutrient dynamics in leaf litter and roots of *Poa ligularis* and *Stipa gynerioides*. J Arid Environ 55:503–514

Morris JW, Benzuidenhout JJ, Furniss PR (1982) Litter decomposition, In: Huntley BJ, Walker BH (eds) Ecology of tropical savannas. Springer, Berlin, pp 535–554

Mtambanengwe F, Kirchmann H (1995) Litter from a tropical savanna woodland (Miombo): chemical composition and C and N mineralization. Soil Biol Biochem 27:1639–1651

Muniafu M, Kinyamario JI (2007) Soil nutrient content, soil moisture and yield of Katumani maize in a semi-arid area of Kenya. Afr J Environ Sci Technol 4:81–85

Nadiaye EL, Sandeno JM, McGrath D, Dick RP (2000) Integrative biological indicators for detecting change in soil quality. Am J Altern Agric 15:26–36

Nagarajan M, Sundaramoorthy S (2000a) Effect of *Prosopis cineraria* (Linn.) Druce on microbial biomass and soil C, N in arid agroforestry systems of western Rajasthan, India. Ann Arid Zone 39:431–438

Nagarajan M, Sundaramoorthy S (2000b) Effect of cultivation and tree canopy on available nutrients in a traditional agroforestry system. Indian J Agrofor 2:99–100

Nair PKR (1995) An introduction to agroforestry. Kluwer, Dordrecht
Nkongolo NV, Caron J, Gauthier F, Yamada M (2000) Organic wastes for improving soil physical properties and enhancing plant growth in container substrates. J Crop Prod 3:97–112
Nord FF (1964) The formation of lignin and its biochemical degradation. Geochim Cosmochim Acta 28:1507–1521
Oglesby KA, Fownes JH (1992) Effect of chemical composition on nitrogen mineralization from green manures of seven tropical leguminous trees. Plant Soil 143:127–132
Palm CA (1995) Contribution of agroforestry trees to nutrient requirements of intercropped plants. Agrofor Syst 30:105–124
Palm CA, Rowland AP (1997) A minimum data set for characterization of plant quality for decomposition. In: Cadisch G, Giller GE (eds) Driven by nature: plant litter quality and decomposition, CAB International, Wallingford, UK, pp 379–392
Palm CA, Sanchez PA (1990) Decomposition and nutrient release patterns of the leaves of three tropical legumes. Biotropica 22:330–338
Palm CA, Sanchez PA (1991) Nitrogen release from the leaves of some tropical legumes as affected by their lignin and polyphenolic contents. Soil Biol Biochem 23:83–88
Parkinson D, Coleman DC (1991) Methods for assessing soil microbial populations, activity and biomass. Agric Ecol Environ 34:3–33
Paul EA, Clark FE (1996) Soil microbiology and biochemistry. Academic, London
Paustian K, Agren GI, Bosatta E (1997) Modeling litter quality effects on decomposition and soil organic matter dynamics. In: Cadisch G, Giller GE (eds) Driven by nature: plant litter quality and decomposition. CAB International, Wallingford, UK, pp 313–335
Pellinen J, Väisänen E, Salkinoja-Salonen M, Brunow G (1984) Utilization of dimeric lignin model compounds by mixed bacterial cultures. Appl Microbiol Biotechnol 20:77–82
Pellinen J, Jokela J, Salkinoja-Salonen M (1987) Degradability of different lignins by bacteria. Holzforschung 41:271–276
Persson J, Otabbong E (1994) Fertility of cultivated soils. In: Soil fertility and regulating factors, Report 4337 Swedish EPA, Stockholm, pp 9–70
Purohit U, Mehar SK, Sundaramoorthy S (2002) Role of *Prosopis cineraria* on the ecology of soil fungi in Indian desert. J Arid Environ 52:17–27
Rao VMB, Singh SP (1993) Crop responses to organic sources of nutrients in dryland conditions. In: Somani LL (ed) Recent advances in dryland agriculture. Scientific, Jodhpur, pp 287–304
Rao YS, MacDicken KG (1991) Foreword. In: Mallick W, Rao YS, MacDicken KG (ed) Agroforestry in Asia and the Pacific. RAPA, Bangkok, pp 1–2
Raychaudhuri SP, Agarwal RR, Daita NP, Gupta SP, Thomas PK (1963) Soils of India. ICAR, New Delhi
Rayner ADM, Boddy L (1988) Fungal communities in the decay of wood. Adv Microb Ecol 10:115–166
Rhoades CC (1997) Single-tree influence on soil properties in agroforestry: lessons from natural forest and savannah ecosystems. Agrofor Syst 35:71–94
Richle DE, McBrayer JF, Ausmus BS (1975) Ecological energetics of decomposer invertebrates in deciduous forest and total respiration budget. In: Venek J (ed) Progress of soil zoology. Slovak Academy of Sciences, Prague, pp 283–292
Rosenbrock P, Buscot F, Munch JC (1995) Fungal succession and changes in the fungal degradation potential during the initial stage of litter decomposition in a black alder [(*Alnus glutinosa*) Gaertn. (L.)] forest. Eur J Soil Biol 31:1–11
Roy BB, Sen AK (1968) Soil map of Rajasthan. Ann Arid Zone 7:1–14
Santhy P, Velusamy MS, Murugappan V, Selvi D (1999) Effect of inorganic fertilizers and fertilizer-manure combination on soil physiochemical properties and dynamics of microbial biomass in an Inceptisol. J Indian Soc Soil Sci 47:479–482
Sasson A (1972) Microbial life in arid environments – prospectus and achievements. Ann Arid Zone 11:67–86

Satpathy B, Behera N, Dash MC (1982) Microbial population, biomass and activity in some tropical soils of Orissa, India. Biol Bull India 4:150–157
Saxena SK (1997) Traditional agroforestry systems in western Rajasthan. In: Gupta JP, Sharma BM (eds) Agroforestry for sustained productivity in arid regions. Scientific, Jodhpur, pp 21–30
Schaefer D, Steinberger Y, Whitford WG (1985) The failure of nitrogen and lignin control of decomposition in a North American desert. Oecologia 65:382–386
Schimel JP, Hattenschwiler S (2007) Nitrogen transfer between decomposing leaves of different N status. Soil Biol Biochem 39:1428–1436
Schlesinger WH (1985) Decomposition of chaparral shrub foliage. Ecology 66:1353–1359
Schlesinger WH, Pilmanis AM (1998) Plant–soil interactions in deserts. Biogeochemistry 42:169–187
Schlesinger WH, Raikes JA, Hartley AE, Cross AF (1996) On the spatial pattern of soil nutrients in desert ecosystems. Ecology 77:364–374
Scholes RJ, Archer SR (1997) Tree grass interaction in savannahs. Annu Rev Ecol Syst 28: 517–544
Schroth G (2003) Decomposition and nutrient supply from biomass. In: Schroth G, Sinclair FL (eds) Trees, crops and soil fertility: concepts and research methods, CAB International, UK, pp 131–150
Seeling B, Jungk A (1996) Utilization of organic phosphorous in calcium chloride extract of soil by barley plants and hydrolysis of acid and alkaline phosphatases. Plant Soil 178:179–184
Sen DN (1982) Ecological approaches to Indian weeds. Geobios, Jodhpur, India
Seneviratne G (2000) Litter quality and nitrogen release in tropical agriculture: a synthesis. Biol Fertil Soils 31:60–64
Shankar V, Kumar S (1987) Grazing resources of Jaisalmer: ecology and development planning with special reference to Sewan Grasslands. CAZRI Monograph 28, CAZRI, Jodhpur
Shaw MR, Harte J (2001) Control of litter decomposition in a sub alpine meadow-sagebrush steppe ecotone under climate change. Ecol Appl 11:1206–1223
Singh JS, Gupta SR (1977) Plant decomposition and soil respiration in terrestrial ecosystems. Bot Rev 43:449–527.
Singh RP, Daulay HS, Singh KC (1981) Effect of systems of planting pearl millet on the yield, total productivity, moisture use and monetary returns. Indian J Agric Sci 51:409–416
Soil Survey Staff (1992) Keys to soil taxonomy, 6th edn. USDA, Soil Conservation Service, Washington DC
Sorensen LH (1975) The influence of clay on the rate of decay of amino acid metabolites synthesized in soil during decomposition of cellulose. Soil Biol Biochem 7:171–177
Speir TW (1977) Studies on a climosequence of soil in tussock grasslands. 11. Urease, phosphatase and sulphatase activities of topsoils and their relationships with other properties including plant available sulphur. N Z J Sci 20:159–166
Speir TW, Ross DJ (1976) Studies on a climosequence of soil in tussock grasslands. 9. Influence of age of *Chinochloa rigida* on enzyme activities. N Z J Sci 19:389–396
Sundaramoorthy S (1987) Weed ecology and competitive ability of certain weeds in cultivated fields. PhD Thesis, JNV University, Jodhpur
Swift MJ (1976) Species diversity and structure of the microbial communities in terrestrial habitats. In: Anderson JM, Maafadgin A (eds) The role of terrestrial and aquatic organisms in decomposition processes. Blackwell, Oxford, pp 185–222
Swift MJ, Heal OW, Anderson JM (1979) Decomposition in terrestrial ecosystems. Studies in ecology, vol 5. Blackwell, Oxford
Szott LT, Fernandez ECM, Sanchez PA (1991) Soil plant interaction in agroforestry system. Forest Ecol Manage 45:127–152
Tarafdar JC (1989) Use of electrofocussing technique for characterizing the phosphatases in the soil and root exudates. J Indian Soc Soil Sci 37:393–395
Tarafdar JC, Chhonkar PK (1979) Phosphatase production by micro-organisms isolated from diverse types of soils. Zenterblatt Bakteriol 134:119–124

Tarafdar JC, Jungk A (1987) Phosphatase activity in the rhizosphere and its relation to the depletion of soil organic phosphorus. Biol Fertil Soils 3:199–204
Tarafdar JC, Rao AV (1990) Effect of manures and fertilizers on phosphateses in the rhizosphere of arid crops. Pol J Soil Sci 23:81–85
Tarafdar JC, Rao AV (1992) Decomposition of tree leaves in arid soils at different moisture levels. J Tree Sci 11:140–143
Tarafdar JC, Meena SC, Kathju S (2001a) Influence of straw size on activity and biomass of soil microorganisms during decomposition. Eur J Soil Biol 37:157–160
Tarafdar JC, Yadav RS, Meena SC (2001b) Comparative efficiency of acid phosphatase originated from plant and fungal sources. J Plant Nutr Soil Sci 164:279–282
Tarafdar JC, Yadav RS, Niwas R (2002) Relative efficiency of fungal intra- and extracellular phosphatases and phytase. J Plant Nutr Soil Sci 165:17–20
Teklay T, Malmer A (2004) Decomposition of leaves from two indigenous trees of contrasting qualities under shaded-coffee and agricultural land-uses during the dry season at Wondo Genet, Ethiopia. Soil Biol Biochem 36:777–786
Teklay T, Nordgren A, Nyberg G, Malmer A (2007) Carbon mineralization of leaves from four Ethiopian agroforestry species under laboratory and field conditions. Appl Soil Ecol 35: 193–202
Tewari JC, Bohra MD, Harsh LN (1999) Structure and production function of traditional extensive agroforestry systems and scope of intensive agroforestry in Thar desert. Indian J Agrofor 1:81–94
Tian G, Brussaard L, Kang BT (1995) Breakdown of plant residues with contrasting chemical compositions under humid tropical conditions – decomposition and nutrient release. Soil Biol Biochem 27:277–280
Upadhyay VP, Singh JS (1989) Pattern of nutrient immobilization and release in decomposing forest litter in central Himalaya. India J Ecol 77:127–146
Vanlauwe B, Nwoke OC, Sanginga N, Merckx R (1996) Impact of residues quality on the C and N mineralization of leaf and root residues of three agro forestry species. Plant Soil 183:221–231
Vanlauwe B, Diels J, Sanginga N, Merckx R (1997) Residue quality and decomposition: an unsteady relationship? In: Cadisch G, Giller GE (eds) Driven by nature: plant litter quality and decomposition. CAB International, Wallingford, UK, pp 157–166
Wagner GH, Wolf DC (1999) Carbon transformations and soil organic matter formation. In: Sylvia DM, Fuhrmann JJ, Hartel PG, Zuberer DA (eds) Principles and applications of soil microbiology. Prentice Hall, NJ, pp 218–258
Wardle DA (2002) Communities and ecosystems: linking the aboveground and belowground components. Monographs in population biology. Princeton University Press, Princeton, NJ
Whitford WG (2002) Ecology of desert systems. Academic, San Diego
Whitford WG, Steinberger Y, Mackay W, Parker LW, Freckman D, Wallwork JA, Weems D (1986) Rainfall and decomposition in the chihuahuan desert. Oecologia 68:512–515
Wieder RK, Lang GE (1982) A critique of the analytical methods used in examining decomposition data obtained from litter bags. Ecology 63:1636–1642
Young A (1991) Change and constancy: an analysis of publications in agroforestry systems, vols 1–10. Agrofor Syst 13:195–202
Young JC, Insam H (1991) Microbial biomass and relative contributions of bacteria and fungi in soil beneath tropical rain forest, Hainan Island, China. J Trop Ecol 7:385–395
Zinke PJ (1962) The patterns of influence of individual forest trees on soil properties. Ecol 43: 130–133
Zvyagintsev DG (1994) Vertical distribution of microbial communities in soils. In: Ritz K, Dighton J, Giller KE (eds) Beyond the biomass. Wiley/Blackwell, Chichester, pp 29–37

Chapter 6
Aspects of Mycorrhizae in Desert Plants

Martha E. Apple

Abstract Mycorrhizal symbioses are critical to desert plants since they face the challenges of scarce, sporadic precipitation, nutrient deficiencies, intense solar radiation, and the high temperatures found in hot deserts. Deserts are covering increasingly more of the Earth's surface area as desertification increases globally. Mycorrhizal desert plants have a greater chance of survival in the harsh desert environment. Desert plants form mycorrhizae with endomycorrhizal arbuscular fungi and with ectomycorrhizal fungi. Both form extensive networks of hyphae in the soil, and glomalin, a glycoprotein produced by hyphae of arbuscular mycorrhizal fungi, is crucial in soil structure and carbon storage. Mycorrhizal desert plants are important in agriculture, ecosystem biology, and conservation of the deserts.

6.1 Introduction

Desert plants face the challenges of scarce, sporadic precipitation and nutrient deficiencies, as well as those of the intense solar radiation, high temperatures, and the great diurnal ranges in temperature often found in desert ecosystems. The percentage of the Earth's surface covered by deserts is increasing along with the human population, and hot deserts are globally the most common type of climate, covering 14.2% of the Earth's land area (Peel et al. 2007).

6.1.1 Deserts

While deserts are stereotypically portrayed as vast seas of sand dunes or mountainous regions with Saguaro cacti and vultures, in fact there are many different kinds of

M.E. Apple
Department of Biological Sciences, Montana Tech of the University of Montana, Butte, Montana, MT 59701, USA
e-mail: mapple@mtech.edu

K.G. Ramawat (ed.), *Desert Plants*,
DOI 10.1007/978-3-642-02550-1_6, © Springer-Verlag Berlin Heidelberg 2010

deserts with varying patterns of temperature and precipitation. Landforms, altitude, and life forms vary with deserts. The African Sahara exemplifies a hot desert, while the American Great Basin Desert is a temperate desert. The driest place on earth is the Atacama Desert of Chile, and surprisingly, Antarctica is considered to be an extremely cold desert. According to the Koppen-Geiger climate classification (Peel et al. 2007), deserts are regions where, on average, annual precipitation is less than 250 mm, with semi-arid regions receiving 250–500 mm. Deserts exist where potential evapotranspiration exceeds precipitation. The diurnal temperature variation in deserts is pronounced, with high daytime and low nighttime temperatures (Woodward 2003).

6.1.2 Mycorrhizae

In arid environments, mycorrhizal symbioses are an essential and quite often critical factor in the survival of plants (Titus et al. 2002). Mycorrhizal desert plants have a greater likelihood of survival in these harsh environments, in part because of the enhanced nutrient acquisition and availability of water provided to the plants by the mycorrhizal fungi. Specifically, mycorrhizal plants are more robust, have increased uptake of phosphorus, nitrogen, calcium, magnesium, and iron (Bohrer et al. 2003; Smith and Read 1997; Titus et al. 2002), greater drought resistance (Davies et al. 1992), and greater tolerance to stresses such as salinity, acidity, and heavy metals. They are stronger competitors and are more resistant to fungal infections (West 1997) and herbivory by insects (Gange and Bower 1997; Titus et al. 2002) than their non-mycorrhizal counterparts. It is important to understand mycorrhizal symbioses in desert plants because of their importance in agriculture, conservation and reclamation (White et al. 1989), ecology and biodiversity.

Although mycorrhizal symbioses approach ubiquity in the plant kingdom, they do not reach it, since not all plants are mycorrhizal. However, among the angiosperms, arbuscular mycorrhizal associations form in 83% of the dicotyledonous species and 79% of the monocotyledonous species tested so far. Mycorrhizae are not found in the Brassicaceae, the source of cruciferous vegetables such as broccoli and cabbage (Smith and Read 1997). In the Australian deserts, 73% of the plants tested were mycorrhizal (O'Conner et al. 2001), and most perennial plants of arid zones are mycorrhizal (Titus et al. 2002). Therefore, mycorrhizae is an important enough aspect of life in desert plants to warrant considerable attention, especially in the contexts of ecosystem studies, horticultural work, and biotechnological and agricultural activities in the desert and in semi-arid lands.

6.1.3 Mycorrhizal Fungi

The fungi in mycorrhizal symbioses benefit because they obtain a fraction of the photosynthate produced by their host plants. This fraction varies and can be considerable. On average, it is between 10% and 30% of the net carbon fixed by the host plant (Finlay and Soderstrom 1992; Allen et al. 2003). While this percentage represents a fair amount of carbon lost by the plant, the benefits conferred on the plant by the fungus outweigh the effects of the carbon losses. In fact, roots generally lose some carbon through the process of root exudation. The quantity of carbon lost varies but ranges from 1% to 40%, and this exuded carbon influences microbial activity in the rhizosphere and nutrient acquisition by the rhizosphere (Dilkes et al. 2004; Meharg and Killham 1994; Whipps 1990). Therefore, carbon lost through root exudation can be thought of as an important expenditure by the plant in terms of gaining vital nutrients such as N and P, without which the plant could not survive.

Plants living in the desert form mycorrhizal symbioses with vesicular- or arbuscular-mycorrhizal (VAM or AM) fungi, and ectomycorrhizal (ECM) fungi (Mejstrik and Cudlin 1983). In VAM, which is an endomycorrhizal association, the fungal partner of the plant-fungal mycorrhizal symbiosis forms vesicles, which indicate carbon storage (Titus et al. 2002), and tree-like arbuscules, which form from hyphal tips that penetrate the root's cortical cell walls but not the plasmalemma (Allen 2007). Arbuscules resemble small trees with extensive lateral branching, and they are the sites of exchange of materials between the fungus and the plant, so an abundance of arbuscules suggests an intensely active symbiosis (Titus et al. 2002). In contrast, ECM fungi form a mantle of hyphae that surrounds the root and a network of hyphae called the Hartig net (Hartig 1840) around the peripheral cells of the root, with some hyphal cells entering the root between cortical cells (Smith and Read 1997).

6.1.4 Taxonomy of Desert Mycorrhizal Fungi

Taxonomically, the AM fungi fall into the Glomales, an order of obligate biotrophic fungi characterized by the intraradical arbuscules that they use to obtain carbon from their host plants (Smith and Read 1997; Walker and Trappe 1993). Important genera of desert mycorrhizal fungi include *Acaulospora*, *Archaeospora*, *Glomus* and *Paraglomus* (Shi et al. 2006). Spores of AM fungi isolated from the volcanic desert of Mount Fuji, Japan, and analyzed with fragment and sequence analysis of the large ribosomal subunit RNA gene belonged to the *Glomaceae*, *Acaulosporaceae*, and *Gigasporaceae* (Wu et al. 2007).

Gigaspora sp. produce gigantic spores. For example, the spores of *G. margarita* are on average 321 μm in diameter. Spores of *Glomus* sp. are generally smaller, with *G. luteum* producing spores with an average diameter of 164 μm.

ECM fungal associations are formed by fungi in the Basidiomycotina when the fruiting bodies are epigeous (underground) and by Basidiomycotina, Ascomycotina or Zygomycotina if the fruiting bodies are hypogeous (aboveground) (Smith and Read 1997). The ECM fungi include many easily recognizable families, which include the Amanitaceae and the Boletaceae, and in some cases, economically important families such as the Hellvelaceae, which includes the edible and choice morels and the Tuberaceae, which includes the much sought after truffles of haute cuisine.

6.1.5 Initiation of Arbuscular Mycorrhizal Symbioses

Arbuscular mycorrhizal symbioses are initiated when the fungus contacts the root epidermis and differentiates into an appressorium, which it uses to enter the root. Once the appressorium is in the cortex, it sends intercellular hyphae through the apoplastic pathway in the cell wall. Next, the hyphae grow into the space between the cell wall and the plasmalemma, where they form extensively branched arbuscules (Gianinazzi-Pearson et al. 2006). Once the arbuscule forms, the host plant's cortical cells undergo a proliferation of organelles and membranes, resulting in an extensive interface between the plant and fungal cells (Smith and Smith 1990). Fungal hyphae grow and extend outward into the soil to form a network, which then invades other roots, acquires nutrients, and forms spores. None of this happens without preliminary signaling between the fungus and the host plant, which sends out root exudates containing signaling molecules, including strigolactones, which stimulate transcription and hyphal growth in the fungus. Once the appressoria contact the epidermal cell surfaces, new apoplastic compartments form in the cells where the hyphae penetrate (Smith and Smith 1990). Host plants upregulate genes involved in responses to mycorrhizal fungi following contact with fungal exudates but it is not necessary for host plant cells to have actual contact with the fungal cells themselves (Parniske 2004).

6.2 Economics, Agriculture, and Mycorrhizae in the Desert

Economically useful mycorrhizal fungi grow in the deserts of the world. One example is an edible and prized hypogeous ECM fungus, the *Helianthemum* sp. associated desert truffle, *Terfezia pfeilii*, which grows in the Kingdom of Saudi Arabia (Diez et al. 2002; Hashem and Al-Obaid 1996). The desert truffle has a broad range of hosts (Molina et al. 1992) and it may in turn have an economically important host plant, *Citrullus lanatus*, the watermelon, with which it forms non-vesicular arbuscular endomycorrhizae. The watermelon can be cultivated in the desert and in semi-arid regions, but not without irrigation (Kagan-Zur et al. 1999). Correct cultivation of the watermelon may ensure economically gainful harvests of both the desert truffle and the watermelon's edible fruit.

Generally, the plants in a mycorrhizal association have an economic importance that is immediately and apparently greater than that of their fungi. However, the increase in biomass in crop plants that often occurs as a result of mycorrhizal symbioses tells of the underlying economic importance of the fungi. Western Rajasthan, India, is an arid region where the crop plant *Vigna aconitifolia*, the moth bean, was grown with and without inoculations of VAM fungi. Of the fungi tested, *Glomus fasciculatum* was found to be the most effective VAM fungus, as was evident by the significantly increased N, P, K, Zn and Cu concentrations and biomass at crop maturity (Tarafdar and Kumar 1996).

Agriculture in the desert is very important in terms of securing an abundant global food supply, and its importance will increase as desertification becomes more extensive. The outcome of agricultural efforts can be improved by mycorrhizal symbioses in crop plants. Greenhouses are a key means of producing crops in deserts and in semi-arid regions, and inoculation of the soil with AM fungi has been demonstrated to be important in preventing the stunting of plants that results after greenhouse soils are disinfected prior to planting the crops (Koltai et al. 2008). Cost effective and easy to use methods such as fabricated bead-based AM fungal (AMF) inoculum for use in inoculating the soil in greenhouses and in field trials will likely increase the use of inoculation with AMF in agriculture of deserts and semi-arid regions (Koltai et al. 2008).

6.2.1 Disease Resistance

Mycorrhizae are effective in increasing resistance to disease in desert crop plants. Date palm seedlings, *Phoenix dactylifera*, treated with mycorrhizal spores of *Glomus monticolus*, *Glomus deserticola*, and *Glomus clarum* and wild AMF from a date grove in Morocco were more resistant to Bayoud Disease (Jaiti et al. 2007), a severe fungal disease caused by *Fusarium* that has killed over 15 million date palms in Morocco and Algeria over the last century. As a result of these deaths, desertification is increasing and farmers have abandoned their land and moved to urban centers (Zaid et al. 2002). If inoculation with *Glomus* sp. can indeed save the date palm, then there will be many positive repercussions.

6.2.2 Endemic Mycorrhizal Plants as Food

Besides the usefulness of inoculating with AM fungi in increasing the likelihood of production of economically profitable crops, naturally mycorrhizal endemic desert plants can be evaluated for their suitability as crop plants. One example is *Solanum centrale*, the Australian bush tomato, which is also known as the desert raisin. The fruit of this native plant is eaten whole, dried, or ground by the indigenous people of Australia, a group of citizens who can experience financial and cultural gains by successfully cultivating this potentially commercial desert crop plant. Spores of *Glomus* sp. form AM mycorrhizae with the fine roots of *S. centrale*, even in very dry conditions (Dennett 2006).

6.3 Water

Water is either continuously or intermittently scarce in the desert. Both AM and ECM fungi are important in transmitting water to the host plants because of their extensive hyphal networks that extend outwork into the soil beyond the rhizosphere

(Allen 2007). In AM fungi, these networks can extend outward approximately 6 cm from the root, and they form fan-like, branching networks that provide over 100 cm of hypha per fungal infection point. Because these infection points are placed at millimeter intervals along roots (Fitter 1991), a 10 mm segment of root would have an average of 1,000 cm of hypha fanning outward into the soil (Friese and Allen 1991; Allen et al. 2003; Allen 2007). Networks forming from arterial hypha have the potential to greatly expand the network originating from the infection points (Bago et al. 1998).

In ECM symbioses, individual or clustered root tips are surrounded by a mantle of ECM fungi that can act as a physical barrier to water loss. Aggregations of hyphae called chords, and sometimes individual hyphal elements, can extend outward from the root tips to great distances of several meters, as in the case of *Pisolithus* sp. and, in one instance, 20 m from the host tree (Allen 1991). These mycelia can form interconnections that can transport C or N between individual plants (Allen 2007; He et al. 2006).

6.4 Ecological Considerations

6.4.1 Succession

Mycorrhizal plants are important in early successional stages in the colonization of volcanic deserts (Peter 2003; Nara et al. 2003; Wu et al. 2007), as in the example of *Laccaria* and *Inocybe*, two ECM fungal genera, colonizing their pioneering host plant, *Salix reinii*, on a volcanic desert on Mount Fuji, Japan. ECM fungi form common mycorrhizal networks (CMNs), which enhanced nutrient acquisition and seedling growth in *Salix reinii* seedlings at Mount Fuji (Nara 2006). Mycorrhizal pioneer plants are found as colonizers on volcanoes of other continents, as was the case following the 1980 eruption of Mount St. Helens in the state of Washington in the northwestern United States (Allen 1988). Water drains quickly out of pumice, but the small pieces of pumice at Mount St. Helens hold water and AM hyphae formed bridges between the pumice particles (Allen 2007). These bridges are likely very important in water conductivity, since it increased in mycorrhizal lupines, *Lupinus lepidus*, that were the first plants to colonize Mount St. Helens after the eruption (Allen et al. 1992).

6.4.2 Physiognomy and Mycorrhizae

Ephemeral annuals, biennials, long-lived perennial herbaceous plants, shrubs, and trees are all found in deserts. Just as these types of plants have different life strategies, they also differ in terms of their dependence on mycorrhizal symbioses. Collier et al. (2003) found that annuals in the Chihuahuan Desert of North America had significantly less infection with AM fungi and thinner roots than their longer-lived, thicker rooted perennial counterparts. A similar pattern was found in the Mojave Desert, where annuals were either nonmycorrhizal or had little AM hyphae, while perennial plants were colonized with AM fungi (Titus et al. 2002). Interestingly, a non-native tree, *Tamarix ramosis*, (which by definition is perennial) was non-mycorrhizal in the Mojave desert (Titus et al. 2002), although *Tamarix* sp. are mycorrhizal in a desert riparian forest along the Tarim River in China (Yang et al. 2008). In the Sonoran desert, VAM was high in plants that colonized zones without nurse plants (Carrillo-Garcia et al. 1999), and it increased the stabilization of windblown soil landing under canopies of nurse plants.

6.4.3 Seedling Establishment

The likelihood of seedling establishment in desert plants can be enhanced by mycorrhizal fungi, although the harsh edaphic factors sometimes overrule the potential benefits of mycorrhizae. For example, in cactus seedlings of the Sonoran desert of Baja, Mexico, favorable edaphic factors were more important in promoting seedling establishment than VAM fungi and, although the VAM inoculum density was the same under mesquite trees and in adjacent barren areas, the soil under mesquite trees had greater water holding capacity and higher N and C (Bashan et al. 2000). However, ectomycorrhizal established pioneer shrubs can promote seedling establishment in new seedlings of the same species and of other early successional species. At a volcanic desert on Mt. Fuji, Japan, new seedlings of *Salix reinii*, a willow, *Betula ermani*, a birch, and *Larix kaempferi*, a larch, were more successful when growing under established shrubs of *Salix reinii*, a pioneer species, and had ECM fungi of the same species as the established, early successional *Salix reinii* (Nara and Hogetsu 2004).

6.4.4 Phenology

The phenology of desert plants is linked tightly to the availability of water and there are corresponding variations in the extent of mycorrhizal colonization with seasonal variations in precipitation and, hence, availability of water (Apple et al. 2005) as observed in two Mojave desert shrubs, the evergreen *Larrea tridentata*, also known as the creosote bush, and the drought deciduous bursage, *Ambrosia dumosa*. In *L. tridentata*, the mycorrhizal roots are most prevalent in the upper 10 cm of soil (Allen 1991; Staffeldt and Vogt 1974). Levels of hyphae, arbuscles, and vesicles in mycorrhizal roots can increase or decrease with the seasons (Titus et al. 2002), in evergreen and drought deciduous Mojave desert shrubs.

The mycorrhizal inoculum potential (MIP) is an index of the quantity of inoculum in the soil and it is measured by the percent AM fungal colonization in corn

(*Zea mays*) roots. MIP can vary with season, spatially, and with environmental variables (Titus and Del Moral 1998; Titus et al. 2002; Moorman and Reeves 1979).

The successful establishment of mycorrhizal symbioses is related to MIP and to the phenology and patterns of root growth in desert plants. Because inoculum must be present in order for mycorrhizal symbioses to become established, and because MIP is variable, then the percent colonization will vary with MIP. Likewise, mycorrhizal establishment can drop during periods of rapid root growth that outpace the speed at which spores and hyphae can colonize the roots. Root anatomy and root branching patterns vary with species and are therefore important factors in the establishment of mycorrhizal symbioses.

6.4.5 *Salinity*

The salinity of the soil is a very important edaphic factor in determining whether mycorrhizal symbioses become established. Halophytic, or salt-loving, plants have the ability to tolerate salt, and AM colonization occurred in halophytes of semi-arid northeastern Iran with high salinity (16 dS m^{-1}) but was inhibited when salinity became very high (45 dS m^{-1}), (Asghari et al. 2008). Very importantly, AM in salty soils can increase salt tolerance and eventual crop yield (Asghari et al. 2005). It is important to realize that spores of mycorrhizal fungi are sensitive to salt, and high salinity can inhibit spore germination and subsequent hyphal growth of AMF (Juniper and Abbott 2006).

6.5 Mycorrhizal Helper Bacteria

Microbes known as mycorrhizal helper bacteria (MHB) stimulate the development of mycorrhizal symbioses in AM and ectomycorrhizal fungi (Frey-Klett et al. 2007). Proposed mechanisms for the stimulation of the development of mycorrhizal symbioses by MHB include the production of root cell wall softening enzymes by bacteria, enhancement of the recognition of the fungus and the root, nutritional enhancement of fungal growth, favorable changes of soil properties, and factors that trigger the germination of fungal propagules (Garbaye 1994; Smith and Read 1997). Not only are MHBs influential in development, they are apparently indispensable in such vital processes as nitrogen fixation, nutrient mobilization from soil minerals into the plant, and the ever-continuing necessity of protecting roots against pathogens (Frey-Klett et al. 2007). It is important to realize that soil bacteria, as well as mycorrhizal fungi and plants, are important in establishing mycorrhizal symbioses. Species-specific interactions among MHBs, mycorrhizal fungi, and plants all play roles in determining the functionality of mycorrhizal symbioses. Some bacteria found in the cytoplasm of AM fungi are likely related to *Burkholderia cepacia*, although other bacterial genera are found as MHBs, and may contribute to the metabolism of nitrogen because of their involvement with the synthesis of amino acids. The roots, the mycorrhizal fungi and its extraradical hyphae and the helper bacteria all form the mycorrhizosphere, through which carbon flows outward from the roots into the soil. The mycorrhizosphere is especially important in water relations of desert plants because it flattens the gradient between cellular water and what is often very dry soil. Thus, the plants are not as prone to dehydration if they have an intact mycorrhizosphere.

MHB are found as part of the mycorrhizal symbioses of plants grown agriculturally in deserts and in native plants of semi-arid regions. It is likely that native plants of deserts also have MHBs as part of their mycorrhizal symbioses. Cells of the bacterial genera *Bacillus*, *Pseudomonas*, and *Arthrobacter* were found along with ten AM species (*Acaulospora* sp., *A. laevis*, *A. spinosa*, *Gigaspora* sp., *Gi. ramisporophora*, *Glomus* sp., *Gl. aggregatum*, *Gl. ambisporum*, *Gl. sinuosum*, and *Scutellospora biornata*) in a South American grassland, and the diversity of the AM fungi decreased with altitude (Lugo et al. 2008). MHBs can live intracellularly in mycorrhizal fungi. After inoculation of *Gigaspora decipiens* spores with the bacterium *Burkholderia sp.*, TEM was used to demonstrate that *Burkholderia* cells colonized the interior of the *G. decipiens* spores. *Burkholderia* attached end-on or side-on to both spores and hyphae of *G. decipiens*, apparently via fibrillar structures (Levy et al. 2003).

When *Bacillus* sp. were inoculated along with the AM fungi *Glomus mosseae* and *Glomus intraradices*, in drought-stressed lettuce, *Lactuca sativa*, two key fungal metabolites, succinate dehydrogenase (SDH) and alkaline phosphatase (ALP), increased their metabolic activities (Vivas et al. 2003). In fact, *Bacillus* sp. stimulated the development of *Glomus intraradices* development. The enhancement of such important physiological parameters as photosynthetic rate, water use efficiency (WUE), and stomatal conductance in the lettuce plants varied with the species-specific interactions of the AM fungi with *Bacillus* sp. Even so, the quantities of living and physiologically active AM fungi increased with co-inoculation of the helper bacterium (Vivas et al. 2003). Since lettuce is an important crop species, the process of co-inoculation with helper bacteria has important implications in the agriculture of deserts and semi-arid lands.

6.5.1 Ectomycorrhizal Associated Bacteria

Kozdrój et al. (2007) found that ectomycorrhizal associated bacteria (EMAB) *Pseudomonas putida* or *Bacillus cereus* along with inoculation of *Amanita rubescens* or *Hebeloma sinapizans* protected *Pinus sylvestris* seedlings from Cd(II) damage when grown in an industrial desert area polluted with heavy metals. In these *P. sylvestris* seedlings, ectomycorrhization by *A. rubescens* increased with dual inoculation with both species of EMAB, and the seedlings were more vigorous in terms of root and shoot lengths and biomass than if they had not been inoculated. Interestingly, the roots of the pine seedlings accumulated more (56 μg g^{-1} to

72 μg g^{-1} dry weight) of Cd(II) when inoculated with the EMAB *P. putida* than with *B. cereus* (Kozdrój et al. 2007). Helper bacteria along with ECM fungi are especially important in sequestering metals from polluted sites and in soils with nutrient deficiencies, as in the desert. ECM fungi sequester metals by binding heavy metals to negatively charged sites on cell walls, immobilization with metallothioneins and polyphosphates, concentration of metals in the slime layer surrounding the hyphae, and sequestration in vacuoles (Kozdrój et al. 2007; Leyval et al. 1997).

6.6 Glomalin

On an even smaller scale than the already small but very important scale of the MHB, there is glomalin, a very sticky iron-containing glycoprotein produced by the hyphae of VAM fungi in the Glomales (Wright et al. 1996). The concentration in the soil of this abundant soil glycoprotein varies, but can range from 1.08 to 4.91 mg^{-1} cm^{-3} soil (Rillig et al. 2003), and in undisturbed soils, it can account for approximately 15–20% of the organic carbon. Because glomalin is very important in the formation and stability of soil aggregates, it can be thought of as the glue of the soil (Wright and Upadhyaya 1998). Soil aggregates are in turn important in stabilizing the soil and increasing its water holding capacity. Where glomalin levels are higher, the soil is more permeable with greater infiltration of water, root development and microbial activity are enhanced, and the soil is more resistant to erosion (Wright and Upadhyaya 1998).

6.6.1 Glomalin and Agriculture

Land-use practices are very influential in determining the quantity of glomalin in the soil (Rillig et al. 2003). Glomalin content was lower in an agricultural field with a corn, *Zea mays*, and soybean, *Glycine max*, rotation than in a nearby forest (Rillig et al. 2003), and Wright et al. (1999) found that soils contained more glomalin after an agricultural transition from tilling to no-tilling. Hyphal networks are more likely to remain intact without the mechanical disruption of tilling (Wright et al. 1999). The concentration of glomalin will not increase if crops from the Brassicaceae are grown, since they do not form mycorrhizae and the presence of glomalin is dependent upon hyphae.

6.6.2 Glomalin, Ecosystems, and Carbon

Glomalin is found in the desert, although in lower levels than in more productive ecosystems such as the tropical rainforest and temperate forests (Treseder and

Turner 2007) where net primary production (NPP) is greater. Comparisons of glomalin levels across biomes show that deserts had an average value of 0.079 ± 0.038 g/kg soil of immunoreactive soil protein (ISRP), which is a measure of glomalin. Although glomalin levels were lower, on average, than in all other ecosystems tested, the abundance of AM was 0.36 km colonized root/m, a value that was slightly higher than the abundance of AM in temperate forests and tropical rainforests, which both had 0.31 km colonized root/m (Treseder and Turner 2007).

Globally, glomalin is of significant and current importance because of the increases in soil aggregates and glomalin found with elevated CO_2 at two experimental annual grassland ecosystems in Jasper Ridge in Northern California (Rillig et al. 1999). Since glomalin can remain in the soil for decades, it is an important pool of soil carbon (Rillig et al. 2003). Therefore, agricultural practices that contribute to increases in glomalin also influence the quantity of carbon in the soil. This is very important in desert soils, which generally have a low percentage of organic material.

6.7 Conclusion

In closing, mycorrhizal symbioses are of crucial importance in desert ecosystems because they are important for the survival of native and agricultural plants, because they are a key factor in the successful cultivation of economically profitable desert plants and fungi, and because their presence has a great impact on carbon dynamics in the soil.

References

Allen MF (1988) Re-establishment of mycorrhizae following severe disturbance: comparative patch dynamics Allen MF (1991) The ecology of mycorrhizae. Cambridge University Press, New York

Allen MF (2007) Mycorrhizal fungi: highways for water and nutrients in arid soils. Vadose Zone J 6:291–297 of a shrub desert and a subalpine volcano. Proc R Soc Edinburgh 94:63–71

Allen MF, Crisafulli C, Friese CF, Jeakins SL (1992) Reformation of mycorrhizal symbioses on Mount St. Helens, 1980–1990: interactions of rodents and mycorrhizal fungi. Mycol Res 98:447–453

Allen MF, Swenson W, Querejeta JL, Egerton-Warburton LM, Treseder KK (2003) Ecology of mycorrhizae: a conceptual framework for complex interactions among plants and fungi. Annu Rev Phytopathol 41:271–303

Apple ME, Thee CI, Smith-Longozo VL, Cogar CR, Wells CE, Nowak RS (2005) Arbuscular mycorrhizal colonization of *Larrea tridentata* and *Ambrosia dumosa* roots varies with precipitation and season in the Mojave Desert. Symbiosis 39:131–136

Asghari HA, Marschner P, Smith SE, Smith FA (2005) Growth response of *Atriplex nummularia* to mycorrhizal inoculation at different salinity levels. Plant Soil 273:245–256

Asghari HR, Amerian MR, Gorbani H (2008) Soil salinity affects arbuscular mycorrhizal colonization of halophytes. Pak J Biol Sci 11:1909–1915

Bago B, Ascon-Aquilar C, Piche Y (1998) Architecture and developmental dynamics of the external mycelium of the arbuscular mycorrhizal fungi *Glomus intraradices* grown under monoxenic conditions. Mycologia 90:52–62

Bashan Y, Davis EA, Carrillo-Garcia A, Linderman RG (2000) Assessment of VA mycorrhizal inoculum potential in relation to the establishment of cactus seedlings under mesquite nurse-trees in the Sonoran Desert. Appl Soil Ecol 14:165–175

Bohrer G, Kagan-Zur V, Roth-Berjerano N, Ward D, Beck G, Bonifacio E (2003) Effects of different Kalahari-desert VA mycorrhizal communities on mineral acquisition and depletion from the soil by host plants. J Arid Environ 55:193–208

Carrillo-Garcia A, Leon de la Lux JL, Bashan Y, Bethlenfalvay GJ (1999) Nurse plants, mycorrhizae, and plant establishment in a disturbed area of the Sonoran Desert. Restor Ecol 7:321–335

Collier SC, Yarnes CT, Herman RP (2003) Mycorrhizal dependency of Chihuahuan Desert plants is influenced by life history strategy and root morphology. J Arid Environ 55:223–229

Davies FT, Potter JR, Linderman RG (1992) Mycorrhiza and repeated drought exposure affect drought resistance and extra radical hyphae development of pepper plants independent of plant size and nutrient content. J Plant Physiol 139:289–294

Dennett A (2006) Underground structures and mycorrhizal associations of *Solanum centrale* (the Australian bush tomato), Honours Thesis, Plant Breeding Institute, University of Sydney, Sydney http://www.desertknowledgecrc.com.au. Cited 9 February 2009

Diez J, Manjon JL, Martin F (2002) Molecular phylogeny of the mycorrhizal desert truffles, *Terfezia* and *Tirmania*, host specificity and edaphic tolerance. Mycologia 94:247–259

Dilkes NB, Jones DL, Farrar J (2004) Temporal dynamics of carbon partitioning and rhizodeposition in wheat. Plant Physiol 134:706–715

Finley R, Soderstrom B (1992) Mycorrhiza and carbon flow to the soil. In: Allen MF (ed) Mycorrhizal functioning: an integrated plant fungal process. Chapman and Hall, New York, pp 134–160

Fitter AH (1991) Costs and benefits of mycorrhizas: implications for functioning under natural conditions. Experimentia 47:350–362

Frey-Klett P, Garbaye J, Tarkka M (2007) The mycorrhiza helper bacteria revisited. New Phytol 176:22–36

Friese CF, Allen MF (1991) The spread of VA mycorrhizal fungal hyphae in the soil: inoculum types and external hyphal architecture. Mycologia 83:409–418

Gange AC, Bower E (1997) Interactions between insects and mycorrhizal fungi. In: Gange AC, Brown VK (eds) Multitrophic interactions in terrestrial systems. Blackwell, Oxford, pp 115–130

Garbaye J (1994) Helper bacteria: a new dimension to the mycorrhizal symbioses. New Phytol 128:197–210

Gianinazzi-Pearson V, Maldonado-Mendoza I, Lopez-Meyer M, Weidmann S, Harrison MJ (2006) Arbuscular mycorrhiza. In: *Medicago truncatula* Handbook, Noble Foundation. Available via http://www.noble.org/MedicagoHandbook/ArbuscularMycorrhiza Cited 9 February 2009

Hartig T (1840) Vollstandige Naturgeschichte der Forstlichen Kulturpflanzen Deutschlands. Forstner'sche, Berlin

Hashem AR, Al-Obaid AM (1996) Mineral composition of soil and wild desert truffles in Saudi Arabia. J King Saud Univ Sci 8:5–10

He X, Bledsoe CS, Zasoski RJ, Southworth D, Horwath WR (2006) Rapid nitrogen transfer from ectomycorrhizal pines to adjacent ectomycorrhizal and arbuscular mycorrhizal plants in a California oak woodland. New Phytol 170:143–151

Jaiti F, Meddich A, El Hadrami I (2007) Effectiveness of arbuscular mycorrhizal fungi in the protection of date palm (*Phoenix dactylifera* L.) against Bayoud disease. Physiol Mol Plant Pathol 71:166–173

Juniper S, Abbott LK (2006) Soil salinity delays germination and limits growth of hyphae from propagules of arbuscular mycorrhizal fungi. Mycorrhiza 16:371–379

Kagan-Zur V, Kuang J, Tabak S, Taylor FW, Roth-Bejerano N (1999) Potential verification of a host plant for the desert truffle *Terfezia pfeilii* by molecular methods. Mycol Res 103:1270–1274

Koltai H, Meir D, Shlomo E, Resnick N, Ziv O, Wininger S, Ben-Dor B, Kapulnik Y (2008) Exploiting arbuscular mycorrhizal technology in different cropping systems under greenhouse conditions in semi-arid regions. Acta Hortic 797:223–228

Kozdrój J, Piotrowska-Seget Z, Krupa P (2007) Mycorrhizal fungi and ectomycorrhiza associated bacteria isolated from an industrial desert soil protect pine seedlings against Cd(II) impact. Ecotoxicology 16:449–456

Levy A, Chang BJ, Abbott LK, Kuo J, Harnett G, Inglis TJJ (2003) Invasion of spores of the arbuscular mycorrhizal fungus *Gigaspora decipiens* by *Burkholderia* spp. Appl Environ Microbiol 69:6250–6256

Leyval C, Turnau K, Haselwandter K (1997) Effect of heavy metal pollution on mycorrhizal colonization and function: physiological, ecological and applied aspects. Mycorrhizae 7:139–153

Lugo M, Ferrero M, Menoyo E, Estévez M, Siñeriz F, Anton A (2008) Arbuscular mycorrhizal fungi and rhizospheric bacteria diversity along an altitudinal gradient in South American Puna grassland. Microb Ecol 55:705–713

Meharg AA, Killham K (1994) A critical review of labeling techniques used to quantify rhizosphere carbon flow. Plant Soil 166:55–62

Mejstrik VK, Cudlin P (1983) Mycorrhiza in some desert plant species in Algeria. Plant Soil 71:363–366

Molina R, Massicotte H, Trappe JM (1992) Specificity phenomena in mycorrhizal symbiosis: community-ecological consequences and practical implications. In: Allen MF (ed) Mycorrhizal functioning. Chapman and Hall, London, pp 357–423

Moorman T, Reeves FB (1979) The role of endomycorrhizae in revegetation practices in the semi-arid west. II. A bioassay to determine the effect of land disturbance on endomycorrhizal populations. Am J Bot 66:14–18

Nara K (2006) Ectomycorrhizal networks and seedling establishment during early primary succession. New Phytol 169:178

Nara K, Hogetsu T (2004) Ectomycorrhizal fungi on established shrubs facilitate subsequent seedling establishment of successional plant species. Ecology 85:1700–1707

Nara K, Nakaya H, Wu B, Zhou Z, Hogetsu T (2003) Underground primary succession of ecyomycorrhizal fungi in a volcanic desert on Mount Fuji. New Phytol 159:743–756

O'Conner PJ, Smith SE, Smith FA (2001) Arbuscular mycorrhizal associations in the southern Simpson Desert. Aust J Bot 49:493–499

Parniske M (2004) Molecular genetics of the arbuscular mycorrhizal symbioses. Curr Opin Plant Biol 7:414–421

Peel MC, Finlayson BL, McMahon TA (2007) Updated world map of the Koppen-Geiger climate classification. Hydrol Earth Syst Sci 11:1633–1644

Peter M (2003) Volcanic deserts and primary succession: when and how do mycorrhizal fungi participate? New Phytol 159:534–536

Rillig MC, Wright SF, Allen MF, Field CB (1999) Rise in carbon dioxide changes soil structure. Nature 400:628

Rillig MC, Ramsey PW, Morris S, Paul EA (2003) Glomalin, an arbuscular-mycorrhizal fungal soil protein, responds to land-use change. Plant Soil 253:293–299

Shi Z, Zhang L, Feng G, Christie P, Tian C, Li X (2006) Diversity of arbuscular mycorrhizal fungi associated with desert ephemerals growing under and beyond the canopies of Tamarisk shrubs. Chin Sci Bull 51:132–139

Smith SE, Read DJ (1997) Mycorrhizal symbioses, 2nd edn. Academic, London

Staffeldt EE, Vogt KB (1974.) Mycorrhizae of the desert. Biome Research Memo, Utah State University, Logan, UT:7

Smith SE, Smith FA (1990) Structure and function of the interfaces in biotrophic symbioses as they relate to nutrient transport. New Phytol 114:1–38
Tarafdar JC, Kumar P (1996) The role of vesicular arbuscular mycorrhizal fungi on crop, tree and grasses grown in an arid environment. J Arid Environ 34:197–203
Titus JH, Del Moral R (1998) Vesicular-arbuscular mycorrhizae influence Mount St. Helens pioneer species in greenhouse experiments. Oikos 81:495–510
Titus JH, Titus PJ, Nowak RS, Smith SD (2002) Arbuscular mycorrhizae of Mojave desert plants. West N Am Nat 62:327–334
Treseder KK, Turner KM (2007) Review and analysis: glomalin in ecosystems. Soil Sci Soc Am J 71:1257–1266
Vivas A, Marulanda A, Ruiz-Lozano J, Barea J, Azcón R (2003) Influence of a *Bacillus* sp. on physiological activities of two arbuscular mycorrhizal fungi and on plant responses to PEG-induced drought stress. Mycorrhizae 13:249–26
Walker CM, Trappe JM (1993) Names and epithets in the Glomales and Endogonales. Mycol Res 97:339–344
West NE (1997) Interactions between arbuscular mycorrhizal fungi and foliar pathogens: consequences for host and pathogen. In: Gange AC, Brown VK (eds) Multitrophic interactions in terrestrial systems. Blackwell, Oxford, pp 79–89
Whipps JM (1990) Carbon economy. In: Lynch JM (ed) The rhizosphere. Wiley, New York, pp 59–98
White JA, Munn LC, Williams SE (1989) Edaphic and reclamation aspects of vesicular-arbuscular mycorrhiza in Wyoming red desert soils. Soil Sci Soc Am J 53:86–90
Woodward SL (2003) Biomes of Earth: terrestrial, aquatic, and human-dominated. Greenwood, Abingdon, UK
Wright SF, Upadhyaya A (1998) A survey of soils for aggregate stability and glomalin, a glycoprotein produced by hyphae of arbuscular mycorrhizal fungi. Plant Soil 198:97–107
Wright SF, Franke-Snyder M, Morton JB, Upadhyaya A (1996) Time-course study and partial characterization of a protein on hyphae of arbuscular mycorrhizal fungi during active colonization of roots. Plant Soil 181:193–203
Wright SF, Starr JL, Paltineanu IC (1999) Changes in aggregate stability and concentration of glomalin during tillage management transition. Soil Sci Soc Am J 63:1825–1829
Wu B, Hogetsu T, Isobe K, Ishii R (2007) Community structure of arbuscular mycorrhizal fungi in a primary successional volcanic desert on the southeast slope of Mount Fuji. Mycorrhiza 17:495–506
Yang Y, Chen Y, Li W (2008) Arbuscular mycorrhizal fungi infection in desert riparian forest and its environmental implications: a case study in the lower reach of Tarim River. Prog Nat Sci 18:983–991
Zaid A, de Wet PF, Djerbi M, Oihabi A (2002) Chapter XII: Diseases and pests of date palm. In: Zaid A (ed) Date palm cultivation FAO plant production and protection paper 156, http://www.fao.org/documents. Cited 9 February 2009

Chapter 7
Anatomical Variations in the Woody Plants of Arid Areas

Shumin Yang, Ikuo Furukawa, and Zehui Jiang

Abstract Variations in secondary xylem characteristics in 13 xerophytic species growing in an arid sandy region in China are described and compared in detail from an ecological perspective. All species showed similar wood structure (except *Haloxylon ammodendron* – rayless), obvious growth ring boundaries (except in *H. ammodendron*, *Tamarix mongolica* and *Zygophyllum xanthoxylon* – sometimes discontinuous), ring to semi-ring-porosity, a simple perforation plate, alternate intervessel pitting, non-septate fibres, paratracheal confluent axial parenchyma, helical thickenings and heterocellular rays. However, some quantitative differences in rays and vessels between species were observed. Rays are uniseriate in *Salix psammophila*, one- to two-seriate in *Tamarix mongolica* and *Hippophae rhamnoides*, two- to five-seriate in *Ammopiptanthus mongolicus*, *Lespedeza bicolor*, *Z. xanthoxylon*, *Nitraria tangutorum*, *Elaeagnus angustifolia*, and *Calligonum mongolicum*, and generally four- to ten-seriate in the other three species. This quantitative study of anatomical characteristics revealed that secondary xylem cells have high adaptability to desert conditions. Vessels with small diameter, either solitary or grouped in multiples, very short elements and minute pits are among the responses to demand for greater water transport capacity, and the appearance of such features in the xylem of arid zone species is interpreted as a strategy for conductive safety. *T. mongolica* and *A. mongolicus* had narrower vessels and higher vessel frequency than the other 11 species, which could lead to lowered vulnerability and mesomorphy value. Thus it was considered that the anatomical features of *T. mongolica* and *A. mongolicus* were more likely to suffer from water stress than those of other species. Fibre length and vessel element length were measured and analysed, with their horizontal variations showing either

S. Yang and Z. Jiang (✉)
Research Institute of Wood Industry, Chinese Academy of Forestry, Beijing 100091, China
e-mail: Shangke620@hotmail.com

I. Furukawa
Faculty of Agriculture, Tottori University, Koyama, Tottori 680-8553, Japan

K.G. Ramawat (ed.), *Desert Plants*,
DOI 10.1007/978-3-642-02550-1_7,

decreasing, increasing, considerable fluctuation or constant tendency with age. Fibre length is less than 900 μm, which is defined as "short range" according to IAWA Committee classification. There was significant difference both in among-tree and within-tree fibre length and vessel element length except in *A. mongolicus*.

7.1 Introduction

Drylands (arid, semi-arid and dry sub-humid areas) cover more than 40% of the Earth's surface and are distributed in about 100 countries. In such regions, land degradation-desertification has become a serious problem. China is one of the most severely desertified countries in the world, and the area affected has reached approximately 50% of the national territory. Desertification rehabilitation through planting grass, shrubs and trees that are found to have been successful in reducing environmental degradation is an effective and economical technique. Such species are typical plants in desert environments and have been the focus of many studies over several decades. These plants live in arid environments subject to drought and often intense sunlight, thus many such plants have evolved a number of modifications that minimise water loss through transpiration or evaporation of water from plant surfaces.

Several taxonomic and systematic studies of selected species have been carried out based on fruit, pollen and leaves, while studies on wood anatomy are quite scarce (Cao and Zhang 1991; Fahn et al. 1986; Schweingruber 1990; Zhang and Cao 1990, 1993). There are no published references on the wood anatomy of *Haloxylon ammodendron*, which lacks rays. Most studies on rayless species have focussed mostly on different families or other genera or species in Chenopodiaceae (Rajput 2001). Recently, this genus has been classified as rare and endangered in China, and therefore more studies have been initiated. As yet, there is no detailed description of the wood of *H. ammodendron* although there are some pictures and brief descriptions scattered in books or articles (Schweingruber 1990). It is difficult to find any description of the wood anatomy of selected species of *Tamarix austromongolica*, although Fahn (1958) described the anatomy of the xylem and annual rhythm of development, as well as the seasonal changes in starch content in other *Tamarix* species. Published articles on Leguminosae focus on other genera (Gasson 1999a, 1999b), and Cao and Zhang (1991) and Zhang and Cao (1993) reported the anatomy of secondary xylem of a few examined species. There is published information on the wood anatomy of all the examined genera in Zygophyllaceae and in Polygonaceae, but the selected species have not yet been described. Previous studies have included data on the secondary xylem of Elaeagnaceae (Fahn et al. 1986; Schweingruber 1990; Zhang and Cao 1990) based on light microscopy examination. Jansen et al. (2000) reported the wood anatomy of some members of the Elaeagnaceae, including *Hippophae rhamnoides*, commenting on vestured pits, helical thickening and systematic relationships with other families. A brief study of *Salix* can be found in Xu (2006).

Additionally, previous studies on wood anatomy of other species have focussed mainly on general description (Schweingruber 1990). However, quantitative data for wood have not been presented in the relevant literature. This chapter concentrates on 13 xerophytic species growing in an arid sandy region in China. It proved difficult to find information in the literature on the horizontal variations of cell dimensions of the selected 13 species. Here, we present preliminary information of the quantitative wood anatomy of these selected species and discuss the relationship between anatomical characteristics of secondary xylem and adaptability to arid climate, as well as the relationship between variations in the characteristics of fibre length and the dimensions of cambial fusiform cells (Yang and Furukawa 2003, 2006; Yang et al. 2007a, 2007b, 2008). Furthermore, horizontal variations in vessel element pattern length and fibre length were studied. The information obtained through this study will be useful in selecting and introducing suitable species to control desert expansion.

7.2 Materials and Methods

7.2.1 Site Description

The selected species were collected from Dengkou Psammophytes and Zhongtan Station, located on the border of the Ulan Buh Desert, which lies in a temperate region in the middle of Asia representing a transition zone between desert and dry steppe with a semi-arid continental monsoon climate. Mean annual precipitation is more than 130 mm, and annual evaporation is 2,351.9 mm. Vegetation is poor and sparse.

7.2.2 Methods

Thirteen representative species from six families were examined (Table 7.1). Five healthy trees of each species with approximately the same diameter, measured on a straight part of the trunk, were chosen, and then felled using a bow saw. Two discs (2–3 cm thick) from each tree were taken at a height of 20–30 cm above the ground, and dried under natural conditions. Some of the discs were immediately fixed in formalin-acetic-alcohol. Wood samples were softened in 5% glycerin solution, subsequently sectioned with a sliding microtome moving on transverse, radial and tangential surfaces of the discs. Thin sections were stained with safranine, dehydrated in a graded alcohol series and mounted in Canada balsam for light microscope examination. Small blocks exposing transverse, radial and tangential surfaces were prepared according to Exley et al. (1977) for scanning electron microscope observations. Macerated wood pieces were prepared by soaking in

Table 7.1 Variation of selected wood anatomical characters of 13 species

Family / species	Characteristic[a]										
	1	2	3	4	5	6	7	8	9	10	11
Family Chenopodiaceae											
Holoxylon ammodendron (C.A.Mey.) Bge	–[b]	26	87	53	301	2	Ray-less	Ray-less	Ray-less	–[b]	–[b]
Family Tamaricaceae											
Tamarix austromongolica Naka	45	58	108	51	746	2	875	5	4–10	1.29	139.2
Family Leguminosae											
Hedysarum scoparium Fisch. et Mey	84	41	161	17	786	3.3	972	5	4–10	0.49	78.6
Caragana korshinskii Kom	85	44	126	18	610	3.3	435	6	4–10	0.52	65.2
Lespedeza bicolor Turcz	79	43	130	44	668	1.8	466	8	2–5	0.54	70.8
Ammopiptanthus mongolicus Cheng f.	122	20	144	23	682	2.2	1176	12	2–5	0.16	23.6
Family Zygophyllaceae											
Nitraria tangutorum Bobr.	106	42	150	31	404	1.9	521	8	2–5	0.40	59.4
Zygophyllum xanthoxylon (Bge.) Maxim	110	49	138	65	468	2	981	4	2–5	0.45	61.5
Tetraena mongolica Maxim	221	31	93	38	278	1.9	88	10	1–2	0.14	13.0
Family Elaeagnaceae											
Hippophae rhamnoides L.	113	37	164	63	841	1.5	165	16	1–2	0.33	53.7
Elaeagnus angustifolia L.	73	60	182	69	801	2.8	229	5	1–10	0.82	149.6
Family Polygonaceae											
Calligonum mongolicum Turcz	–[b]	59	152	79	568	2.2	388	7	1–5	–[b]	–[b]
Family Salicaceae											
Salix psammophila C. Wang et Ch. Y. Yang	97	53	341	37	641	4.3	218	11	1–3	0.55	186.3

[a] *1* Vessel number per mm^2, *2* tangential vessel diameter (μm), *3* vessel element length (μm), *4* solitary vessels (%), *5* fibre length (μm), *6* intervessel pit diameter (μm), *7* ray height (μm), *8* number of rays per mm, *9* ray width, *10* vulnerability ratio, *11* Mesomorphy ratio

[b] Vessel frequency is not measured in *Calligonum* and *Holoxylon* because they are ring porous

Jeffrey's solution and mounted in glycerin-jelly. Quantitative data are based on 25 measurements of vessel element length and 50 of fibre length. Terminology and methodology follow the IAWA list of microscopic features for hardwood identification (IAWA Committee.1989).

7.3 Results

7.3.1 Wood Anatomical Variation in Secondary Xylem Cells

Slightly undulating growth ring boundaries are generally distinct, marked by marginal parenchyma bands or flattened fibres, sometimes discontinuous in *Holoxylon ammodendron*, *Tetraena mongolica* and *Zygophyllum xanthoxylon* (Figs. 7.1, 7.2). Wood is somewhat ring- to semi-ring-porous (Fig. 7.3); vessel frequency is 45–221/mm^2. Vessel outline is round, oval or slightly irregular in cross-section; some vessels have short tails, with a tangential diameter of 20–60 μm; vessel

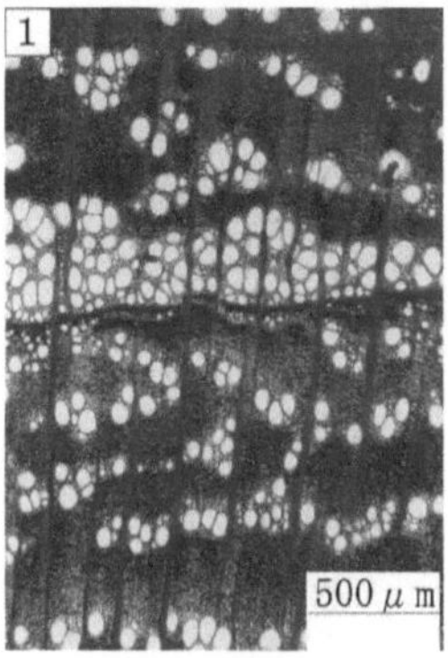

Fig. 7.1 Light microscopy photographs. *Caragana korshinskii*, transverse section (TS). Wavy growth ring boundary is distinct, marked by parenchyma bands

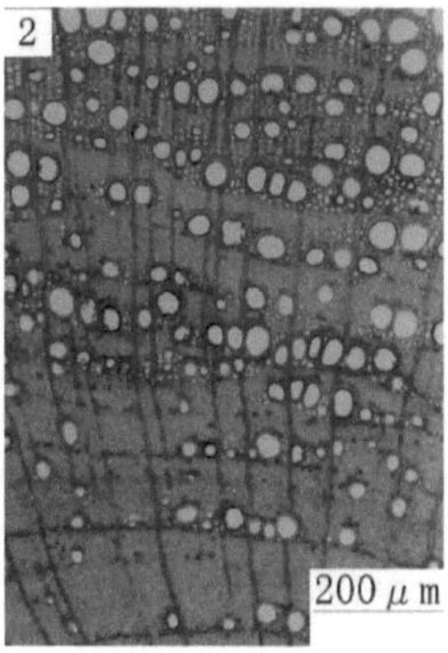

Fig. 7.2 *Tetraena mongolica* (TS). Growth ring boundaries are distinct, occasionally continuous

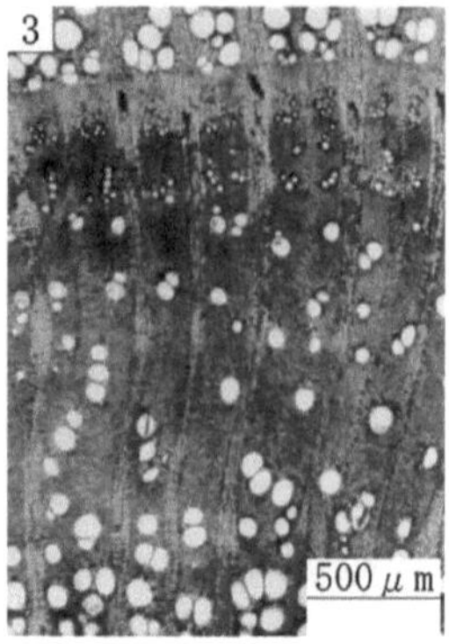

Fig. 7.3 Light microscopy photographs. *Tamarix austromongolica* (TS). Wood is semi-ring porous. Vessels are solitary or in tangential multiples. Gummy contents present in vessels

Fig. 7.4 Light microscopy photographs. *Holoxylon ammodendron*, tangential longitudinal section (TLS). Rays are absent, libriform wood fibre present

elements are 87–341 μm long. Vessels are 17–79% solitary; the rest are in tangential multiples of two–four or in very small clusters; while in *Caragana korshinskii* and *Hedysarum scoparium*, vessels are in multiples or clusters (Fig. 7.1); vessels are arranged in tangential-to-oblique bands, or seemingly dendritic patterns, in Leguminosae and *Nitraria tangutorum*. Gummy contents are present in vessels in the Leguminosae, *Tamarix austromongolica* and *N. tangutorum* (Fig. 7.3).

No rays were observed in *Holoxylon ammodendron* (Fig. 7.4). Rays are uniseriate in *Salix psammophila*, one- to two-seriate in *Tetraena mongolica* and *Hippophae rhamnoides* (Figs. 7.5–7.6); uniseriate, larger rays are two- to five-seriate in *Ammopiptanthus mongolicus*, *Lespedeza bicolor*, *Z. xanthoxylon*, *Nitraria tangutorum*, *Elaeagnus angustifolia*, and *Calligonum mongolicum*, infrequently uniseriate (Fig. 7.7). Larger rays are commonly four- to ten-seriate in *Tamarix austromongolica*, *Caragana korshinskii* and *Hedysarim scoparium*, occasionally with sheath cells, but narrower rays and sometimes even uniseriate rays are also detected (Fig. 7.8). Rays distend at the growth ring boundary in *Hedysarim scoparium*; ray height varies from 88 to 1,176 μm, and frequency is 4–16/mm; body ray cells are composed of procumbent cells with mostly 2–6 rows of square marginal cells, but square with upright marginal cells, rarely procumbent cells in *L. bicolor*,

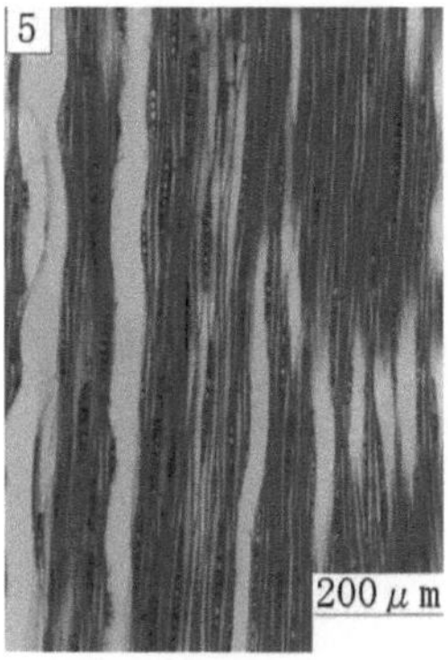

Fig. 7.5 Light microscopy photographs. *Salix psammophila* (TLS). Rays are uniseriate

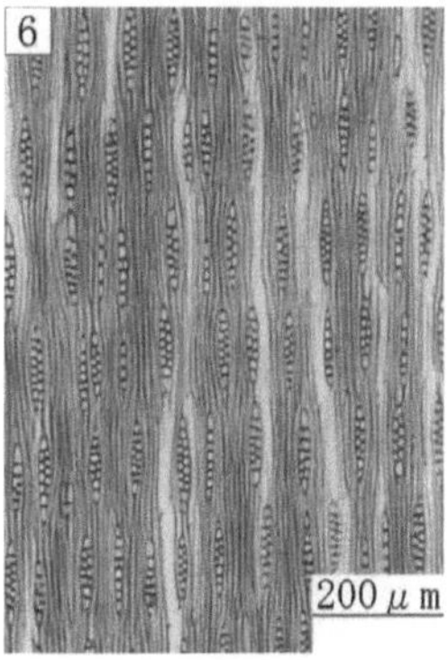

Fig. 7.6 Light microscopy photographs. *Hippophae rhamnoides* (TLS). Axial parenchyma cells, rays and vessel elements are storied. Rays are uniseriate, occasionally biseriate

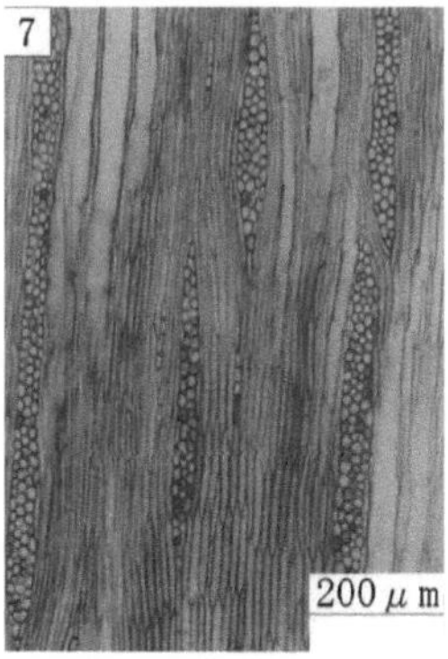

Fig. 7.7 Light microscopy photographs. *Zygophyllum xanthoxylon*, rays two- to five-seriate, infrequently uniseriate

A. mongolicus and *Z. xanthoxylon*, and mainly procumbent in *Caragana korshinskii, Tetraena mongolica, N. tangutorum, E. angustifolia* and *S. psammophila*; (Figs. 7.9–7.10); pits – round, oval or elongated – are present on the tangential wall and radial wall of ray cells. Infrequent sheath cells are present in ray cells in

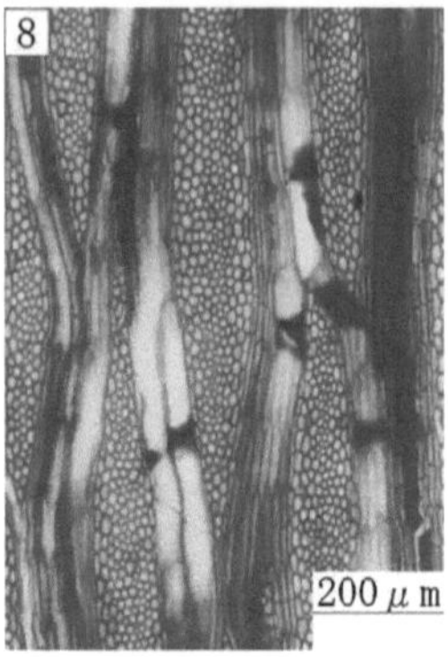

Fig. 7.8 Light microscopy photographs. *Ammopiptanthus mongolicus* (TLS). Larger rays are commonly four- to ten-seriate

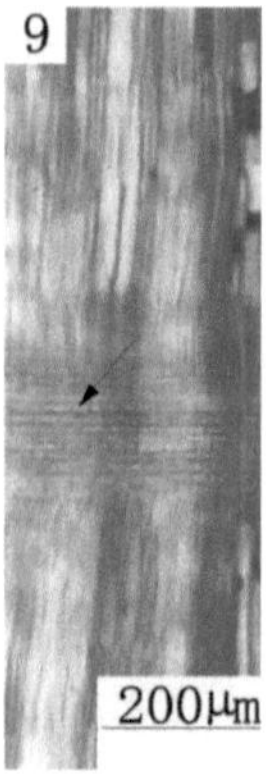

Fig. 7.9 Light microscopy photographs. *Nitraria tangutorum*, radial longitudinal section (RLS). Rays consist of procumbent cells, with square marginal cells

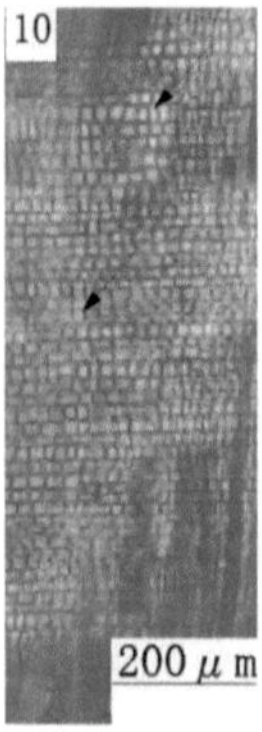

Fig. 7.10 Light microscopy photographs. *Lespedeza bicolor* (RLS). Rays consist of procumbent and square cells, with upright marginal cells

Tamarix austromongolica, *Caragana korshinskii*, *E. angustifolia* and *Hippophae rhamnoides*. Vascular tracheids have distinct spiral thickening in *A. mongolicus*, *Hedysarum scoparium*, *E. angustifolia* and *Hippophae rhamnoides*.

Axial parenchyma is absent or extremely rare in *E. angustifolia* and *Hippophae rhamnoides*, while abundantly present as predominantly paratracheal to scanty paratracheal in the other species, frequently confluent in wide tangential bands or marginal bands; fusiform parenchyma is present with 2–8 cells or more cells per parenchyma strand (Fig. 7.7). Axial parenchyma cells, vascular tracheid, rays and vessel elements are irregularly storied (Figs. 7.7–7.8), and the vessel wall surface is covered with warts in *N. tangutorum*.

Non-septate fibres of around 278–841 μm long are present with simple pits, mainly thick-walled in Zygophyllaceae; the fibre tracheid has round, oval, elongated or irregular bordered pits with slit-like inner apertures (Fig. 7.11).

Intervessel pitting is alternate to opposite, occasionally diffuse, some vestured (Figs. 7.12–7.14); the shape of vessel pits is round to oval or coalescent; pit apertures are round, oval, elongated or slit-like or partly coalescent, enclosed within pit borders 1–4 μm in diameter. The shape and size of vessel-ray pitting are similar to those of intervessel pitting (Fig. 7.15). Vessel-parenchyma pits have thin vesture, and their pit apertures are oval to elongated in *A. mongolicus* and *Calligonum mongolicum* (Fig. 7.16).

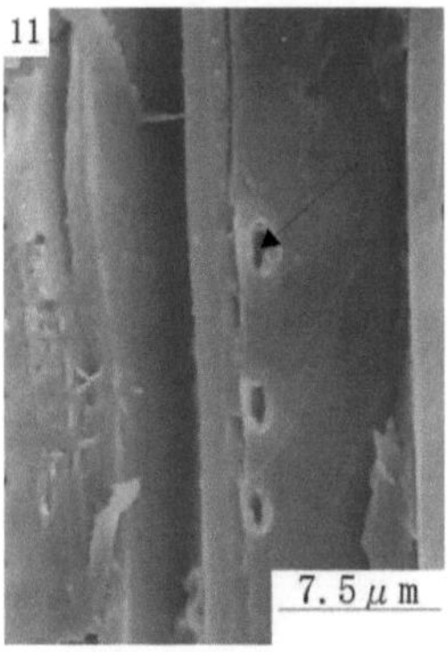

Fig. 7.11 Scanning electronic microscope photographs. *Tamarix austromongolica* (TLS). Pit on fibre tracheid

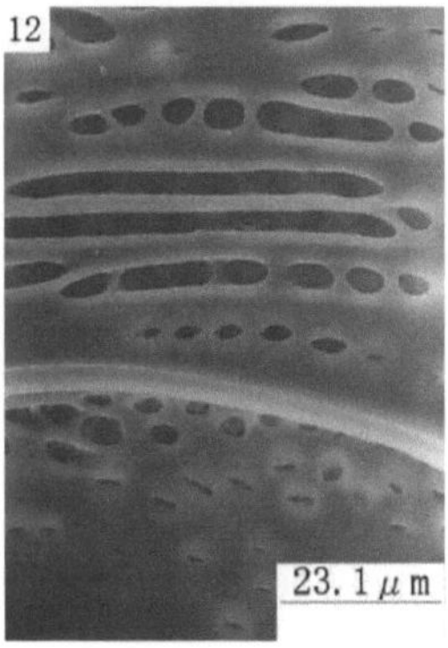

Fig. 7.12 *Elaeagnus angustifolia* (RLS). Alternate intervessel pitting. Pit apertures oval, slit-like, or coalescent

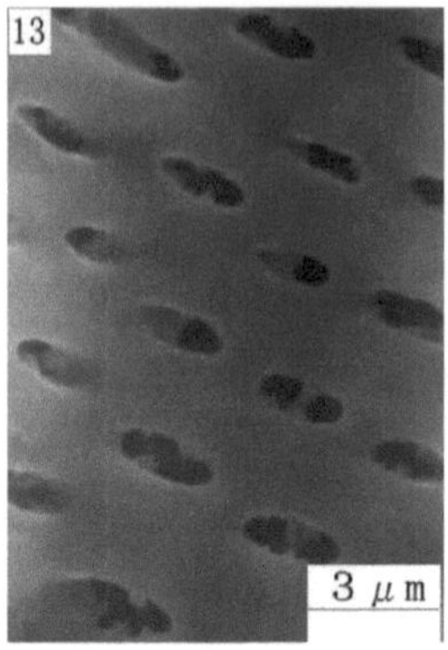

Fig. 7.13 *Calligonum mongolicum* (RLS). Intervessel pitting is alternate with small vestures

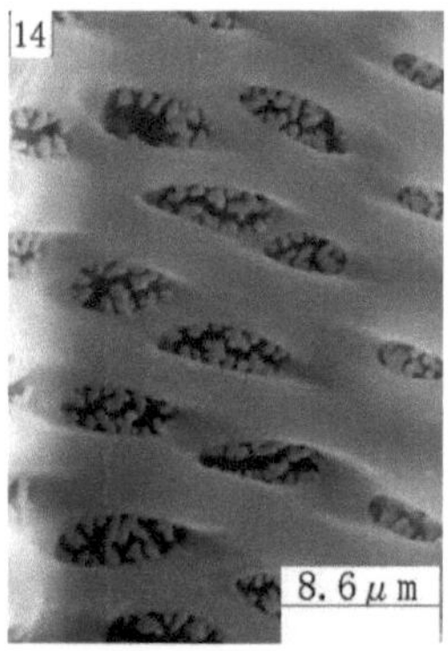

Fig. 7.14 *A. mongolicus* (TLS). Branched vestures present full of intervessel pit canal

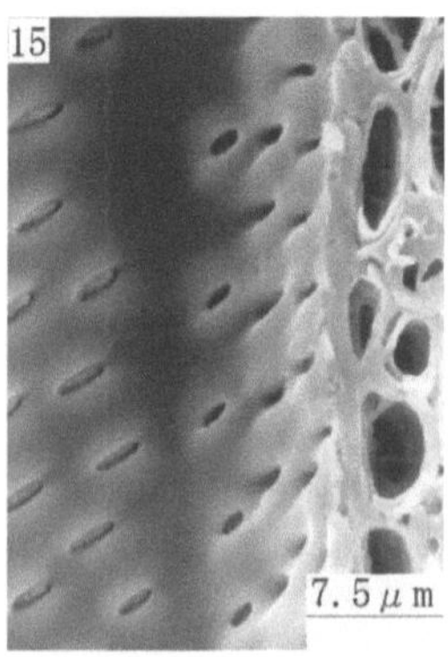

Fig. 7.15 *Caragana korshinskii* (TLS). Vessel-ray pitting is similar to intervessel pitting

Well-developed helical thickenings are present in narrower vessels and only faint spirals exist throughout the body of wide ones (Fig. 7.17).

Perforation plates are exclusively simple, and are present in oblique to almost horizontal end walls (Fig. 7.17).

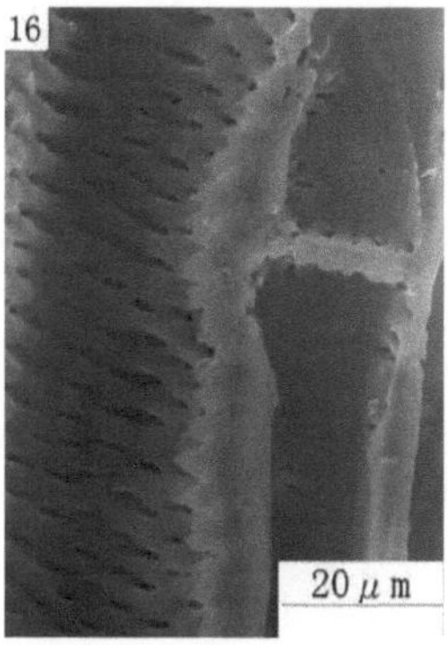

Fig. 7.16 *Calligonum mongolicum* (TLS). Vessel-parenchyma pitting is similar to intervessel pitting, alternate

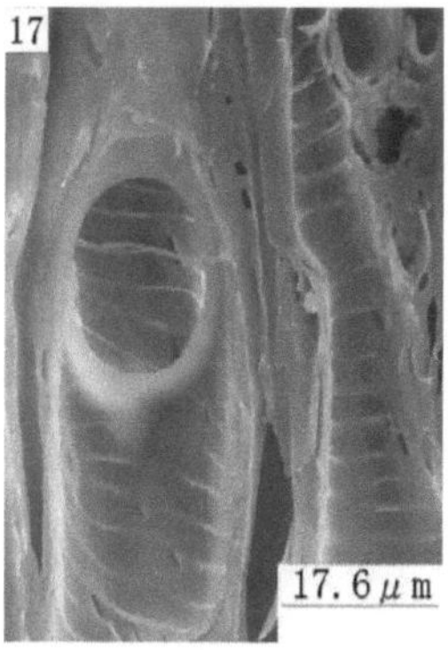

Fig. 7.17 *Hippophae rhamnoides* (TLS). Helical thickening is present in narrow vessel elements; perforation plate is simple

Crystals are present in axial parenchyma and in fibre in *Holoxylon ammodendron* and *L. bicolor*, and in ray cells in *Tamarix austromongolica*, *Hedysarum scoparium*, *A. mongolicus*, *N. tangutorum* and *Z. xanthoxylon* (Figs. 7.18–7.19).

7.3.2 Ecological Perspectives on the Variations in Wood Anatomy

There are larger variations in tangential vessel diameter, ranging from 15–110 μm within-tree and 20–60 μm between-tree. These few differences in wood anatomy among species are listed in Tables 7.1 and 7.2. It is clear that vessel element length and tangential vessel diameter are very small in all 13 species, mostly less than 150 and 50 μm, respectively (except Tamarix, which has a vessel diameter of 58 μm), in agreement with earlier observations in other xerophytes (IAWA committee 1989).

All the species examined were grown under the same environmental conditions and in the same locality; nevertheless, there are functionally significant variations

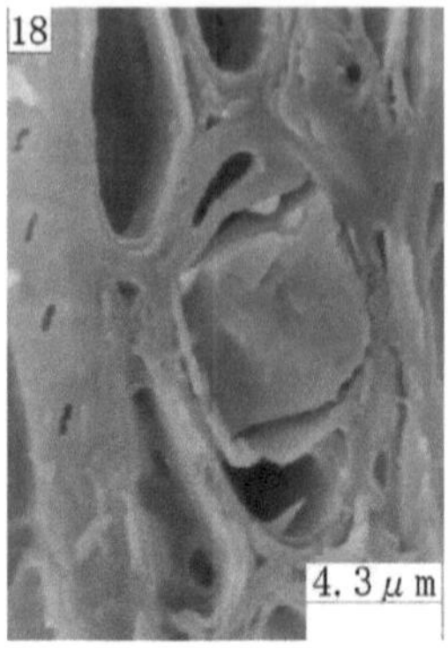

Fig. 7.18 *A. mongolicus* (TLS). Crystals present in ray cells

Fig. 7.19 *Holoxylon ammodendron*. Crystals present in fibres

Table 7.2 Qualitative variation of selected wood anatomical characters of 13 species

Family and species	Anatomical character[a]										
	1	2	3	4	5	6	7	8	9	10	11
Holoxylon ammodendron	±[b]	+	+	+	+	–	+	+	–	+	±
Tamarix austromongolica	+	+	±	–	+	–	+	+	–	+	+
Hedysarum scoparium	+	+	+	–	+	–	+	+	–	+	+
Caragana korshinskii	+	+	+	–	+	–	+	±	±	+	+
Lespedeza bicolor	+	+	+	–	+	–	+	+	–	±	+
Ammopiptanthus mongolicus	+	+	+	–	+	–	+	+	–	+	+
Nitraria tangutorum	±	+	+	–	+	–	+	±	±	+	±
Zygophyllum xanthoxylon	±	+	+	–	+	–	+	+	–	+	±
Tetraena mongolica	±	+	+	–	+	–	+	±	±	+	±
Hippophae rhamnoides	+	+	+	+	–	+	–	+	–	+	+
Elaeagnus angustifolia	+	+	+	+	–	+	–	–	+	+	±
Calligonum mongolicum	+	+	+	+	+	–	+	+	–	+	±
Salix psammophila	+	+	–	+	+	–	+	–	+	–	±

[a] *1* Distinct growth ring boundaries, *2* simple perforation plate, *3* vestured pit, *4* intervessel pitting alternate, *5* intervessel pitting alternate-opposite, *6* parenchyma scanty, *7* parenchyma paratracheal, *8* rays heterogeneous, *9* rays homogeneous, *10* helical thickening, *11* storied structure
[b] + Character present, ± character infrequently or irregularly present, – character absent

of characteristics among eight of these species. *Tetraena mongolica* and *A. mongolicus* had narrower vessels and larger vessel frequency than the other species, indicating a slightly higher safety and conductive efficiency (Table 7.1). Thus, *T. mongolica* and *A. mongolicus* have larger vulnerability and mesomorphy indices than the other species, indicating that, according to Carlquist (1977), these two species are better adapted to more xeric conditions or seasonal drought than the other six species. Additionally, other features, such as exclusively simple perforation plate and helical thickening, are general trends for species confined to dry sites.

7.3.3 Horizontal Variations in Vessel Element Length and Fibre Length

Vessel element length and fibre length were plotted from pith to bark on the basis of sample measurements taken from every ring. According to the IAWA Committee (1989), fibre length is considered short if it is less than 900 μm. The mean length values and the results of the analysis of variance are summarised in Table 7.1. Statistically, there is a significant difference in within-tree vessel element length and fibre length in all species except *A. mongolicus* among all 13 species (Table 7.3).

The results of horizontal variation with ring number are shown in Fig. 7.20. No readily apparent pattern was found in either vessel element length or fibre length. Instead, three variation tendencies in vessel element length could be discerned:

- vessel element length decreased slightly with ring number from the pith in *Holoxylon ammodendron* and *N. tangutorum*;
- vessel element length exhibited an irregular increasing or decreasing curve in *Tetraena mongolica* and *Z. xanthoxylon*;
- within the stem, vessel element length remains more or less constant from pith to bark (all other species).

Three patterns for fibre length were found in the 13 species:

- fibre length increases slightly as the diameter from the pith increases (Fig. 7.20);
- fibre length decreases slightly with ring number from the pith (*N. tangutorum*);
- the variation trend of fibre length is neither gradual nor constant, i.e. shows irregularly increasing or decreasing curves (*Tetraena mongolica* and *Z. xanthoxylon*).

This pattern is very common in trees and is similar to previous findings on horizontal variations (Tayor 1968, 1979; Furukawa et al. 1983a, 1983b, 1989; Stringer and Olson 1987).

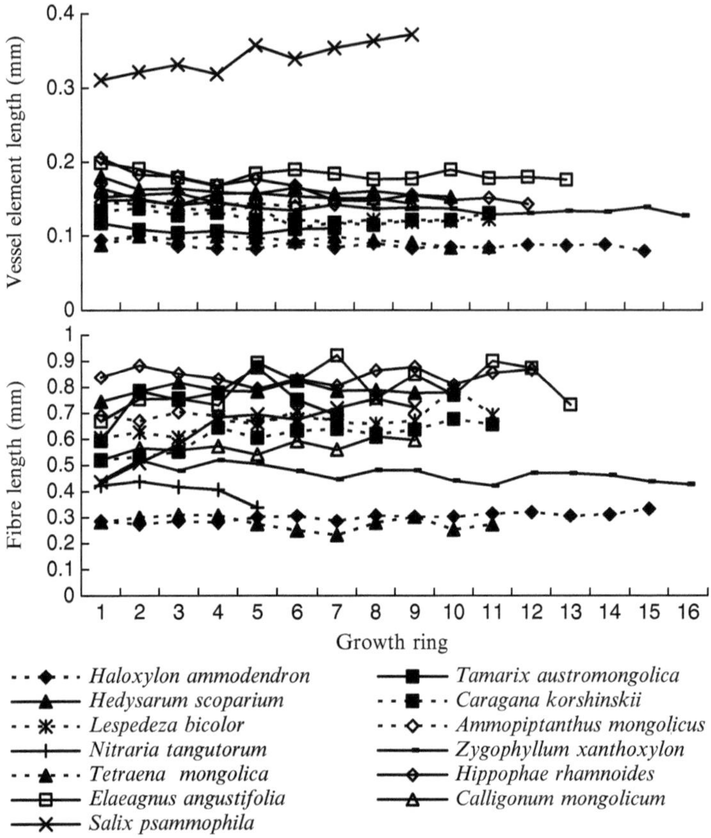

Fig. 7.20 Variation in vessel element length and fibre length with growth rings from pith to bark

7.4 Discussion and Conclusions

7.4.1 Comparison of Wood Anatomy and Diagnostic Value

The secondary xylem of *Holoxylon ammodendron* in Chenopodiaceae is similar to that of another species in the same genus – *Holoxylon articulatum* – in which rays are absent and conspicuously included phloem and storied structures are present; distinct helical thickening is also present, which differs from other species (Schweingruber 1990).

Tamarix species within Tamaricaceae cannot be distinguished according to available information. Their wood structures are similar to those of other *Tamarix* species, such as *T. boveana*, *T. gallica*, *T. parviflora*, *T. aphylla*, *T. jordainis* var.

Table 7.3 Analysis of variance (ANOVA) of vessel element length and fibre length

Index	Vessel element length *F* value	Fibre length *F* value
Holoxylon ammodendron	5.5**	12.8**
Tamarix austromongolica	84.8**	464.3**
Hedysarum scoparium	4.3**	6.3**
Caragana korshinskii	8.2**	34.2**
Lespedeza bicolor	11.9**	26.1**
Ammopiptanthus mongolicus	0.54^{ns}	0.8^{ns}
Nitraria tangutorum	3.3*	34.1**
Zygophyllum xanthoxylon	2.1**	4.0**
Tetraena mongolica	2.7**	11.6**
Hippophae rhamnoides	2.0*	4.0**
Elaeagnus angustifolia	405.5**	1051.5**
Calligonum mongolicum	126.1**	333.1**
Salix psammophila	5.8**	176.1**
Among species	342.4**	151.3**

** Significant at 1% level, * significant at 5% level, *ns* not significant

negevensis, and *T. gallica* var. *maris mortu* (Schweingruber 1990), thus wood anatomical features are very similar throughout this family.

Four species of Leguminosae have storied vessel elements, vascular tracheid, ray and parenchyma. The most obvious differences in anatomical characteristics are ray and vessel number and arrangement, and helical thickening. Vessels in *A. mongolicus* are frequent and small in diameter, while there is little difference among the other three species. Rays are abundant and well developed in Leguminosae, and larger rays are two- to five-seriate in *Ammopiptanthus* and *Lespedeza* and four- to ten-seriate in the other two species. Ray cells in *Caragana korshinskii* are only procumbent, being square to slightly upright, rarely procumbent, in the other three species. No distinct helical thickening was present in *L. bicolor*, but thickening was present in both narrow and wider vessels in the other three species. The four Leguminosae species can be distinguished from each other by these differing anatomical characteristics.

The significant characteristics of the three species of Zygophyllaceae are growth ring boundary and ray type and size. Complete growth ring boundaries are present in *N. tangutorum*; however, these are sometimes discontinuous in *Tetraena mongolica* and *Z. xanthoxylon*. Rays are uniseriate in *T. mongolica* and two- to five-seriate in the other two species. Ray cells are mainly procumbent in *Nitraria* and *T. mongolica* and procumbent or square in *Z. xanthoxylon*.

Typical distinct characters of the two species in Elaeagnaceae are ray type and width, and storied structure. In *Hippophae rhamnoides*, there is a distinct storied structure in rays, axial parenchyma cells and vessel elements, but the structure is irregularly storied in *E. angustifolia*. The rays are one- to two-seriate, heterogeneous in *Hippophae rhamnoides*, with two distinct classes in *E. angustifolia* (one- to two-seriate or more than four-seriate, homogeneous).

Calligonum has distinct helical thickening, one- to five-seriate rays, distention at the growth ring boundary, abundant axial parenchyma, which differentiates this genus easily from other genera in Polygonaceae (Schweingruber 1990).

Salix has uniseriate, developed rays.

All species tend towards ring- to semi-ring porous, and the vessel elements have simple perforation plates. The intervessel and vessel-ray pits are similar in size, shape and arrangement. The fibres are nonseptate thick-walled, with pit simple to minutely bordered. Storied structures are present.

Anatomically, there are suites of features that delimit the species in each family. Differences among species are confined mostly to rays, axial parenchyma, helical thickening, growth ring boundaries and quantitative features. Axial parenchyma is typically confluent, forming irregular bands or seemingly marginal bands of one cell wide. Ray type and size are different in other species. Helical thickening is not clear in *Lespedeza* and *Salix*, but is present in the others. A distinct growth ring boundary is present in all species examined, but was sometimes discontinuous in *Haloxylon*, *Zygophyllum* and *Tetraena*. These differences in qualitative characters have a diagnostic value at the family, genus or species level, since these features were present in all samples studied. In most cases where more than one sample per species was studied, these features were constantly absent, present, or faint, which has a diagnostic value on the species level.

7.4.2 Ecological Wood Anatomy Perspectives and Significant Functions

Wood or secondary xylem provides a complex tissue for water transport, mechanical strength, and for metabolic processes such as storage and mobilisation of reserve carbohydrates and lipids (Carlquist 1988; Zimmermann and Brown 1977; Zimmermann 1983). The relationship between wood structure and environmental characteristics has long been acknowledged and has been further confirmed in recent decades. Variations in the hydraulic architecture of xylem, as well as functional and ecological wood anatomy have been discussed extensively with regards ecologically diverse genera and families or of entire woody floras (Baas 1973, 1982; Baas and Zhang 1986; Baas et al. 1983; Carlquist 1975, 1977; Carlquist and Hoekman 1985; Chalk 1983; Lindorf 1997; Van Den Oever et al. 1981; Zhang and Cao 1990; Zhong et al. 1992).

Special emphasis has been placed on analysis of the correlation of some ecological parameters such as temperature and water availability, as well as geographical variables including latitude and altitude, with quantitative anatomical features, e.g. pore diameter, vessel frequency, the degree of vessel grouping, vessel element length, and intervessel pit size. In general, the tendency is for vessel members to become shorter and narrower as the aridity increases to prevent collapse of vessels under high negative pressures, and towards vessel grouping in arid environments

(Baas and Zhang 1986; Carlquist and Hoekman 1985; Fahn et al. 1986; Zhong et al. 1992). Spiral thickening in vessels or fibres is restricted largely to flora at higher latitudes, and more frequent, narrower vessels, shorter vessel members and fibres, and low rays also appear to be associated with higher latitudes (Alves and Angyalossy-Alfonso 2000; Baas 1973; Baas et al. 1988; Chen et al. 1993; Deng and Baas 1990; Van Den Oever et al. 1981) and, to a lesser extent, also with altitude rather than the tropical lowland (Van Den Oever et al. 1981).

The anatomical features of both xylem and leaves of species occurring in habitats subject to high water stress show adaptations correlated to the environmental extremes they experience (Metcalfe and Chalk 1983). Leaf surface reduction with thick cuticles, photosynthesis by green stems and cutinisation of the outer walls in leaf epidermis enables plants to withstand dry climatic periods (Lindorf 1997). Ecological and evolutionary trends in vessel diameter, perforation plate type, vessel frequency, vessel member length, total vessel length, and fibre type have all been discussed in terms of their input to the safety and efficiency of water transport (Zimmermann and Brown 1977). Mainly solitary vessels, or groups of only a few narrow and numerous vessels, could lead to greater conductive safety because it renders the inactivation of any vessel less harmful by enabling water transport to be transferred to an adjacent vessel (Carlquist 1984). Indeed, small-diameter vessels, either solitary or grouped in multiples, very short elements and minute pits can be seen as a response to demand for greater water transport capacity, and the appearance of such features in the xylem of arid zone species has been interpreted as a strategy for conductive safety in several previously published articles (Baas 1976; Baas and Carlquist 1985; Baas et al. 1988; Carlquist 1977, 1980, 1984, 1988; Carlquist and Hoekman 1985; Fahn et al. 1986; Zimmermann and Brown 1977; Zweypfenning 1978).

Both efficiency (or maximal conductivity) and safety are related strongly to vessel diameter and vessel frequency. Increased vessel diameter increases efficiency of water conduction dramatically, while at the same time decreasing safety. However, ring-porosity and the presence of different vessel size classes in general are of importance for the combined efficiency and safety of xylem sap transport at different times in or throughout growing seasons (Baas and Schweingruber 1987). The gradually decrease in vessel diameter from earlywood to latewood allows for optimal transport efficiency by wide vessels and provides great conductive safety through the narrow latewood vessels (Baas and Schweingruber 1987; Baas et al. 1988; Ellmore and Ewers 1985; Zimmermann 1982). For example, vessel elements tend to be shorter and narrower and more frequent in *Tetraena mongolica* and *A. mongolicus* than in other species (see Table 7.1), which could lead to greater conductive safety.

The terms "vulnerability" (V; mean vessel diameter divided by the mean number of vessels per square millimetre) and "mesomorphy" (M; vulnerability multiplied by mean vessel element length) were proposed by Carlquist (1977, 1992) to express the conductive safety and efficiency, respectively, within xylem. The value of vulnerability and mesomorphy indices could reflect a range of ecological factors (Baas 1982; Carlquist 1988). Lower values of V and M (less than 1 and 50, respectively) indicate species to be xeromorphic. Higher indices ($V > 1$, $M > 800$) indicate species to be more mesomorphic. Tables 7.1 and 7.2 show the different value of V and M due to the variations in species both within the same family and in different families.

Apart from these quantitative characters, qualitative characters also show ecological correlations. Some narrow vessels in these species have coarse helical thickenings, which increase cell wall strength to withstand high pressures or enlarge the wall surface area to promote water bonding to the surface (Carlquist 1975, 1982). In this study, all species showed helical thickenings, together with vessel groups and tracheids, i.e. features associated with greater conductive safety in arid environments.

Carlquist and Hoekman (1985) pointed out many other alternative strategies that enable plants to survive in water deficit conditions in addition to wood features. Wood structure should be considered as only one xeromorphic characteristic.

7.4.3 Horizontal Variations in Vessel Element Length and Fibre Length

Variations in cell length and volume have long been discussed due to their marked effect on product quality and the utilisation of wood (Zobel and Buijtenen 1989). Fibre length in particular is considered one of the more important indicators of wood quality, and is related to the mechanical strength and longitudinal shrinkage of wood. It has been studied intensively within individual annual rings, from the base to the top of the tree; among different rings within one tree; among species; and even on different sides of a tree in relation to sunlight and temperature (Slunder 1972).

Within-tree patterns differ for each species and for each environment (Zobel and Buijtenen 1989). Previous studies on axial variation in wood fibre length have commonly agreed with the tendency that fibre length increases more often than not up to a point well up the bole, and then decreases (Olson and Carpenter 1985; Ridoutt and Sands 1993; Wilkes 1988; Bhat et al. 1990; Yamada et al. 1985). Other researchers have reported constant length or a decrease in fibre length with height. The most commonly reported pattern of fibre length for both hardwood and softwood is an increase in the horizontal direction with age in rings near the pith, followed by a more gradual increase until a maximum is reached, but there are other trends, e.g. constant cell length, or large fluctuations or a marked decrease with age (Furukawa et al. 1983a, 1983b, 1989; Stringer and Olson 1987; Tayor 1968, 1979). In present study, fibre length and vessel element length of the 13 selected species exhibited the whole range of patterns, i.e. decreasing, increasing, unremarkable fluctuations or nearly constant values, except that in *Holoxylon ammodendron* vessel element is characterised by a slight decrease going up the tree. This trend concurs with the general conclusions of Furukawa and co-workers in their study on 71 hardwood species (Furukawa et al. 1983a, 1983b).

Vessel element length and fibre length is dictated by initial fusiform cambial cell length (Carlquist 1988). Additionally, fibre dimensions are determined by the dimensions of the cambial fusiform cells from which they are derived, and by processes that occur during cell differentiation (Ridoutt and Sands 1993, 1994). In species with non-storied cambium, increase in fibre length is explained on the basis of the increase in the length of cambial initials with increasing cambial age (Heinowicz and Heinowicz 1959). In this study, most species have regular or irregular storied structure, and nearly constant length could result equally from a retarded production from the cambial initials or a lower degree of intrusive growth. The difference in the range of fibre length is attributed to differences in the age of wood and to the different species.

References

Alves ES, Angyalossy-Alfonso V (2000) Ecological trends in the wood anatomy of some Brazilian species. 1. Growth rings and vessels. IAWA J 21(1):3–30

Baas P (1973) The wood anatomical range in *Ilex* (Aquifoliaceae) and its ecological and phylogenetic significance. Blumea 21:193–258

Baas P (1976) Some functional and adaptive aspects of vessel member morphology. In: Baas P, Bolton AJ, Catling DH (eds) Wood structure in biological and technological research. Leiden Bot 3:157–181

Baas P (1982) Systematic, phylogenetic, and ecological wood anatomy. History and perspectives. In: Bass P (ed) New perspectives in wood anatomy. Nijhoff/Junk, The Hague pp 23–58

Baas P, Carlquist S (1985) A comparison of the ecological wood anatomy of the floras of Southern California and Israel. IAWA Bull n s 6:349–353

Baas P, Schweingruber FH (1987) Ecological trends in the wood anatomy of trees, shrubs and climbers from Europe. IAWA Bull n s 8 (3):245–274

Baas P, Zhang XY (1986) Wood anatomy of trees and shrubs from China. Oleaceae. IAWA Bull n s 7(3):195–220

Baas P, Werker E, Fahn A (1983) Some ecological trends in vessel characters. IAWA Bull n s 4:141–159

Baas P, Esser PM, Van Den Western MET, Zandee M (1988) Wood anatomy of Oleaceae, IAWA Bull n s 9:103–182

Bhat KM, Bhat KV, Dhamodaran TK (1990) Wood density and fibre length of *Eucalyptus grandis* grown in Kerala, India. Wood Fibre Sci 22:54–61

Chalk L (1983) The effects of ecological conditions on wood anatomy. In: Metcalfe CR, Chalk L (eds) Anatomy of the dicotyledons, vol 2. Clarendon, Oxford, pp 152–156

Cao WH, Zhang XY (1991) The secondary xylem anatomy of 6 desert plants of *Caragana*. Acta Bot Sin 33(3):181–187

Chen BL, Baas P, Wheeler EA, Wu SM (1993) Wood anatomy of trees and shrubs from China. Magnoliaceae. IAWA J 14:391–412

Carlquist S (1975) Ecological strategies of xylem evolution, University of California Press, Berkeley, CA

Carlquist S (1977) Ecological factors in wood evolution: a floristic approach, Am J Bot 64:887–896

Carlquist S (1980) Further concepts in ecological wood anatomy, with comments on recent work in wood anatomy and evolution. Aliso 9:499–553

Carlquist S (1982) Wood anatomy of Illicium (Illiciaceae): phylogenetic, ecological, and functional interpretation. Am J Bot 69:1587–1598
Carlquist S (1984) Vessel grouping in dicotyledon wood: significance and relationship to imperforate tracheary elements. Aliso 10:505–525
Carlquist S (1988) Comparative wood anatomy. Systematic, ecological, and evolutionary aspects of dicotyledon wood. Springer, Berlin
Carlquist S (1992) Wood anatomy in Solanaceae: a survey. Allertonia 6:279–326
Carlquist S, Hoekman DA (1985) Ecological wood anatomy of the woody southern Californian flora. IAWA Bull n s 6(4):319–347
Deng L, Baas P (1990) Wood anatomy of trees and shrubs from China. Theaceae. IAWA Bull n s 11:337–378
Ellmore GS, Ewers FW (1985) Hydraulic conductivity in trunk xylem of elm, *Ulmus Americana*. IAWA Bull n s 6:303–307
Exley RR, Meylan BA, Butterfield BG (1977) A technique for obtaining cut surfaces on wood samples prepared for the scanning electron microscope, J Microsc 110:75–78
Fahn A (1958) Xylem structure and annual rhyme of development in trees and shrubs of desert. I. *Tamarix aphylla*, *T. jordanis* var. *negevensis*, *T. gallica* var. *maris mortui*. Trop Woods 109:81–94
Fahn A, Werker E, Baas P (1986) Wood anatomy and identification of trees and shrubs from Israel and adjacent regions. Israel Academy of Science, Jerusalem, Israel
Furukawa I, Sekoguchi M, Sakuno T, Kishimoto J (1983a) Wood quality of small hardwoods. 2. Horizontal variations in the length of fibres and vessel elements in seventy-one species of small hardwoods. Hardwood Res 2:104–134
Furukawa I, Nakayama T, Sakuno T, Kishimoto J (1983b) Wood quality of small hardwoods. 3. Horizontal variations in the length of fibres and vessel elements in trees with storeyed and non-storeyed wood. Bull Fac Agric Tottori Univ 35:42–49
Furukawa I, Fukutani S, Kishimoto J (1989) Characteristics of the variation of wood quality within Keyaki (*Zelkova aerrata* Makino) trees – horizontal variations of ring width, fiber length, vessel element length, specific gravity and longitudinal compression strength. Hardwood Res 5:197–206
Gasson P (1999a) Wood anatomy of *Exostyles venusta* (Swartzieae, Papilionoideae, Leguminosae). IAWA J 20(1):59–66
Gasson P (1999b) Wood anatomy of the tribe Dipterygeae with comments on related Papilionoid and Caesalpinioid Leguminosae. IAWA J 20(4):441–455
Heinowicz A, Heinowicz Z (1959) Variations of length of vessel members and fibres in the trunk of *Robinia pseudoacacia*. Acta Soc Bot Pol 28:453–460
IAWA Committee (1989) IAWA list of microscopic features for hardwood identification, IAWA Bull n s 10:219–332
Jansen S, Piesschaert F, Smets E (2000) Wood anatomy of Elaeagnaceae, with comments on vestured pits, helical thickenings, and systematic relationships. Am J Bot 87(1):20–28
Lindorf H (1997) Wood and leaf anatomy in *Sessea corymbiflora* from an ecological perspective. IAWA J 18(2):157–168
Metcalfe CR, Chalk L (1983) Anatomy of the dicotyledons. Clarendon, Oxford
Olson JR, Carpenter SB (1985) Specific gravity, fibre length, and extractive content of young *Paulownia*. Wood Fibre Sci 17 (4):428–438
Rajput KS (2001) Occurrence of rayless secondary xylem in some Indian herbaceous species. Isr J Plant Sci 49:221–227
Ridoutt BG, Sands R (1993) Within-tree variation in cambial anatomy and xylem cell differentiation in *Eucalyptus globules*. Trees 8:18–22
Ridoutt BG, Sands R (1994) Quantification of the process of secondary xylem fibre development in *Eucalyptus globules* at two height levels. IAWA J 15:417–424
Schweingruber FH (1990) Anatomy of European woods. Haupt, Stuttgart, Germany

Slunder ER (1972) Variation in wood specific gravity of yellow poplar in the southern Appalachians. Wood S 5:132–138
Stringer JW, Olson JR (1987) Radial and vertical variation in stem properties of juvenile black locust (*Robinia pseudoacacia*). Wood Fibre Length 19(1):59–67
Tayor FW (1968) Variations in the size and proportions of wood elements in Yellow-Poplar trees. Wood Sci Technol 2:153–165
Tayor FW (1979) Property variation within stems of selected hardwoods growing in Mid-south. Wood Sci 11(3):193–199
Van Den Oever L, Baas P, Zandee M (1981) Comparative wood anatomy of symplocos and latitude and altitude of provenance. IAWA Bull n s 2(1):3–24
Wilkes J (1988) Variations in wood anatomy within species of Eucalyptus. IAWA Bull n s 9:13–23
Xu F, Jones Gwynn LL, Sun RC (2006) Fibre morphology and anatomical structure of sandlive willow (*Salix psammophila*). Chem Ind For Prod 26(1):91–94
Yamada M, Sakuno T, Furukawa I, Kishimoto J (1985) Wood quality of small hardwoods (χ) within tree variation of fibre length in Niseakashia (*Robinia pseudoacacia* L). Hardwood Res 3:107–120
Yang SM, Furukawa I (2003) Anatomical features of a rayless woody xerophyte (*Haloxylon ammodendron*). Sand Dune Res 49(3):99–104
Yang SM, Furukawa I (2006) Ecological wood adaptation and horizontal variations of vessel element and fibre length of *Calligonum mongolicum.* Chin For Sci Technol 5(2):17–21
Yang SM, Jiang ZH, Ren HQ, Furukawa I (2007a) Quantitative characteristics of xylem cells and variation in vessel element length and fibre length for 13 psammophytes. Bull Bot Res 27 (5):601–606
Yang SM, Jiang ZH, Ren HQ (2007b) Study on anatomical property of xylem cells of the woody xerophytes grown in China. Acta Bot Borea Occident Sin 27(8):1507–1516
Yang S M, Jiang ZH, Ren HQ, Furukawa I (2008) Ecological anatomy characteristics of secondary xylem cells of two xerophytes in Elaeagnaceae. Sci Silv Sin 44(2):106–110
Zhang XY, Cao WH (1990) Studies on the secondary xylem anatomy of *Hippophae rhamnoides* under different habitats. Acta Bot Sin 3(12):909–915
Zhang XY, Cao WH (1993) The ecological secondary xylem anatomy of 7 desert species of *Leguminosae*. Acta Bot Sin 35(12):929–935
Zhong Y, Baas P, Wheeler EA (1992) Wood anatomy of trees and shrubs from China. Ulmaceae. IAWA Bull n s 13(4):419–453
Zimmermann MH, Brown CL (1977) Trees: structure and function. Springer, New York
Zimmermann MH (1982) Functional xylem anatomy of angiosperm trees. In: Bass P (ed) New perspectives in wood anatomy. Nijhoff/Junk, The Hague, pp 59–70
Zimmermann MH (1983) Xylem structure and ascent of sap. Springer, Berlin
Zobel BJ, Buijtenen JP (1989) Wood variation, its causes and control. Springer, Berlin
Zweypfenning RCVJ (1978) A hypothesis on the function of vestures pits. IAWA Bull 1:13–15

Chapter 8
Diversity and Conservation in the Cactus Family

Pablo Ortega-Baes, Silvia Sühring, Jesús Sajama, Ezequiel Sotola, Mariana Alonso-Pedano, Silvia Bravo, and Héctor Godínez-Alvarez

Abstract Cacti are conspicuous elements of the Western hemisphere. They have long attracted attention due to their peculiar biology, and have maintained close relationships with local settlers. Cacti are used as food, fodder, medicinal and ornamental plants, and as a source of wood. The greatest diversity of this family is recorded in Mexico, with 586 species, followed by Brazil, Argentina, Bolivia and Peru. These countries are also the richest in endemic species. The underlying factors affecting diversity patterns are varied and include temperature, rainfall and geological history. Cacti are threatened by loss and degradation of habitat, and illegal collection. Though it is commonly stated that many species are endangered, most have not yet been evaluated. Due to the growing impact of human activities on biodiversity it is necessary to increase conservation actions to preserve this interesting group of plants, especially in South American countries where knowledge and conservation of cacti are still incipient.

8.1 Introduction

Cacti have attracted special interest due to their particular traits, such as a spinose succulent body that stores water and photosynthesizes with a CAM metabolism. Such traits allow them to live in extreme environments and tolerate water stress (Gibson

P. Ortega-Baes (✉), J. Sajama, E. Sotola, M. Alonso-Pedano, and S. Bravo
Laboratorio de Investigaciones Botánicas (LABIBO), Facultad de Ciencias Naturales, Universidad Nacional de Salta, Av. Bolivia 5150, 4400, Salta, Argentina
e-mail: portega@miranda.ecologia.unam.mx; ortega_baes@yahoo.com.ar

S. Sühring
EEA-Salta, Instituto Nacional de Tecnología Agropecuaria (INTA), Ruta Nac. 68 km 172, (4403), Cerrillos, Salta, Argentina

H. Godínez-Alvarez
UBIPRO, FES-Iztacala, UNAM, Av. de los Barrios 1, Los Reyes Iztacala, Tlalnepantla, 54090, Estado de México, Ap. Postal 314, Iztacala, México

K.G. Ramawat (ed.), *Desert Plants*,
DOI 10.1007/978-3-642-02550-1_8,

and Nobel 1986; Anderson 2001). The species of this family can have cylindrical, globular or flat (cladode) stems. These traits and plant architecture determine different life forms, such as arborescent, columnar, globular, barreliform, and articulated (Fig. 8.1; Gibson and Nobel 1986; Terrazas-Salgado and Mauseth 2002).

Flower and fruit traits have allowed cacti to interact with the different animals on which they depend for seed production and seed dispersal. Flowers are associated with different pollinators, such as bats, birds and insects. Fruits (usually fleshy) are eaten by birds, mammals, reptiles and insects, which also disperse seeds (Gibson

Fig. 8.1 Life forms in the Cactaceae. **a** Barriliform (*Denmoza rodacantha*), **b** globular (*Echinopsis ancistrophora*), **c** epiphyte (*Pfeiffera iantothele*), **d** articulated or opuntioide (*Opuntia quimilo*), and **e** columnar (*Echinopsis terscheckii*)

and Nobel 1986; Godínez-Alvarez et al. 2002; Pimienta-Barrios and del Castillo 2002; Godínez-Alvarez 2004).

The first phases of the cactus life cycle are decisive since only a small proportion of the annual seed production will germinate, and an even smaller fraction of seedlings will survive the first year. Precipitation is the main factor affecting seedling emergence and survival. It has been found that spatial and temporal variations in precipitation explain establishment patterns in cacti. However, other factors such as the presence of nurse plants, favorable climatic conditions in fall and winter, types of soil, and human activities such as overgrazing can also affect this process (Pierson and Turner 1998; Godínez-Alvarez et al. 2003).

8.2 Taxonomic Diversity

Cactaceae is endemic to the Western hemisphere, except for *Rhipsalis baccifera*, which is also found in Africa and Asia. It comprises 124 genera and 1,427 species (Table 8.1; Hunt 2006). These figures have varied in last decade (Hunt 1999, 2006; Anderson 2001), with a remarkable reduction of valid species (Hunt 2006).

Recent classifications of Cactaceae recognize four subfamilies: Pereskioideae, Maihuenioideae, Opuntioideae and Cactoideae (Anderson 2001; Wallace and Gibson 2002; Nyffeler 2002). Pereskioideae, including species of *Pereskia* genus, has 17 species distributed in both hemispheres. Maihuenioideae is represented by

Table 8.1 Taxonomic diversity in the Cactaceae (based on Hunt 2006). Threatened species according to IUCN Red List (IUCN 2008)

Subfamily	Tribe	Number of species	Number of endemic species (%)	Threatened species
Pereskioideae		17	10 (59)	1 (6)
Maihueniodeae		2	1 (50)	–
Opuntioideae		186	119 (64)	7 (4)
	Austrocylindropuntieae	12	6 (50)	–
	Pterocacteae	9	9 (100)	–
	Tephrocacteae	15	11 (73)	1 (7)
	Cylindropuntieae	55	30 (55)	–
	Opuntieae	95	64 (67)	6 (6)
Cactoideae		1222	982 (88)	96 (8)
	Calymmantheae	1	1 (100)	–
	Hylocereeae	66	36 (55)	–
	Cereeae	141	113 (80)	23 (16)
	Trichocereeae	279	234 (84)	11 (4)
	Notocacteae	133	104 (78)	–
	Rhipsalideae	62	49 (79)	5 (8)
	Browningieae	20	14 (70)	1 (5)
	Pachycereeae	169	131 (78)	3 (2)
	Cacteae	351	300 (86)	53 (15)
All Family		1427	1113 (80)	104 (7)

only one genus (*Maihuenia*), with two species distributed in Argentina and Chile (Anderson 2001; Hunt 2006).

Opuntioideae has 186 species distributed in the American continent. These species are articulated cacti with flat, cylindrical and globular stem-segments (Gibson and Nobel 1986; Terrazas-Salgado and Mauseth 2002). The latest classification of Opuntioideae includes five tribes: Austrocylindropuntieae, Pterocacteae, Tephrocacteae, Cylindropuntieae and Opuntieae (Wallace and Dickie 2002). The first three are distributed exclusively in South America, while Cylindropuntieae is distributed mainly in Mexico and the United States. Tribe Opuntieae is distributed across the American continent and has the highest number of species and endemism of this subfamily (Anderson 2001; Hunt 2006). The highest proportion of endemic species is registered in the tribe Pterocacteae, where 100% of the species are exclusive to Argentina.

Cactoideae has the highest diversity with 1,222 species, which are widely distributed in America. This subfamily also has the highest number of endemic species (political endemism: species restricted to only one country) totaling 982 species, which represents 88% of the species in this taxon (Table 8.1; Hunt 2006). Its species have stems with ribs or tubercles and areoles without glochids (Gibson and Nobel 1986). Nine tribes are recognized in this subfamily: Calymmanthea, Hylocereeae, Cereeae, Trichocereeae, Notocacteae, Rhipsalideae, Browningieae, Pachycereeae, and Cacteae (Anderson 2001). Some of them have diversified in the Northern hemisphere (i.e., Pachycereeae, Cacteae) and others in the Southern hemisphere (i.e., Trichocereeae, Notocacteae; Gibson and Nobel 1986; Wallace and Gibson 2002). Cacteae has the highest number of species with 351, followed by Trichocereeae with 279 species. Both tribes have the highest number of endemic species (300 and 234, respectively). All tribes have over 70% endemic species, except for Hylocereeae (Table 8.1). The tribe Calymmantheae is endemic to Peru (Anderson 2001; Hunt 2006).

8.3 Diversity Patterns on a Global Scale

Cacti are distributed in every ecoregion of the Americas, even in the rainforests. However, the highest diversity is concentrated in arid and semi-arid regions located between 35º north and south latitudes, and from sea level to 5,000 m (Oldfield 1997; Anderson 2001).

Four main diversity centers have been recognized: (1) Mexico and the southwestern United States; (2) the Central Andes (Southern Ecuador, Peru, Bolivia, Northeast Chile and Northwest Argentina); (3) East Brazil; and (4) West and South Brazil, Paraguay, Uruguay and South and East Argentina (Oldfield 1997; Boyle and Anderson 2002).

Knowledge on cactus diversity patterns on the global and regional scales has improved recently due to increasing interest in cactus conservation (Hernández and Godínez 1994; Hunt 1999, 2006; Gómez-Hinostrosa and Hernández 2000; Hernández et al. 2001, 2004; Guzmán et al. 2003; Pin and Simon 2004; Martínez-Avalos and Jurado 2005). However, this improvement has been asymmetric among the American countries, and South American diversity has been the least studied.

In terms of cactus species, Mexico is the most diverse country with 586 species. Brazil, Argentina, Bolivia, the United States and Peru have over 100 species (Fig 8.2a; Ortega-Baes and Godínez-Alvarez 2006; Godínez-Alvarez and Ortega-Baes 2007). Mexico has the highest number of endemic species (Fig. 8.2b; Ortega-Baes and Godínez-Alvarez 2006; Godínez-Alvarez and Ortega-Baes 2007), while Brazil, Argentina, Bolivia and Peru have more than 90 endemic species (Fig. 8.2b). It is important to note that the countries with the highest diversity also have the highest proportion of endemic species, reaching over 50% in Mexico, Brazil, Peru, Chile, Bolivia and Argentina. A positive relationship between species richness and endemism has been reported by Ortega-Baes and Godínez-Alvarez (2006). Similarly, it has been shown that the species richness and endemism of cacti in American countries may be explained by their area (i.e., countries with greater area have higher diversity). However, diversity in some countries (Mexico, Argentina, Peru, Bolivia, Chile, Paraguay, and Costa Rica) is higher than expected (Ortega-Baes and Godínez-Alvarez 2006).

The countries with the highest diversity in the Cactus family also have the highest diversity in the subfamily Cactoideae (Fig. 8.3a). The same pattern is registered for endemic species (Fig. 8.3b). Mexico, the United States and Argentina have the highest number of species in the subfamily Opuntioideae (Fig. 8.3c). Mexico and Argentina have the highest endemism in this subfamily (Fig. 8.3d).

8.4 Factors Explaining Cactus Diversity

Cactus diversity (i.e., α diversity or species richness, and β diversity or species turnover) can be explained by different environmental factors such as temperature and precipitation. However, these factors restrict species distribution differently depending on their life form (Mourelle and Ezcurra 1996, 1997).

According to this idea, α diversity of columnar cacti is high in regions with the greatest number of frost-free days (Mourelle and Ezcurra 1996). Frost damages the apical meristems, reducing apical growth and stem diameter. The cactus diversity of articulated species increases as the amount of annual rainfall in summer and the annual mean temperature increase (Mourelle and Ezcurra 1996). These are wide-range cacti with high tolerance to environmental factors; however, they seem to be associated with hot regions and summer rainfall. Globular cactus diversity is associated strongly with summer annual rainfall. Furthermore, it has been indicated that globular diversity is related to substrate characteristics such as microsites and soil rockiness (Mourelle and Ezcurra 1996).

Apart from these environmental factors, there is a positive relationship between species richness and region aridity. Thus, species richness increases together with aridity (Godínez-Alvarez and Ortega-Baes 2007). Geological history is another

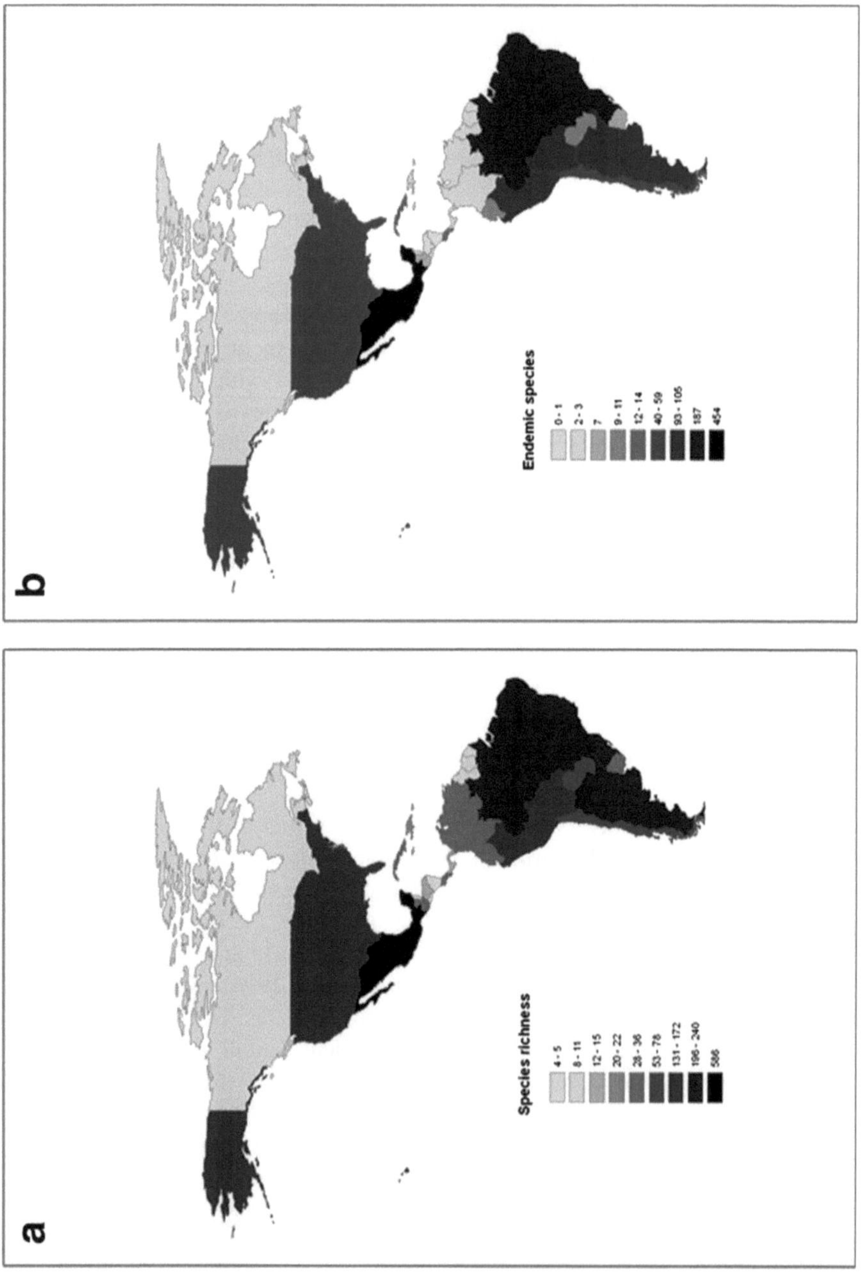

Fig. 8.2 Global cactus diversity. **a** Species richness, **b** endemism

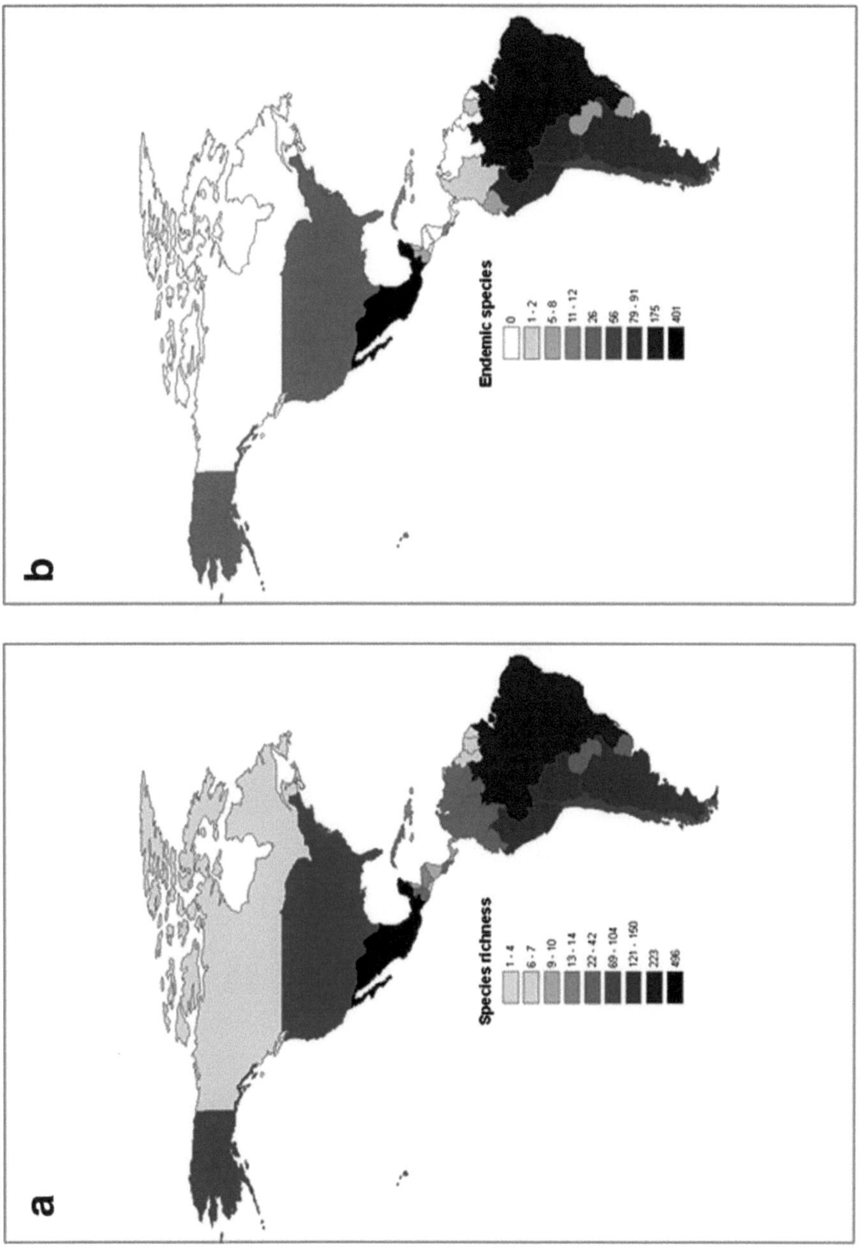

Fig. 8.3 Diversity patterns of subfamilies Cactoideae and Opuntioideae. **a** Species richness of Cactoideae, **b** endemism of Cactoideae

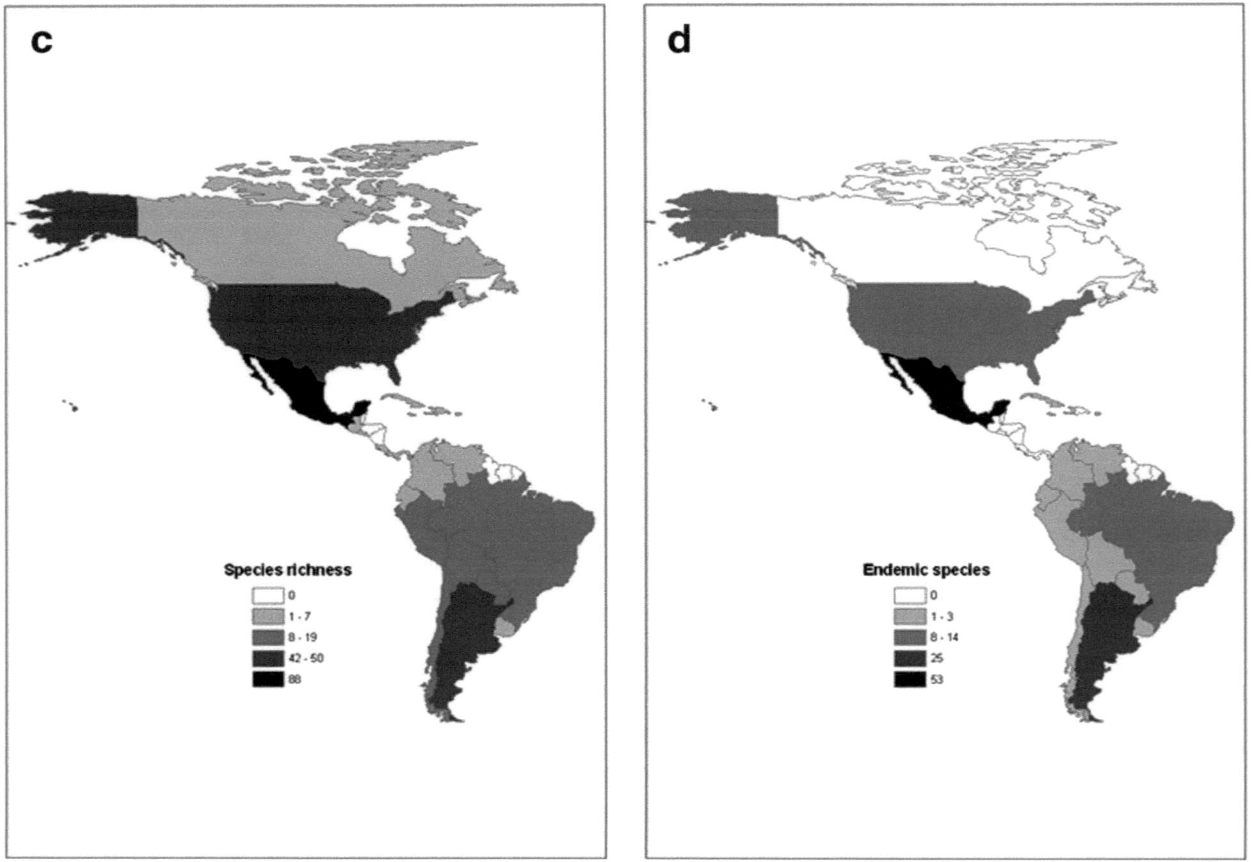

Fig. 8.3 (continued) **c** species richness of Opuntioideae, and **d** endemism of Opuntioideae

factor that could explain spatial patterns of cactus diversity in some regions. For example, the Prepuna and the Monte biogeographic regions in Argentina, and the Chihuahuan Desert in Mexico have a high diversity, probably because they functioned as refuges during the Pleistocene glaciations (Hernández and Bárcenas 1995; Mourelle and Ezcurra 1996).

High cactus diversity in some regions may be explained by a relatively high level of species turnover or β diversity (Goettsch and Hernández 2006). This pattern could be related to environmental heterogeneity (Mourelle and Ezcurra 1997). In arid and semi-arid regions, variations in environmental factors such as temperature, precipitation, topography, and soil produce a wide variety of habitats that are occupied by cactus species with different environmental requirements. Nevertheless, β diversity varies according to cactus life form. Columnar and articulated cacti have a comparatively lower species turnover than globular cacti (Mourelle and Ezcurra 1997). This could be explained by the fact that fruits of columnar and articulated cacti are eaten by birds and bats. These animals have great vagility, increasing the seed dispersal capacity of these cacti. Globular cacti have a lower seed dispersal capacity because their fruits are eaten by reptiles and insects (Godínez-Alvarez 2004), or they show passive dispersal (Peters et al. 2009). These latter traits would lead to a narrower habitat size (Mourelle and Ezcurra 1997; Goettsch and Hernández 2006; Hernández et al. 2008).

8.5 Threatened Species

Cacti are threatened by loss and degradation of habitat, and illegal collection (Oldfield 1997; Boyle and Anderson 2002). It has been shown that land-use change is the most important factor causing biodiversity loss. Acute disturbances (e.g., clearing forest to turn it into cropland) can affect species in different ways. Some species may be threatened by loss of available habitat, while other species may be favored by disturbance. On the other hand, low level disturbances (e.g., extensive goat grazing) also affect biodiversity because, in the long run, they become chronic disturbances (Cuarón 2000; Martorell and Peters 2005; Broennimann et al. 2006).

Arid and semi-arid lands have been modified since European colonization by the expansion of agriculture and cattle farming, urbanization, foresting and mining. For example, the Chaco forest in Argentina has lost 85% of its original extension due to the expansion of agriculture, negatively affecting regional biodiversity (Zak et al. 2004).

Cattle grazing has moved towards more arid zones due to agricultural expansion (Vilela et al. 2009; Villagra et al. 2009). This is one of the most important production activities in the arid regions of America. It has been estimated that over 270,000 km^2 of the Monte biogeographic province in Argentina are used for extensive cattle raising (goats, cattle, and sheep). This activity modifies both landscape structure and species biodiversity because it reduces plant cover and increases bare soil patches (Villagra et al. 2009).

Fig. 8.4 Cactus use as wood. **a** *Echinopsis atacamensis* individual cut for wood. **b** Illegally collected wood of *E. atacamensis* in Seclantás (Salta, Argentina). **c** Handcrafts made from *E. atacamensis*. **d** Door made from *E. atacamensis*

Cacti are exposed to pressure from collectors due to their value (Oldfield 1997; Anderson 2001) as ornamental plants, medicinal plants, food and wood (Figs. 8.4, 8.5; Casas and Barbera 2002; Inglese et al. 2002; Nerd et al. 2002; Nefzaoui and Ben Salem 2002; Saénz Hernández et al. 2002; Wright et al. 2007). They are collected from their natural habitat (Figs. 8.4a, 8.5a), except for a few species that are cultivated. The ornamental species (Fig. 8.5), especially the globular cacti, suffer great pressure from illegal collecting. Although many species are cultivated and techniques of species propagation have been developed (Anderson 2001), illegal collecting is still a significant threat (Oldfield 1997).

A way to control illegal trade at an international level is through the Convention on International Trade of Endangered Species (CITES). This organization regulates international trade by including the species in Appendices. Some cactus species are included in Appendix I (CITES 2008) because their trade is banned. In addition, all cacti are included in Appendix II due to the fact that their trade must be regulated to avoid possible threat. Despite this regulation, trade in cultivated cacti acts as a front for the trade in wild collected plants (Oldfield 1997). Consequently, more rigorous

Fig. 8.5 Ornamental use of cacti. **a** Illegally collected individuals of *Echinopsis ancistrophora* in Vaqueros (Salta, Argentina). **b** Trade wild of collected cactus. **c** Ornamental cultivated species. **d** Domestic garden with cactus species

control of international and national trade is necessary. Some countries, such as Argentina, do not regulate cactus trade, allowing cactus exchange among regions and border countries (A. Cavalli and P. Ortega Baes, unpublished data).

The Red List of Threatened Species of the International Union for the Conservation of Nature (IUCN) provides a species categorization based on the relative risk of extinction at global scale. It includes at least 104 cactus species (7% of all species) as vulnerable to extinction (IUCN 2008). The subfamily Cactoideae is the best represented with 96 species. Tribes Cereeae and Cacteae have the greatest number of categorized species, while other highly diverse tribes are either not represented (e.g., Notocacteae) or have a low number of species in this category (e.g., Pachycereeae and Trichocereeae). Only seven species from the subfamily Opuntioideae are represented in the IUCN Red List. Most of the categorized species belong to the tribe Opuntieae (Table 8.1).

Brazil has the highest proportion of threatened species on a global scale with 18%, followed by Mexico (10%), Ecuador (9%) and Peru (7%). Most countries either lack endangered species or have a very low percentage, including countries with high diversity such as Argentina, Bolivia, Chile, the United States and Paraguay. Brazil, Mexico, Ecuador, and Puerto Rico have more categorized species than expected according to their richness, while the remaining countries have a smaller percentage (P. Ortega-Baes et al. unpublished data). These results show that the proportion of cactus species categorized as endangered is low. This is particularly critical for certain taxonomic groups (e.g., tribes Notocacteae, Pachycereeae and Trichocereeae) and countries (e.g., Bolivia, Argentina). Consequently, it is necessary to evaluate the conservation status of non-represented taxonomic groups as well as the cactus diversity of countries with low representation in IUCN Red List.

Population evaluation and demographic studies are the best ways to determine the conservation status of cactus species. However, these are scarce and skewed towards North American species (Godínez-Alvarez et al. 2003). This situation has become more severe during the last few years (Valverde et al. 2004; Godínez-Alvarez and Valiente-Banuet 2004; Clark-Tapia et al. 2005; Esparza-Olguín et al. 2005; León de la Luz 2005; Medel-Narvaez et al. 2006; Valverde and Zavala-Hurtado 2006; Jiménez-Sierra et al. 2007; Mandujano et al. 2007a, 2007b; Martínez-Ávalos et al. 2007). As stated, South American species must be studied (Godínez-Alvarez et al. 2003), emphasizing: (1) taxonomic groups with low representation on IUCN Red List, (2) cactus species distributed in regions that have been seriously impacted by human activity, and (3) cactus species that are economically important since they are exposed to intensive collecting.

8.6 Conservation Strategies

The first step towards setting geographic priorities is to select important regions at a global scale, and then to establish smaller-scale priorities. In this context, countries or political territories with outstanding biological diversity have been identified (Caldecott et al. 1996; Mittermeier et al. 1997; Sarukhán and Dirzo 2001). Through this approach, 24 countries have been selected as priorities for cactus conservation, with Mexico, Argentina, Peru, Brazil and Bolivia being the most important (Ortega-Baes and Godínez-Alvarez 2006).

Only Mexico has formally determined the most important areas for cactus conservation (Godínez-Álvarez and Ortega-Baes 2007). In this country, eight states representing 80% of its cactus diversity (Baja California Sur, Coahuila, Hidalgo, Jalisco, Nuevo León, Oaxaca, San Luis Potosí, and Sonora; Godínez-Alvarez and Ortega-Baes 2007) have been selected as the most important for cactus conservation. Following the same approach, eight provinces (Salta, La Rioja, Entre Ríos, Córdoba, Misiones, Catamarca, San Juan and Neuquén) were selected as a priority for the same conservation target in Argentina (P. Ortega-Baes et al. unpublished data).

Mexico is one of the few countries where several regional studies on cactus diversity have been carried out. For instance, studies in the Chihuahuan Desert have provided important information on cactus diversity and specific conservation schemes for the different areas of this region (Hernández and Bárcenas 1995, 1996; Gómez-Hinostrosa and Hernández 2000; Hernández et al. 2001). However, these types of studies are rare in other American regions, making it difficult to define *in situ* conservation at a regional level.

When there is little data on regional diversity, surrogates for biodiversity can be used to select conservation areas. These areas are frequently selected assuming that they will represent a broad proportion of biodiversity (Howard et al. 1998; Reyers et al. 2000). In this context, the lack of cactus diversity data to select conservation areas could be solved using a cactus species group. For example, the diversity of one subfamily could be used. Figure 8.6 shows the performance of subfamilies

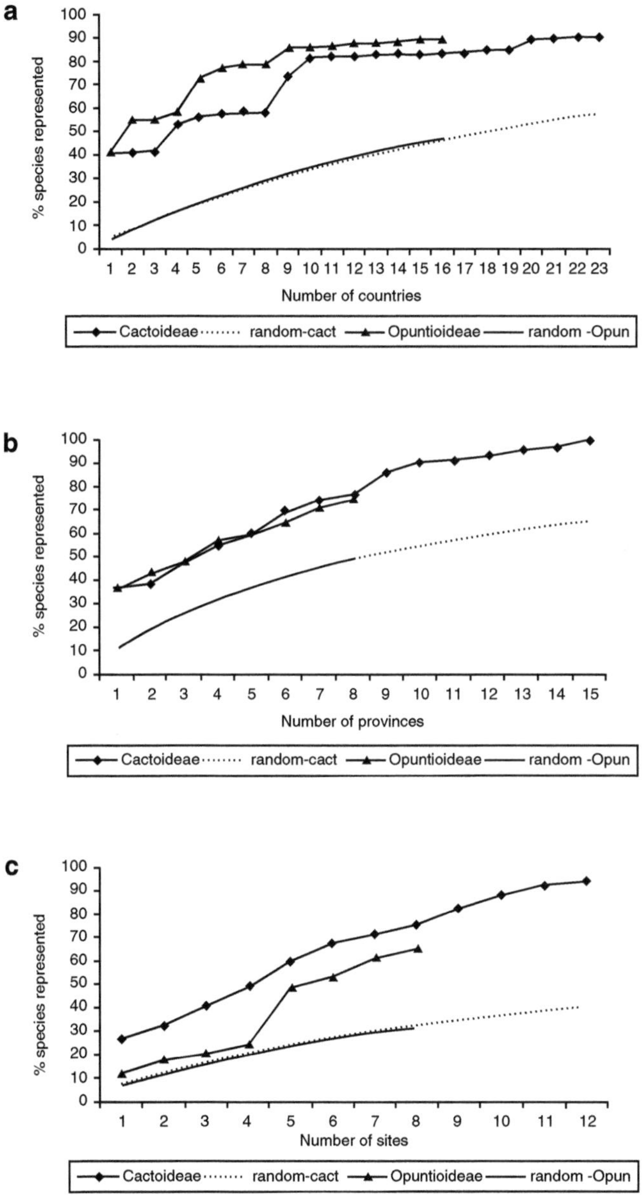

Fig. 8.6 Percentage of cactus species represented by subfamilies Cactoideae and Opuntioideae. **a** Western hemisphere, **b** Argentina, and **c** Calchaquí valley (Argentina, Salta)

Cactoideae and Opuntioideae as surrogates for cactus diversity at the global (Western hemisphere; Fig. 8.6a), country (Argentina; Fig. 8.6b) and regional (Calchaquí valley, Argentina; Fig. 8.6c) scales. It can be noted that, at the three scales of analysis, Cactoideae performs better than Opuntioidea, whose performance decreases from the global to the regional scale. This could be due to the effects of the number of species on surrogates' performance (Manne and Williams 2003). Additionally, cactus can be used as a surrogate for other plants and animals because they are a focal group for conservation. In Argentina, priority provinces for cactus conservation represent a high proportion of dicot diversity (P. Ortega-Baes et al. unpublished data). This idea should be tested using smaller cells.

Ex situ conservation aims are to protect and maintain genetic resources of endemic, endangered and economic species of the world. Seed banks and botanical gardens are the most common *ex situ* strategies for plant conservation. Seed banks allow long-term conservation of representative samples of genetic diversity in reduced spaces (Gold et al. 2004). The Millennium Seed Bank (Royal Botanic Gardens Kew, England) and other international institutions orchestrate the Millennium Seed Bank Project (MSBP), the main goal of which is seed conservation of more than 10% of the plant species in the world ($\sim$ 24,000 species) by 2010. MSBP has produced different research projects, e.g., the *Cactus Seed Biology Monograph*, in which Argentina, Chile, México, Peru, the United States and United Kingdom take part. The objective of this project is to produce and disseminate all available information on the seed biology of the Cactaceae (Seal et al. 2009).

8.7 Concluding Remarks

Cacti are conspicuous elements of arid and semi-arid regions of the Western Hemisphere. These plants have had a historical relationship with local people, but are now exposed to harsh pressure due to habitat conversion and illegal collecting. Global and regional efforts aimed at their conservation have proved insufficient. This is a consequence of the increase of land demand for human activities and the deep socio-economical crisis in Latin American countries, especially in arid lands.

Some actions to improve the conservation of this interesting plant group should be directed at: (1) evaluating implemented conservation strategies; (2) evaluating the conservation status of cactus species in highly threatened ecoregions; (3) evaluating the conservation status of collected species; (4) making regional inventories; (5) conducting studies on cactus biology; (6) determining the human impact on cactus diversity and population; (7) implementing educational programs; (8) increasing governmental control of the cactus trade; (9) organizing training and educational programs for people involved in cactus collection and trade; (10) promoting cooperation among countries, especially bordering ones; (11) developing and enforcing national legislation; (12) expanding protected area systems; and (13) including cactus species in seed banks.

Cactus conservation demands society's commitment, especially that of governments, which must assume this challenge in order to guarantee this legacy for future generations.

Acknowledgments We wish to thank the editor, Dr. K.G. Ramawat for his invitation to contribute to this volume. The authors' research activity was funded by CIUNSalta (Grant N° 1666) and FONCYT (Grant N° 36822).

References

Anderson EF (2001) The cactus family. Timber, Portland

Boyle TH, Anderson E (2002) Biodiversity and conservation. In: Nobel PS (ed) Cacti. Biology and uses. University of California Press, Los Angeles, pp 125–141

Broennimann O, Thuillerw W, Hughesz G, Midgley GF, Alkemade JMR, Guisan A (2006) Do geographic distribution, niche property and life form explain plants' vulnerability to global change? Glob Change Biol 12:1079–1093

Caldecott JO, Jenkins MD, Johnson TH, Groombridge B (1996) Priorities for conserving global species richness and endemism. Biodivers Conserv 5:699–727

Casas A, Barbera G (2002) Mesoamerican domestication and diffusion. In: Nobel PS (ed) Cacti. Biology and uses. University of California Press, Los Angeles, pp 125–141

CITES (2008) http://www.cites.org/eng/app/E-Jul01.pdf

Clark-Tapia R, Mandujano MC, Valverde T, Mendoza A, Molina-Freaner F (2005) How important is clonal recruitment for population maintenance in rare plant species?: the case of the narrow endemic cactus, *Stenocereus eruca*, in Baja California, Mexico. Biol Conserv 124:123–132

Cuarón AD (2000) Effects of land-cover changes on mammals in a neotropical region: a modeling approach. Conserv Biol 14:1676–1692

Esparza-Olguin L, Valverde T, Mandujano MC (2005) Comparative demographic analysis of three *Neobuxbaumia* species (Cactaceae) with differing degree of rarity. Popul Ecol 47: 229–244

Gibson AC, Nobel PS (1986) The cactus primer. Harvard University Press, Cambridge, MA

Godínez-Alvarez H (2004) Pollination and seed dispersal by lizards: a review. Rev Chil Hist Nat 77:569–577

Godínez-Alvarez H, Ortega-Baes P (2007) Mexican cactus diversity: environmental correlates and conservation priorities. Bol Soc Bot Mex 81:81–87

Godínez-Alvarez H, Valiente-Banuet A (2004) Demography of the columnar cactus *Neobuxbamia macrocephala*: a comparative approach using population projection matrices. Plant Ecol 174:109–118

Godínez-Alvarez H, Valiente-Banuet A, Rojas-Martínez A (2002) The role of seed dispersers in the population dynamics of the columnar cactus *Neobuxbaumia tetetzo*. Ecology 83:2617–2629

Godínez-Alvarez H, Valverde T, Ortega-Baes P (2003) Demographic trends in the Cactaceae. Bot Rev 69:173–203

Goettsch B, Hernández H (2006) Beta diversity and similarity among cactus assemblages in the Chihuahuan Desert. J Arid Environ 65:513–528

Gold G, León-Lobos P, Way M (2004) Manual de recolección de semillas de plantas silvestres para la conservación a largo plazo y restauración ecológica. Instituto de Investigaciones Agropecuarias. Centro Regional De Investigación INTIHUASI (ed) La Serena, Chile

Gómez-Hinostrosa C, Hernández H (2000) Diversity, geographical distribution, and conservation México. Biodivers Conserv 9:403–418

Guzmán U, Arias S, Dávila P (2003) Catálogo de cactáceas mexicanas. UNAM/CONABIO, México
Hernández HM, Bárcenas RT (1995) Endangered cacti in the Chihuahuan Desert: I. Distribution patterns. Conserv Biol 9:1176–1188
Hernández HM, Bárcenas RT (1996) Endangered cacti in the Chihuahuan Desert: II. Biogeography and conservation. Conserv Biol 10:1200–1209
Hernández HM, Godínez H (1994) Contribución al conocimiento de las cactáceas mexicanas amenazadas. Acta Bot Mex 26:33–52
Hernández HM, Gómez-Hinostrosa C, Bárcenas RT (2001) Diversity, spatial arrangement, and endemism of Cactaceae in the Huizache area, a hot spot in the Chihuahuan Desert. Biodivers Conserv 10:1097–1112
Hernández H, Gómez-Hinostrosa C, Goettsch B (2004) Check list of Chihuahuan Desert Cactaceae. Harv Pap Bot 9:51–68
Hernández HM, Goettsch B, Gómez-Hinostrosa C, Arita HT (2008) Cactus species turnover and diversity along a latitudinal transect in the Chihuahuan Desert Region. Biodivers Conserv 17:703–720
Howard PC, Viskanic P, Davenport TRB, Kigenyi FW, Baltzer M, Dickinson C, Lwanga JS, Matthews RA, Balmford A (1998) Complementarity and the use of indicator groups for reserve selection in Uganda. Nature 394:472–475
Hunt D (1999) CITES. Cactaceae checklist. Royal Botanic Gardens Kew and International Organization for Succulent Plant Study, Kew, UK
Hunt D (2006) The new cactus lexicon. Remous, Milborne Port, UK
Inglese P, Basile F, Schirra M (2002) Cactus pear fruit production. In: Nobel PS (ed) Cacti. Biology and uses. University of California Press, Los Angeles, pp 163–183
IUCN (2008) IUCN Red List of Threatened Species. http://www.iucnredlist.org
Jiménez-Sierra C, Mandujano MC, Eguiate LE (2007) Are population of the candy barrel cactus (*Echinocactus platyacanthus*) in the desert of Tehuacan, Mexico at risk? Population projection matrix and life table response analysis. Biol Conserv 135:278–292
León de la Luz JL (2005) Evaluation of the conservation status of *Morangaya pensilis* (Cactaceae), little known endemic monotypic genus of southern Baja California, Mexico. Oryx 39:219–222
Mandujano MC, Golubov J, Huenneke LF (2007a) Effect of reproductive modes and environmental heterogeneity in the population dynamics of a geographically widespread clonal desert cactus. Popul Ecol 49:141–153
Mandujano MC, Verhulst JAM, Carrillo-Angeles IG, Golubov J (2007b) Population dynamics of *Ariocarpus scaphirostris* Bödeker (Cactaceae): evaluating the status of a threatened species. Int J Plant Sci 168:1035–1044
Manne LL, Williams PH (2003) Building indicator groups based on species characteristics can improve conservation planning. Anim Conserv 6:291–297
Martínez-Ávalos JG, Jurado E (2005) Geographic distribution and conservation de Cactaceae from Tamaulipas Mexico. Biodivers Conserv 14:2483–2506
Martínez-Ávalos JG, Golubov J, Mandujano MC, Jurado E (2007) Causes of individual mortality in the endangered star cactus *Astrophytum asterias* (Cactaceae): the effect of herbivores and disease in Mexican populations. J Arid Environ 71:250–258
Martorell C, Peters E (2005) The measurement of chronic disturbance and its effects on the threatened cactus *Mammillaria pectinifera*. Biol Conserv 124:197–207
Medel-Narvaez A, Leon de la Luz JL, Freaner-Martinez F, Molina-Freaner F (2006) Patterns of abundance and population structure of *Pachycereus pringlei* (Cactaceae), a columnar cactus of the Sonoran Desert. Plant Ecol 187:1–14
Mittermeier RA, Gil PR, Mittermeier CG (1997) Megadiversidad. CEMEX, México
Mourelle C, Ezcurra E (1996) Species richness of Argentine cacti: a test of biogeographic hypotheses. J Veg Sci 7:667–680
Mourelle C, Ezcurra E (1997) Differentiation diversity of Argentine cacti and its relationship to environmental factors. J Veg Sci 8:547–558

Nefzaoui A, Ben Salem H (2002) Forage, fodder, and animal nutrition. In: Nobel PS (ed) Cacti. Biology and uses. University of California Press, Los Angeles, pp 199–210

Nerd A, Tel-Zur N, Mizrahi Y (2002) Fruits of vine and columnar cacti. In: Nobel PS (ed) Cacti. Biology and uses. University of California Press, Los Angeles, pp 185–197

Nyffeler R (2002) Phylogenetic relationships in the cactus family (Cactaceae) based on evidence from TRNK/MATK and TRNL-TRNF sequences. Am J Bot 89:312–326

Oldfield S (1997) Cactus and succulent plants: status survey and conservation action plan. IUCN/SSC cactus and succulent specialist group. International Union for Conservation of Nature and Natural resources, Gland, Switzerland, and Cambridge, UK

Ortega-Baes P, Godínez-Alvarez H (2006) Global diversity and conservation priorities in the Cactaceae. Biodivers Conserv 15:817–827

Peters EM, Martorell C, Ezcurra E (2009) The adaptive value of cued seed dispersal in desert plants: seed retention and release in *Mammillaria pectinifera* (Cactaceae), a small globose cactus. Am J Bot 96:537–541

Pierson EA, Turner RM (1998) An 85-year study of Saguaro (*Carnegiea gigantea*) demography. Ecology 79:2676–2693

Pimienta-Barrios E, del Castillo RF (2002) Reproductive biology. In: Nobel PS (ed) Cacti. Biology and uses. University of California Press, Los Angeles, pp 163–183

Pin AB, Simon J (2004) Guía ilustrada de los cactus de Paraguay. Secretaría del Ambiente, Paraguay

Reyers B, van Jaarsveld AS, Krüger M (2000) Complementarity as a biodiversity indicator strategy. Proc R Soc Lond, Ser B 267:505–513

Sáenz-Hernández C, Corrales-García J, Aquino-Pérez G (2002) Nopalitos, mucilage, fibber, and cochineal. In: Nobel PS (ed) Cacti. Biology and uses. University of California Press, Los Angeles, pp 211–234

Sarukhán J, Dirzo R (2001) Biodiversity-rich countries. In: Levin SA (ed) Encyclopedia of biodiversity. Academic, San Diego, pp 419–436

Seal CE, Flores J, Ceroni Stuva A, Dávila Aranda P, León-Lobos P, Ortega-Baes P, Galíndez G, Aparicio-González MA, Castro Cepero V, Daws MI, Eason M, Flores Ortiz CM, del Fueyo PA, Olwell P, Ordoñez C, Peñalosa Castro I, Quintanar Zúñiga R, Ramírez Bullón N, Rojas-Aréchiga M, Rosas M, Sandoval A, Stuppy W, Ulian T, Vázquez Medrano J, Walter H, Way M, Pritchard HW (2009) The cactus seed biology database. Royal Botanic Gardens, Kew

Terrazas-Salgado T, Mauseth JD (2002) Shoot anatomy and morphology. In: Nobel PS (ed) Cacti. Biology and uses. University of California Press, Los Angeles, pp 125–141

Valverde PL, Zavala-Hurtado JA (2006) Assessing the ecological status of *Mammillaria pectinifera* Weber (Cactaceae), a rare and threatened species endemic of the Tehuacán-Cuicatlán Region in Central Mexico. J Arid Environ 64:193–208

Valverde T, Quijas S, López-Villavicencio M, Castillo S (2004) Population dynamics of *Mammilaria magnimamma* Haworth. (Cactaceae) in a lava-field in central Mexico. Plant Ecol 170:167–184

Vilela A, Bolkovic ML, Carmanchahi P, Cony M, de Lamo D, Wassner D (2009) Past, present and potential uses of native flora and wildlife of the Monte Desert. J Arid Environ 73 238–243

Villagra PE, Defosse GE, del Valle HF, Tabeni S, Rostagno M, Cesca E, Abraham E (2009) Land use and disturbance effects on the dynamics of natural ecosystems of the Monte Desert: implications for their management. J Arid Environ 73:202–211

Wallace RS, Dickie SL (2002) Systematic implications of chloroplast DNA sequence variation in the Opuntioideae. In: Hunt D, Taylor N (ed) Studies in the Opuntioideae (Cactaceae). Hunt, UK, pp 9–24

Wallace RS, Gibson AC (2002) Evolution and systematics. In: Nobel PS (ed) Cacti. Biology and uses. University of California Press, Los Angeles, pp 125–141

Wright CI, Van-Buren L, Kroner CI, Koning MMG (2007) Herbal medicines as diuretics: a review of the scientific evidence. J Ethnopharmacol 114:1–31

Zak MR, Cabido M, Hodgson JG (2004) Do subtropical seasonal forests in the Gran Chaco, Argentina, have a future? Biol Conserv 120:589–598

Part B
Reproductive Biology

Chapter 9
Reproductive Biology of Some Gum-Producing Indian Desert Plants

Rajesh Tandon, K.R. Shivanna, and H.Y. Mohan Ram

Abstract This chapter covers the reproductive biology of *Acacia senegal*, *Butea monosperma*, *Boswellia serrata*, *Commiphora wightii* and *Sterculia urens* – important gum and gum-resin yielding tree species that grow in the deserts and dry deciduous forests of India. These species have been overexploited and there is a need for conservation of their germplasm to ensure their availability on a sustainable basis. *Acacia*, *Butea* and *Boswellia* bear bisexual flowers. *Commiphora* exhibits trioecy with predominantly female plants with only two male and one andromonoecious plants out of 1,185 plants scored. *Sterculia* exhibits cryptic monoecy; morphologically, the plants are andromonoecious with male and bisexual flowers, but the pollen grains in bisexual flowers are sterile. Flowers of *Acacia*, *Sterculia* and *Boswellia* are of generalised type and are pollinated by honey bees, particularly *Apis dorsata* and *A. indica*, while those of *Butea* are of typical bird syndrome and are pollinated by sunbirds and also the three-striped squirrel. No pollination occurs under field conditions in *Commiphora* and manual pollination does not result in pollen tube growth into the ovary. *Boswellia* and *Butea* show typical self-incompatibility. *Acacia* and *Sterculia* exhibit late-acting self-incompatibility; although pollen tubes reach the embryo sacs following selfing, no seeds develop. *Commiphora* exhibits non-pseudogamous apomixis, characterised by degeneration of the egg, adventive nucellar polyembryony and autonomous endosperm formation. Seed set under field conditions is generally low in most of these species. This is

R. Tandon
Department of Botany, University of Delhi, Delhi 110 007 India
e-mail: rjtnd@rediffmail.com

K.R. Shivanna (✉)
Ashoka Trust for Research in Ecology and the Environment, Royal Enclave, Srirampura, Jakkur Post, Bangalore 560 064 India
e-mail: shivanna@atree.org

H.Y. Mohan Ram
194, SFS DDA Flats, Mukherji Nagar, Delhi 110 009 India
e-mail: hymohanram@gmail.com

K.G. Ramawat (ed.), *Desert Plants*,
DOI 10.1007/978-3-642-02550-1_9,

partly due to the limiting amount of compatible pollen. In *Sterculia* and *Boswellia*, although seeds germinate readily below the canopy of the parent trees, there is hardly any seedling recruitment in forests. The relevance of reproductive features in conservation, and the need for sustainable utilisation of these important species are highlighted.

9.1 Introduction

Nearly 20% of the Earth's surface is occupied by cold and hot deserts (Mohan Ram and Gupta 1997). Thar, the Indian hot desert, covers about 180,000 km^2 in the state of Rajasthan and 45,000 km^2 in the Rann of Kutchh in western Gujarat, and extends to Pakistan. Deserts are generally considered to support a limited number of species. In reality, however, they support a considerable diversity of plants and animals (Mohan Ram and Gupta 1997). According to Khan and Frost (2001), the Thar desert supports 682 plant species including 107 grasses and 200 species of medicinal plants. Commonly termed xerophytes, desert plants grow under harsh environmental conditions where annual precipitation is very low – generally $<$125 mm. Desert plant species show many adaptations that allow them to harvest the limited amount of available water and conserve it. Several desert species extend to dry deciduous forests (Bhandari 1978). Desert plant species have evolved effective strategies to reproduce under adverse environmental growth conditions. In spite of the large number of species growing in Indian deserts, the reproductive biology of only very few species has been studied in detail. Furthermore, excessive exploitation of desert plants for food, feed, firewood, construction and medicine, and harsh adverse ecological conditions, have pushed many desert species into the category of rare/threatened/endangered (RET) plants. A sound knowledge of their reproductive biology is essential not only for their sustainable utilisation and genetic improvement, but also for developing effective conservation measures.

Many plant species of economic importance grow in the Indian deserts. Some yield valuable economic products. Of these, gum and oleo gum-resin yielding species are important, as India is a large exporter of gums. Only a limited number of species have been projected as crops for gum production because gum collection falls largely under the jurisdiction of State Forests Departments. Generally, gum exudates are obtained from wild trees, and constitute an important component of non-wood forest products (NWFP). Secretion of gum/gum-resin requires wounding of the bark. The traditional techniques used for tapping are so crude and destructive that they cause the death of the tree (Nair et al. 1995). Owing to the enormous demand for gum/gum-resin, these species are ruthlessly harvested. Our field studies in the dry deciduous forests of Western Madhya Pradesh have shown that there is hardly any seedling recruitment in these species. There is an urgent need to conserve the germplasm of gum and gum-resin yielding species in order to ensure their sustained viability.

Our research group at the University of Delhi has been engaged in establishing various aspects of the reproductive biology of some of the gum and gum-resin yielding species that occur in arid and semi-arid deserts. In this chapter, we present our findings on the reproductive biology of some of these species. A literature survey indicates that information on the reproductive biology of other species is fragmentary at best.

9.2 Gum and Gum-Resin Yielding Plants

Among the several gum yielding plants in India, only gum ghatti (*Anogeissus latifolia*) and gum karaya (*Sterculia urens*) predominate in international trade. Some of the other sources of gum in India include *Acacia senegal*, *Acacia seyal*, *Azadirachta indica*, *Bauhinia* spp, *Boswellia serrata*, *Butea monosperma*, *Commiphora wightii*, *Cochlospermum religiosum*, *Lannea coromandelica*, *Leucaena leucocephala*, *Prosopis juliflora*, *P. spicigera*, *Sesbania grandiflora*, and *Terminalia* spp (Anonymous 1986). Some of these can be suitably raised to crop level under agroforestry for sustainable utilisation. *Cyamopsis tetragonoloba* is a cultivated annual forage crop that also holds a premier position in the seed gum trade (Anonymous 1986). Barring a handful of initial reports, in-depth research focussing on the pollination biology of gum/resin yielding plants has not been performed.

Our group has investigated the reproductive biology of five gum/gum-resin yielding species: *Acacia senegal* (Tandon et al. 2001), *Boswellia serrata* (Sunnichan et al. 2005), *Butea monosperma* (Tandon et al. 2003), *Commiphora wightii* (Gupta et al. 1996, 1998) and *Sterculia urens* (Sunnichan et al. 1998, 2004).

9.3 Phenology

Arborescent desert plants are generally deciduous and remain defoliated for various durations. They usually exhibit seasonal and synchronised flowering behaviour. However, non-seasonal rains may sometimes induce flowering. For example, in *Acacia pycnantha* (Buttrose et al. 1981), well irrigated-trees are known to flower throughout the year.

The plants of *Commiphora* remain leafless for most of the year. Sparse flowering also occurs throughout the year, with the result that some flowers and fruits are present all year round even when individual plants occur at a considerable distance from one another. In general, there are two flowering peaks – April/May and October/November. The flower buds and the new flush of leaves appear simultaneously. A study carried out in six populations (Gupta 1996) showed that there were not many male plants in the wild. The two male plants at the Central Arid Zone Research Institute (CAZRI), Jodhpur, showed that the peak of male flowering was recorded 1 month earlier. Thus, only a few male flowers were available during the female phase at the population level. Fruits mature by June/July and January/February.

Trees of *Acacia senegal* (Tandon et al. 2001) remain partially or completely leafless for about 4 months from February–March until mid-June. Leaves appear on the branches of the previous year by the end of June, followed by the emergence of new shoots. Flowering is initiated in July with the onset of monsoon. The inflorescences develop only from the axils of leaves of newly developed shoots and come to occupy a peripheral position on the canopy. Fruit development is initiated during the peak of the flowering season and continues until the end of September. Fruits attain their full size in 20–25 days, and reach maturity by January. Shedding of leaves starts in January, and by February and March the trees become totally leafless but the pods are retained. The pods dehisce during March but seeds remain attached to the pod walls. Fruits/seeds are dispersed by strong winds and rain during June–August.

In *Butea monosperma*, the trees usually remain completely leafless for almost 5–6 months (January–mid-June). The peak of flowering occurs during March/April. Blooming of flowers in *Butea* during the deciduous state of trees probably enhances the floral display, as many birds, including the legitimate pollinator – the purple sunbird – and a few nectar robbers such as the White-eye, visit the flowers (Tandon et al. 2003). Trees remain in bloom for up to 2 months. Fruiting begins from the 2nd week of April and mature fruits (single-seeded samara; Augsperger 1989) dislodge from the trees during May/June and are not carried too far from the parent plants. Although a large number of young fruits begin to form, a considerable number are shed or consumed by the langurs and birds, such as parakeets, which invade the trees. Thus, only a small percentage of fruit reaches maturity.

In *Sterculia urens* and *Boswellia serrata* (Sunnichan et al. 2004, 2005), the plants are leafless for most of the year. The white trunk of *Sterculia urens* appears as a prominent feature. Both species produce leaves for about 3 months during the monsoon season (June–September). The trees shed their leaves before the beginning of winter (October–November) and initiate flowering. The peak of flowering occurs in the winter months (December–February). The fruits mature during early summer. Leaflessness during most of the year may be an adaptation to reduce transpiration to conserve water loss.

9.4 Floral Morphology and Sexuality

Acacia, *Butea* and *Boswellia* bear bisexual flowers. In all species of *Acacia* investigated, pollen grains occur in compound units called polyads. The number of pollen grains in a polyad varies from 4 to 64 among the various species. However, the number of pollen grains in a polyad is constant within a particular species. An occurrence of 16 pollen grains in a polyad is most prevalent.

In *A. senegal* (Tandon et al. 2001), the flowers are creamy-white and pentamerous, borne spirally on a spicate inflorescence. There are over 100 stamens in every

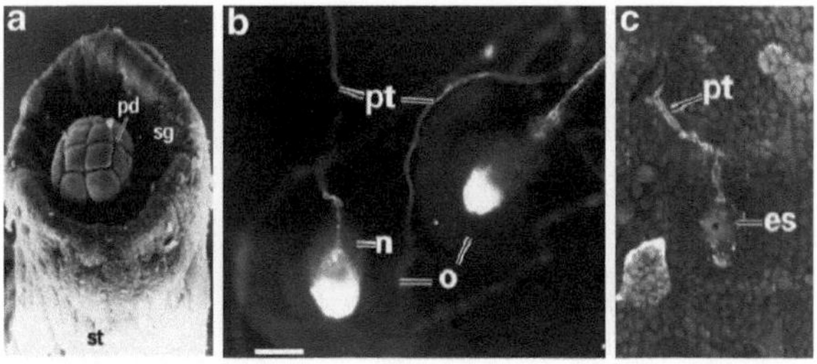

Fig. 9.1a–c *Acacia senegal*: morphology of stigma and pollen and late acting self-incompatibility. **a** Scanning electron micrograph of a pollinated stigma (*sg*) and part of the style (*st*). The cup-shaped stigma and a polyad (*pd*) are clearly seen. **b** Fluorescence micrograph of cleared ovules 28 h after self-pollination. The fluorescing pollen tubes (*pt*) have grown through the micropyle of the ovules (*o*) and have entered the embryo sac. *n* Nucellus. **c** Longitudinal section of ovary after self-pollination showing pollen tube (*pt*) entry into the embryo sac (*es*). The ovary was embedded in glycolmethacrylate, cut at 4 μm and stained with aniline blue. *Bars* **a** 16.14 μm, **b** 750 μm, **c** 625 μm (Tandon et al. 2001; reproduced with permission from the Botanical Journal of the Linnaean Society)

flower. Each anther is eight-loculed and each locule bears one polyad containing 16 pollen grains. The mature pollen grains are two-celled and are rich in lipids. The stigma is cup-shaped (Fig. 9.1a) and the receptive surface is confined to the concave surface of the stigma cup, which is wet and non-papillate. The length of the style varies among the three populations studied in Delhi and Rajasthan. However, the dimensions of other floral organs are more or less uniform in all populations. The style is solid, with a core of transmitting tissue with intercellular spaces filled with extracellular matrix. The unilocular and superior ovary bears, on average, 10.18 $\pm$ 1.09 ovules interspersed with a large number of placental papillae. The papillae secrete an extracellular substance containing proteins and insoluble polysaccharides. The pollen:ovule ratio varies between 1,264:1 and 1,513:1.

Flower colour of *Butea monosperma* has been variously described as scarlet, red and orange. However, according to the colour chart of the Royal Horticultural Society (Anonymous 1942), the colour of *Butea* flowers is "Indian Orange". There are reports of the rare occurrence of white and yellow flower-bearing trees (Sanjappa 1987). The functional floral morphology of these colour variants would be interesting to investigate; it is likely that they might attract different pollinators. Trees that bear Indian-orange-coloured flowers are borne on fascicles of paniculate racemes that bear, on average, nearly 150 flowers. Anthesis under Delhi conditions occurs between 0900 and 1030 hours, and the flower lasts 1–2 days. Nectar secreted by five nectaries located at the base of the ovary accumulates in the calyx cup. The average number of pollen grains produced by a flower is 52,775 $\pm$ 1698. About 37% of the pollen grains are sterile. Each ovary bears five or six ovules. The pollen: ovule ratio is 9,719:1.

Flowers of *Boswellia serrata* (Sunnichan et al. 2005) are pinkish-white and are borne on recemose inflorescences at the ends of branches. There are ten stamens in two whorls of five each. The stamens of the outer whorl are longer (4.2 $\pm$ 0.22 mm)

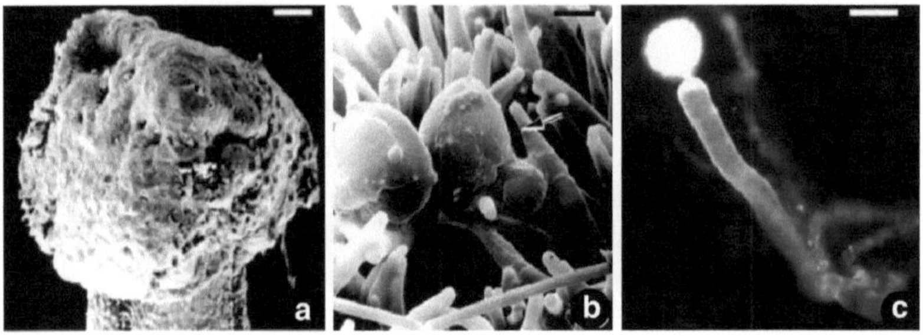

Fig. 9.2a–c *Boswellia serrata*: details of the stigma and inhibition of selfed pollen tubes. **a** Scanning electron micrograph of the stigma on the day of anthesis. The stigma is covered with a copious amount of exudate, which has obscured the visibility of the papillae. **b** As in **a**, but the stigma was fixed in FAA (5 ml 37% formaldehyde, 5 ml glacial acetic acid, 90 ml 70% ethanol) to remove the exudate; the papillae and pollen grains are clear. **c** Fluorescence micrograph of self-pollinated stigma to show normal pollen germination but inhibition of pollen tube in the stigma region following its entry. Note the narrow isthmus between pollen grain and the tube. *Bars* **a** 110 μm, **b** 23 μm, **c** 38 μm (Sunnichan et al. 2005, reproduced by permission from the Botanical Journal of the Linnean Society)

than those of the inner whorl (1.1 ± 0.33 mm). The anthers are dithecous and adnate. Anthers dehisce longitudinally along the theca and release the pollen grains. On average, a flower produces 10,044 ± 1,259 starch-filled pollen grains. About 85% of the fresh pollen grains are viable. The pistil is well demarcated into ovary, style and stigma. The ovary is superior and trilocular; each locule bears a single ovule. Placentation is axile. The pollen:ovule ratio is 3,348:1. The stigma is capitate and is of the wet papillate type (Fig. 9.2a, b). The papillae are unicellular. When the stigma surface reaches the receptive stage, it is covered with a copious amount of exudate and stains positively for insoluble polysaccharides and lipids. Non-specific esterases are also localised. The style is hollow with three flattened stylar canals filled with a secretion product.

9.4.1 Trioecy

In *Commiphora wightii* (Gupta et al. 1996), nearly all the plants are functionally female. Out of 1,185 plants covering six locations in Rajasthan and Madhya Pradesh that were screened for sexuality, it turned out that only two were male and one andromonoecious; all the remaining plants were females. The sexuality of individual plants did not vary during the study period (1992–1995). The male flowers bear stamens arranged in two whorls of four each, with a centrally located minute pistillode. Stamens from male flowers produce fertile pollen grains. The female flowers have a superior ovary on a raised nectary disc surrounded by staminodes arranged in two whorls of four each. Sections of staminodes showed early disintegration of the sporogenous tissue. The bisexual flowers in a solitary andromonoecious plant bear fertile stamens as well as a fully developed pistil. We have recently found that solitary plants cultivated as far away as Bangalore, and a considerably long distance from natural populations, set copious amount of fruits, validating our earlier findings (Gupta et al. 1998).

9.4.2 *Cryptic Monoecy*

All trees of *Sterculia urens* bear morphologically distinguishable male and bisexual flowers (Sunnichan et al. 2004). Their sexuality can be classified as andromonoecious. The percentage of bisexual flowers is very low and ranges from 0.3 to 7.8% in different trees. Male flowers show ten stamens with their filaments fused at the base to form an androphore. No vestigial carpels are present in male flowers. Anthers are dithecous and extrorse. The male flowers open between 0900 and 1130 hours on sunny days, and between 1200 and 0200 hours on cloudy days. The anthers dehisce 1–2 h after anthesis and disperse their pollen soon after. On average, each male flower produces 5,160 $\pm$ 1,141 pollen grains. The pollen grains are two-celled at the time of shedding and are rich in starch (Fig. 9.3a). A high proportion of pollen grains (86.75 $\pm$ 5.89%) are fertile. Pollen from freshly dehisced anthers showed 83.85% viability (Fig. 9.3b) as determined by the fluorescein diacetate test (Heslop-Harrison and Heslop-Harrison 1970).

Bisexual flowers also produce ten stamens similar to those in male flowers. Although anthers of bisexual flowers produce pollen grains that are comparable in number, size and morphology with those produced by male flowers, the anthers do not dehisce unless there is some mechanical disturbance. Also, unlike mature pollen from male flowers, which accumulate insoluble polysaccharides, pollen grains from

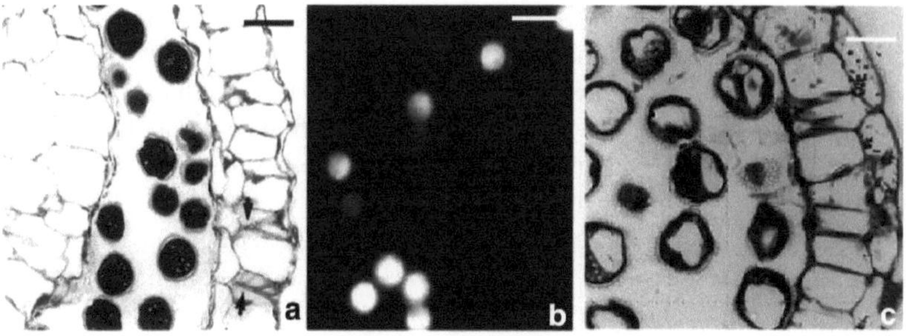

Fig. 9.3a–c *Sterculia urens*: pollen development in male and bisexual flowers. **a** Part of a transection of a mature anther lobe from a male flower stained with I_2KI before dehiscence to show turgid pollen grains filled with reserve materials. **b** Fluorescence micrograph of pollen grains stained with fluorescein diacetate (FDA). Pollen grains are viable as revealed through bright fluorescence. **c** Transection of anther lobe from bisexual flower. Pollen grains are highly vacuolated. *Bars* **a** 12 μm, **b** 42 μm, **c** 30.6 μm (Sunnichan et al. 2004, reproduced with permission from Plant Systematics and Evolution)

bisexual flowers are highly vacuolated without any reserve material (Fig. 9.3c). They do not take up acetocarmine stain. The fluorescein diacetate test also confirmed that the pollen grains in bisexual flowers are non-viable. Further, these pollen grains failed to germinate on receptive stigmas following controlled pollinations. The pistil in a bisexual flower consists of well-demarcated stigma, style and ovary. The stigma is five-lobed and reaches the middle level of the perianth tube at the time of anthesis. Although the stigma appears dry under the light microscope, a thin and irregular secretion is seen under the scanning electron microscope. At the time of anthesis, proteins and non-specific esterases were localised on the surface of the stigma. The gynoecium is pentacarpellary and apocarpous. On average, each carpel bears 6.03 ± 1.02 ovules. The stigma is receptive for 12 h after anthesis.

Taken together, results from acetocarmine staining, fluorescein diacetate test and pollination of receptive stigmas with pollen from bisexual flowers confirm that the pollen grains of ‘bisexual’ flowers are sterile, and that the morphologically ‘bisexual’ flowers are in fact functionally female. Thus, at the operational level, flowers of *S. urens* exhibits cryptic monoecy.

9.5 Pollination Biology

Pollination has been studied in many species of *Acacia* (Kenrick et al. 1987). Although a wide variety of flower foragers are known to visit acacias, bees belonging to Apoideae are the most common vectors in many Australian (Bernhardt 1989) and African (Coetzee 1955) populations of *Acacia*. Some other confirmed pollinators of *Acacia* species are birds (Knox et al. 1985; Ford and Forde 1976), beetles (see Bernhardt 1989) and, unbelievably, even giraffes (*A. nigrescens*; du Toit 1992). In *A. senegal* (Tandon et al. 2001), nectar and pollen grains are the floral rewards for visiting insects. Over 25 insect species, especially bees, wasps and butterflies, visit the flowers of *A. senegal*, and most of them carry pollen load on their bodies. However, the maximum pollen load is found on the giant Asian honeybee (*Apis dorsata*). Pollen grains are presented sternotribically on the body of insects when the latter crawl all over the open flowers of a spike. Foraging activity starts around 0700 hours and reaches a maximum between 1100 hours and 1300 hours, and again around 1600 hours and lasting for 40–60 min. Bees were the most frequent, active and effective foragers.

The cup-shaped stigmas of *Acacia* species show a dimensional adaptation to receive one polyad each (Kenrick and Knox 1989). In *A. retinodes*, up to 80% of the open pollinated flowers received one polyad (Knox and Kenrick 1983). In *A. senegal* (Tandon et al. 2001), over 90% of the open pollinated flowers showed lodging of a single polyad in the stigmatic cup (Fig. 9.1a), although instances of two polyads aligned vertically in the stigma were observed. Knox and Kenrick (1983) reported a correlation between the number of pollen grains in a polyad and the number of ovules in different species of *Acacia*. The ovule number in a particular species is invariably lower than the number of pollen grains occurring in a polyad of that species. Thus, in most species, pollination efficiency is maximised by a single act of pollination since one polyad has the potential to fertilise all the ovules in an ovary.

In *Commiphora wightii* (Gupta 1996), no animals were observed visiting male, female or bisexual flowers. Microscopic examination of stigmas of over 450 flowers collected from female plants growing at different sites showed no pollen grains. Thus, there is no effective pollinator under natural conditions in this species. Hand-pollinated pistils showed germination of a limited number of pollen grains. However, the pollen tubes were confined mostly to the stigma surface or inside the stigma, but they never entered the ovary. Pollen tubes were often branched and showed thick callose accumulation. Microscopic examination of the histological details of the pistil (Gupta et al. 1998) in *C. wightii* revealed some features that are distinctly different from those species that support normal pollen tube growth. The style is typically solid, with two strands of transmitting tissue that traverse the entire length of the style. The cells of the transmitting tissue are isodiametric in transverse as well as longitudinal section and do not form longitudinal files of elongated cells as reported for other taxa. The presence of intercellular proteins amidst the cells of the transmitting tissue is a regular feature in all solid-styled species. However, in *C. wightii*, proteins could not be localised in the intercellular matrix of the transmitting tissue. The authors suggested that the changed orientation of the cells of the transmitting tissue, and the absence of proteins in the intercellular matrix could be the main causes for the failure of the pistil to support pollen tube growth.

In the flowers of *Sterculia urens*, anthers of both male and bisexual flowers are bright yellow, produce abundant pollen, and might serve to attract insects (Sunnichan et al. 2004). As no nectar is produced, pollen grains appear to be the only floral reward for insect visitors. Careful and prolonged observations showed limited insect activity even during the peak of flowering. *Apis indica* was the only floral visitor observed, visiting both male and 'bisexual' flowers on the day of anthesis. Interestingly, no bees visited 'bisexual' flowers when their anthers were removed without disturbing other floral organs, indicating that the anthers of bisexual flowers are the floral attractants. Foraging activity starts around 1000 hours in the morning and reaches its maximum between 1130 and 1230 hours. There was hardly any insect activity in the afternoon. A pollen load of 412 $\pm$ 15 was found distributed on the head, thorax and abdomen of the bees. Non-functional pollen grains have been reported in many other androdioecious species, particularly in species whose flowers lack nectar and are pollinated exclusively by bees (Cane 1993). Many investigators have suggested that the non-viable pollen in the pistillate flowers may act as a reward for pollinators (Kaplan and Mulcahy 1971; Bawa and Beach 1981; Sullivan 1984).

The possibility of wind pollination was studied by hanging glass slides smeared with glycerine jelly at various heights around the canopy of the trees and examining them for the presence of pollen grains after 48 h. No pollen grains were found on any of the slides, ruling out wind as an agency of pollination. The efficiency of pollination in *S. urens* was low. Only about 56% of flowers contained pollen load. Amongst the stigmas that contained pollen grains, only 36% showed 30 or more pollen grains (equal to the number of ovules) and the remaining 20% showed less than 30 pollen grains.

Flowers of *B. serrata* offer both nectar and pollen as rewards to floral visitors. Several insects visited the flowers (Sunnichan et al. 2005). *Apis dorsata* and *A. cerana* were the only active foragers. Maximum activity of the bees was observed between 1200 and 13.00 hours. Bees visited only freshly opened flowers and avoided those that had opened the previous day. During foraging activity, a large number of pollen grains was transferred to the body parts of the bees. The pollen loads recorded on *A. dorsata* and *A. cerana* were, respectively, 714 ± 187 and 472 ± 98. Some individuals of *Oxycetonia versicolor* and *Mylabiris* sp. visited the flowers occasionally, but carried no pollen load. No wind pollination occurred in this species either.

The bright Indian-Orange-coloured flowers of *B. monosperma* attract a variety of birds (Tandon et al. 2003). The structure of the flower complies with ornithophilous pollination (Fig. 9.4a, b). However, detailed observations have demonstrated that, besides the purple sunbirds (*Nectarinia asiatica*) with gently curved bills (Fig. 9.4a) that fit the size and curvature of the corolla, the three-striped squirrel (*Funambulus tristiatus*; Fig. 9.4c) also brings about successful pollination (Tandon et al. 2003). These studies lend support to the earlier findings of Waser et al. (1996) and Johnson and Steiner (2000) that any given flower may have more than one pollinator.

Fig. 9.4a–c Pollination in *Butea monosperma*. **a** A female purple sunbird (*Nectarinia asiatica*) foraging the flower from the legitimate side (opened margin of the fused keel petals). **b** A White-eye robbing the nectar collected in the cup-like cavity of the calyx. **c** The three-striped squirrel (*Funambuus tristiatus*) also consumes nectar from the legitimate site and brings about pollination by transferring pollen onto the stigma when its head (inset) and snout come in contact with the stigma. (Tandon et al. 2003, reproduced with permission from the Annals of Botany)

9.6 Breeding System

Boswellia serrata shows a typical self-incompatible response (Sunnichan et al. 2005). In cross-pollinated pistils, pollen grains germinate profusely and pollen tubes grow through the style and enter the ovary. The pollen tubes are narrow, show no abnormalities and effect fertilisation resulting in fruit and seed set. Although pollen germination is high also in the self-pollinated pistils, the pollen tubes are inhibited soon after their entry into the stigmatic region (Fig. 9.2c). Self pollen tubes are larger than those in cross-pollinated pistils and often show swellings at their tips. Soon after the emergence of the tube through the narrow germ pore, the lumen of the tube enlarges to form a clear 'isthmus'. This condition is not observed in cross-pollinated pollen tubes. The pollen grain and the pollen tube at the isthmus region show accumulation of callose. Formation of an isthmus is an unusual feature not reported in other self-incompatible taxa. None of the selfed flowers set fruits, and they abscised within 3 days.

Butea monosperma shows a weak self-incompatibility, as geitonogamous self-pollinations result in some (~5.25%) fruit set (Tandon et al. 2003). The index of self-incompatibility (ISI; Zapata and Arroyo 1978) was 0.19, which approximates that of a species with complete self-incompatibility (0.20). Weak self-incompatibility is believed to ensure reproductive assurance in the species through geitonogamy. The latter is likely to be facilitated by squirrels as they exhibit territoriality, while sunbirds are likely to promote xenogamy by their ability to carry pollen over longer distances.

9.6.1 Late-Acting Self-Incompatibility

Both *Acacia senegal* and *Sterculia urens* show late-acting self-incompatibility. The breeding system of many species of *Acacia* has been investigated. The Australian species of *Acacia* are self-incompatible and pollen tube inhibition is delayed until their entry into the nucellar tissue of the ovule (Kenrick et al. 1986). *A. senegal* (Tandon et al. 2001) is also strictly self-incompatible. Both light- and electron-microscopic studies have shown clearly that the site of incompatibility is located not in the nucellus but in the embryo sac; the self-pollen tubes enter the embryo sacs (Fig. 9.1b, c) but fail to induce seed set. This type of incompatibility has been termed 'late-acting' or 'ovarian self-incompatibility' (see Seavey and Bawa 1986) and has been reported in many other tree species.

An interesting observation made during the manual pollination of *Acacia senegal* (Tandon et al. 2001) is the difference in the response of flowers to autogamy and geitonogamy. There was a total failure of fruit set through autogamous self-pollination in contrast to 6% fruit set in geitonogamous self-pollination. Tree breeders can, therefore, use geitonogamy as an effective means of developing pure lines in this species. The genetic basis of such a response is unclear.

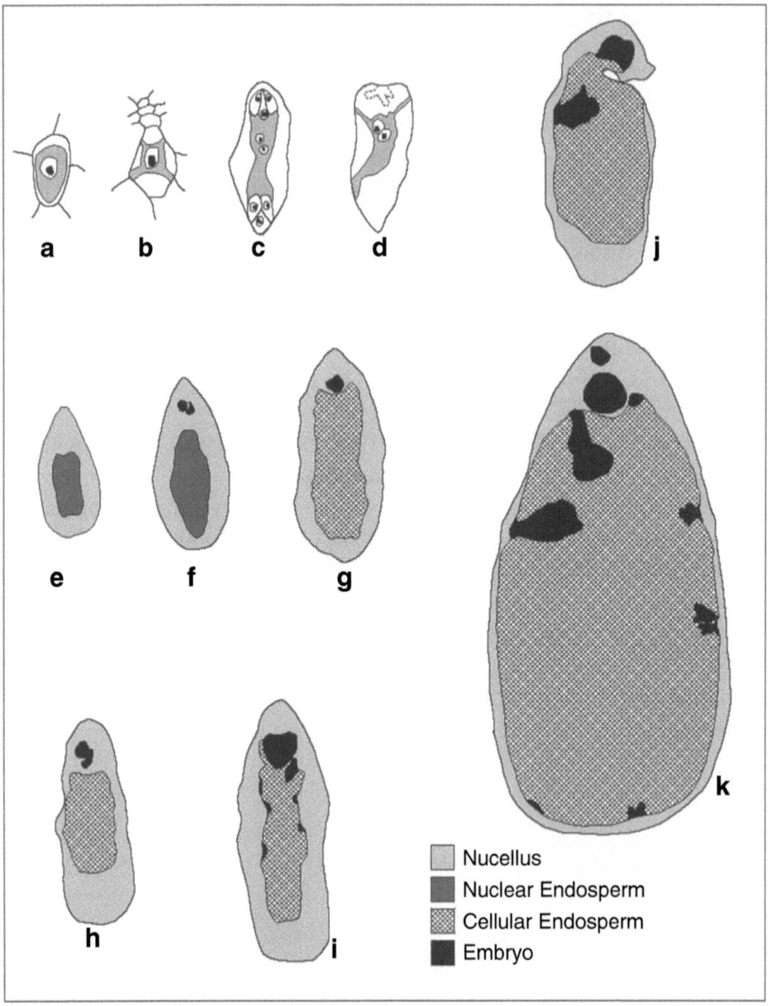

Fig. 9.5 Pseudogamous nucellar polyembryony in *Commiphora wightii*. Diagrammatic representation of the development of embryo sac, endosperm and nucellar embryos (based on dissections and serial sections). **a** Megaspore mother cell. **b** A megaspore tetrad with a functional chalazal megaspore and three degenerating micropylar megaspores. **c** Organized (*Polygonum* type) embryo sac. **d** Embryo sac at a later stage showing juxtaposed polar nuclei and remnants of degenerated egg apparatus. The antipodals have been resorbed. **e** Free nuclear endosperm (20 days after anthesis, DAA). **f** Two young embryos in the micropylar part of the nucellus (25 DAA). **g–k** Later stages reconstructed from serial sections to show cellular endosperm and differentiation and development of numerous nucellar embryos, especially towards the micropyle (Gupta et al. 1996, reproduced with permission from Annals of Botany)

In *S. urens*, which shows cryptic monoecy, manual geitonogamous self-pollinations and xenogamous cross-pollinations were carried out to ascertain the breeding system. Both geitonogamous and xenogamous pollinations supported high pollen germination.

Pollen tubes grew through the style and entered the ovules in both types of pollinations. However, none of the geitonogamously pollinated flowers set fruits. Only xenogamous pollinations resulted in fruit set (Sunnichan et al. 2004).

9.6.2 Apomixis

As pointed out earlier, in *Commiphora wightii*, all female plants produced seeded fruits in spite of the absence of male plants or effective pollinators. These features indicate that the species is apomictic. Detailed embryological studies (Fig. 9.5) confirmed the presence of apomixis (Gupta et al. 1996). The ovule in *C. wightii* is anatropous, bitegmic and crassinucellate. An embryo sac develops deep in the nucellus in young ovules. However, it degenerates partially or completely at anthesis. In the former, the egg apparatus and the antipodal cells degenerate. Even in the absence of double fertilisation, the polar nuclei fuse to give rise to the free nuclear endosperm in a large number of ovules. Some of the nucellar cells situated above the embryo sac become richly cytoplasmic with a conspicuous nucleus, and they develop into nucellar embryos. Several nucellar embryos are initiated in the micropylar region of the nucellus. The endosperm eventually becomes completely cellular. During seed maturation, the nucellus and endosperm are consumed and the 2–6 nucellar embryos fill up the entire seed. The cotyledons are massive and contorted. Mature fruits generally show twin embryos. Thus, the following features show clearly that *C. wightii* is an obligate, non-pseudogamous apomict: (1) formation of seeded fruits from female flowers with no access to pollen, (2) failure of pollen tubes to enter the ovule after hand pollination, (3) occurrence of nucellar polyembryony, and (4) autonomous development of endosperm.

9.7 Pollination Efficiency and Fruit Set

In *C. wightii* (Gupta et al. 1996), there is no pollination and the seeds contain 2–6 nucellar embryos that are true to the mother plant. The fruits are red when fully ripe and are typically two chambered drupes. Only about 50% of the fruits contain mature seeds; the remaining fruits are either empty or contain aborted or shrivelled seeds. Shrivelled seeds may be the result of lack of endosperm development, or early degeneration of the endosperm in some ovules.

In the other three species studied, the average fruit production through open pollination was very low (0.36% in *Acacia*, 2.6–10% in *Boswellia* and 0.7–3.2% in *Sterculia*). In *B. serrata* (Sunnichan et al. 2005), the stigmas of >90% of the flowers had pollen grains. In *A. senegal* (Tandon et al. 2001), pollination efficiency was about 35%, and in *S. urens* (Sunnichan et al. 2004) it was 56%. Availability of a large number of flowers on each tree facilitates the visits of insects to more flowers of the same tree in succession. As the species is strictly self-incompatible, neither autogamous nor geitonogamous self-pollinations results in fruit set. It is, therefore, expected that the number of flowers that receive a sufficient amount of xenogamous pollen is likely to be much lower than that recorded for pollination efficiency. Recruitment of cross pollen through insect visits is likely to be infrequent. Thus, insufficient cross-pollen is likely to represent the main constraint to fruit set in all three species mentioned above, as has been reported in several other species (Knight et al. 2005).

Although manual pollination increased fruit set significantly in all these species, it was still low (maximum ~30% in *Acacia*, ~22% in *Butea*, ~20% in *Boswellia* and ~16% in *Sterculia*). These results indicate that, apart from the limitation of cross-pollination, there are other constraints to fruit set. In the absence of leaves on the trees during the partial or entire period of flowering and fruiting, trees have to utilise limited resources stored in the stem for fruit and seed loading. Therefore, the availability of resources is likely to be another constraint on fruit set. Similar resource limitations on the development of all fertilised ovules into seeds has been reported in several other species (Sutherland 1980; Udovic 1981; Sutherland and Delph 1984).

9.8 Seed Germination and Seedling Establishment

The fruit of *A. senegal* is a light brown and strap-shaped pod. The mature pods collected from five populations in Rajasthan exhibited variation in their average length (5.41–8.09 cm), the average number of seeds borne per fruit (3.92–5.02), the percentage of seed abortion (5.55–23.90) and the extent of infestation by bruchids (14.75–38.55%; Tandon 1997). Mature fruits wither and dehisce on the canopy or are dispersed as diaspores. Seed germination occurs during the subsequent monsoon. Many seeds drop under the canopy (~4.53/m^2), but only a few seedlings emerge (~0.23/m^2) during monsoon. Like many other leguminous seeds with hard testa, the seeds of *A. senegal* exhibit seed-coat dormancy. Under laboratory conditions, ablation of the hard seed coat with the help of a nail-clipper proved more effective than either scalding or scarification (mechanical and chemical). Under natural conditions, new seedlings and trees become established along shallow seasonal drains with gravel and stones, which suggest that seeds of *A. senegal* become scarified while being dispersed with water (ombrohydrochory; Gutterman 1994) and the seedlings establish along the margins of the drain.

The fruit of *Boswellia* is a simple, dry, dehiscent, septicidal capsule (Sunnichan et al. 2005). All three ovules in the ovary develop into winged seeds, bringing the ovule:seed ratio to 1:1. The average flower:fruit ratio was 18.35:1.

In *Sterculia*, the mature fruit is a dry dehiscent follicle covered with dense, rough and pointed trichomes. Out of a total of 30 ovules borne in each ovary, 8–26 ovules develop into seeds, each with a solitary embryo (Sunnichan et al. 2004).

Fruit of *Commiphora* is a two-chambered drupe with a bright yellow pseudoaril around the endocarp found in all species of *Commiphora* and *Bursera* (Gillett 1980).

In *Commiphora wightii*, nearly 50% of the fruits contain abortive and shrivelled seeds (Gupta 1996). Seed germination tests in Petri dishes showed that the developmental stage of the fruit is an important factor for optimal germination, as only 10% of the seeds from dried fruits germinated. The germination response from seeds extracted from fruits with deep red exocarp was nearly 50%. Germination in soil required a 3-day pre-sowing treatment in water before planting. Seeds and embryos of guggul germinate readily under in vitro conditions on Murashige and Skoog's medium (Murashige and Skoog 1962).

In *Sterculia urens*, freshly harvested seeds showed 100% germination, which was reduced to 70% in 10 months, 60% in 18 months and 42% in 2 years. Nearly 20% of the seeds were able to germinate even after 3 years of storage (Sunnichan 1998). The seedlings grow normally when raised in earthen pots kept in the botanical garden for up to 4 months, after which the shoots of all the seedlings dry up. However, an underground tuber develops before the shoots dry up. The tuber perennates in the soil during the ensuing winter and summer months. During the next monsoon (July–August) each tuber puts out a new shoot, which grows vigorously for about 4 months. The tuber enlarges during this period and the shoot dries up again during September–October. This cycle continues for one more year before a permanent shoot develops in the 4th year.

Seed germination and seedling establishment follow the same pattern as those noted in potted plants in botanical gardens as also under natural forest conditions. A large number of vigorously growing seedlings appear under the tree canopy during the monsoon. The shoots of the seedlings invariably die back during September–October. Tubers of different sizes are found when the soil below the canopy is dug up. However, no permanent shoots (4th year shoots) were recorded in the forest during the study, which lasted 4 years. According to local people, the tubers are dug up from the forest soil, and are cooked and eaten as food or pounded and used to relieve constipation and to facilitate childbirth. Wild herbivores such as boars also dig up the tubers and consume them. Harsh drought conditions (such as leafless trees and exposure of the soil to the sun) prevailing during summer may also result in the desiccation or death of some of the tubers.

Seeds of *Boswellia serrata* retained viability for up to only 5 months (Sunnichan 1998). The highest percentage germination recorded was 45. Scarification of seeds increased germination only marginally (51%). However, soaking of seeds for 1 h in gibberellic acid solution (2×10^{-4}) before sowing increased the percentage of germination to 72.

Seedling establishment in *B. serrata* (Sunnichan 1998) also follows the same pattern as in *S. urens*. Seedlings raised in pots grow vigorously for 4–5 months. With the onset of winter, the shoots dry up and the seedlings perennate by means of an underground tuber. Each tuber produces a new shoot during the next monsoon; the cycle of dying of the shoot and its regeneration is repeated for 2 more years before the permanent shoot emerges in the 4th year. In the forest, a large number of seedlings are seen germinating under the canopy of adult trees in the study sites. These follow a similar pattern of death of juvenile shoots and survival by means of tubers. In *B. serrata* also, there was no recruitment of young trees in the forests.

The reasons for lack of recruitment are not clear. This is a serious concern, as both *S. urens* and *B. serrata* populations are dwindling and no practical measures are being taken up by the foresters.

9.9 Concluding Remarks

Apart from generating comprehensive data on the reproductive biology of some of the important gum and oleo gum-resin yielding species occurring in India, our investigations have highlighted their reproductive strategies as well as features relevant to their conservation and sustainable utilisation. Apomixis in *Commiphora wightii* has some interesting features. Unlike many other apomictic species, this plant does not require pollination stimulus. Also, most recorded instances of apomictic plants require endosperm developed by semigamy (fusion of one of the sperm cells with the fused polar nucleus) for development of apomictic embryos (Hanna 1991; Asker and Jerling 1992; Koltunow 1993). In *C. wightii*, however, endosperm development is autonomous and does not require pseudogamy. As pollination is not essential, it is presumed that the endosperm is diploid.

By evolving pseudogamous obligate apomixis, *C. wightii* has eliminated the need for pollination. Autonomous apomixis and polyembryony in *C. wightii* could be a strategy for reproduction and survival in the absence of male plants and effective pollination. This is unique because pollination is a requirement for the development of semigamous endosperm development in many other apomicts. Thus, the progeny obtained in *Commiphora* from seeds is clonal. This knowledge has applications in cloning of high oleo gum-resin yielding plants. Instead of the presently practiced cumbersome and sub-optimal method of propagation through stem cuttings, it would be convenient to use seeds.

The presence of two male and one andromonoecious plants in the populations screened suggests the possibility of occurrence of sexual plants in *Commiphora wightii*. It would be worthwhile to screen a larger number of populations in other locations to identify sexual plants. This would then make *C. wightii* an ideal system for understanding the genetics of apomixis. As the development of endosperm is a prerequisite for commercial exploitation of apomixis in cereals, *C. wightii* with an autonomous endosperm development would be especially useful in approaches aimed at introducing genes that control apomixis into cereals. As far as we are aware, no such gene is available in *Arabidopsis thaliana*, although the identification of a mutant that causes apomeosis has been successfully used to induce the production of a non-reduced egg, and even produce a triploid that retains heterozygosity and possibly hybrid vigour (Ravi et al. 2008).

The three sexually reproducing species, *Acacia senegal*, *Sterculia urens* and *Boswellia serrata*, are strictly self-incompatible. Although this compromises, to some extent, the ability of the seeds to develop, it maintains genetic variability in the population. Also, all three species produce leaves profusely during the monsoon period, achieving robust growth and storage of sufficient photosynthates for reproductive process during the ensuing months. Defoliation during part or entire reproductive period is obviously an adaptation to conserve water and survive under water stress conditions.

Our studies have shown that a reliable method of conservation and sustainable utilisation of *S. urens* and *B. serrata*, two of the overexploited tree species from the semi-arid regions of India, is to raise plantations for commercial tapping. Seedling establishment in both these species follows a unique pathway of tuber formation and perennation. It is convenient to raise seedlings in nurseries until permanent shoots emerge, and then transplant them to plantation locations. Trees of both species grow in areas where agriculture is not possible. It should, therefore, be possible to raise plantations through cooperative efforts for organised and sustainable tapping. *Boswellia serrata* is the only non-coniferous tree that yields oleo-resin (pines are the main source) in India. The resin of *B. serrata* is used for relieving rheumatic pains and is in great demand in both local and export markets. Unwisely, foresters have cut down huge quantities of Salai wood for paper making (and sold it also to mills at throwaway prices as they did with bamboos). The high value of this plant was realised only recently because Malaysia imports and sells Salai guggul at high prices.

As the trees of both *S urens* and *B. serrata* are self-incompatible, they are highly heterozygous. Seeds are not suitable for raising the progeny from plus trees. However, clonal progeny from plus trees would be advantageous for raising plantations. But there are no reports of induction of rooting in stem cuttings in either of these species. Sunnichan (V.G. Sunnichan, unpublished data) attempted to induce rooting of stem cuttings by treatment with plant growth regulators without any success. Nevertheless, we have been able to standardise effective methods for micropropagation of *S. urens* through shoot tip and nodal cuttings from mature trees (Sunnichan et al. 1998). Application of this technique would facilitate propagation of selected trees. Attempts need to be made to develop methods for micropropagation of *B. serrata* also using explants from mature trees.

Tree breeding programmes involve selection of high yielding plants or plus trees, raising seed orchards, selection of elite trees and their production. But as these involve time, trained personnel and resources, our suggestion is to bring together procedures that would not only assure availability of valuable trees but will also employ the results of basic research to ensure a multidisciplinary approach by sharing knowledge and planning coordinated attempts for better utilisation of plant resources of deserts and semi-arid regions.

Another approach towards the sustainable utilisation of these species is to devise an improved tapping method to reduce injury to the trees. Development of a modified technique to tap gum/gum-resin has been reported in *C. wightii* (Bhatt et al. 1989) and *S. urens* (Nair et al. 1995). This technique significantly improves gum yield and promotes early wound healing with least damage to the tree. The modified technique needs to be tested for long-term use, and standardisation for large-scale commercial application. Effective management of tree populations on a sustainable basis would provide employment and assured income to the tribal and rural people living in and around the forests.

References

Anonymous (1942) Horticultural colour chart II, vols I and II. Wilson and Royal Horticultural Society, p 155

Anonymous (1986) The useful plants of India. Publication and Information Directorate. Council of Scientific and Industrial Research, New Delhi

Asker S, Jerling L (1992) Apomixis in plants. CRC, Boca Raton

Augspurger CK (1989) Morphology and aerodynamics of wind-dispersed legumes. In: Stirton CH, Zarucchi JL (eds) Advances in legume biology. Monogr Syst Bot Mo Bot Gard 29:451–466

Bawa KS, Beach JH (1981) Evolution of sexual systems in flowering plants. Ann Mo Bot Gard 68:254–274

Bernhardt P (1989) Floral ecology of Australian acacias. In: Stirton CH, Zarucchi JL (eds) Advances in legume biology. Monogr Syst Bot Mo Bot Gard 29:263–282

Bhandari MM (1978) Flora of the Indian desert. Scientific, Jodhpur

Bhatt JR, Nair MNB, Mohan Ram HY (1989) Enhancement of oleo-gum resin production in *Commiphora wightii* by improved tapping technique. Curr Sci 58:349–357

Buttrose MS, Grant WJR, Sedgley M (1981) Floral development in *Acacia pycnantha* Benth. in Hook. Aust J Bot 29:385–395

Cane JH (1993) Reproductive role of sterile pollen in cryptically dioecious species of flowering plants. Curr Sci 65:223–225

Coetzee JA (1955) The morphology of *Acacia* pollen. S Afr J Sci 52:23–27

Du Toit JT (1992) 'Winning by a neck'. Nat Hist 8:29–33

Ford HA, Forde N (1976) Birds as possible pollinators of *Acacia pycnantha*. Aust J Bot 24:793–795

Gillett JB (1980) *Commiphora* (Burseraceae) in South America and its relationship to *Bursera*. Kew Bull 34:569–587

Gupta P (1996) Reproductive biology of guggul (*Commiphora wightii*) with special emphasis on apomixis and polyembryony. PhD Thesis, University of Delhi

Gupta P, Shivanna KR, Mohan Ram HY (1996) Apomixis and polyembryony in guggul plant, *Commiphora wightii*. Ann Bot 78:67–72

Gupta P, Shivanna KR, Mohan Ram HY (1998) Pollen–pistil interaction in a non-pseudogamous apomict, *Commiphora wightii*. Ann Bot 81:589–594

Gutterman Y (1994) Strategies of seed dispersal and germination in plants inhabiting deserts. 60:373–425

Hanna WW (1991) Apomixis in crop plants – cytogenetic basis and role in plant breeding. In: Gupta PK, Tsuchiya T (eds) Chromosome engineering in plants: genetics, breeding, evolution, part A. Elsevier, Amsterdam, pp 229–242

Heslop-Harrison J, Heslop-Harrison Y (1970) Evaluation of pollen viability by enzymatically induced fluorescence; intracellular hydrolysis of fluorescein diacetate. Stain Technol 45:115–120

Johnson SD, Steiner KE (2000) Generalization versus specialization in plant pollination systems. Trends Ecol Evol 15:140–143

Kaplan SM, Mulcahy DL (1971) Mode of pollination and floral sexuality in *Thalictrum*. Evolution 25:659–668

Kenrick J, Knox RB (1989) Pollen-pistil interaction in Leguminosae (Mimosoideae). In: Stirton CH, Zarucchi JL (eds) Advances in legume biology. Monogr Syst Bot Mo Bot Gard 29:127–156

Kenrick J, Kaul V, Williams EG (1986) Self-incompatibility in *Acacia retinodes*: site of pollen tube arrest in the nucellus. Planta 169:245–250

Kenrick J, Bernhardt P, Marginson R, Beresford G, Knox RB, Baker I, Baker HG (1987) Pollination-related characteristics in mimosoid legume *Acacia terminalis*. Plant Syst Evol 157:49–62

Khan TI, Frost S (2001) Floral biodiversity: a question of survival in the Indian Thar Desert. Environmentalist 21:231–236

Knight TM, Steets JA, Vamosi JC, Mazer SJ, Burd M, Campbell DR, Dudash MR, Johnston MO, Mitchell RJ, Ashman Tia-Lynn (2005) Pollen limitation of plant reproduction: pattern and process. Annu Rev Ecol Evol Syst 36:467–497

Knox RB, Kenrick J (1983) Polyad function in relation to the breeding system of *Acacia*. In: Mulcahy D, Ottaviano E (eds) Pollen: biology and implication for plant breeding. Elsevier Biomedical, New York, pp 411–417

Knox RB, Kenrick J, Bernhardt P, Marginson R, Beresford G, Baker I, Baker HG (1985) Extrafloral nectaries as adaptation for bird pollination in *Acacia terminalis*. Am J Bot 72:1185–1196

Koltunow AM (1993) Apomixis: embryo sacs and embryos formed without meiosis or fertilization in ovules. Plant Cell 5:1425–1437

Mohan Ram HY, Gupta P (1997) Plant life under extreme environments. Curr Sci 72:306–315

Murashige T, Skoog F (1962) A revised medium for rapid growth and bioassays with tobacco tissue cultures. Physiol Plant 15:473–497

Nair MNB, Shivanna KR, Mohan Ram HY (1995) Ethephon enhances karaya gum yield and wound healing response: a preliminary report. Curr Sci 69:809–810

Ravi M, Marimuthu MP, Siddiqi I (2008) Gamete formation without meiosis in *Arabidopsis*. Nature 451:1121

Sanjappa M (1987) Revision of the genera *Butea* Roxb. Ex Willd. and *Meizotropis* Voigt. (Fabaceae). Bull Bot Surv India 29:199–225

Seavey SR, Bawa KS (1986) Late-acting self-incompatibility in Angiosperms. Bot Rev 52:195–219

Sullivan JR (1984) Pollination biology of *Physalis viscosa* var. *cinerascens* (Solanaceae). Am J Bot 71:815–820

Sunnichan VG (1998) Reproductive biology of two forest trees of India: *Sterculia urens* (gum karaya) and *Boswellia serrata* (salai guggul). PhD Thesis, University of Delhi

Sunnichan VG, Shivanna KR, Mohan Ram HY (1998) Micropropagation of gum karaya tree (*Sterculia urens* Roxb.) through multiple shoot formation and somatic embryogenesis. Plant Cell Rep 17:951–956

Sunnichan VG, Mohan Ram HY, Shivanna KR (2004) Floral sexuality and breeding system in gum karaya tree, *Sterculia urens*. Plant Syst Evol 244:201–218

Sunnichan VG, Mohan Ram HY, Shivanna KR (2005) Reproductive biology of *Boswellia serrata*, the source of salai guggul, an important gum-resin. Bot J Linn Soc 147:73–82

Sutherland S (1980) Energy limited fruit set in a paniculate agave: a test of the bateman principle. Bull Ecol Soc Am 61:105

Sutherland S, Delph LF (1984) On the importance of male fitness in plants: patterns of fruit set. Ecology 65:1093–1104

Tandon R. (1997) Reproductive biology of two leguminous trees: *Acacia senegal* and *Butea monosperma*. PhD Thesis, University of Delhi

Tandon R, Shivanna KR, Mohan Ram HY (2001) Pollination biology and the breeding system of *Acacia senegal*. Bot J Linn Soc 135:251–262

Tandon R, Shivanna KR, Mohan Ram HY (2003) Reproductive biology of *Butea monosperma* (Fabaceae). Ann Bot 92:715–723

Udovic D (1981) Determinants of fruit set in *Yucca whipplei*: reproductive expenditure vs pollinator availability. Oecologia 48:389–399

Waser NM, Chittka L, Price MV (1996) Generalization in pollination systems, and why it matters. Ecology 77:1043–1060

Zapata TR, Arroyo MTK (1978) Plant reproductive ecology of a secondary deciduous tropical forest in Venezuela. Biotropica 10:221–230

Chapter 10
Reproductive Biology of Cactaceae

María del Carmen Mandujano, Israel Carrillo-Angeles, Concepción Martínez-Peralta, and Jordan Golubov

Abstract Floral biology in the Cactaceae represents a new field of research, with only 2% of approximately 2,000 species having been studied. Studies on breeding systems cover functional and morphological floral traits of sexual expression, while research on mating (hybridization) systems includes experiments on pollination, morphological and functional floral traits vis-à-vis estimation of out-crossing rates, and inbreeding depression. Most cacti are hermaphroditic with some exceptions of dioecy. Herkogamy and dichogamy seem to be common, and are coupled with self-incompatibility and inbreeding depression as mechanisms to avoid selfing; these traits are important in guiding the evolution of mating (fertilization) systems from mixed to mainly either out-crossing or selfing in all three subfamilies. We found no clear pattern between mating systems and genetic diversity. The impressive variety of fertilization and breeding systems, as well as the genetic diversity within Cactaceae highlights the complex evolution of this family and the plasticity of their reproductive response to the spatially and temporally unpredictable habitats in which they occur. This chapter reviews information on the floral biology, pollinators and genetics of Cactaceae, covering about 70 references: 36% on genetics, 43% on different aspects of pollination ecology, and 21% on diverse subjects with limited descriptions.

M.C. Mandujano (✉), I. Carrillo-Angeles, and C. Martínez-Peralta
Departamento de Ecología de la Biodiversidad, Laboratorio de Dinámica de Poblaciones y Evolución de Historias de Vida, Instituto de Ecología, Universidad Nacional Autónoma de México, Apartado Postal 70-275, Ciudad Universitaria UNAM, C.P, 04510 México, D.F
e-mail: mcmandu@miranda.ecologia.unam.mx

J. Golubov
Laboratorio de Ecología, Sistemática y Fisiología Vegetal, Departamento El Hombre y Su Ambiente-CBS-Universidad Autónoma Metropolitana-Xochimilco, Calzada del Hueso 1100, Col. Villa Quietud, 04960 México, D.F
e-mail: jgolubov@gmail.com

K.G. Ramawat (ed.), *Desert Plants*,
DOI 10.1007/978-3-642-02550-1_10,

10.1 Introduction

Reproduction is a vital process in the life cycle of organisms. Plants produce flowers in order to reproduce, and syngamy becomes the link between present genotypes and future generations. The fertilization process of plant species, mediated by breeding systems, determines the mode of transmission of the genetic information from one generation to the next, which ultimately establishes the genetic structure of wild populations. As reproductive biology is under genetic control and subject to natural selection, investigating the fertilization process and its genetic potential (Brown 1990; Waser 1983) will provide insights into its evolution. Although research into the fundamentals of reproductive biology in plants began 200 years ago, the theoretical foundations are relatively new and have only recently been included in empirical studies. In particular, reproductive biology of the cactus family is a relatively unexplored field, and, due to the variety of life forms, photosynthetic pathways, floral displays and diversity, research has concentrated on their taxonomy, morphology and physiology (Gibson and Nobel 1986; Nobel 1988; Table 10.1).

Six plant families are well represented in the Americas: Compositae (Asteraceae), Leguminosae (Fabaceae), Poaceae, Orchidaceae, Cactaceae and Rubiaceae (Rzedowski 1991). Of these families, the Cactaceae, native to the New World, is particularly important because of its high species diversity and endemism (84% are endemic; Arias et al. 2005a). The lack of information on these species regarding their reproductive biology in their natural habitat is now a major concern because many species are included in national (e.g., the Mexican red list with 239 species), and international [59 species in the International Union for Conservation of Nature and Natural Resources (IUCN) red list, and 64 species in the Convention on International Trade in Endangered Species (CITES) Appendix I] endangered species lists, requiring urgent evaluation of population status (Lüthy 2001; Arias et al. 2005a).

The Cactaceae display different sexual strategies for successful reproduction, and several theories have attempted to explain the origin and causes of such diversity based on the facts that plants are sessile, modular, many are cosexual and need to develop strategies in order to reach mates (Lovett-Doust and Lovett-Doust 1988; Waser 1983). Sprengel (1793), Knight (1799) and later Darwin (1876) acknowledged that outcrossed progeny performed better than selfed progeny, and suggested that the decreased success of selfed progeny might determine floral traits that promote outcrossing. However, Darwin (1859) also documented that some species display features that are especially adapted for self-fertilization, an observation that led to the reproductive assurance hypothesis: self-fertilization will evolve only when it is necessary or under specific conditions to ensure reproduction (Darwin 1859; Lloyd 1979). Fisher proposed a different approach, in which a gene causing self-fertilization will increase in frequency in each generation and will rapidly become fixed unless opposed by other mechanisms (Holsinger 1992). According to the genetic proposal, a selfer has, on average, three successful

Table 10.1 Flower traits, life forms, number of genera, species and main geographic distribution by subfamilies of Cactaceae. *SHFT* Short floral tube, *LFT* long floral tube. Flowers are bisexual, unless otherwise mentioned. Symmetry: *R* radial or actinomorphous, *Z* bilateral or zigomorphous. Size indicates the smallest to the largest flower present in the subfamily. Individual sexual unit: *B* bisexual flower, *U* unisexual flower (*UM* morphological or *UF* functional). Inflorescence refers to the number of sessile flower per areole. Nomenclature and life forms according to Anderson (2001), Guzmán et al. (2003), and Mauseth (2006). Hybrids and non-confirmed species not included

Subfamily	Genera (*n*)	Species (*n*)	Life form	Geographic distribution	Metabolism	Flower traits						Inbreeding depression
						Shape	Individual sexual unit	Symmetry	Color of perianth segments	Corolla aperture (cm)	Inflorescence	
Pereskioideae	1	17	Treelike or shrub-like with long-lasting leaves	Lowland neotropics from southern Mexico, Caribbean region, Central America, and to northern Argentina	CAM, C3	Floral cup, SHFT	B	R	Red, pink to white	0.5–7	Clusters or solitaries	No data
Maihuenioideae	1	2	Shrub-like as short cushions	Chile, Southern Andes throughout Patagonia, Argentina	C3 ?	Floral cup, SHFT	B	R	Yellow, white	2.5–4	Solitary	No data
Opuntioideae	15	341	Treelike, shrub-like	From Canada throughout South America and Caribbean	CAM, C3	Floral cup, SHFT	B, UM, UF	R	Yellow, pink, red, green off-white	0.5–11	Solitary	0.79 ± 0.29
Cactoideae	105	1,458	Treelike, shrub-like cylindrical, expansive			Floral cup, SHFT, LFT	B, UM, UF	R, Z	Yellow, pink, magenta-blue, purple, orange, red, white	0.5–37	Clusters or solitaries	0.83 ± 0.25
Total	122	1,818										0.82 ± 0.26

[a]Undetermined

gametes, two as an ovule and pollen parent to the selfed progeny, and one as pollen parent to the outcrossed progeny of another individual. In contrast, an outcrosser will have only two successful gametes, one each as ovule parent and as pollen parent. This 3:2 advantage is the source of the automatic selection often associated with self-fertilization (Holsinger 1992). The genetic model gave way to a new paradigm in which species can evolve either to outcrossing or towards selfing mating systems with attributes that avoid outcrossing and inbreeding depression as the main force hindering selfing (Charlesworth and Charlesworth 1987; Schemske and Lande 1985). Following Knight and Darwin, most of the research on plant mating systems over the past century has focused on the fitness consequences of selfing, documenting the magnitude of inbreeding depression, elucidating its genetic basis, or describing its evolutionary consequences. Among the list of plants traits that have been suggested to have evolved to discourage selfing are the breeding systems that can be found at the level of a single flower (i.e., herkogamy, dichogamy or self-incompatibility), individual (e.g., gynoecious or androecious plants) or population (e.g., hermaphrodite or monoecious individuals; Richards 1997).

In this chapter, we present evidence of the origin of the mating and breeding systems in the Cactaceae, following both ecological and phylogenetic approaches (Barrett and Eckert 1990; Barrett et al. 1996). Breeding systems covers all the functional and morphological floral traits reported in the cactus family, while mating systems will include pollination experiments, morphological and functional floral traits as well as genetic estimations of outcrossing rates. We first present the origin of the subfamilies and phylogenetic relationships within the Cactaceae and then explore the presence of different floral traits that will enable us to determine functional aspects of the mating systems and examine this family from an ecological, morphological, and physiological perspective. The focus of the chapter will include the traditional selfing vs outcrossing paradigm (Barrett 2003), introducing floral traits, floral behavior and interactions with pollinators and the relation of these factors with the evolution of both mating and breeding systems. In addition, we explore how clonal reproduction and life form may affect mating patterns, potentially leading to evolutionary transitions between mating systems in some subfamilies. Finally, we compare the genetic diversity of cacti species with their reproductive biology in order to provide evidence of the relationship between mating system and genetic diversity.

10.2 The Cactus Family

10.2.1 Evolution and Systematics

In America, several succulent families, e.g., the Agavaceae (Good-Avila et al. 2006; Eguiarte et al. 1994) and Cactaceae (Wallace 1995; Nyffeler 2002; Arias et al. 2005b), have experienced recent diversification. The Cactaceae have been shown to have evolved from the Caryophyllales of the Andes as a monophyletic group (Wallace 1995; Hershkovitz and Zimmer 1997; Nyffeler, 2002), with a subsequent radiation of species towards North and South America, a diversification that has led to the formation of specific groups such as *Copiapoa*, *Echinopsis* and *Maihuenia* in South America, and *Opuntia*, *Mammillaria* and *Ariocarpus* in North America (Anderson 2001), associated mainly with the arid and semi-arid environments characterized by low and variable annual rainfall, high evaporation and contrasting temperatures (Gibson and Nobel 1986). The most recent classification of the family includes four subfamilies, 121–127 genera, and close to 2,000 species (Table 10.1; Anderson 2001; Wallace and Gibson 2002; Hunt et al. 2006). Molecular phylogenetic analyses of this family have been plagued by low statistical resolution and under- or over-representation of specific groups (e.g., Nyffeler 2002; Arias et al. 2003, 2005a, 2005b; Butterworth and Wallace 2004; Edwards et al. 2005). However, general patterns are fairly consistent across studies, i.e., a monophyletic origin, (but see Edwards et al. 2005, who suggest a paraphyletic evolution), Pereskioideae as a basal group, and consistent evidence of the Opuntioideae and Cactoideae as independent groups. Apart from these general patterns, there is little evidence that can be used to subdivide Maihuenioideae as a subfamily, or to define lower taxonomic levels within each tribe. This suggests that the best approach for future phylogenetic studies in the Cactaceae should consider smaller groups and should include a biogeographic component as well as life history attributes. This small-scale approach should generate adequate phylogenies that can then be used to determine the adaptive value of other components, disentangling phylogenetic effects (Harvey and Pagel 1991). This is especially important when we consider that recently derived species differ only in reproductive attributes (Carson and Templeton 1984).

Unfortunately, the cactus family is very well represented in national and international endangered species lists (Lüthy 2001; Arias et al. 2005a). The vulnerability of the Cactaceae is due largely to a combination of factors (among others, low growth and recruitment rates, long life cycles, limited distribution, small population size; Valverde et al. 2004; Esparza-Olguín et al. 2005) that, coupled with habitat loss and illegal collection of individuals, negatively impact many populations (Mandujano et al. 2007; Martínez-Ávalos 2007).

10.2.2 The Cactus Flower

The basic structure of Cactaceae flowers follows that of dicotyledons, but there are significant differences from the typical flower (Bravo-Hollis 1978). The cactus flower is bisexual, with an inferior ovary with the exception in some species of *Pereskia*. They are sessile and solitary. Usually, a single flower is produced from each axillary bud (i.e., areole) – commonly from areoles near the apex – but position, color, shape, size and behavior can vary greatly (Bravo-Hollis 1978;

Gibson and Nobel 1986; Anderson 2001; Table 10.1). Interestingly, despite the diversity in floral characteristics, they are highly uniform from a structural and anatomical point of view (Fig. 10.1). The flowers are partially enclosed by shoot or stem tissue (the pericarpel), where the carpels are found, which may have areoles (Mauseth 2006). For example, in the Opuntioideae these areoles can give rise to new clonal plants by pseudoviviparity (Mauseth 2006; Palleiro et al. 2006), bracts (small modified leaves), spines or trichomes, like in the genus *Selenicereus*, *Leptocereus*, *Echinopsis* and *Ferocactus*. In some genera of cacti, such as *Mammillaria, Rhipsalis* and *Ariocarpus*, the pericarpel can be completely devoid of areoles (Bravo-Hollis 1978; Fig. 10.1). The pericarpel is often referred to as the floral tube, but it is composed of more than the parts of the perianth. The floral tube can be longitudinally elongated, as in *Nopalea*, *Polaskia* and *Stenocereus*, or short, as in *Grusonia*, *Opuntia*, *Mammillaria*, *Maihuenia* and *Pereskia* (Fig. 10.1). Flowers are typically actinomorphic, but a few species, such as those of *Cleistocactus*, *Disocactus*, *Hylocereus* and *Selenicereus* are zygomorphic (Gibson and Nobel 1986), but may be found in some species of other genera (e.g., *Brasilicereus phaeacanthus*). The flowers have numerous stamens, usually inserted on the inner part of the floral tube, and produce copious amounts of pollen. Nectar is secreted by a disc or along the basal portion of the hypanthium, and in some species it is possible to recognize a well-defined nectar chamber. The gynoecium consists of an ovarian cavity, which contains numerous ovules, and a style that bears a multi-lobed stigma (Table 10.1). Mature stigmas have been reported as either wet (with a free-flowing secretion) or dry (with a hydrated, proteinaceous extracuticular layer or pellicle, but no free-flowing secretion). This condition is correlated with the type of self-incompatibility (SI) system and these chemically complex secretions, which can change with flower age, are thought to be important in pollen capture or germination (Table 10.1; Heslop-Harrison and Shivanna 1977; Boyle 1997). There are a very few examples with unisexual flowers caused by the atrophy of the gynoecium or androecium during early development of floral parts, e.g., *Mammillaria dioica* (Sánchez-Carbajal 2007).

10.3 Breeding Systems

Breeding systems include all the aspects of sexual expression in plants that affect the relative contribution to the next generation (Dafni 1992). Plants exhibit a huge variation in breeding systems and the cactus family is no exception (Tables 10.1, 10.2) even considering the limited number of studies. Approximately 90% of angiosperms are hermaphrodites and 10% display different breeding systems (Dellaporta and Calderon-Urrea 1993). Anderson (2001) includes ca. 2,000 species in the cactus family (Table 10.1), and less than 5% of these species have had their breeding system, pollination syndromes or some aspect of their reproductive biology determined (Table 10.1). In those studies that have been published, around 20 species with flowers that display dysfunctional arrangements of sexual

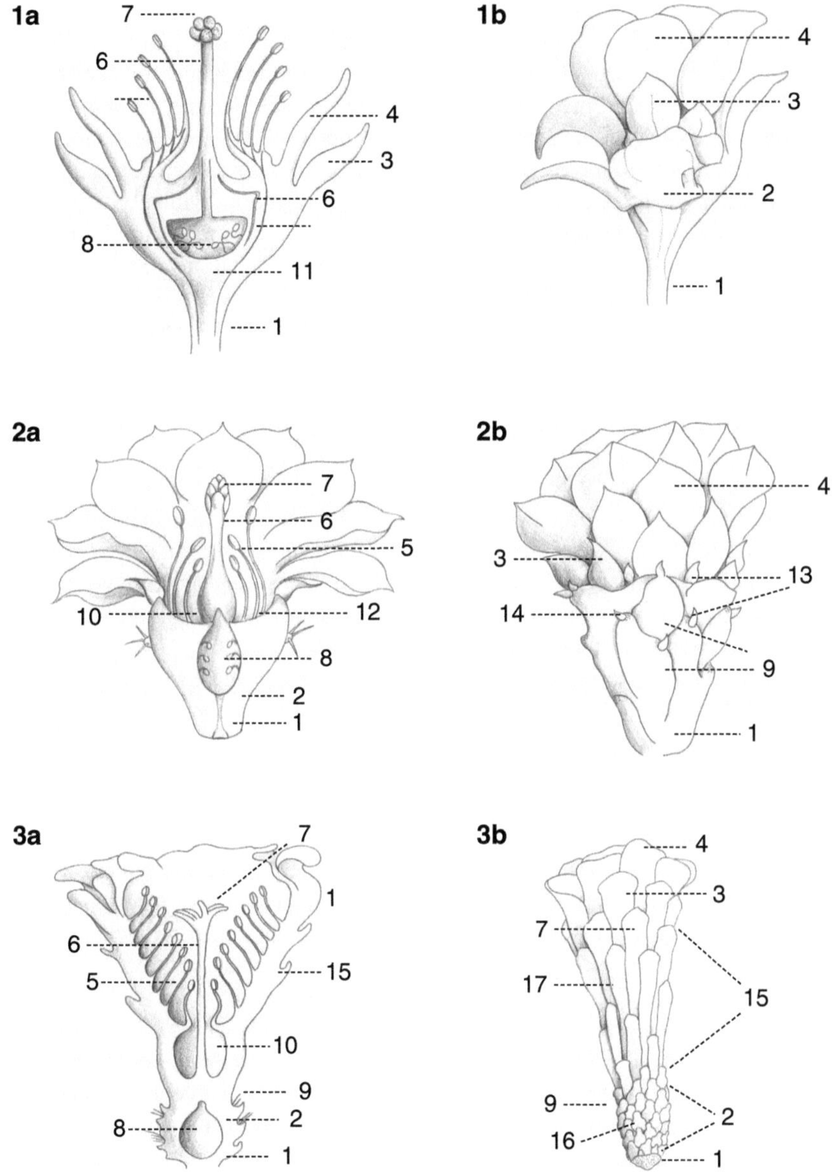

Fig. 10.1 Main floral types in the Cactaceae: *1 Pereskia*, *2 Opuntia*, and *3 Stenocereus* flowers. Schematic representation of each type (*1a*, *2a*, *3a*) and external view (*1b*, *2b*, *3b*) are shown to compare similarities and differences in structures. Common structures in the three types: *1* peduncle (gradually shorter from *Pereskia* to *Stenocereus*), *2* pericarpel, *3* external perianth segments, *4* internal perianth segments, *5* stamens, *6* pistil, *7* stigma lobes and *8* ovarian cavity. Structures differing between floral types: *9* podarium and *10* nectar chamber in *Opuntia* and *Stenocereus*; *11* hypanthium in *Pereskia*; *12* wide receptacular zone where stamens and perianth segments insert, *13* rudimentary leaves and *14* woolly areoles in *Opuntia*; *15* long receptacular tube, *16* scales and *17* decurrent podarium in *Stenocereus*

expressions, being functionally male or female, have been described (Strittmatter et al. 2002). Variations in sexual expression at an individual plant (e.g., monoecious, dioecious), morphological (herkogamy, distyly and heterostyly) or physiological (self-incompatibility, dichogamy) level of the female or male sexual function (breeding systems sensu Neal and Anderson 2005) are mechanisms that have been proposed to have evolved to promote outcrossing, which is achieved in different ways. Plant sexuality varies at the level of the individual unit (flower), individual plant or population. A large proportion of cacti species are hermaphrodites, with perfect flowers, and the few reported species that have separate sexual expressions or a dysfunctional sexual expression are considered dioecious (i.e., populations having separate male and female plants) as in *Opuntia stenopetala* (Bravo-Hollis 1978; Orozco-Arroyo 2002), *M. dioica* (Sánchez-Carbajal 2007), and *Echinocereus coccineus* (Powell 1995). Some populations are gynodioecious (plants having hermaphrodite flowers and plants with female flowers) such as *Consolea spinosissima* (=*O. spinosissima*, Strittmatter et al. 2002), *Pachycereus pringlei* (Fleming et al. 1998), and *Mammillaria blossfeldiana* (Rebman 2001), and trioecy (i.e., plants with male flowers, plants with female blossoms and plants with hermaphrodite flowers) can also be found in *Opuntia robusta* and *P. pringlei* (Pimienta-Barrios and del Castillo 2002). Unisexuality appears to have evolved independently several times within the cactus family, resembling the derived condition reported for other angiosperms (Webb 1979).

In monoclinous species, selfing or sexual interference can be avoided or reduced (see Webb and Lloyd 1986) by a spatial separation of the anthers and stigmas (i.e., herkogamy), or a temporal separation (i.e., dichogamy) of these two pollination surfaces (i.e., pollen is produced when the stigma of the same flower is not receptive). If pollen is produced before the stigma of the same flower is receptive, protandrous flowers are generated; while protogyny describes flowers that have stigmas that can accept foreign or xenogamous pollen before the pollen from the same flower or plant. Herkogamy is apparently frequent, with differing degrees of separation between anthers and style. For example, much longer styles than anthers have been found in *Ariocarpus fissuratus* (C. Martínez-Peralta and M.M. Mandujano, unpublished data), *Opuntia imbricata* (McFarland et al. 1989), *Pilosocereus lanuginosus*, *P. moritzianus* (Nassar et al. 1997), *Stenocereus queretaroensis* (Ibarra-Cerdeña et al. 2005), the genus *Melocactus*, *Nopalea* (Anderson 2001), *Peniocereus striatus* and *P. greggii* (Raguso et al. 2003; Table 10.2). There are two classes of pollination systems in which individual flowers are herkogamous and bisexual, having two or three different floral arrangements, either on the same or on different individuals. One of these classes, heterostyly, occurs when the floral morphs are found on separate plants and differ reciprocally in style and stamen lengths (Richards 1997). These spatial differences between flowers or plants can be maintained by the reciprocal morphs, usually genetically linked to a strong SI (Webb and Lloyd 1986), but some species that display heterostyly are self-compatible (Yeo 1975; Charlesworth and Charlesworth 1979). Heterostyly has been reported in *Opuntia robusta*, in which plants with female flowers have short

Table 10.2 Life form, reproductive traits and estimation methods of outcrossing rates from 70 studies on cacti species. Life form: *T* Treelike, *C* columnar, *Cy* cylindrical, *E* expansive, *G* globose, *S* shrub-like; Subfamily-tribe: *C Cac* Cactoideae-Cacteae, *C Cer* -Cereeae, *C Hyl* -Hylocereeae, *C Pac* -Pachycereeae, *C Rhp* -Rhipsalideae, *C Trc* -Trichocereeae, *Opt* Opuntioideae, *Per* Pereskioideae; Floral longevity: *D* diurnal, *N* nocturnal. Compatibility: *SC* self-compatibility, *SI* self-incompatibility, *GSI* gametophytic self-incompatibility; Mating system: *A* Agamospermy, *M* Mixed, *O* Outcrossing. Outcrossing rate estimation method: *CE* crossing experiments, *MM* molecular marker. *H* Herkogamy, *D* Dichogamy; Nectar: + presence of nectar. Nomenclature according to Anderson (2001). ? Undetermined, *Some flowers can be N or D

Species	Life form	Subfamily-tribe	Floral longevity (days)	Compatibility	Mating system morphological-functional	Inbreeding depression	Outcrossing rate estimation method	H	D	Nectar	Pollination syndrome	Reference
Ariocarpus fissuratus (Engelm.) Schum.	G	C Cac	1–2 D	SC	O	Fruit set	?[a]	?	?	+	Melittophily	Martínez-Peralta 2007
Astrophitum asterias (Zucc.) Lem.	G	C Cac	1 D	SI–SC	M	Fruits (0.745), seeds (0.893)	CE	?	?	?	Melittophily	Martínez-Ávalos 2007; Strong and Williamson 2007
Carnegia gigantea (Englem.) Britton & Rose	T	C Pac	?	?	?	?	MM	?	?	?	Chiropterophily	Hamrick et al. 2002
Cephalocereus totolapensis (Bravo & MacDoug.) Buxb.	C	C Pac	?	?	?	?	MM	?	?	?	Chiropterophily	Palleiro 2008
Cereus horrispinus Backeberg	T	C Cer	1 N	SI	O	?	CE	?	?	?	Chiropterophily	Nassar et al. 1997
Creus repandus (L.) Miller	T	C Cer	1 N	SI	O	?	CE, MM	?	?	?	Chiropterophily	Nassar et al. 1997, 2003
Consolea spinosissima (Miller) Lem.	S	Opt	1–2 D	SI	A	No	CE	?	?	+	Mirmecophily	Negrón-Ortiz 1998
Cylindropuntia bigelovii (Engelm.) Knuth	S	Opt	?	?	?	?	?	?	?	?	Melittophily	Hernández-Rosas 2006
Cylindropuntia imbricata (Haworth) Knuth	S	Opt	?	?	O	?	?				Melittophily	McFarland et al. 1989

(*continued*)

Table 10.2 (continued)

Species	Life form	Subfamily-tribe	Floral longevity (days)	Compatibility	Mating system morphological-functional	Inbreeding depression	Outcrossing rate estimation method	H	D	Nectar	Pollination syndrome	Reference
Echinocactus platyacanthyus Link. et Otto	G	C Cac	2–3 D	SC	M	Fruits, seeds	MM	?	?	?	Melittophily	Jiménez-Sierra 2008
Echinomastus erectrocentrus (Coult.) Britton & Rose	Cy	C Cac	1–2 D	SI	O	No	CE	?	?	+	Melittophily	Johnson 1992
Echinopsis chamaecereus Friedrich & Rowley	S	C Trc	2–3 D	SI	O	?	?	?	?	?	?	Boyle and Idnurm 2001
Escontria chiotilla (Weber ex Schum.) Rose	C	C Pac	1(D)	SI	O	No	CE	No	No	+	Two pollinators	Oaxaca-Villa et al. 2006
Facheiroa squamosa (Gürcke) Braun et Esteves	T	C Trc	?	?	?	?	MM	?	?	?	?	Moraes et al. 2005
Ferocactus cylindraceus (Engelm.) Orcutt	G	C Cac	2 D	SC	O	Fruit set, seeds/fruit	CE	?	?	+	Melittophily	McIntosh 2002
Ferocactus histrix (DC.) Linds.	G	C Cac	4–6 D	?	O	?	?	?	?	+	Melittophily	del Castillo 1994
Ferocactus robustus (Pfeiff.) Britton & Rose	G	C Cac	2 D	SI	M	Fruits, seeds, germination	?	?	?	+	Melittophily	Piña 2000
Ferocactus wislizenii (Engelm.) Britton & Rose	G	C Cac	2 D	SC	O	Fruit set, seeds/fruit	CE	?	?	+	Melittophily	McIntosh 2002
Grusonia bradtiana (Coult.) Britton & Rose	S	Opt	1 D	SC	M	Fruits, seeds	?	No	?	+	Melittophily	Plasencia-López 2003

Hatiora gaertneri (Regel) Barthlott	E	C Rhp	?	GSI	O	?	?	?	?	?	?	Boyle 2003
Hatiora rosea (Lagerheim) Barthlott	E	C Rhp	?	GSI	O	?	?	?	?	?	?	Boyle 2003
Hatiora x graeseri Barthlott ex Hunt	E	C Rhp	?	GSI	O	?	?	+	?	?	?	Boyle 2003
Hylocereus polyrhizus (Weber) Britton & Rose	E	C Hyl	?	SC	M	?	?	?	?	?	?	Lichztenzveig et al. 2000
Hylocereus undatus (Haw.) Britton & Rose	E	C Hyl	?	SC	M	?	CE	?	?	?	Two pollinators	Lichztenzveig et al. 2000
Lophocereus schotii (Engelm.) Britton & Rose	T	C Pac	1 N	SI	O	?	CE, MM	+	?	+	Phalaenophily	Fleming and Holland 1998; Parker and Hamrick 1992
Mammillaria grahamii Engelm.	G	C Cac	1–2 D	SI	O	No	CE	?	?	+	Melittophily	Bowers 2002
Marginatocereus marginatus (DC.) Backeb.	C	C Pac	1 D	SI	O	No	?	+	?	+	?	Dar et al. 2006
Melocactus concinnus Buining & Brederoo	G	C Cer	?	?	?	?	MM	+	?	?	Ornithophily	Mota et al. 2006
Melocactus curvispinus Pfeiff.	G	C Cer	1 D	SC	M	Fruit set	MM	+	?	+	Ornithophily	Nassar and Ramírez 2004; Nassar et al. 2001
Melocactus paucispinus Heimen & Paul	G	C Cer	?	SC	M	Fruits	MM	?	?	?	Ornithophily	Mota et al. 2006
Neobuxbaumia macrocephala (Weber ex Schum.) Dawson	C	C Pac	1 N	SI	O	No	CE	?	?	+	Chiropterophily	Valiente-Banuet et al. 1997

(*continued*)

Table 10.2 (continued)

Species	Life form	Subfamily-tribe	Floral longevity (days)	Compatibility	Mating system morphological-functional	Inbreeding depression	Outcrossing rate estimation method	H	D	Nectar	Pollination syndrome	Reference
Neobuxbaumia mezcalaensis (Bravo) Backeb.	T	C Pac	1 N	SI	O	No	CE	?	?	+	Chiropterophily	Valiente-Banuet et al. 1997
Opuntia basilaris Engelm. & Bigelow	S	Opt	?	?	?	?	?	+	+	?	?	Grant and Grant 1979
Opuntia compressa (Salisb.) MacBr.	S	Opt	1 D	SC	M	?	?	?	?	No	?	Grant et al 1979
Opuntia discata Griffiths	S	Opt	1 D	SC	M	?	?	?	?	No	?	Grant et al. 1979
Opuntia lindheimeri Englem.	S	Opt	1 D	SC	M	?	?	?	?	No	?	Grant et al. 1979
Opuntia littoralis (Engelm.) Cockerell	S	Opt	?	?	?	?	?	?	?	+	?	Grant and Grant 1979
Opuntia macrocentra Engelm.	S	Opt	1 D	SC	M	Fruits (0.07), seeds (0.53)	CE	?	?	?	Melittophily	M.M. et al. unpublished data
Opuntia microdasys (Lehm.) Pfeiff	S	Opt	1 D	SI	O	?	CE	?	?	?	Melittophily	Piña et al. 2007
Opuntia phaeacantha Engelm.	S	Opt	1 D	SC	M	No	CE	?	?	No	?	Grant et al. 1979; Osborn et al. 1988
Opuntia polyacantha Haw.	S	Opt	1 D	SI	O	?	CE	?	?	?	Melittophily	Osborn et al. 1988
Opuntia rastrera Weber	S	Opt	1 D	SC	M	Fruit set, germination, seedlings, seeds/fruit	CE	+	?	+	Melittophily	Mandujano et al. 1996; Plasencia-López 2008
Opuntia robusta Wendl. ex Pfeiff.	S	Opt	1–2 D	SC	M	No	?	?	?	+	Melittophily	del Castillo and González-Espinosa 1988

Pachycereus pecten-aboriginum (Engelm.) Britton & Rose	C	C Pac	1 N	SI	O	?	CE	?	?	?	Two pollinators	Molina-Freaner et al. 2004
Pereskia guamacho Weber	T	Per	1 D	SI	O	?	MM	?	?	?	Melittophily	Nassar et al. 2002
Pilosocereus aureispinus (Buining & Brederoo) Ritter	S	C Cer	N	?	?	?	MM	?	?	?	Chiropterophily	Moraes et al. 2005
Pilosocereus lanuginosus (L.) Byles & Rowley	T	C Cer	1 N	SI	O	?	MM	?	?	?	Chiropterophily	Nassar et al. 2003
Pilosocereus machrisii (Dawson) Backeb.	S	C Cer	N	?	?	?	MM	?	?	?	Three pollinators	Moraes et al. 2005
Pilosocereus moritzianus (Otto.) Byles & Rowley	T	C Cer	1 N	SC	M	?	CE	?	?	?	Chiropterophily	Nassar et al. 1997
Pilosocereus royenii (L.) Byles & Rowley	T	C Cer	1 N	SC	M	Seeds	CE	?	?	+	Chiropterophily	Rivera-Marchand and Ackerman 2006
Pilosocereus vilaboensis (Diers et Esteves) Braun	S	C Cer	N	?	?	?	MM	?	?	?	Chiropterophily	Moraes et al. 2005
Polaskia chende (Gosselin) Gibson & Horak	T	C Pac	1 D	SI	O	?	CE	?	?	+	Melittophily	Cruz and Casas 2002
Polaskia chichipe (Gosselin) Backeb.	T	C Pac	1 D*	SC	M	Seeds	MM	?	?	+	Two pollinators	Otero-Arnaiz et al. 2003
Praecereus euchlorus (Weber) Taylor	E	C Cer	?	?	?	?	MM	?	?	?	Chiropterophily	Moraes et al. 2005
Pterocereus gaumeri (Britton & Rose) McDoug. & Miranda	S	Opt	1 N	SI	O	?	CE	?	?	?	Chiropterophily	Méndez et al. 2005

(*continued*)

Table 10.2 (continued)

Species	Life form	Subfamily-tribe	Floral longevity (days)	Compatibility	Mating system morphological-functional	Inbreeding depression	Outcrossing rate estimation method	H	D	Nectar	Pollination syndrome	Reference
Schlumbergera russelliana (Hook.) Britton & Rose	E	C Rhp	?	GSI	O	?	?	?	?	?	?	Boyle 1997
Schlumbergera truncata (Haworth) Moran	E	C Rhp	?	GSI	O	?	?	?	?	?	?	Boyle 1997, 2003
Schlumbergera x buckleyi (Bukley) Tjaden	E	C Rhp	?	GSI	O	?	?	?	?	?	?	Boyle 2003
Selenicereus megalanthus (Schum. ex Vaupel) Moran	E	C Hyl	?	SC	M	?	CE	?	?	?	Two pollinators	Lichztenzveig et al. 2000
Stenocereus eruca (Brandegee) Gibson & Horak	Cy	C Pac	1 N	SI	O	No	MM	?	?	+	Two pollinators	Clark-Tapia and Molina-Freaner 2004
Stenocereus griceus (Haworth) Buxbaum	T	C Pac	1 N	?	?	?	MM	?	?	?	Chiropterophily	Hamrick et al. 2002; Nassar et al. 2003
Stenocereus gummosus (Engelm. ex Brandegee) Gibson & Horak	S	C Pac	N	SI	?	?	MM	?	?	?	Phalaenophily	Clark-Tapia 2000
Stenocereus queretaroensis (Weber) Buxbaum	T	C Pac	1 N*	SI	O	?	CE	?	?	+	Chiropterophily	Ibarra-Cerdeña et al. 2005
Stenocereus stellatus (Pfeiff.) Riccob.	C	C Pac	1 N	SI	O	?	CE	?	?	+	Chiropterophily	Casas et al. 1999
Stenocereus thurberi (Engelm.) Buxbaum	C	C Pac	?	?	?	?	MM	?	?	?	Two pollinators	Hamrick et al. 2002

Thelocactus hastifer (Werderm. & Boed.) Knuth.	G	C Cac	?	?	?	?	?	?	?	?	?	Ramírez-Corona 2000
Thelocactus tulensis (Poselg.)Britton & Rose	G	C Cac	?	?	?	?	?	?	?	?	?	Ramírez-Corona 2000
Turbinicarphus horripilus (Lem.) Vác. John & Riha	G	C Cac	2 N	?	O	?	?	?	?	+	?	Matías-Palafox 2007

styles and male flowers have long styles with an atrophied stigma (del Castillo and González-Espinoza 1988).

There are other peculiar examples in which herkogamy can be found in some populations. In *Peniocereus greggi*, SI is present regardless of herkogamy (Raguso et al. 2003) as well as in *Echinocactus platyacanthus* which is self-compatible (Jiménez-Sierra 2008). In addition, the presence of heteromorphic flowers may occur during the reproductive season. For example, flowers of *Opuntia lindheimeri*, *O. discata* and *O. phaeacantha* that are produced during the flowering peak display herkogamy and are allogamous, whereas those produced at the end of the floral season are both homogamous and autogamous (Grant et al. 1979). There are also examples of cacti that do not show herkogamy, e.g., *Opuntia microdasys* (Piña et al. 2007), *O. rastrera* (Mandujano et al. 1996), *Stenocereus griseus, Subpilocereus repandus* and *Subpilocereus horrispinus* (Nassar et al. 1997), which has been associated with SI.

Some cacti species have been reported to be dichogamous with cases of protandry (Table 10.2). As pollen is easier to detect, determinations are based on observations of pollen release from anthers, and no attempts have been made to quantify pollen viability or stigma receptivity, which causes pollen to adhere and germinate. In addition, low fruit set in hand pollination experiments can be related to the timing of stigma receptivity, leading to unreliable results. In general, dichogamy is an almost unexplored trait; only four of the published reproductive biology studies specifically addressed the presence of dichogamy, despite the fact that it determines the optimal state of male and female floral functions. The evidence suggests that temporal separation is short in some species (hours), for example the stigma of *Hylocereus* spp. becomes receptive 3 h after it sheds its pollen (Pimienta-Barrios and del Castillo 2002) – a pattern also found in *Pilosocereus royenii* (Rivera-Marchand and Ackerman 2006). In other instances, separation between sexual functions within a flower can be longer (e.g., days in *F. histrix*, del Castillo 1994) and the species may have stigma lobes that remain closed until the onset of receptivity. Unfortunately, in most cases, dichogamy is not easy to determine. Species that open stigma lobes are *F. histrix* and *E. platyacanthus*, while others species have stigma lobes that never open, even though there are receptive (e.g., *O. rastrera*, *Grusonia bradtiana*, and *O. microdasys*, in which receptivity was determined with pollen adherence). Dichogamy can also be a plastic trait. Flowers of *A. trigonus* may last for 1–3 days; pollen is released soon after the flower opens on the first day, but pollen can adhere to the stigmatic surface of the same flower only at the time of floral closure. By the 2nd day, self-pollen is depleted from anthers, and stigmas remain receptive to foreign pollen (M.M., personal observation). Climatic conditions may also affect stigma receptivity, as in some *Ariocarpus* species cold days prevent stigmas from being pollinated in both manual pollination and open pollination treatments (C.M.-P., personal observation). Lack of dichogamy is often related to difficulties in determining stigma receptivity because standard techniques are usually destructive (i.e., enzymatic reactions for peroxidase enzymes or esterases, collecting pistil for pollen tube growth). Given the endangered status of many species, destructive methods are rarely, if at all, used whilst other methods involve complex experimental designs. The oldest and probably most complex non-destructive method is to hand-pollinate flowers at different times of day, or over a period of days, to evaluate fruit and seed production (Kearns and Inouye 1993). This is definitely a field that deserves attention and sound experimental research to accurately describe the duration and function of female sexual expression in cacti flowers.

10.3.1 Self Incompatibility

Outbreeding and inbreeding have important effects on the progeny formed, thus syngamy is affected by gamete identity and relatedness. Perhaps the most important strategy to select mates in angiosperms is the presence of an SI system. In plants bearing genetic SI, self-fertilization and inbreeding are prevented by the gene products of the *S*-locus, which preclude reproduction between individuals sharing SI alleles (de Nettancourt 1997, 2001). Perfect flowers display SI to avoid self-fertilization, which involves a cell–cell recognition system between maternal and parental genotypes that is used to regulate the acceptance or rejection of pollen landing on the stigma of the same species or in the germination/inhibition of pollen tube growth along the style (Lovett-Doust and Lovett-Doust 1988; Franklin-Tong and Franklin 2003; McClure and Franklin-Tong 2006). There are two major classes of SI at the genetic level: gametophytic SI (GSI) and sporophytic SI (SSI). GSI is so-called because the incompatibility phenotype of the pollen is determined by its haploid (gametophytic) genotype, whereas with SSI the pollen exhibits the incompatibility phenotype of its diploid (sporophytic) parent (Franklin-Tong and Franklin 2003). SI differs in the evolution of plant families – close to 60% of angiosperms display some kind of SI, and at least 68 plant families include members with SI (de Nettancourt 2001; Hiscock and Tabah 2003; Ferrer and Good-Avila 2007). Self-incompatibility is by far the most effective mechanism used by plants to prevent self-fertilization and consequent inbreeding. Other mechanisms such as herkogamy, dichogamy and different floral arrangements at the individual or population level do not prevent inbreeding (Wyatt 1983; Dafni 1992) as mediated or autonomous self-pollination remains possible (Barrett 2003). The latter strategies act more as promoters of pollen exchange than as barriers to selfing (Lloyd and Webb 1986; Webb and Lloyd 1986). This subtle difference explains why plants that display both dichogamy and herkogamy can also have SI (e.g., *Schlumbergera* show SI; Boyle 1997). However, some species, like *Opuntia excelsa* (Bullock 1985) and *Opuntia microdasys* (Piña et al. 2007), are homogamous and display SI. Self-incompatibility is not a discrete trait; plant populations may vary from being strictly SI to showing intra- or inter-population variation in the strength of SI. This is known as partial or pseudo self-compatibility/SI (Levin 1996; Ferrer and Good-Avila 2007) – a characteristic that has not been addressed in the cactus family. The ultimate goal of SI is to avoid selfing through the rejection of pollen grains coming from the same flower (autogamy), flowers from the same plant (geitonogamy), or from genetically related plants. Self-incompatibility has been classified according to floral morphology within populations (homomorphic and heteromorphic) and the type of recognition (SSI and GSI; de Nettancourt 1997). Homomorphic systems refer to populations in which all flowers display the same floral morphology (e.g., *Ariocarpus fissuratus*; C.M.-P. and M.M., unpublished data), while in heteromorphic species individuals within a population differ in floral morphology (i.e., distyle or tristyle; e.g., *Opuntia robusta*; del Castillo and González-Espinosa 1988). In addition to these divisions it has been suggested that a wet stigma surface and bi-nucleated pollen leads to GSI while a dry stigma surface and tri-nucleate pollen leads to SSI (de Nettancourt 1997; Heslop-Harrison and Shivanna 1977). In the cactus family, SI systems have been barely studied; however, some authors have suggested that SI is widespread among the Cactaceae (Strong and Williamson 2007), and preliminary evaluations suggest SI in 28 out of 98 genera from three subfamilies: Pereskioideae, Opuntioideae and Cactoideae (Boyle 1997). However, our review of the literature revealed only three formal reports in which the SI system is determined with specialized techniques. The results show GSI, with tri-nucleated pollen and dry stigmas (Table 10.2) – traits that in general have been associated with SSI. GSI is controlled by a single multi allelic locus in *Schlumbergera truncata*, *S. russelliana*, *S.* x *buckleyi* (Boyle 1997, 2003), *Echinopsis chamaecereus* (Boyle and Idnurm 2001), *Hatiora rosea*, *H. gaertneri* and *H.* x *graeseri* (Boyle 2003).

Other authors have proposed the presence of SI as responsible for null fruit set in self-pollination experiments (Table 10.2) in *Ferocactus cylindraceus* (McIntosh 2002), *Echinomastus erectocentrus* (Johnson 1992), *Stenocereus eruca* (Clark-Tapia and Molina-Freaner 2004), *Pereskia guamacho* (Nassar et al. 2002), *Trichocereus pasacana* (Badano and Schlumpberger 2001), *Astrophytum asterias* (Strong and Williamson 2007), *Neobuxbaumia mezcalensis* and *N. macrocephala* (Valiente-Banuet et al. 1997), *Hylocereus polyrhizus* (Lichtenzveig et al. 2000), *Mammillaria grahamii* (Bowers 2002), *Lophocereus schottii* (Fleming and Holland 1998; Holland and Fleming 1999), *Stenocereus stellatus* (Casas et al. 1999), *S. eruca* (Clark-Tapia and Molina-Freaner 2004), *S. queretaroensis* (Ibarra-Cerdeña et al. 2005), *Pterocereus gaumeri* (Méndez et al. 2005), *Escontria chiotilla* (Oaxaca-Villa et al. 2006), and the revision of 55 taxa by Ross (1981). There are also several informal reports suggesting that SI is widespread in the cactus family from collectors and gardeners who observed low or null fruit for several species of *Mammillaria*, *Astrophytum* and *Ariocarpus* under cultivation. However, it is important to bear in mind that, without the genetic determination of *S*-alleles, pollen tube growth experiments, and protein determination of cell–cell reactions, it is impossible to tell if a null fruit set is due to the presence of SI or to extreme inbreeding depression (Uyenoyama 1993). Evidence in any case suggests that outcrossing may be favored in the cactus family because of high inbreeding depression or the presence of partial or complete SI among the species. This confounding aspect can be related to bottlenecks that reduce population size. For example, *A. asterias* has a remnant population that has been described as having SI (Strong and Williamson 2007); however, Martínez-Ávalos (2007) conducted similar controlled pollination assays in a much denser population and found ca. 30% fruit set in autogamy treatments, suggesting partial SI. Delayed self-pollination has been detected in *Grusonia bradtiana* where cross-pollen tubes start growing during the first hours, and first 12 h after floral anthesis, while self-pollen grains did not grow until 24 h after flower closure (Plasencia-López 2003). Clearly, more detailed studies are needed to establish trends in the breeding systems of the Cactaceae.

10.4 Mating Systems

Mating system studies primarily address genetic issues associated with inbreeding depression or statistical characterization of mating patterns. The patterns of transmission of genes through one mating cycle reflect the mating actually taking place in the population and summarize the mating behavior of the plant population as a whole. The models can specify differences between different hierarchical levels among or within species, populations or individuals. Mating systems include three broad categories that differ from random mating: predominantly outcrossing, mixed selfing and outcrossing, and predominantly selfing (but see Brown 1990). Populations of many species experience a significant amount of selfing and outcrossing (mixed mating, Barrett 2003), and therefore plant mating systems should be considered as a continuum of possibilities that allows plants to deal with the varying pollinator/resource availability and environmental heterogeneity that characterizes deserts in order to produce offspring.

Models of mating system evolution predict that a large number of species will be predominantly selfing, but many examples show that selfing is widespread in annuals and less frequent in perennial species (Barrett and Eckert 1990; Barrett et al. 1996). However, studies have shown that selfing can evolve in annual plant species as well as in perennial species in the same taxonomical family (see Ferrer and Good-Avila 2007). Members of the cactus family are all perennials and non-herbaceous (Mauseth 2006), but we can elaborate hypotheses based on life cycle duration. In general, we can associate life cycle duration with life form (i.e., globose, expansive, columnar, cylindrical, shrub-like and treelike; Gibson and Nobel 1986; Anderson 2001; Fig. 10.2), which in turn is correlated with other life history components such as size at first reproduction, number and size of offspring, rates of growth, survival, etc (Mandujano et al. 2001, 2007; Rosas-Barrera and Mandujano 2002). For example, treelike, columnar and shrub-like species (similar to perennial trees) like *Carnegiea gigantea*, *Pereskia* spp. or *Opuntia rastrera*, which have long life cycles (thousands of years), many seeds per fruit, and large and fast-growing seedlings in comparison with globose species like *Ariocarpus scaphirostris*, *A. asterias* and *Mammillaria hernandezii*, which have short life cycles, small and slow-growing seedlings (Rosas-Barrera and Mandujano 2002). Following this scheme, we divided the family into long-lived species and short-lived species. Using this characterization, we contrast the mating systems, expecting more outcrossing species in long-lived species (Fig. 10.1). The outcrossing rates (t) were estimated using two sources of information. First, indirect estimations were

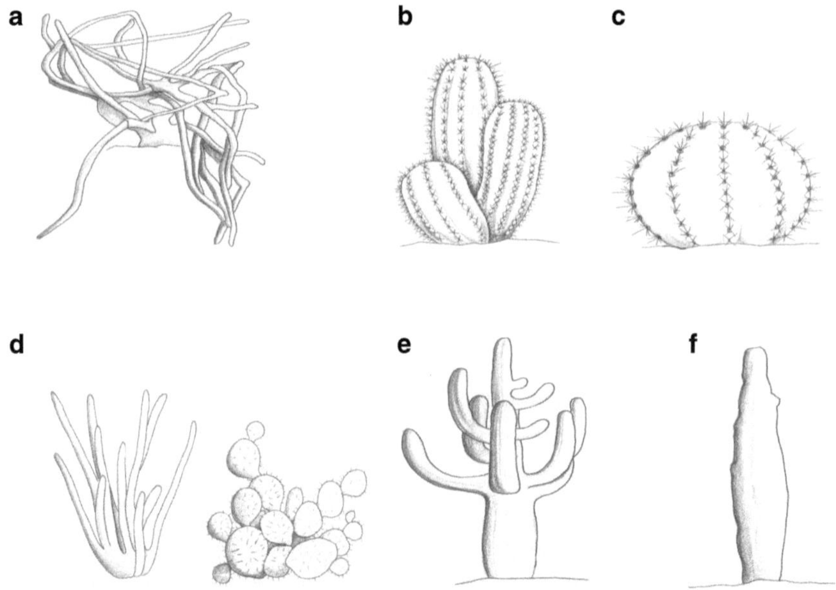

Fig. 10.2a–f Principal life forms in the Cactaceae. **a** Expansive (e.g., *Rhipsalis baccifera, Pseudorhipsalis ramulosa*). **b** Cylindrical (e.g., *Ferocactus histrix, Echinocereus laui*). **c** Globose (e.g., *Ortegocactus macdougallii, Melocactus curvispinus, Mammillaria* spp.). **d** Shrub-like (e.g., *Opuntia basilaris, Ferocactus robustus, Echinopsis cephalomacrostibas*). **e** Tree-like (e.g., *Pereskia sacharosa, Pachycereus weberi, Myrtillocactus geometrizans*). **f** Columnar (e.g., *Carnegiea gigantea, Cephalocereus columna-trajani, Trichocereus pachanoi*)

obtained from population genetic studies (some sources specifically reported t but in other cases it was estimated from the reported heterozygosity; Brown 1990):

$$F = (H_e - H_o)/H_e \tag{10.1}$$

$$t = (1 - F)/(1 + F) \tag{10.2}$$

where F is the fixation index, and H_o, H_e are the observed and expected heterozygosity (genetic diversity; Brown 1990; Hamrick and Godt 1989).

The second source of information was direct quantification of the mating system through pollination experiments. In these cases we estimated the outcrossing rate (t_e) using the fruit set or seed set (depending on reported information available) for selfing (w_s) and outcrossing treatments (w_x; following our unpublished observations):

$$t_e = 1 - s \tag{10.3}$$

where s is the selfing rate estimated by:

$$s = (d_r - w_X)(ws_s) / w_X - w_s) \quad (10.4)$$

The selfing rate is a function of $ws_s = w_s / (w_X + w_s)$, which represents the proportion of selfing relative to the total plant fitness reached by both selfing and outcrossing, inbreeding depression $\delta = 1/(w_X - w_s)$ (see below) and $d_r = w_X + (w_X - w_s)$ is the discounting rate (i.e., the fraction of outcrossing that the plant lost by selfing).

Substituting the values ws_s and d_r in Eq. 10.4, it can be seen that s and t_e can be obtained directly as:

$$s = w_s / (w_X + w_s) \quad (10.5)$$

$$t_e = w_X / (w_X + w_s) \quad (10.6)$$

Outcrossing rates vary from 0 to 1, i.e., 0 in species with selfing, 1 in species with outcrossing and 0.5 for species with mixed mating systems. So values > 0.5 tend towards outcrossing and < 0.5 to selfing (Barrett and Eckert 1990; Barrett et al. 1996; Table 10.2).

Our revision includes 70 study cases, 36 studies report population genetics, but only 21 of these provided enough information to estimate t. On the other hand, 30 studies conducted pollination experiments and included the results of fruit set from selfing and outcrossing treatments to estimate t_e. We pooled the sample by subfamily, pollination syndrome and life form to explore mating system patterns (Fig. 10.1).

Pereskioideae is found among the species leaning towards outcrossing. We found that selfing is more widespread in two clades of the family tree, Cactoideae and Opuntioideae; however, some species, like *O. microdasys* and *S. eruca*, have SI and are found as complete outcrossers. Interestingly, species that have been reported in these subfamilies display high levels of inbreeding depression at the level of seed and fruit set. Patterns are clearer when segregating species by tribe; within Cactoideae, Pahcycereeae are mainly outcrossers, but a few species, which display diurnal and nocturnal flowers, tend toward selfing. Species of tribe Cereeae show mostly mixed mating systems, tending towards selfing, with a few representatives having outcrossed mating systems. Long-lived tree-like and shrub-like species are usually outcrossers, similar to other perennial species (Barrett and Eckert 1990). According to our expectations, species with shorter lifespans (e.g., globose) are not outcrossers, and tended towards mixed and selfing mating systems while long-lived species appear to be mainly outcrossers. Shrubby species, which are also long-lived, can be found having a mixed mating system with a tendency towards selfing, and a few species are obligate outcrossers (Fig. 10.3a, b). Effective population size could possibly reduce the possibilities of evolving towards outcrossing. Several globose species face reduced population size or are rare species (*sensu* Rabinowitz). Autogamy in these cases can be seen, as proposed by Darwin (1876), as an insurance against environmental variability and lack of pollinators.

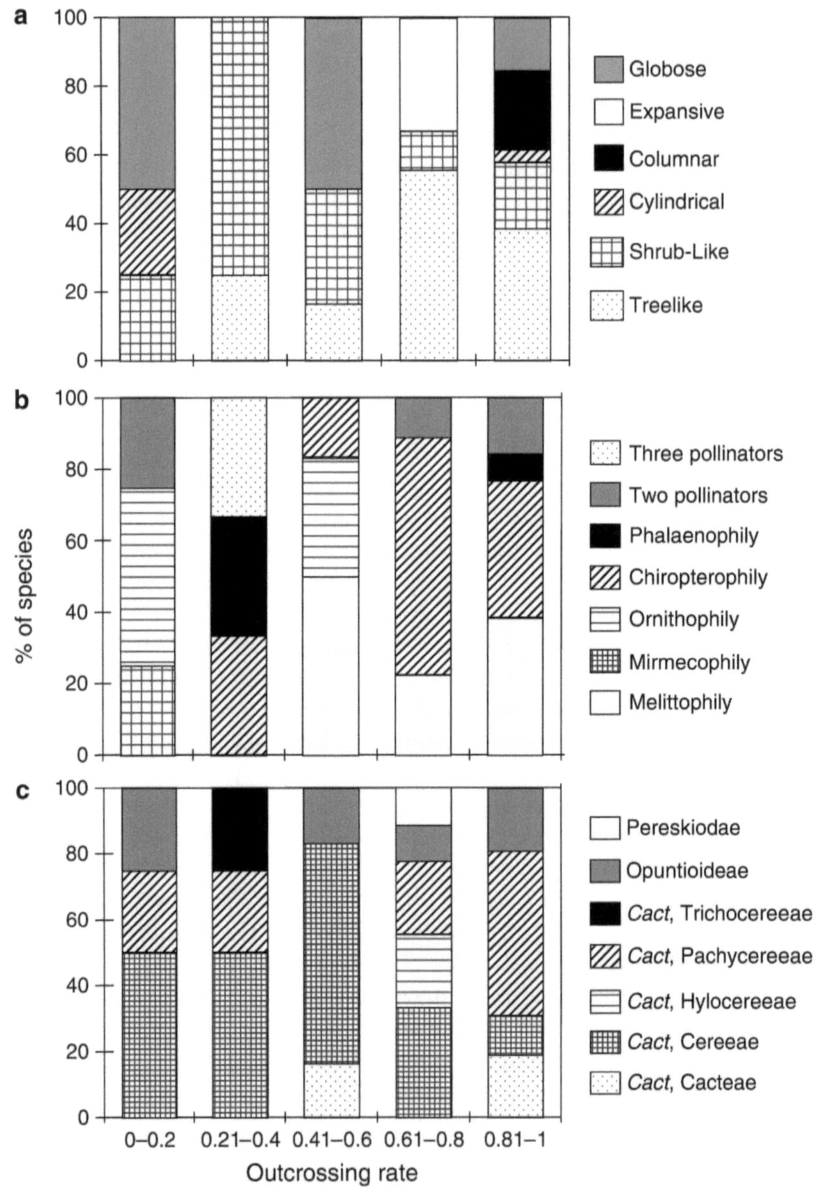

Fig. 10.3 Relationships between outcrossing rate and **a** life form, **b** pollination syndrome, **c** subfamily, tribe for 49 species of Cactaceae. *Cact* Cactoideae. See text for further details

10.4.1 Inbreeding Depression

The harmful effects of inbreeding have been noticed since Darwin (1859) – effects that can appear during early development or later in the life cycle (Holsinger 1992). The relative performance of selfed and outcrossed progeny can be inferred from genetic data or can be measured directly for some traits such as ovule fertilization, fruit set, seed set, seedling survival, juvenile survival, age at first reproduction, reproductive effort of produced offspring, etc. In all cases, the main goal is to assess the average fitness effect of inbreeding in the population (Holsinger 1992). Population inbreeding depression is defined as:

$$\delta = 1 - w_s/w_x \tag{10.7}$$

where w_s is the average fitness of selfed progeny and w_x is the average fitness of outbred progeny (Charlesworth and Charlesworth 1987; Holsinger 1992).

Selfing will purge populations of deleterious alleles, so that only low levels of inbreeding depression are to be expected in partially self-fertilizing species. Models of selfing vs outcrossing have included inbreeding depression as an explicit parameter. Lloyd (1979) showed that inbreeding depression is a critical parameter determining the outcome of selection that affects selfing rate. Outcrossing is favored when inbreeding depression > 0.5, whereas selfing is favored and would tend towards fixation when the value is < 0.5. Breeding systems in cosexual species are thought of as measures to avoid the negative effects of inbreeding. Early inbreeding depression is invoked in partially selfed species in which authors report reduced output in selfing treatments in comparison with outcrossing treatments. This reduction can be observed at the level of fruit set (e.g., *Ariocarpus fissuratus*, Martínez-Peralta 2007; and *Melocactus curvispinus*, Nassar and Ramírez 2004), seed production (e.g., *Ferocactus robustus*, Piña 2000; *Ferocactus cylindraceus* and *F. wislizeni* McIntosh 2002; *Polaskia chichipe*, Otero-Arnaiz et al. 2003; *Pilosocereus royenii*, Rivera-Marchand and Ackerman 2006), germination percentage (e.g., *Ferocactus robustus*, Piña 2000; *Hylocereus undatus*; and *Selenicereus megalanthus*, Lichtenzveig et al. 2000) or seedling survival (e.g., *Opuntia rastrera*, Mandujano et al. 1996).

Based on the few studies that report or provide ecological information to estimate δ, average inbreeding depression in the cactus family is 0.82 ± 1 SD 0.26 (*n*= 27). This high level of inbreeding depression suggests that outcrossing will be favored within the cactus family. A close approximation, segregating species by subfamily, indicates similar patterns within Opuntioideae and Cactoideae (no data was found for Pereskioideae). We also explored the variance in inbreeding depression for tribes; the highest inbreeding depression was found for Pachycereeae (δ = 0.97 ± 1 SD 0.07) in which outcrossing mating systems and SI predominate, and the lowest in Hylocereeae (two species) and Cereeae (δ = 0.57 and 0.61, respectively), in which herkogamy and dichogamy have been recorded. Large variances are associated with inbreeding depression, which suggests the strength of harmful effects due to selfing in some cases and in others inbreeding has not been detected.

10.5 Pollination Syndromes

Cactus flowers are animal-pollinated (Pimienta-Barrios and del Castillo 2002). However, few species are able to set fruit in the absence of floral visitors (Table 10.2), for example *Opuntia macrocentra* (M.M. et al. unpublished data) and *Melocactus curvispinus* (Nassar and Ramírez 2004). Pimienta-Barrios and del Castillo (2002) proposed a close relationship between flower traits and pollinators, based on flower longevity, time of aperture, color, shape and nectar. In our review of the literature, we found a close match between these traits, but several species do not follow this trend. Floral visitors of cactus blossoms include bats, birds, bees (native and introduced), wasps, moths, beetles, grasshoppers and ants (Table 10.2). Cacti species have been thought of as a group specialized in animal pollination because floral traits are well adapted to animal behavior (Gibson and Nobel 1986; Mandujano et al. 1996; Valiente-Banuet et al. 1996, 2002). Pollination mediated by bats and moths is frequent in columnar species (Badano and Schlumpberger 2001) as in *Stenocereus griseus*, *Pilosocereus moritzianus*, *Subpilocereus repandus*, *Subpilocereus horrispinus* (Nassar et al. 1997), *N. macrocephala* (Valiente-Banuet et al. 1997), *S. eruca* (Clark-Tapia and Molina-Freaner 2004), *Pachycereus pringlei*, *C. gigantea* and *L. schottii* (Fleming et al. 2001). These species are night blooming, with white large flowers that produce copious amounts of nectar and pollen as floral rewards. Some species, however, can be also open diurnally with efficient visitors (Molina-Freaner et al. 2004). Bee-pollinated species are common in Opuntioideae and Pereskioideae (e.g., *P. guamacho*, Nassar et al. 2002) and solitary bees (mainly *Diadasia* and *Lithurge*) have been especially suggested oligolectic to *Opuntia* (Mandujano et al. 1996; Reyes-Agüero et al. 2006; Piña et al. 2007). Other columnar species can be also pollinated by bees (*Polaskia chende*; Cruz and Casas 2002), and bees are important visitors to globose cacti (e.g., *Echinomastus erectocentrus*; *A. asterias*; Johnson 1992; Strong and Williamson 2007). There is no clear pattern according to mating system but some tendencies can be proposed. Bee-pollinated species are found along the whole outcrossing gradient (selfing to outcrossing). Bat-pollinated species are located in the range of mixed towards outcrossing mating, but a few show a tendency towards selfing. On the other hand, species pollinated by birds are found skewed towards a mixed mating system with a tendency towards selfing. Non-specialized species tend towards selfing and species visited by ants appear as selfers (Fig. 10.3b). However, it is important to bear in mind that our findings are exploratory because few studies provide the list of floral visitors and information to estimate outcrossing rates ($n = 49$). Pollination syndromes provide great utility in understanding the mechanisms of floral diversification. Our conclusions are based on the importance of organizing pollinators into functional groups according to presumed similarities in the selection pressures they exert. For example, *C. gigantea* and *Stenocereus thurberi* in the Sonoran desert have modified the timing of anthesis and pollination system from a specialized *chiropterophily*, to generalist systems that allow nocturnal and diurnal floral visitors. In contrast, the sister species *Neobuxbaumia mezcalaensis* and *Stenocereus martinezii* have maintained the original bat pollination system (Fleming et al. 2001). A second possibility is associated to specificity of floral visitor but attributed to flower size or color and functional herkogamy, which increases pollination efficiency by more often attracting pollinators; this can be one route towards diversification in *Opuntia*. In general, flowers of the *Opuntia*, *Consolea* and *Cylindropuntia* show a clear match with melittophily (Rebman and Pinkava 2001; Reyes-Agüero et al. 2006). Areas with high bee species richness correlate with the presence of Opuntioideae (Reyes-Agüero et al. 2006), and variations in flower size match the size of bees that function as effective pollinators (Grant and Hurd 1979; Reyes-Agüero et al. 2006). Environmental heterogeneity, association with different kinds of pollinators, and limited gene flow among cacti populations are probably responsible for the diversification of some species, because they modify florescence periods depending on factors such as rainfall and temperature, among others, that function as barriers to gene flow. Species can live in the same environment, yet show very limited gene flow due to different floral periods. For example, in *Cylindropuntia spinosior* in Arizona, whose lowland populations start flowering before mountain populations, total blooming lasting from April to June (Grant and Grant 1971).

Some visitors that have been observed are considered nectar thieves with undetermined function. Beetles (e.g., Nitidulidae) are commonly found in Opuntioideae flowers, which may favor selfing (Mandujano et al. 1996), while other beetles (e.g., Tenebrionidae and Meloidae), ants and grasshoppers consume floral parts or complete flowers by florivory, which can ultimately reduce plant fitness. Such observations are common but there is a need for further research to completely understand the role of all floral visitors.

10.6 Clonal Reproduction

Aside from the wide variation in sexual reproductive strategies in the cactus family, several species, mainly within Opuntioideae and Cactoideae, are able to generate new offspring by clonal propagation (i.e., ramets), e.g., *Stenocereus eruca* (Clark-Tapia et al. 2006), *Lophocereus schottii* (Parker and Hamrick 1992), *Ferocactus robustus* (Carrillo-Angeles 2006), *Opuntia rastrera* (Mandujano et al. 1996, 2001), *Opuntia microdasys* (Piña et al. 2007), *Cylindropuntia* spp. (Anthony 1954; Rebman and Pinkava 2001), *O. echios* (Hicks and Mauchamp 1999), *O. microdasys* (Palleiro et al. 2006), among others. The literature on the evolution of clonal propagation and its effects on population ecology, genetics, and recently on mating system and breeding systems, is constantly increasing. In the context of this chapter, addressing clonal propagation is relevant because the pattern of clonal growth can influence mating systems dramatically, favoring either outcrossing or selfing. Some patterns of clonal growth favor an intermingled distribution of ramets of different genets, increasing the probability of outcrossing. Similarly, the aggregation of ramets can result in higher floral display in one genet and the attraction of a higher number of pollinators who carry pollen of other genets (Charpentier 2002). In contrast, proximity between ramets of the same genet can result in fitness decrease both at genet and ramet level by geitonogamic crosses, particularly in obligate xenogamous species (Handel 1985). In self-compatible species, this type of cross can encourage the expression of deleterious alleles, give rise to offspring with low fitness, and reduce the availability of pollen for outcrossing (Holsinger 1992; Charpentier 2002), whereas in self-incompatible and semi-incompatible species the costs of geitonogamy are associated with saturation of the stigma with incompatible pollen, blocking of the styles and abortion of self-fertilized ovules (Charpentier 2002). Four studies have explored the effects of clonality on mating systems within the Cactaceae (*Stenocereus eruca*; Clark-Tapia et al. 2006; *Opuntia rastrera*; Plasencia-López 2008; *Opuntia microdasys*; Piña 2000; and *Ferocactus robustus*; Carrillo-Angeles 2006). All studies revealed a significant reduction in female fecundity when pollination occurred between ramets located at short distances, while genetic data showed high levels of similarity at those distances. The reduction in female fecundity is apparently a consequence of geitonogamy and inbreeding depression. Carrillo-Angeles (2006) studied the fruit and seed set of focal individuals located in different genetic neighborhoods of *Ferocactus robustus*, suggesting that individual ramet fitness decreases as a complex function of genetic diversity in the neighborhood with the amount and genotypes. The effect of clonality on sexual reproduction is a complex, and growing, field of research that promises interesting findings on the evolution of cactus floral traits and may provide useful links to demographic processes.

10.7 Genetic Diversity

The evolutionary potential of species relies on genetic diversity, and the ability of species to adapt to environmental change depends on the extent of genetic diversity. Mating systems affect the genetic realization of species from one generation to the next, and it has been proposed that selfing species will have low levels of genetic diversity in comparison with outcrossing or mixed mating species. However, evidence indicates similar levels of variation among species regardless of the mating system (Hamrick and Godt 1989). Theory suggests that outbreeding species will produce more heterozygous progeny, which in turn will be fitter, an argument that permeates all levels of biological studies. Information of the extent and distribution of genetic variation in Cactaceae has many practical applications, from how to collect and maintain genetically representative samples in order to conserve genetic diversity in *ex situ* collections (e.g., in botanical gardens or nurseries) to planning in situ management strategies (Jiménez-Sierra 2008). In general, genetic diversity for germplasm collections or cultivated species is low because popular ornamental cacti as well as cultivated prickly pears are grown asexually (e.g. *Echinopsis chamaecereus*, several species of *Rhipsalis* and *Schlumbergera*, *Opuntia ficus-indica*, among others; Boyle and Anderson 2002).

In the first review of genetic diversity of cacti species, Boyle and Anderson (2002) found four genetic studies performed on wild cacti species. To date, with 25 studies, we found that genetic diversity of the Cactaceae is slightly above average ($\%P = 62.98$, $A = 2.18$ and $H_o = 0.16$). Within the Cactaceae, and even within tribes, no clear pattern can be seen between genetic diversity and outcrossing rate. For example, similar percentages of polymorphic loci and heterozygosity can give a self, mixed or outcrossed mating system. This apparent decoupling may be due to recent severe population bottlenecks that have reduced genetic diversity, and the fact that the outcrossing rate t is an indirect estimate that depends largely on values of H. This high variability suggests that the mating system has changed throughout the phylogeny of the Cactaceae from the ancestral condition of *Pereskia* (an outcrosser with average genetic diversity) to extreme selfers.

10.8 Conclusions

Many attributes are associated with the evolution of breeding and mating systems in the cacti family, e.g., changes in corolla size, relative maturation times of anthers and stigmas, herkogamy, floral color, size and time of anthesis, floral visitors, SI systems, inbreeding depression, life span and life form, and they all determine the relative success of selfed and outcrossed gametes. Breeding and mating systems in the cactus family show a tendency towards outcrossing for long-lived and selfing for short-lived species. Outcrossing has long-term advantages as the population can maintain high levels of genetic diversity and preserve the possibility of producing diverse progeny (Wyatt 1983; Charlesworth and Charlesworth 1987), but species are then condemned to depend upon pollinator services. The data supports the strong dependence of cacti species on pollinator services even though a small proportion does not need pollinators. On the other hand, selfing may limit genetic diversity, but species ensure seed production with the possible coupled effect of reduced inbreeding depression of selfers (Charlesworth and Charlesworth 1987; Barrett and Eckert 1990). Species with a selfing mating system can persist over long time periods while environmental conditions remain stable – a condition that may not be met for endangered species. On the other hand, mixed mating systems promote recombination and also ensure reproduction, but face inbreeding depression and a reduction in fitness (Charlesworth and Charlesworth 1987; Barrett and Eckert 1990). We are clearly in need of detailed studies that can determine the reasons behind variation in herkogamy and the subtle effects of dichogamy. SI studies seem to be largely anecdotal and difficult to differentiate from inbreeding depression, which seems to be very high in the Cactaceae. Overall, reproductive biology of the Cactaceae is a developing flower waiting to be pollinated.

Acknowledgments We thank Francisco Molina-Freaner, Luis Eguiarte, Mariana Rojas-Aréchiga, Judith Márquez and Gisela Aguilar Morales for comments and fruitful discussions on the manuscript, sourcing some obscure results, enthusiasm for floral biology in succulent species, and their friendship. Hugo Altamirano helped with figures and in the search for lost references. Leonardo Guerrero Aguilera drew all the figures depicting flower structures and life forms. The chapter is dedicated to all those who love cacti flowers. This paper was financed partially by CONACYT-SEMARNAT 0350, PAPIIT IN205007 to M.M., and CONACyT CB-62390, CONACyT CB 83790, and BBVA-BIOCON 04–084 to J.G.

References

Anderson EF (2001) The cactus family. Timber, Portland, OR

Anthony M (1954) Ecology in the Opuntiae in the Big Bend region of Texas. Ecology 35:334–347

Arias S, Terrazas T, Cameron K (2003) Phylogenetic analysis of *Pachycereus* (Cactaceae, Pachycereeae) based on chloroplast and nuclear DNA sequences. Syst Bot 28:547–557

Arias S, Guzmán U, Mandujano MC, Soto M, Golubov J (2005a) Las especies Mexicanas de Cactáceas en riesgo de extinción: una comparación entre los listados NOM-ECOL-2001 (México), la lista roja (UICN) y CITES. Cact Suc Mex 50:100–125

Arias S, Terrazas T, Arreola-Nava HJ, Vázquez-Sánchez M, Cameron KM (2005b) Phylogenetic relationships in *Peniocereus* (Cactaceae) inferred from plastid DNA sequence data. J Plant Res 118:317–328

Badano E, Schlumpberger B (2001) Sistema de cruzamiento y estimaciones en la eficiencia de polinización sobre *Trichocereus pasacana* (Cactaceae) en dos poblaciones del Noroeste Argentino. Gayana Bot 58:115–122

Barrett SCH (2003) Mating strategies in flowering plants: the outcrossing-selfing paradigm and beyond. Philos Trans R Soc Lond B 358:991–1004

Barrett SCH, Eckert CG (1990) Variation and evolution of mating systems in seed plants. In: Kawano S (ed) Biological approaches and evolutionary trends in plants. Academic, London, pp 229–254

Barrett SCH, Harder LD, Worley AC (1996) The comparative biology of pollination and mating in flowering plants. Philos Trans R Soc Lond B 351:1271–1280

Bowers JE (2002) Flowering patterns and reproductive ecology of *Mammillaria grahamii* (Cactaceae), a common, small cactus in the Sonoran Desert. Madroño 49:201–206

Boyle TH (1997) The genetics of self-incompatibility in the genus *Schlumbergera* (Cactaceae). J Hered 88:209–214

Boyle TH (2003) Identification of self-incompatibility groups in *Hatiora* and *Schlumbergera* (Cactaceae). Sex Plant Reprod 16:151–155

Boyle TH, Anderson EF (2002) Biodiversity and conservation. In: Nobel PS (ed) Cacti: biology and uses. University of California Press, Berkeley, pp 125–141

Boyle TH, Idnurm A (2001) Physiology and genetics of self-incompatibility in *Echinopsis chamaecereus* (Cactaceae). Sex Plant Reprod 13:323–327

Bravo-Hollis H (1978) Las cactáceas de México, vol 1. Universidad Nacional Autónoma de México, Mexico

Brown AHD (1990) Genetic characterization of plant mating systems. In: Brown AHD, Clegg MT, Kahler AL, Weir BS (eds) Plant population genetics, breeding, and genetic resources. Sinauer, Sunderland, MA, pp 143–162

Bullock SH (1985) Breeding systems in the flora of a tropical deciduous forest in Mexico. Biotropica 17:287–301

Butterworth CA, Wallace RS (2004) Phylogenetic studies of *Mammillaria* (Cactaceae) insights from chloroplast sequence variation and hypothesis testing using the parametric bootstrap. Am J Bot 91:1086–1098

Carrillo-Angeles IG (2006) Distribución espacial de clones de *Ferocactus robustus*: consecuencias sobre la reproducción sexual. MSc Thesis, Universidad Nacional Autónoma de México, Mexico

Carson HL, Templeton AR (1984) Genetic revolutions in relation to speciation phenomena: the founding of new populations. Annu Rev Ecol Syst 15:97–131

Casas A, Valiente-Banuet A, Rojas-Martínez A, Dávila P (1999) Reproductive biology and the process of domestication of the columnar cactus *Stenocereus stellatus* in Central Mexico. Am J Bot 86:534–542

Charlesworth D, Charlesworth B (1979) The evolutionary genetics of sexual systems in flowering plants. Proc R Soc Lond B 205:513–530

Charlesworth D, Charlesworth B (1987) Inbreeding depression and its evolutionary consequences. Annu Rev Ecol Syst 18:237–268

Charpentier A (2002) Consequences of clonal growth for plant mating. Evol Ecol 15:521–530

Clark-Tapia R (2000) Estructura genética de dos cactáceas columnares del Desierto Sonorense: *Stenocereus gummosus* y *S. eruca* (Cactaceae). MSc Thesis, Universidad Nacional Autónoma de México, Mexico

Clark-Tapia R, Molina-Freaner F (2004) Reproductive ecology of the rare clonal cactus *Stenocereus eruca* in the Sonoran desert. Plant Syst Evol 247:155–164

Clark-Tapia R, Alfonso-Corrado C, Mandujano MC, Molina-Freaner F (2006) Reproductive consequences of clonal growth in *Stenocereus eruca*, a rare clonal cactus of the Sonoran Desert. Evol Ecol 20:131–142

Cruz M, Casas A (2002) Morphological variation and reproductive biology of *Polaskia chende* (Cactaceae) under domestication in Central Mexico. J Arid Environ 51:561–576

Dafni A (1992) Pollination ecology. Oxford University Press, Oxford

Dar S, del Coro Arizmendi M, Valiente-Banuet A (2006) Diurnal and nocturnal pollination of *Marginatocereus marginatus* (Pachycereeae: Cactaceae) in Central Mexico. Ann Bot 97:423–427

Darwin C (1859) On the origin of species. Oxford University Press, Oxford, UK

Darwin C (1876) The effects of cross and self-fertilisation in the vegetable kingdom. Adamant Media. Boston, MA

De Nettancourt D (1997) Incompatibility in angiosperms. Sex Plant Reprod 10:185–199

De Nettancourt D (2001) Incompatibility and incongruity in wild and cultivated plants. Springer, Berlin

Del Castillo R (1994) Polinización y otros aspectos de la biología floral de *Ferocactus histrix*. Cact Suc Mex 39:36–42

Del Castillo R, González-Espinosa M (1988) Una interpretación evolutiva del polimorfismo sexual de *Opuntia robusta* (Cactaceae). Agrociencia 71:184–196

Dellaporta SL, Calderon-Urrea A (1993) Sex determination in flowering plants. Plant Cell 5:1241–1251

Edwards E, Nyffeler JR, Donoghue MJ (2005) Basal cactus phylogeny: implications of *Pereskia* (Cactaceae) paraphyly for the transition to the cactus life form. Am J Bot 92:1177–1188

Eguiarte LE, Duvall MR, Learn GH Jr, Clegg MT (1994) The systematic status of the Agavaceae and Nolinaceae and related Asparagales in the monocotyledons: an analysis based on the rbcL gene sequence. Bol Soc Bot Mex 54:35–56

Esparza-Olguín L, Valverde T, Mandujano MC (2005) Comparative demographic analysis of three *Neobuxbaumia* species (Cactaceae) with differing degree of rarity. Popul Ecol 47:229–245

Ferrer MM, Good-Avila SV (2007) Macrophylogenetic analyses of the gain and loss of self-incompatibility in the Asteraceae. New Phytol 173:401–414

Fleming TH, Holland JN (1998) The evolution of obligate pollination mutualisms: senita cactus and senita moth. Oecologia 114:368–375

Fleming TH, Maurice S, Hamrick JL (1998) Geographic variation in the breeding system and the evolutionary stability of trioecy in *Pachycereus pringlei* (Cactaceae). Evol Ecol 12:279–289

Fleming T, Sahley C, Holland J, Nassar J, Hamrick J (2001) Sonoran desert columnar cacti and the evolution of generalized pollination systems. Ecol Monogr 71:511–530

Franklin-Tong VE, Franklin FCH (2003) The different mechanisms of gametophytic self-incompatibility. Philos Trans R Soc Lond B 358:1025–1032

Gibson AC, Nobel PS (1986) The cactus primer. Harvard University Press, Boston, MA

Good-Avila S, Souza V, Gaut SB, Eguiarte LE (2006) Timing and rate of speciation in *Agave* (Agavaceae). Proc Natl Acad Sci USA 103:9124–9129

Grant V, Grant KA (1971) Dynamics of clonal microspecies in cholla cactus. Evolution 25:144–155

Grant V, Grant KA (1979) Pollination of *Opuntia basilaris* and *O. littoralis*. Plant Syst Evol 132:321–325

Grant V, Hurd P (1979) Pollination of the southwestern opuntias. Plant Syst Evol 133:15–28

Grant V, Grant KA, Hurd PD Jr (1979) Pollination of *Opuntia lindheimeri* and related species. Plant Syst Evol 132:313–320

Guzmán U, Arias S, Dávila P (2003) Catálogo de Cactáceas Mexicanas. CONABIO UNAM, Mexico

Hamrick JL, Godt MJ (1989) Allozyme diversity in plant species. In: Brown AHD, Clegg MT, Kahler AL, Weir BS (eds) Plant population genetics, breeding and germplasm resources. Sinauer, Sunderland, MA, pp 43–63

Hamrick JL, Nason JD, Fleming TH, Nassar JM (2002) Genetic diversity in columnar cacti. In: Fleming TH, Valiente-Banuet A (eds) Evolution, ecology and conservation of columnar cacti and their mutualists. University of Arizona Press, Tuscon, AZ, pp 122–133

Handel SN (1985) The intrusion of clonal growth patterns on plant breeding system. Am Nat 125:367–384

Harvey PH, Pagel MD (1991) The comparative method in evolutionary biology. Oxford University Press, Oxford

Hernández-Rosas L (2006) Variación genética y producción de semillas en poblaciones de *Cylindropuntia bigelovii* con diferente ploidía. MS Thesis, Universidad Nacional Autónoma de México, Mexico

Hershkovitz MA, Zimmer EA (1997) On the evolutionary origins of the cacti. Taxon 46:217–232

Heslop-Harrison Y, Shivanna KR (1977) The receptive surface of the angiosperm stigma. Ann Bot 41:1233–1258

Hicks D, Mauchamp A (1999) Population structure and growth patterns of *Opuntia echios* var. *gigantea* along an elevation gradient in the Galápagos Islands. Biotropica 32:235–243

Hiscock SJ, Tabah DA (2003) The different mechanisms of sporophytic self-incompatibility in flowering plants. Philos Trans R Soc Lond B 358:1037–1045

Holland JN, Fleming TH (1999) Mutualistic interactions between *Upiga virescens* (Pyralidae), a pollinating seed-consumer, and *Lophocereus schottii* (Cactaceae). Ecology 80:2074–2084

Holsinger KE (1992) Ecological models of plant mating systems and the evolutionary stability of mixed mating systems. In: Wyatt R (ed) Ecology and evolution of plant reproduction. Chapman & Hall, New York, pp 169–191

Hunt D, Taylor N, Charles G (2006) The cactus lexicon. DH Books, Milborne Port, UK

Ibarra-Cerdeña CL, Iñiguez-Dávalos, Sánchez-Cordero V (2005) Pollination ecology of *Stenocereus queretaroensis* (Cactaceae), a chiropterophilous columnar cactus, in a tropical dry forest of Mexico. Am J Bot 92:503–509

Jiménez-Sierra CL (2008) Estudios sobre la biología y demografía de *Echinomastus platyacanthus* Link et Otto, en Zapotitlán de las Salinas, Puebla. PhD Thesis, Universidad Nacional Autónoma de México, Mexico

Johnson R (1992) Pollination and reproductive ecology of Acuña cactus, *Echinomastus erectrocentrus* (Cactaceae). Int J Plant Sci 153:400–408
Kearns CA, Inouye DW (1993) Techniques for pollination biologists. University Press of Colorado, Niwot, CO
Knight TA (1799) An account of some experiments on the fecundation of vegetables. Philos Trans R Soc Lond 195–204
Levin DA (1996) The evolutionary significance of pseudo-self-fertility. Am Nat 148:321–332
Lichtenzveig J, Abbo S, Nerd A, Tel-Zur N, Mizrahi Y (2000) Cytology and mating systems in the climbing cacti *Hylocereus* and *Selenicereus*. Am J Bot 87:1058–1065
Lloyd DG (1979) Some reproductive factors affecting the selection of self-fertilization in plants. Am Nat 113:67–79
Lloyd DG, Webb CJ (1986) The avoidance of interference between the presentation of pollen and stigmas in angiosperms. I. Dichogamy. N Z J Bot 24:135–162
Lovett-Doust J, Lovett-Doust L (1988) Plant reproductive ecology. Oxford University Press, New York
Lüthy JM (2001) The Cacti of CITES. Appendix I. CITES identification manual. CITES, Federal Veterinary Office Switzerland, Botanical Garden of the University of Berne, IOS & Sukulent-Sammlung Zürich, Bern
Mandujano MC, Montaña C, Eguiarte L (1996) Reproductive ecology and inbreeding depression in *Opuntia rastrera* (Cactaceae) in the Chihuahuan desert: why are sexually derived recruitments so rare. Am J Bot 83:63–70
Mandujano MC, Montaña C, Franco M, Golubov J, Flores-Martínez A (2001) Integration of demographic annual variability in a clonal desert cactus. Ecology 82:344–359
Mandujano MC, Verhulst JAM, Carrillo-Angeles I, Golubov J (2007) Population dynamics of *Ariocarpus scaphirostris* Bödeker (Cactaceae): evaluating the status of a threatened species. Int J Plant Sci 168:1035–1044
Martínez-Ávalos JG (2007) Estudio demográfico del "star cactus" *Astrophytum asterias* (Lem.) Zucc. (Cactaceae) una especie en riesgo de extinción. PhD Thesis, Universidad Autónoma de Nuevo León, Mexico
Martínez-Peralta C (2007) Biología floral de *Ariocarpus fissuratus* (Engelmann) Schumann (Cactaceae) en Cuatro Ciénegas, Coahuila, México. BSc Thesis, Universidad Nacional Autónoma de México
Matías-Palafox L (2007) Estructura poblacional y biología reproductiva de *Turbinicarpus horripilus* (Lem.) Vác. John & Riha (Cactaceae). MS Thesis, Universidad Autónoma Metropolitana, Mexico
Mauseth JD (2006) Structure–function relationships in highly modified shoots of Cactaceae. Ann Bot 98:901–926
McClure BA, Franklin-Tong V (2006) Gametophytic self-incompatibility: understanding the cellular mechanisms involved in "self" pollen tube inhibition. Planta 224:233–245
McFarland JD, Kevan PG, Lane MA (1989) Pollination biology of *Opuntia imbricata* (Cactaceae) in southern Colorado. Can J Bot 67:24–28
McIntosh ME (2002) Plant size, breeding system, and limits to reproductive success in two sister species of *Ferocactus* (Cactaceae). Plant Ecol 162:273–288
Méndez M, Durán R, Dorantes A, Dzib G, Simá L, Simá P, Orellana R (2005) Floral demography and reproductive system of *Pterocereus gaumeri*, a rare columnar cactus endemic to Mexico. J Arid Environ 62:363–376
Molina-Freaner F, Rojas-Martínez A, Fleming TH, Valiente-Banuet A (2004) Pollination biology of the columnar cactus *Pachycereus pecten-aboriginum* in north-western México. J Arid Environ 56:117–127
Moraes EM, Abreu AG, Andrade SCS, Sene FM, Solferini VN (2005) Population genetic structure of two columnar cacti with a patchy distribution in eastern Brazil. Genetica 125:311–323

Mota SL, Leila EB, Câmara MM, Da Silva SSA (2006) Allozyme diversity and morphometrics of *Melocactus paucispinus* (Cactaceae) and evidence for hybridization with *M. concinnus* in the Chapada Diamantina, North-eastern Brazil. Ann Bot 97:389–403

Nassar JM, Ramírez N (2004) Reproductive biology of the melon cactus, *Melocactus curvispinus* (Cactaceae). Plant Syst Evol 248:31–44

Nassar JM, Ramírez N, Linares O (1997) Comparative pollination biology of Venezuelan columnar cacti and the role of nectar-feeding bats in their sexual reproduction. Am J Bot 84:918–927

Nassar JM, Hamrick JL, Fleming TH (2001) Genetic variation and population structure of the mixed-mating cactus, *Melocactus curvispinus* (Cactaceae). Heredity 87:69–79

Nassar JM, Hamrick JL, Fleming TH (2002) Allozyme diversity and genetic structure of the leafy cactus (*Pereskia guamacho* [Cactaceae]). J Hered 93:193–200

Nassar JM, Hamrick JL, Fleming TH (2003) Population genetic structure of venezuelan chiropterophilous columnar cacti (Cactaceae). Am J Bot 90:1628–1637

Neal PR, Anderson GJ (2005) Are 'mating systems' 'breeding systems' of inconsistent and confusing terminology in plant reproductive biology? or is it the other way around. Plant Syst Evol 250:173–185

Negrón-Ortiz V (1998) Reproductive biology of a rare cactus, *Opuntia spinosissima* (Cactaceae), in the Florida Keys: why is seed set very low? Sex Plant Reprod 11:208–212

Nobel PS (1988) Environmental biology of agaves and cacti. Cambridge University Press, New York

Nyffeler R (2002) Phylogenetic relationships in the cactus family (Cactaceae) based on evidence from *trnk matk* and *trnl-trnf* sequences. Am J Bot 89:312–326

Oaxaca-Villa B, Casas A, Valiente-Banuet A (2006) Reproductive biology in wild and silvicultural managed populations of *Escontria chiotilla* (Cacataceae) in the Tehuacán Valley, Central Mexico. Genet Resour Crop Evolut 53:277–287

Orozco-Arroyo G (2002) Sistema reproductivo de *Opuntia stenopetala* Engelmann (Cactaceae). BSc Thesis, Universidad Nacional Autónoma de México, Mexico

Osborn MM, Kevan PG, Lane MA (1988) Pollination biology of *Opuntia polyacantha* and *Opuntia phaeacantha* (Cactaceae) in southern Colorado. Plant Syst Evol 159:85–94

Otero-Arnaiz A, Casas A, Bartola C, Pérez-Negrón E, Valiente-Banuet A (2003) Evolution of *Polaskia chichipe* (Cactaceae) under domestication in the Tehuacán Valley, Central Mexico: reproductive biology. Am J Bot 90:593–602

Palleiro ND (2008) Estructura poblacional y genética del cactus columnar *Cephalocereus totolapensis* (Bravo et MacDougall) en el estado de Oaxaca. MSc Thesis, Universidad Nacional Autónoma de México, Mexico

Palleiro N, Mandujano MC, Golubov J (2006) Aborted fruits of *Opuntia microdasys* (Cactaceae): insurance against reproductive failure. Am J Bot 93:505–511

Parker KC, Hamrick (1992) Genetic and clonal structure in a columnar cactus, *Lophocereus schottii*. Am J Bot 79:86–96

Pimienta-Barrios E, del Castillo RF (2002) Reproductive biology. In: Nobel PS (ed) Cacti: biology and uses. University of California Press, Berkeley, pp 75–90

Piña HH (2000) Ecología reproductiva de *Ferocactus robustus* en el Valle de Zapotitlán Salinas, Puebla. MSc Thesis, Instituto Politécnico Nacional, Mexico

Piña HH (2007) Biología reproductiva de *Opuntia microdasys* (Lehm.) Pfeiffer en el Desierto Chihuahuense. PhD Thesis, Instituto de Ecología A.C., Mexico

Piña H, Montaña C, Mandujano MC (2007) Fruit abortion in the Chihuahuan-Desert endemic cactus *Opuntia microdasys*. Plant Ecol 193:305–313

Plasencia-López LMT (2003) Biología reproductiva de *Opuntia bradtiana* (Cactaceae) en Cuatro Ciénegas, Coahuila. BSc Thesis, Universidad Nacional Autónoma de México, Mexico

Plasencia-López LMT (2008) Diversidad clonal y cruzas óptimas en *Opuntia rastrera* Weber (Cactaceae). MSc Thesis, Universidad Nacional Autónoma de México, Mexico

Powell AM (1995) Second generation experimental hybridizations in the *Echinocereus* × *lloydii* complex (Cactaceae), and further documentation of dioecy in *E. coccineus*. Plant Syst Evol 196:63–74
Raguso AR, Henzel C, Buchmann SL, Nabhan GP (2003) Trumpet flowers of the sonoran desert: floral biology of *Peniocereus* cacti and sacred *Datura*. Int J Plant Sci 164:877–892
Ramírez-Corona F (2000) Estudio de la variación genética en poblaciones naturales de dos especies endémica y amenazadas de *Thelocactus* spp. (Cactaceae). MSc Thesis. Universidad Nacional Autónoma de México, Mexico
Rebman J (2001) The succulents of Islote Toro, Baja California, Mexico. Cact Suc Mex 46:52–55
Rebman JP, Pinkava DJ (2001) *Opuntia* cacti of North America – an overview. Fla Entomol 84:474–483
Reyes-Agüero JA, Aguirre JR, Valiente-Banuet A (2006) Reproductive biology of *Opuntia*: a review. J Arid Environ 64:549–585
Richards AJ (1997) Plant breeding systems. Chapman & Hall, Cambridge
Rivera-Marchand B, Ackerman JD (2006) Bat pollination breakdown in the Caribbean columnar cactus *Pilosocereus royenii*. Biotropica 38:635–642
Rosas-Barrera MD, Mandujano MC (2002) La diversidad de historias de vida de cactáceas, aproximación por el triángulo demográfico. Cact Suc Mex 47:33–34
Ross R (1981) Chromosome counts, cytology, and reproduction in the Cactaceae. Am J Bot 68:463–470
Rzedowski J (1991) Diversidad y orígenes de la flora fanerogámica de México. Act Bot Mex 14:13–21
Sánchez-Carbajal D (2007) Embriología de *Mammillaria dioica* K. Barandegee (Cactaceae). BSc Thesis, Universidad Nacional Autónoma de México, Mexico
Schemske DW, Lande R (1985) The evolution of self-fertilization and inbreeding depression in plants. II. Empirical observations. Evolution 39:41–52
Sprengel KC (1793) Discovery of the secret of nature in the structure and fertilization of flowers. In: Lloyd DG, Barrett S (eds) Floral biology: studies on floral evolution in animal-pollinated plants. Chapman & Hall, New York, pp 3–43
Strittmatter LI, Negrón-Ortiz V, Hickey RJ (2002) Subdioecy in *Consolea spinosissima* (Cactaceae): breeding system and embryological studies. Am J Bot 89:1373–1387
Strong AW, Williamson PS (2007) Breeding system of *Astrophytum asterias*: an endangered cactus. Southwest Nat 52:341–346
Uyenoyama MK (1993) Genetic incompatibility as a eugenic mechanism. In: Thornhill NW (ed) The natural history of inbreeding and outbreeding. The University of Chicago Press, Chicago, IL
Valiente-Banuet A, Arizmendi MC, Rojas A, Dominguez (1996) Ecological relationships between columnar cacti and nectar-feeding bats in Mexico. J Trop Ecol 11:1–17
Valiente-Banuet A, Rojas-Martínez A, Arizmendi MC, (1997) Pollination biology of two columnar cacti (*Neobuxbaumia mezcalaensis* and *Neobuxbaumia macrocephala*) in the Tehuacan Valley, central Mexico. Am J Bot 84:452–455
Valiente-Banuet A, Arizmendi MC, Rojas-Martínez A, Casas C, Silva C, Dávila P (2002) Biotic interactions and population dynamics of columnar cacti. In: Fleming TH, Valiente-Banuet A (eds) Columnar cacti and their mutualists: evolution, ecology and conservation, The University of Arizona Press, Tucson, AZ, pp 225–240
Valverde T, Quijas S, López-Villavicencio M, Castillo S (2004) Population dynamics of *Mammillaria magnimamma* Haworth (Cactaceae) in a lava-field in central Mexico. Plant Ecol 170:167–184
Wallace RS (1995) Molecular systematic study of the Cactaceae: using chloroplast DNA variation to elucidate cactus phylogeny. Bradleya 13:1–12
Wallace RS, Gibson AC (2002) Evolution and systematics. In: Nobel PS (ed) Cacti: biology and uses. University of California Press, Berkeley, CA

Waser NM (1983) The adaptive nature of floral traits: ideas and evidence. In: Real L (ed) Pollination biology. Academic, Orlando, pp 241–285

Webb CJ (1979) Breeding systems and the evolution of dioecy in New Zealand apioid Umbelliferae. Evolution 33:662–672

Webb CJ, Lloyd DG (1986) The avoidance of interference between the presentation of pollen and stigmas in angiosperms. II. Herkogamy. N Z J Bot 24:163–168

Wyatt R (1983) Pollinator–plant interactions and the evolution of breeding systems. In: Real L (ed) Pollination Biology. Academic, Orlando, pp 51–95

Yeo PF (1975) Some aspects of heterostyly. New Phytol 75:147–153

Chapter 11
Parthenocarpy and Seed Production in Burseraceae

María F. Ramos-Ordoñez, Judith Márquez-Guzmán, and Ma. del Coro Arizmendi

Abstract Seed production is determined by biotic and abiotic factors; however, crop viability can be affected negatively by phenomenon such as parthenocarpy (production of seedless fruits). *Bursera morelensis* is a deciduous species endemic to Mexico that occurs in the Tehuacan Valley and illustrates the strong effect of parthenocarpy. There is high annual variability in the production of fertilized and parthenocarpic fruits. In some years, crop yield is greater, and fruit removal and visits by birds are higher as compared to a year with poor crops. In mast years, parthenocarpy is a factor that helps to attract seed dispersers, which also reduces predation by insects and birds. These advantages of parthenocarpy probably cannot be observed in years with low fruit production but in years with large crops, this phenomenon can help maximize profits by increasing plant reproductive success.

11.1 Introduction

In arid and semi-arid environments, rainfall is often very low and unpredictable (Valiente-Banuet and Ezcurra 1991; van Rheede and van Rooyen 1999), and plants may exhibit masting behavior due to pulses of rain (Haase et al. 1995). Mast seeding is the intermittent synchronous production of large seed crops by a population of plants (Kelly 1994). Annual variability in crop size has several consequences. In years with large crops, the plant is exposed to density-dependent competition, seedling predation or attack by pathogens; conversely, in years with

M.F. Ramos-Ordoñez and Ma. del Coro Arizmendi (✉)
FES Iztacala UNAM, Laboratorio de Ecología UBIPRO, Av. De los Barrios 1, Los Reyes Iztacala, Tlalnepantla, Edo. México, CP 54090, México
e-mail: mafyx@correo.unam.mx; coro@servidor.unam.mx

J. Márquez-Guzmán
Facultad de Ciencias UNAM, Laboratorio de Desarrollo en Plantas, Circuito Exterior s/n, Ciudad Universitaria, Avenida Universidad 3000, CP 04510, México
e-mail: jmg@fciencias.unam.mx

K.G. Ramawat (ed.), *Desert Plants*,
DOI 10.1007/978-3-642-02550-1_11, © Springer-Verlag Berlin Heidelberg 2010

small crops, the probability of recruitment is reduced (Norton and Kelly 1988; Kelly 1994; Haase et al. 1995). Mast years are more efficient than non-mast years due to the increased attraction of seed dispersers, or to the reduction in the individual probability of seed predation (Norton and Kelly 1988). The crop size produced depends on the number of pollinated flowers, the number of fertilized ovules, available resources, and the ability of the parent to provide the necessary resources for development (Stephenson 1981). Moreover, the proportion of nonviable seeds and the causes thereof, which can include as pre-dispersal predation, embryo death after pollination (Silvertown 1980) and fruit development without fertilization or parthenocarpy (Schwabe and Mills 1981), should be considered. Production of high rates of parthenocarpic fruits can be viewed as a long-term disadvantage whereby the plant reduces its reproductive success, even causing its disappearance. However, the cost of producing fruits without seed is low (Zangerl et al. 1991; Ramos-Ordoñez et al. 2008), and it may confer on the plant certain advantages for seed survival (Coetzee and Giliomee 1987; Zangerl et al. 1991; Traveset 1993; Verdú and García-Fayos 1998, 2001; Ramos-Ordoñez et al. 2008).

11.2 Parthenocarpy

Formation of parthenocarpic fruits occurs during the first two stages of development of the fruit (activation of the fruit and cellular division), in which the ovary begins and finishes its growth whereas the ovule stays latent or degenerates (Solomon 1980; Gay et al. 1987; Jordano 1988; Gillaspy et al. 1993; Sato et al. 2001, 2002; Varoquaux et al. 2000; Young et al. 2004). The causes of this phenomenon can be environmental, physiological or genetic (Table 11.1), with

Table 11.1 Environmental, physiological and genetic causes of parthenocarpy

Cause	Reference
Scarcity of resources	Jang and Sheen 1997; Sato et al. 2001
Thermal stress	Sato et al. 2001, 2002; Higashiyama et al. 2003; Young et al. 2004
Hydric stress	Gay et al. 1987; Jordano 1988
Pollen available	Campbell and Halama 1993; Obeso 1993
Damages to reproductive organs	Galil and Eisikowitch 1971; Solomon 1980
Changes in auxin concentration	Nitsch 1950; Rayle and Cleland 1992
Changes in giberellin concentration	Cano-Medrano and Darnell 1997, 1998
Changes in cytokinin concentration	Bohner and Bangerth 1988
Changes in abscicic acid concentration	Aloni 1995; Zacarías et al. 1995
Changes in polyamine concentration	Alabadí et al. 1996; Fos et al. 2003
Polyploidy and errors in gene expression	Mazzucato et al. 1998, 2003; Ampomah-Dwamena et al. 2002; Zohari 2004

Table 11.2 Species of the plants in which parthenocarpy has been reported as a strategy of defense against seed predators. The works are ordered according to the year of publication.

Species	Predator	Reference
Protea repens	Insects	Coetzee and Giliomee 1987
Pastinaca sativa	Insects	Zangerl et al. 1991
Pistacia terebinthus	Insects	Traveset 1993
Protea (several species)	Insects	Wright 1994
Yucca schottii	Insects	Ziv and Bronstein 1996
Juniperus osteosperma	Birds	Fuentes and Schupp 1998
Pistacia lentiscus	Insects and birds	Verdú and Garcia-Fayos 1998, 2001
Bursera morelensis	Insects and birds	Ramos-Ordoñez et al. 2008; Ramos-Ordoñez 2009

the physiological causes being the most studied, mainly in species of commercial importance.

Several studies have shown that, although parthenocarpy reduces crop size, it is also a strategy leading to diminished seed predation by insects and birds (Table 11.2). In some plants, insects lay eggs indiscriminately in seeded and parthenocarpic fruits, thus allowing many fertile seeds to escape predation (Coetzee and Giliomee 1987; Traveset 1993; Wright 1994; Verdú and García-Fayos 1998). Parthenocarpic fruits can also act as a physical or chemical barrier to prevent insects reaching seeded fruits (Zangerl et al. 1991; Ziv and Bronstein 1996; Ramos-Ordoñez et al. 2008). Seed predation by birds is reduced in trees with a greater proportion of parthenocarpic fruits (Fuentes and Schupp 1998; Verdú and García-Fayos 2001; Ramos-Ordoñez 2009). Producing parthenocarpic fruit also enhances attraction and so increases the frequency of seed disperser visits (Ramos-Ordoñez 2009).

11.3 Parthenocarpy in Burseraceae

The family Burseraceae comprises approximately 18 genera and 550 species of tropical trees and shrubs (Stevens 2001). Parthenocarpy has been recorded in only nine species of the genus *Bursera* distributed in the tropical dry forest and xerophilous shrublands of Mexico. Although the genus has more than 100 species, knowledge on parthenocarpy and patterns of fruiting is scarce. For example, *B. fagaroides* exhibits a strong annual variation in fruit production, and fruit removal is related to a spatio-temporal variation (Ortiz-Pulido and Rico-Gray 2000). Nevertheless, no information is available on the variation in this parthenocarpy or its ecological consequences (Verdú and García-Fayos 1998). In *B. simaruba*, seed production ranging from 60,000 to 3,000 has been reported, and no information on parthenocarpy is available (Stevens 1983; Greenberg et al. 1995). In *B. aptera*, *B. grandifolia*, *B. bipinnata*, *B. lancifolia*, *B. copallifera*, *B. glabrifolia*, *B. bicolor*, *B. schlechtendalii* and *B. submoniliformis*, the presence of parthenocarpic fruits has been recorded but other aspects of reproductive biology remain unknown (Bonfil et al. 2007;

Ramos-Ordoñez et al. 2008). Information is available only for *B. morelensis*, where the presence of parthenocarpy was first reported (Verdú and García-Fayos 1998), and described histologically and anatomically (Ramos-Ordoñez et al. 2008) by measuring its ecological consequences (Ramos-Ordoñez 2009).

11.4 *Bursera morelensis*

Red cuajiote (*Bursera morelensis*) is a dioecious tree endemic to Mexico. It is a very showy plant with its crust of intense reddish color, and is frequent and some times dominant in the tropical dry forest and xerophilous shrublands of the eastern sector of the Balsas, the Papaloapan and Tehuantepec rivers basins (Rzedowski et al. 2004, 2005). The studied population was located in the Barranca del Muchil in San Rafael Coxcatlán, in the southeastern portion of the Tehuacan Valley, Puebla, México (18° 12′ and 18° 14′ N; 97° 07′ and 97° 09′ W). Soils on site are very heterogeneous and determine four different vegetation zones (Medina 2000): (1) a first zone dominated by *Fouqueria formosa* Kunt denominated Fouquieral; (2) Cuajiotal, dominated by *Bursera morelensis* Ramírez; (3) Chiotillal, dominated by *Escontria chiotilla* (Weber) Rose; and (4) Cardonal, dominated by the columnar cactus *Pachycereus weberi* (Coulter) Buxb (Medina 2000; Ríos-Casanova et al. 2004). Flowering of *B. morelensis* at this site is annual and happens between April and May. Approximately 1 week after the first rains, the male begin to produce flowers before the females (Rzedowski et al. 2004; Ramos-Ordoñez et al. 2008). The masculine flowers produce large amounts of pollen – approximately 50,000 pollen grains per flower – and flowers are grouped in inflorescences of 12–16 flowers. The pollen grains are spheroidal, and measure 25.6–28.8 $\mu \times$ 24–27.2 μ (Cortés 1998). Female flowers present a tricarpelar ovary, can be solitary or grouped in inflorescences of up to nine flowers (Ramos-Ordoñez 2009). After pollination by bees (*Apis mellifera*), the ovary remains latent for 7–8 months; the fruits, however, reach a definitive size in the first 2 weeks after pollination. At this stage, the fruits are green in color and little by little they become red, with mature fruits with a single seed appearing only in the month of November (Ramos-Ordoñez et al. 2008). Fruit and seed production are linked to the availability of pollen, and the number of flowers that form fruits of final size varies between 29.15% and 73.98%. Nevertheless, in many cases, viability of the seeds varies between 47% and 88% due to parthenocarpy. Parthenocarpic fruits exhibit unusual growth of ovary walls, and the tissues finally squash the latent or developing ovule (Fig. 11.1). Before maturing, parthenocarpic and seeded fruits appear similar, but when mature it is possible to distinguish them by their pattern of dehiscence, as dehiscence is always incomplete in parthenocarpic fruit. Prior to dispersion, fruits are predated by wasps of the superfamily Chalcidoidea, and flies of the family Cecidomyiidae. There is evidence of damage only in parthenocarpic fruits, with the larval states of these insects using the fruit as a growth chamber, feeding on ovary tissues (Ramos-Ordoñez et al. 2008).

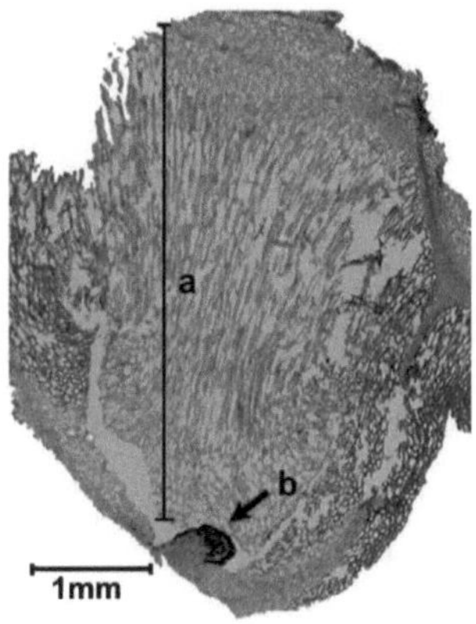

Fig. 11.1 Longitudinal section of a parthenocarpic fruit of *Bursera morelensis*. The walls of the ovary (*a*) fill the locule, crushing the latent ovule (*b*). Micrographs were taken under an optical microscope (Olympus Provis AX70)

11.5 Fruit Crop by *B. morelensis*

Red cuajiote exhibits masting behavior. In a sample of 13 female trees of *B. morelensis* studied by Ramos-Ordoñez (2009) during the years 2005 and 2006, a strong temporal variability in the size of the crop produced was observed (Fig. 11.2). In 2005, the trees produced on average 2,352.62 ± 2,062.45 fruits (± SD), whereas in 2006 they produced 19,358.5 ± 11,528.3 fruits (χ^2 = 1,156, df = 12, P = 0.01). Seed crop size was reduced by varying amounts due to parthenocarpy; the percentage of parthenocarpic fruits produced in 2005 was 9.83 ± 10.72% and in 2006 32.5 ± 10.29 %, with a statistical significant difference between years (χ^2 = 72.9, df = 12, P = 0.01). However, in both periods, the proportion of parthenocarpic fruits produced did not correlate with crop size. Seed production in *B. morelensis* is linked to the availability of pollen (Ramos-Ordoñez et al. 2008), but is also very likely to be affected by temperature and humidity as in *B. microphylla* and other species in arid environments (Maya and Arriaga 1996; Pugnaire et al. 1996). Variations in parthenocarpic proportion may result from stressful environmental conditions (i.e., water or temperature) or the availability of pollen, but it is also necessary to consider hereditary factors (Verdú and García-Fayos 1998, 2001).

Fruits in *B. morelensis* are eaten by several species of migratory birds, and fruit removal is determined mainly by crop size and abundance, and the diet of specific bird species (Ramos-Ordoñez 2009). The number of visits and number of fruits removed per hour of observation is higher in mast years (6.98 visits h^{-1} $tree^{-1}$ and 5.46 fruits h^{-1} $tree^{-1}$ in 2006) than in non-mast years (0.74 visits h^{-1} $tree^{-1}$ and 1.4 fruits h^{-1} $tree^{-1}$ in 2005). To determine the effect of masting on fruit removal, long-term studies are needed (Norton and Kelly 1988).

On the other hand, parthenocarpy has an important role in attracting dispersers, since it has been demonstrated that the number of visits by birds increases with the amount and proportion of parthenocarpic fruit in the trees, as plant attractiveness is

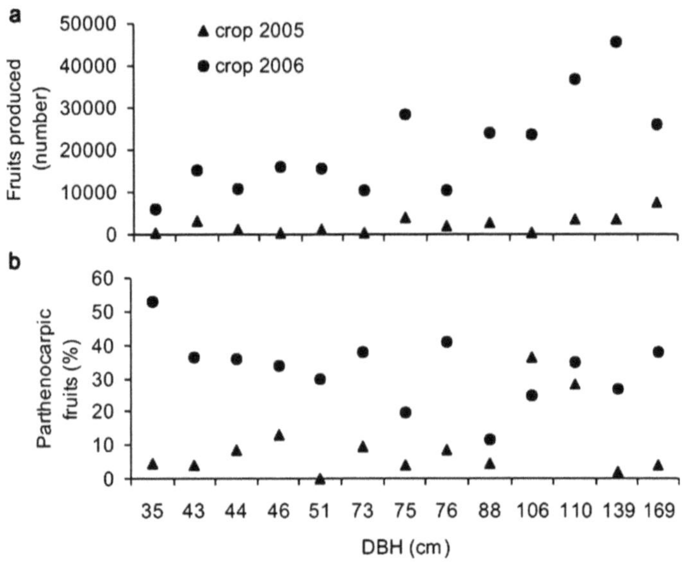

Fig. 11.2 Number of fruits (**a**) and percentage of parthenocarpic fruits (**b**) produced by 13 trees of *Bursera morelensis* during the years 2005 and 2006 in the Tehuacan Valley, México. Trees are arranged in accordance with diameter at breast height (DBH)

increased due the effects of density and color (Ramos-Ordoñez 2009). Before being dispersed, both insects and birds eat the fruits of *B. morelensis*; parthenocarpy thus helps reduce seed predation. The fertile fruits have seeds that develop a layer of calcium oxalate crystals, which is not found in the parthenocarpic fruits (Ramos-Ordoñez et al. 2008). These crystals are toxic to many insects and create a layer of hard tissue (Franceschi and Horner 1980; Molano-Flores 2001; Volk et al. 2002). Larvae of both flies and wasps only occur in fruit without seeds (Ramos-Ordoñez et al. 2008). In 2006, flies and wasp parasitized between 0 and 18% of the fruits, but in 2005 there were no parasites in samples of marked trees. Nevertheless, additional samples taken from non-marked trees showed that 8.95% of fruits were attacked by flies (M.F. Ramos-Ordoñez, unpublished data).

In the case of birds, it was observed that, in trees with the highest proportion of parthenocarpic fruit, predation by granivorous birds decreases (Ramos-Ordoñez 2009). Regardless of crop size, in the absence of migratory dispersers, seeds would be exposed to resident granivorous birds, and the decrease in predation due to parthenocarpy could be minimal. Usually, seeds fall under trees and shrubs that serve as perches for dispersers (Izhaki et al. 1991). After the seeds are shed, secondary removal occurs by rodents, ants and granivorous birds. Vertebrates remove more seeds than ants. At this stage predation of parthenocarpic fruit is not affected because none of these predator groups removes parthenocarpic fruits (M.F. Ramos-Ordoñez, unpublished data). It is likely that the abundant seed availability in mast years reduces post-dispersal seed predation (predator satiation hypothesis; Silvertown 1980; Kelly 1994).

Bursera is a genus that is highly adapted to the ecological and climatic conditions of the dry forest (Becerra 2005), and it is not uncommon in this environment to find a strong variability in the amount of fruit produced annually in response to seasonal conditions. Nevertheless, the reproductive biology in *Bursera* is also linked to biotic seed dispersal, and the advantages of parthenocarpy could decline due to various factors such as changes in the diversity and abundance of species (i.e., birds and insects) that feed on the plant at a specific developmental timepoint, affecting its reproductive success.

Acknowledgments Financial support was provided by DGAPA-PAPIIT No. IN207305 and CONABIO DT006 projects to M.C.A., CONACyT research scholarship to M.F.R.-O. This work is part of the PhD thesis of M.F.R.-O. of the Posgrado en Ciencias Biológicas, Universidad Nacional Autónoma de México.

References

Alabadí D, Agüero MS, Pérez-Amador MA, Carbonell J (1996) Arginase, arginine decarboxylase, ornithine decarboxylase, and polyamines in tomato ovaries: changes in unpollinated ovaries and parthenocarpic fruits induced by auxin or gibberellin. Plant Physiol 112:1237–1244

Aloni R (1995) The induction of vascular tissues by auxin and cytokinin. In: Davies PJ (ed) Plant hormones, physiology, biochemistry and molecular biology, Kluwer, London, pp 531–546

Ampomah-Dwamena C, Morris BA, Sutherland P, Veit B, Yao J (2002) Down-regulation of TM29, a tomato SEPALLATA homolog, causes parthenocarpic fruit development and floral reversion. Plant Physiol 130:605–617

Becerra JX (2005) Timing the origin and expansion of the Mexican tropical dry forest. Proc Natl Acad Sci USA 102:10919–10923, doi:10.1073/pnas.0409127102

Bohner J, Bangerth F (1988) Effects of fruit set sequence and defoliation on cell number, cell size and hormone levels of tomato fruits (*Lycopersicon esculentum* Mill.) within a truss. Plant Growth Regul 7:141–155

Bonfil C, Cajero I, Castellanos C, Healy E, Evans R (2007) Germinación y propagación vegetativa de diversas especies del género *Bursera*. URL: http://www.socbot.org.mx/Congresos/XVII/Resumenes%20Congreso.pdf

Campbell DR, Halama KH (1993) Resource and pollen limitations to lifetime seed production in a natural plant population. Ecol 74:1043–1051

Cano-Medrano R, Darnell RL (1997) Cell number and cell size in parthenocarpic vs pollinated blueberry (*Vaccinium ashei*) fruits. Ann Bot 80:419–425

Cano-Medrano R, Darnell RL (1998) Effect of GA_3 and pollination on fruit set and development in rabbiteye blueberry. HortScience 33:632–635

Coetzee JH, Giliomee JH (1987) Seed predation and survival in the infructescences of *Protea repens* (Proteaceae). South Afr J Bot 53:61–64

Cortés A (1998) Biología reproductiva de *Bursera medranoana* Rzedowski & Ortiz (Burseraceae), una especie de origen híbrido. Tesis de Licenciatura. Universidad Nacional Autónoma de México. México

Fos M, Proaño K, Alabadí D, Nuez F, Carbonell J, García-Martínez JL (2003) Polyamine metabolism is altered in unpollinated parthenocarpic pat-2 tomato ovaries. Plant Physiol 131:359–366

Franceschi VR, Horner HT (1980) Calcium oxalate crystals in plants. Bot Rev 46:361–427

Fuentes M, Schupp E (1998) Empty seeds reduce seed predation by birds in *Juniperus osteosperma*. Evol Ecol 12:823–827

Galil J, Eisikowitch D (1971) Studies on mutualistic symbiosis between syconia and sycophilous wasp in monoecious figs. New Phytol 70:773–787
Gay G, Kerhoas C, Dumas C (1987) Quality of a stress-sensitive *Cucurbita pepo* L. pollen. Planta 171:82–87
Gillaspy G, Ben-David H, Gruissem W (1993) Fruits: a developmental perspective. Plant Cell 5:1439–1451
Greenberg R, Foster MS, Marquez-Valdelamar L (1995) The role of the white-eyed vireo in the dispersal of *Bursera* fruit on the Yucatan Peninsula. J Trop Ecol 11:619–639
Haase P, Pugnaire FI, Incoll LD (1995) Seed production and dispersal in the semi-arid tussock grass *Stipa tenacissima* L. during masting. J Arid Environ 31:55–65
Higashiyama T, Kuroiwa H, Kuroiwa T (2003) Pollen-tube guidance: beacons from the female gametophyte. Curr Opin Plant Biol 6:36–4
Izhaki I, Walton PB, Safriel UN (1991) Seed shadows generated by frugivorous birds in an eastern Mediterranean scrub. J Ecol 79:575–590
Jang JC, Sheen J (1997) Sugar sensing in higher plants. Trends Plant Sci 2:208–214
Jordano P (1988) Polinización y variabilidad de la producción de semillas en *Pistacia lentiscus* L. (Anacardiaceae). An Jard Bot Madr 45:213–231
Kelly D (1994) The evolutionary ecology of mast seeding. Trends Ecol Evol 9:465–470
Maya Y, Arriaga L (1996) Litterfall and phonological patterns of the dominant overstorey species of a desert scrub community in north-western Mexico. J Arid Environ 34:23–35
Mazzucato A, Taddei AR, Soressi GP (1998) The parthenocarpic fruit (pat) mutant of tomato (*Lycopersicon esculentum* Mill.) sets seedless fruits and has aberrant anther and ovule development. Development 125:107–114
Mazzucato A, Olimpieri I, Ciampolini F, Cresti M, Soressi GP (2003) A defective pollen-pistil interaction contributes to hamper seed set in the parthenocarpic fruit tomato mutant. Sex Plant Reprod 16:157–164
Medina JS (2000) Determinación del vigor reproductivo de *Stenocereus stellatus* (Cactaceae) a lo largo de una cronosecuencia edáfica en un abanico aluvial en Coxcatlán, Valle de Tehuacán. Tesis de Licenciatura. Universidad Nacional Autónoma de México. México
Molano-Flores B (2001) Herbivory and concentrations affect calcium oxalate crystal formation in leaves of *Sida* (Malvaceae). Ann Bot 88:387–391
Nitsch JP (1950) Growth and morphogenesis of the strawberry as related to auxin. Am J Bot 37:211–215
Norton DA, Kelly D (1988) Mast seeding over 33 years by *Dacrydium cupressinum* Lamb. (rimu) (Podocarpaceae) in New Zealand: the importance of economies of scale. Funct Ecol 2:399–408
Obeso JR (1993) Selective fruit and seed maturation in *Asphodelus albus* Miller (Liliaceae). Oecologia 93:564–570
Ortiz-Pulido R, Rico-Gray V (2000) The effect of spatio-temporal variation in understanding the fruit crop size hypothesis. Oikos 93:523–528
Pugnaire FI, Haase P, Puigdefábregas J, Cueto M, Incoll LD, Clark SC (1996) Facilitation and succession under the canopy of *Retama sphaerocarpa* (L.) Boiss. in a semi-arid environment in South-east Spain. Oikos 76:455–464
Ramos-Ordoñez MF (2009) Dispersión biótica de semillas y caracterización de frutos de *Bursera morelensis* en el Valle de Tehuacán Puebla. Tesis Doctoral. Universidad Nacional Autónoma de México. México
Ramos-Ordoñez MF, Márquez-Guzmán J, Arizmendi MC (2008) Parthenocarpy and seed predation by insects in *Bursera morelensis*. Ann Bot 102:713–722
Rayle D, Cleland R (1992) The acid growth theory of auxin-induced cell elongation is alive and well. Plant Physiol 99:1271–1274
Ríos-Casanova L, Valiente-Banuet A, Rico-Gray V (2004) Las hormigas del Valle de Tehuacán (Hymenoptera: Formicidae): una comparación con otras zonas áridas de México. Acta Zool Mex 20:37–54

Rzedowski J, Medina R, Calderón G (2004) Las especies de *Bursera* (Burseraceae) en la cuenca superior del Río Papaloapan (México). Acta Bot Mex 66:23–151
Rzedowski J, Medina R, Calderón G (2005) Inventario del conocimiento taxonómico, así como de la diversidad y del endemismo regionales de las especies mexicanas de *Bursera* (Burseraceae). Acta Bot Mex 70:85–111
Sato S, Peet MM, Gardner RG (2001) Formation of parthenocarpic fruit, undeveloped flowers and aborted flowers in tomato under moderately elevated temperatures. Sci Hortic 90:243–254
Sato S, Peet MM, Thomas JF (2002) Determining critical pre- and post-anthesis periods and physiological processes in *Lycopersicon esculentum* Mill. exposed to moderately elevated temperatures. J Exp Bot 53:1187–1195
Schwabe WW, Mills JJ (1981) Hormones and parthenocarpic fruit-set. Hortic Abstr 51:661–698
Silvertown JW (1980) The evolutionary ecology of mast seeding in trees. Biol J Linn Soc 14:235–250
Solomon BP (1980) *Frumenta nundinella* (Lepidoptera: Gelechiidae): life history and induction of host parthenocarpy. Environ Entomol 9:821–825
Stephenson AG (1981) Flower and fruit abortion: proximate causes and ultimate functions. Annu Rev Ecol Syst 12:253–279
Stevens G (1983) *Bursera simaruba* (indio desnudo, jinocuave, gumbo limbo). In: Jansen DH (ed) Costa Rican natural history, The University of Chicago Press, Chicago, pp 201–202
Stevens PF (2001). Angiosperm Phylogeny Website, ver 9, June 2008, http://www.mobot.org/MOBOT/research/APweb/. Cited 12 April 2009
Traveset A (1993) Deceptive fruits reduce insect seed predation in *Pistacia terebinthus* L. Evol Ecol 7:357–361
Valiente-Banuet A, Ezcurra E (1991) Shade as a cause of the association between the cactus *Neobuxbaumia tetetzo* in the nurse plant *Mimosa luisana* in the Tehuacán Valley, Mexico. J Ecol 79:961–971
Van Rheede K, van Rooyen MW (1999) Dispersal biology of desert plants. Springer, Berlin
Varoquaux F, Blanvillain R, Delseny M, Gallois P (2000) Less is better: new approaches for seedless fruit production. Trends Biotechnol 18:233–242
Verdú M, García-Fayos P (1998) Ecological causes, function, and evolution of abortion and parthenocarpy in *Pistacia lentiscus* (Anacardiaceae). Can J Bot 76:134–141
Verdú M, García-Fayos P (2001) The effect of deceptive fruits on predispersal seed predation by birds in *Pistacia lentiscus*. Plant Ecol 156:245–248
Volk GM, Lynch-Holm VJ, Kostman TA, Goss LJ, Franceschi VR (2002) The role of druse and raphide calcium oxalate crystals in tissue calcium regulation in *Pistia stratiotes* leaves. Plant Biol 4:34–45
Wright MG (1994) Unpredictable seed-set: a defence mechanism against seed-eating insects *Protea* species (Proteaceae). Oecologia 99:397–400
Young LW, Wilen RW, Bonham-Smith PC (2004) High temperature stress of *Brassica napus* during flowering reduces micro- and mega-gametophyte fertility, induces fruit abortion, and disrupts seed production. J Exp Bot 55:485–495
Zacarías L, Talon M, Ben-Cheikh W, Lafuente MT, Primo-Millo E (1995) Abscisic acid increases in non-growing and paclobutrazol-treated fruits of seedless mandarins. Physiol Plant 95:613–619
Zangerl AR, Berenbaum MR, Nitao JK (1991) Parthenocarpic fruits in wild parsnip: decoy defence against a specialist herbivore. Evol Ecol 5:136–145
Ziv Y, Bronstein JL (1996) Infertile seeds of *Yucca schottii*: a beneficial role for the plant in the yucca-yucca moth mutualism? Evol Ecol 10:63–76
Zohary D (2004) Unconscious selection and the evolution of domesticated plants. Econ Bot 58:5–10

Part C
Ecophysiological Studies

Chapter 12
Photosynthesis of C_4 Desert Plants

Peixi Su

Abstract There are more than 40 species of C_4 woody plants (including semi-woody) in the deserts of China. These plants are exclusively members of Chenopodiaceae and Polygonaceae; members of other families have not been found. This chapter introduces some of their characteristics, e.g., photosynthetic structure, carbon isotope composition, gas exchange properties, chlorophyll fluorescence, and CO_2 response. *Haloxylon ammodendron* of Chenopodiaceae and *Calligonum mongolicum* of Polygonaceae are the dominant plants in the continental deserts of the Chinese region. Their true leaves are quite reduced, and the cortex of young annual cylindrical shoots is the main photosynthetic tissue. *Salsola arbuscula* and *Salsola collina* are two species in same genus: *S. arbuscula* is woody, *S. collina* is a herb, and the drought resistant capacity of *S. arbuscula* is greater than that of *S. collina*. Their assimilation organs have Kranz anatomy; the three woody species are similar in Kranz anatomy. Compared to herbs, C_4 woody plants have thicker cuticles with a layer of underlying hypodermal cells, and the palisade cells are longer and arranged more densely. Their $\delta^{13}C$ values are between –14‰ and –15‰. In addition, C_4 woody plants have a higher light saturation point, i.e., above 1,600 μmol m^{-2} s^{-1}, the CO_2 compensation point is below 10 μmol mol^{-1}, and their apparent quantum yield is above 0.04 mol mol^{-1}.

P. Su
Laboratory of Plant Stress Ecophysiology and Biotechnology, Linze Inland River Basin Research station, Cold and Arid Regions Environmental and Engineering Research Institute, Chinese Academy of Sciences, Lanzhou 730000, China
e-mail: supx@lzb.ac.cn

K.G. Ramawat (ed.), *Desert Plants*,
DOI 10.1007/978-3-642-02550-1_12, © Springer-Verlag Berlin Heidelberg 2010

12.1 Introduction

Plants can be classified into three groups in terms of their photosynthetic carbon assimilation pathways: C_3, C_4 and CAM (crassulacean acid metabolism). C_3 plants fix atmospheric CO_2 directly in their photosynthetic cells via the enzyme ribulose 1,5-bisphosphate (RuBP) carboxylase/oxygenase (Rubisco). C_4 plants increase the level of CO_2 through phosphoenolpyruvic acid carboxylase (PEPCase), thereby fixing CO_2 via the Calvin cycle in bundle sheath cells. These CO_2-concentrating mechanisms greatly increase photosynthetic efficiency. CAM plants open their stomata at night and fix atmospheric CO_2 through the C_4 pathway, and further process carbon via the C_3 pathway during the day (Winter and Smith 1996; Voznesenskaya et al. 2001).

Plant photosynthesis is a process that converts CO_2 into carbohydrate. The main criteria for a plant to be considered a C_4 plant are a Kranz anatomy, a low CO_2 compensation point, and high $\delta^{13}C$ values, as well as a strong capacity for carbon assimilation and high production. Such plants are thereby well adapted to environments with intensive light and high temperatures. Their light saturation point, photosynthesis rate and water use efficiency are, accordingly, high (Hatch 2002). Voznesenskaya et al. (2001) suggested that Kranz anatomy is not essential for terrestrial C_4 plant photosynthesis, taking *Borszczowia aralocaspica* as an example. Bender (1971) and Smith and Epstein (1971) first suggested using stable carbon isotope techniques to distinguish the photosynthetic CO_2-fixing pathway of plants, and established that the $\delta^{13}C$ value of the C_3 plant is in the range of –23‰ to –32‰, while that of C_4 plants is –6‰ to –19‰. CAM plants have $\delta^{13}C$ values of –10‰ to –22‰, while that of C_3–C_4 intermediate plants is similar to that of C_3 plants. Results obtained by other scholars fall mostly within the above range (Farquhar et al. 1982; Hattersley 1982; Farquhar 1983).

Foliar $\delta^{13}C$ values can reflect the water use efficiency (WUE) associated with plant photosynthetic and transpiration intensity (Lajtha and Michener 1994). Farquhar et al. (1989) believed that foliar $\delta^{13}C$ values could be used to indicate the long-term WUE of plants, and, to a certain extent, $\delta^{13}C$ values are correlated positively with WUE (Marshall and Zhang 1994; Sun et al. 1996).

Desert plants grow in a harsh environment with high temperatures and radiation and are exposed to drought, often for much of the year. Their specific morphological and physiological features frequently reduce water loss and alleviate high-radiation damage to the photosynthetic apparatus. *Haloxylon ammodendron* and *Calligonum mongolicum* are distributed widely in the temperate desert regions of China. Unlike many other desert plants, they have high biomass. *H. ammodendron* has a tree life form or grows as a shrub depending on its age and environment. *C. mongolicum* plants growing on sand dunes have developed a root system with strong sand-binding ability and regenerate well. *C. mongolicum* grows as a large or small shrub depending on soil moisture conditions, and different ecological types have formed in each desert area in response to different growth environments, with heights between 0.3 m and 3.0 m (Su 2007). *Salsola arbuscula* is a woody plant occurring in arid desert regions; *Salsola collina* occurs mostly inside or on the edge of oasis, and both these plants grow in desert areas with better water conditions.

12.2 Photosynthetic Structure of Assimilating Organs

Structure is the basis of function; changes in photosynthetic structure will effect changes in ecophysiological function. The curtailment of leaves is one common characteristic of desert xerophytes. *H. ammodendron* and *C. mongolicum* have an unusual photosynthetic apparatus. True leaves are markedly reduced, and it is the cortex of young annual cylindrical shoots that is the main photosynthetic tissue. *S. arbuscula* leaves are small and succulent; *S. collina* leaves are filamentous.

12.2.1 Photosynthetic Structure of Assimilating Shoots of Haloxylon ammodendron and *Calligonum mongolicum*

The assimilating shoots of *H. ammodendron* are cylindrical and have a layer of epidermal cells, a layer of hypodermal cells, and two layers of chlorenchyma on the stem periphery, as well as an outer layer of palisade cells and an inner layer of bundle sheath cells as shown in Fig. 12.1a, b. The central portion of the shoot is occupied by water storage tissue, with the main vascular bundles located in the center. Central bundles are thus separated from Kranz-type cells by layers of water storage cells. There are some small peripheral bundles that have contact with bundle sheath cells. Some crystal idioblasts are present in palisade cells and water storage tissues (Fig. 12.1b). The palisade cells and bundle sheath cells contain chloroplasts. This Kranz anatomy type is similar to the 'Salsoloid' type (named by Carolin et al. 1975), which is typical for succulent C_4 species of *Salsola* and closely related species (Pyankov et al. 1999).

Assimilating shoots of *C. mongolicum* have a layer of epidermal cells, a layer of hypodermal cells, and two layers of chlorenchyma (an outer layer of palisade cells and an inner layer of bundle sheath cells) on the stem periphery, and many mucilage cells in water storage tissue (Fig. 12.1c, d). Some crystal idioblasts are present in water storage tissue (Fig. 12.1d). The Kranz anatomy is also similar to the Salsoloid type.

12.2.2 Photosynthetic Structure of Leaves of Salsola arbuscula and *Salsola collina*

The cross section of *S. arbuscula* leaves is circular (Fig. 12.2a) or triangular. Most epidermal cells are roughly oval in shape, and differ in size. The outer epidermis

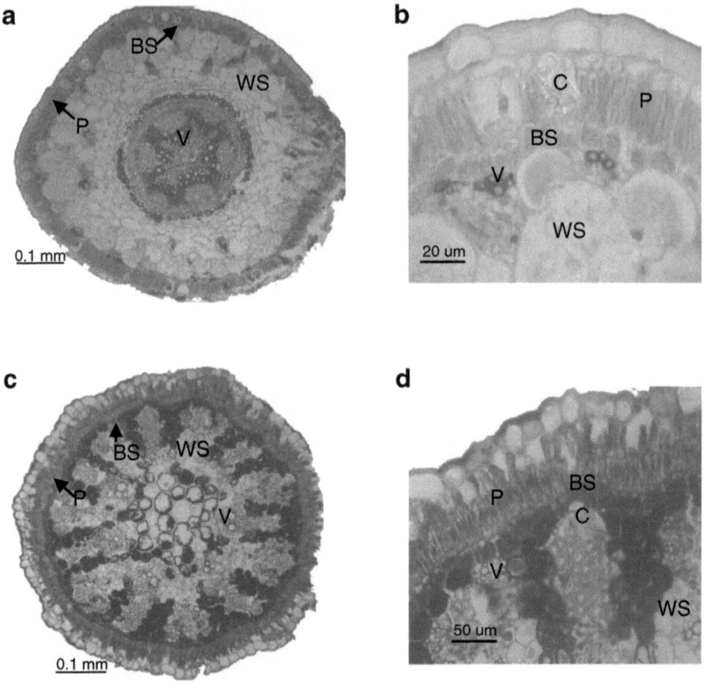

Fig. 12.1a–d Transverse sections of assimilating shoots of *Haloxylon ammodendron* and *Calligonum mongolicum*. **a** Assimilating shoot of *H. ammodendron*. **b** Partial section of *H. ammodendron*. **c** Assimilating shoot of *C. mongolicum*. **d** Partial section of *C. mongolicum*. *P* palisade cells, *BS* bundle sheath cells, *WS* water storage tissue, *V* vascular tissue, *C* crystal idioblast. *Bars* **a**, **c** 0.1 mm; **b** 20 μm; **d** 50 μm

has thicker cuticles, some prominent on the outward surface, with a layer of hypodermal cells under them. Palisade cells are closely arranged (Fig. 12.2b); in the palisade tissue cell layer are vascular bundle sheath cells, which form a Kranz anatomy together with the small vascular bundles beside them and the outer mesophyll cells. In the center of the cross section are large water storage cells and the central vascular bundle. Some crystal cells are present in the palisade cells and epidermal cells.

S. collina leaves are filamentous circular. The cross section in Fig. 12.2c shows the close, orderly arrangement of epidermal cells, with some extending outwards with epidermal hairs. The palisade cells containing the chloroplasts are formed by a layer of elongated cells, and are closely arranged. Inside the palisade cells are vascular bundle sheath cells, inside which are the developed water storage parenchyma cells. Vascular bundles are scattered in storage tissues, and the central vascular bundle is larger. The crystal cluster can be seen in the storage tissues (Fig. 12.2c, d). The palisade tissue, vascular bundle sheaths and vascular bundles form the Kranz anatomy.

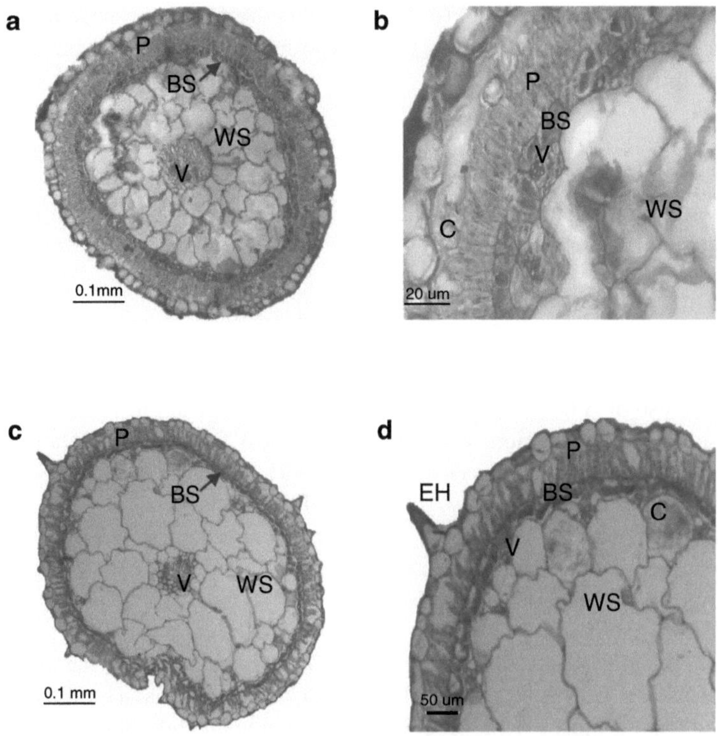

Fig. 12.2a–d Transverse sections of leaves of *Salsola arbuscula* and *Salsola collina*. **a** Leaf of *S. arbuscula*. **b** Partial section of *S. arbuscula*. **c** Leaf of *S. collina*. **d** Partial section of *S. collina*. *P* palisade cells, *BS* bundle sheath cells, *WS* water storage tissue, *V* vascular tissue, *C* crystal idioblast, *EH* epidermal hairs. *Bars* **a**, **c** 0.1 mm; **b** 20 μm; **d** 50 μm

In comparison, *S. collina* leaves have epidermal hairs and more developed water storage parenchyma cells; *S. arbuscula* leaves have thicker cuticles with a layer of hypodermal cells under them, and the palisade cells are longer and arranged more densely.

Kranz anatomy is an important character of C_4 plants, although a study by Voznesenskaya et al. (2001) showed that Kranz anatomy is not essential for terrestrial C_4 plant photosynthesis; however, Kranz anatomy is still the main criteria defining C_4 plants.

12.3 $\delta^{13}C$ Values of Photosynthetic Organs

The carbon isotope discrimination of plants with different carbon assimilation pathways is essentially responsible for variations in $\delta^{13}C$. The difference in $\delta^{13}C$ values is determined mainly by the heredity of the plant, although environmental factors also play a role (Hattersley 1982).

Stable carbon isotope ratios ($\delta^{13}C$) were analyzed by mass spectrometry. All data were reported relative to the PDB[1]. The measured results were determined using the following equation:

$$\delta^{13}C(‰) = \left[\left(R_{\text{sample}} - R_{\text{standard}}\right)/R_{\text{standard}}\right] \times 10^3 \tag{12.1}$$

Carbon isotope discrimination (Δ) was calculated by the following formula (Kloeppel et al. 1998):

$$\Delta = \left(\delta_{\text{air}} - \delta_{\text{plant}}\right)/\left(1 + \delta_{\text{plant}}\right) \tag{12.2}$$

An air $\delta^{13}C$ concentration of –9.23‰ was applied to all calculations (Su et al. 2003). The use of Δ helped focus the analysis on biological questions independent of the source CO_2 variation. Calculation of Δ also facilitated data analysis and direct inter-study comparisons (Körner et al. 1991; Kloeppel et al. 1998).

C_4 woody (and semi-woody) plants in the deserts of China are shown in Table 12.1. There are more than 40 species, concentrating in Chenopodiaceae and Polygonaceae, with no other families having been found. As Table 12.1 shows, the $\delta^{13}C$ values of C_4 desert plants are between –11‰ and –16‰; the Δ values are between 2‰ and 7‰.

The C_4 dicotyledonous Calligonum plants in Polygonaceae are the main woody plants of natural ecosystems in northwestern Chinese deserts. The distribution of *C. mongolicum* is most extensive, being distributed in all eight of the biggest desert areas of China, including the Tarim Basin desert, the Zhunggar Basin desert, the eastern Xinjiang desert, the Chaidamu Basin desert, the Hexi Corridor desert, the Alashan plateau desert, the Ordos Plateau desert, and the western Northeast and the eastern Inner Mongolia deserts. Thus, *C. mongolicum* is the most widely distributed C_4 woody desert plant species. *H. ammodendron* (Chenopodiaceae) is present in all desert areas except the western of Northeast desert and the eastern Inner Mongolia desert, being distributed over a large area of the Zhunggar Basin desert with good natural regeneration. *H. ammodendron* grown solely on precipitation is a shrub or subshrub, with a community coverage of less than 20%; if groundwater is available, it can grow up to subtree size, with relatively high community coverage, which can reach ~30%–40%.

The more mature the leaves or assimilating shoots were, and the later the growth period, the higher the apparent positive correlation of $\delta^{13}C$ values and short-term WUE of desert plants in temperate zones. Thus, at least until the first frost, $\delta^{13}C$ values can be used as an indicator of short-term WUE. If $\delta^{13}C$ values were adopted to represent the long-term WUE of desert plants in temperate zones, it is advised that sampling should be conducted from the last 10 days of August to the last 10 days of September (Su et al. 2005). Table 12.1 presents the results of analysis of samples in

[1]Pee Dee Belemnite carbonate standard, which is a Cretaceous belemnite (*Belemnitella americana*) from the Pee Dee formation of South Carolina

Table 12.1 Stable carbon isotope ratio ($\delta^{13}C$) and discrimination (Δ) of photosynthetic organs of C_4 woody (and semi-woody) plants in the deserts of China

Family and species	Life form	$\delta^{13}C$ (‰)	Δ (‰)	Occurrence (region/s)
Chenopodiaceae				
Aellenia glauca	Semi-shrub	–15.49	6.36	Northern Xinjiang
Anabasis aphylla	Semi-shrub	–14.54	5.39	Xinjiang
Anabasis brevifolia	Semi-shrub	–12.93	3.75	Alashan, Erjina of Inner Mongolia; Western Gansu; Xinjiang
Anabasis elatior	Semi-shrub	–13.52	4.35	Zhunggar Basin in Northern Xinjiang
Anabasis eriopoda	Semi-shrub	–13.02	3.84	Zhunggar Basin in Northern Xinjiang
Anabasis salsa	Semi-shrub	–13.24	4.06	Xinjiang
Anabasis truncata	Semi-shrub	–13.78	4.61	Buerjin, Urumchi in Xinjiang
Atriplex cana	Semi-shrub	–13.91	4.75	Buerjin in Xinjiang
Camphorosma monspeliaca	Semi-shrub	–12.94	3.76	Zhunggar Basin in Northern Xinjiang
Camphorosma lessingii	Semi-shrub	–13.12	3.94	Northern Xinjiang
Haloxylon ammodendron	Shrub or subtree	–14.34	5.18	All desert areas except the western of Northeast and the eastern Inner Mongolia desert
Haloxylon persicum	Shrub	–13.82	4.65	Zhunggar Basin in Xinjiang
Iljinia regelii	Semi-shrub	–11.93	2.73	Zhunggar Basin, Tarim Basin, Turpan Basin in Xinjiang
Kochia prostrata	Semi-shrub	–13.42	4.25	Desert regions in Northeastern, Northern and Northwest China
Salsola arbuscula	Shrub	–14.27	5.11	Inner Mongolia, Ningxia, Gansu, Xinjiang
Salsola orientalis	Semi-shrub	–14.31	5.15	Northern Xinjiang
Salsola passerina	Semi-shrub	–14.82	5.67	Inner Mongolia, Ningxia, Gansu, Xinjiang
Suaeda dendroides	Semi-shrub	–13.28	4.10	Northern Xinjiang
Polygonaceae				
Calligonum alaschanicum	Shrub	–15.66	6.53	Tengger Desert, Kubuqi Desert, East Alashan
Calligonum arborescens	Shrub	–14.23	5.07	Jinghe county in Xinjiang
Calligonum aphyllum	Shrub	–14.11	4.95	Ili region in Xinjiang
Calligonum caput-medusae	Shrub	–15.17	6.03	Northern Xinjiang

(continued)

Table 12.1 (continued)

Family and species	Life form	$\delta^{13}C$ (‰)	Δ (‰)	Occurrence (region/s)
Calligonum chinense	Shrub	–14.42	5.27	Jiuquan to Linze in Hexi Corridor in Gansu
Calligonum colubrinum	Shrub	–14.31	5.15	Qitai county in Xinjiang
Calligonum cordatum	Shrub	–14.34	5.18	Botanical Garden in the Middle Taklimakan Desert
Calligonum densum	Shrub	–14.65	5.50	Botanical Garden in the Middle Taklimakan Desert
Calligonum ebi-nuricum	Shrub	–14.58	5.43	Ebinur Lake in Xinjiang
Calligonum gobicum	Shrub	–12.97	3.79	Gaotai, Linze county in Hexi Corridor in Gansu
Calligonum jimunainum	Shrub	–13.63	4.46	Jimunai in Xinjiang
Calligonum junceum	Shrub	–15.06	5.92	Turpan Basin, Tuokexun in Xinjiang
Calligonum juochiangense	Subshrub	–14.38	5.23	Ruoqiang in east of Tarim Basin
Calligonum klementzii	Shrub	–15.07	5.93	Fukang, Qitai county in Xinjiang, Dunhuang in west of Gansu
Calligonum kozlovi	Shrub	–14.41	5.26	Chaidamu Basin in Qinghai
Calligonum kuerlese	Shrub	–13.75	4.58	Kolar, Luntai county in Xinjiang
Calligonum leucocladum	Shrub	–15.16	6.02	Northern Xinjiang
Calligonum mongolicum	Shrub	–14.49	4.32	Eastern Xinjiang, Western Inner Mongolia, Western Gansu, Ningxia
Calligonum potanini	Shrub	–13.61	4.44	Anxi, Linze county in Hexi Corridor in Gansu
Calligonum pumilum	Subshrub	–14.69	5.54	Shanshan, Yiwu in east of Xinjiang
Calligonum roborowskii	Shrub	–14.92	5.78	West of Linze in Hexi Corridor in Gansu, Tarim Basin in Xinjiang
Calligonum rubicundum	Shrub	–15.13	5.99	Erqis River Region in Xinjiang and Buerjin Desert
Calligonum zaidamense	Shrub	–15.32	6.18	Chaidamu Basin in Qinghai and Linze in Hexi Corridor in Gansu

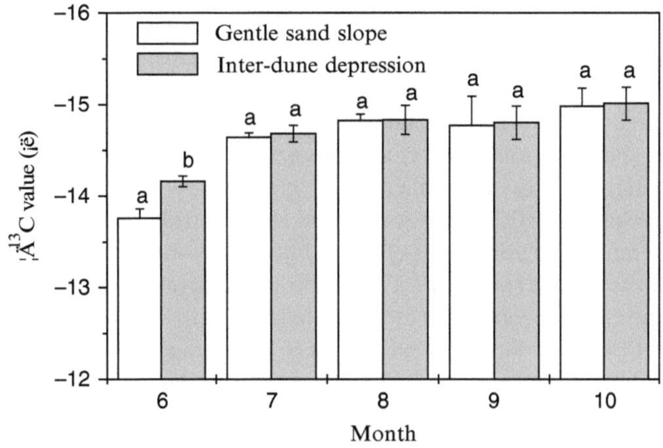

Fig. 12.3 Comparison of $\delta^{13}C$ values of assimilating shoots of *Calligonum mongolicum* in two habitats. Different lower case letters indicate significant difference at the 0.05 *P*-level in the same month, while the same lower case letter indicates a non-significant difference at the 0.05 *P*-level

this period. The Δ of plants with different carbon assimilation pathways essentially determines the variation in $\delta^{13}C$ values; Δ originates mainly from physicochemical processes in which the uptake of $^{13}CO_2$ is much slower than that of $^{12}CO_2$.

The carbon isotope composition of plants differs under different conditions and at different growth stages as a result of the combined action of plant species and environmental factors. The $\delta^{13}C$ value of assimilating shoots of *Calligonum mongolicum* growing in inter-dune depressions was smaller than that of shoots growing in gentle sand slopes (Fig. 12.3). On average, the $\delta^{13}C$ value of plants growing on gentle sand slopes during June to October was –14.59 ± 0.22‰, while that of plants growing in inter-dune depressions in the same period was –14.79 ± 0.07‰. The $\delta^{13}C$ values in different months also differed. There was a significant difference ($P < 0.05$) in $\delta^{13}C$ values on gentle sand slopes between June and other months, but no significant difference in the months from July to October. The $\delta^{13}C$ values indicated that *C. mongolicum* growing in inter-dune depressions and on gentle sand slopes differed significantly in terms of WUE in the early growth stage, but were not significantly different in the middle and late growth stages. The $\delta^{13}C$ values of *Haloxylon ammodendron* increased with decreasing soil moisture content, but the difference was not significant. The $\delta^{13}C$ values of the C_4 desert plant *Salsola passerina* from different habitats also showed notable differences (Su and Yan 2008). Variance analysis indicated significant differences among abandoned land, floodplain, Gobi mountain approaches and desert, with higher values in desert areas than in abandoned land ($P < 0.01$). Multiple comparison analysis showed that the $\delta^{13}C$ values of plants sampled from floodplains and Gobi mountain approaches were not significantly different ($P > 0.05$). When abandoned land and desert conditions were compared, however, their $\delta^{13}C$ values were found to be significantly different ($P < 0.01$).

12.4 Diurnal Course of Gas Exchange

Photosynthesis is one of the physiological processes of plants most sensitive to environmental changes. The diurnal course of CO_2 exchange is different in different seasons, some environmental conditions are more conducive to enhancing plant photosynthetic capacity than others, and plants show different photosynthetic acclimation characteristics under stress. The Hexi Corridor region in northwestern China has a temperate arid desert climate; in July, the midday photon flux density (PFD) can exceed 2,000 μmol m^{-2} s^{-1} and the air temperature can exceed 40°C, and in August the climatic conditions are suitable for the growth of desert plants. In late August, the PFD from 08:00 to 18:00 hours is, on average, 500 μmol m^{-2} s^{-1} lower and the average air temperature is 5°C lower than in late July, while the relative humidity (RH) of the air is 12% higher than in late July.

A typical hot sunny day in late July was selected on which to measure the diurnal course of gas exchange of in vivo assimilating shoots of *Haloxylon ammodendron* and *Calligonum mongolicum* under natural field conditions. The diurnal courses of net photosynthetic rate (Pn) of the two desert plants showed a bimodal mode, but their maximum Pn values were clearly different. The first peak Pn of 36.1 μmol CO_2 m^{-2} s^{-1} and 47.1 μmol CO_2 m^{-2} s^{-1} for *H. ammodendron* and *C. mongolicum* occurred at 13:00 and 11:00 hours, respectively, with a second peak value occurring at 16:00 and 15:00 hours (Fig. 12.4a). Daily mean net photosynthetic rates (mean values from 08:00 to 18:00 hours) for *H. ammodendron* and *C. mongolicum* were 18.0 ± 3.7 and 25.4 ± 4.6 μmol CO_2 m^{-2} s^{-1}, respectively.

In late August, mixed communities of *S. arbuscula* and *S. collina* were chosen, and their gas exchange characteristics observed. Under the same habitat, the diurnal course of their photosynthetic rates was as shown in Fig. 12.4b. The maximum Pn in *S. arbuscula* appeared at 12:00 hours with 22.2 μmol CO_2 m^{-2} s^{-1}, slightly reduced at 14:00 hours; the second peak could be seen at 16:00 hours with 20.1 μmol CO_2 m^{-2} s^{-1}, and the daily average value (08:00–18:00 hours) was 15.7 μmol CO_2 m^{-2} s^{-1}. Pn of *S. collina* increased gradually from a low at 08:00 hours, with the maximum appearing at 14:00 hours with 35.1 μmol CO_2 m^{-2} s^{-1}, followed by a significant reduction; the daily average value was 21.5 μmol CO_2 m^{-2} s^{-1}. According to *t* tests, from 08:00 to 14:00 hours, Pn of *S. collina* was significantly higher than that of *S. arbuscula* ($P < 0.05$), until after 16:00 hours, when the difference was no longer significant. The diurnal changes of Pn in *S. collina* showed a unimodal pattern, while those of *S. arbuscula* showed atypical bimodal changes, which may be due to the lessened high-temperature stress in late August, and there was no obvious photosynthetic midday depression phenomenon as seen in the same diurnal course of photosynthetic rate of *C. mongolicum* in late July.

Generally, the diurnal courses of plant photosynthesis under natural conditions can be divided into two types: a unimodal type with a high peak in the morning, and a bimodal type with two peaks, one in the morning and the other in the afternoon. The interval between the two peaks is called the "midday depression of photosynthesis" (Schulze and Hall 1982).

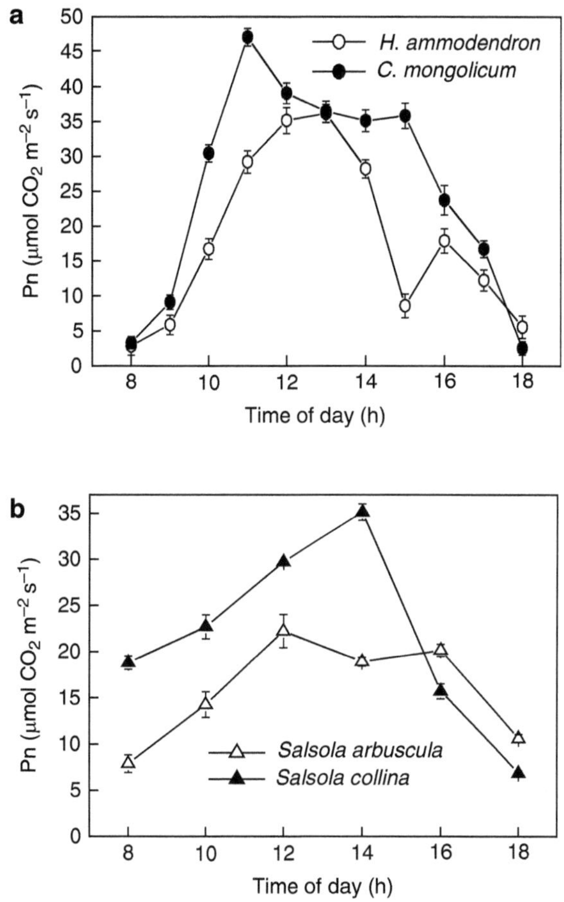

Fig. 12.4 Diurnal course of net photosynthetic rate (Pn) of *Haloxylon ammodendron*, *Calligonum mongolicum*, *Salsola arbuscula* and *Salsola collina*. *Bars* SE

From Fig. 12.5a, it can be seen that the diurnal variations in intercellular CO_2 concentration (Ci) of *H. ammodendron* and *C. mongolicum* tended generally to exhibit a concave pattern, but showed a convex pattern to different degrees around noon, e.g., around 15:00 hours for *H. ammodendron*. From the comparison shown in Fig. 12.4a, it can be seen that the photosynthetic rate is very much lowered at this point, showing a low valley. In the meantime, the stomatal limitation value (Ls) is also significantly reduced (Fig. 12.5b). At 14:00 hours, the Ci values of *C. mongolicum* were higher than those of adjacent points (Fig. 12.5a), and the photosynthetic rate at this point was also lower than that before and after this time (Fig. 12.4a), although with less significance. Concurrently, the Ls values were also correspondingly reduced (Fig. 12.5b).

According to the view of stomatal limitation value put forward by Farquhar and Sharkey (1982), the basic criteria for identifying the major factor causing decline in photosynthesis is the change trends of Ci and Ls. When the direction of the changes in photosynthetic rate and Ci is the same and Ls rises, the reduction in

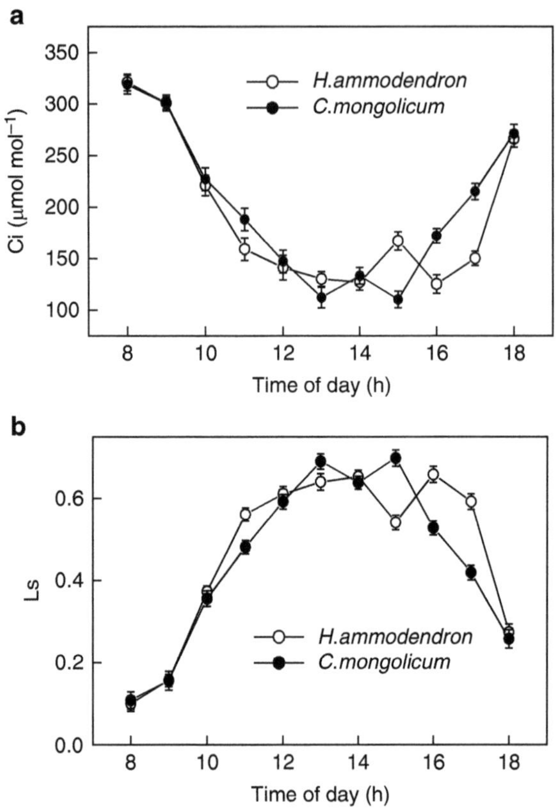

Fig. 12.5 Diurnal course of intercellular CO_2 concentration (Ci) and stomatal limitation value (Ls) of *Haloxylon ammodendron* and *Calligonum mongolicum*

photosynthetic rate is caused mostly by stomatal factors. If Ci and photosynthetic rate change in opposite directions and Ls is reduced, the reduction in photosynthetic rate is caused by non-stomatal factors. From the results of this study (Figs. 12.4a, 12.5), it can be seen that the reduction in photosynthetic rate for *H. ammodendron* and *C. mongolicum* can be attributed to non-stomatal factors.

12.5 Diurnal Changes in Chlorophyll Fluorescence

The measurement of chlorophyll fluorescence is a useful tool in quantifying the effect of stress on photosynthesis (Krause and Weis 1991; Schreiber et al. 1994). Under environmental conditions without stress, the photochemical efficiency of PS II (Fv/Fm) changes little – many species showing optimal values of around 0.83 (Johnson et al. 1993). When the light energy absorbed by photosynthetic organs exceeds the energy used and transferred, Fv/Fm decreases significantly. This is a good indicator of photoinhibition (Oquist et al. 1992; Demmig-Adams and Adams 1992). Photoinhibition occurs whenever the absorbed light energy exceeds the capacity of the plants to use the trapped energy through photosynthetic electron transport (Jia and Lu 2003). Previously, the term photoinhibition was used almost synonymously with damage to PS II. However, it has now been demonstrated that photoinhibition can result not only from some form of "damage" to PS II but also from an increase in thermal energy dissipation, which is a photoprotective process and does not represent damage (Demmig-Adams and Adams 1992). Short-term photoinhibition is not due to photodamage since this aggravated photoinhibition can be rapidly and fully reversed; large reversible decreases in Fv/Fm are compensated by proportional increases in non-photochemical processes related to photoprotection, under long-term photoinhibition the decreased Fv/Fm may be associated partly with protective processes (Jia and Lu 2003).

From analysis of the reasons underlying the reduction of Pn of *H. ammodendron* and *C. mongolicum* under high temperatures in late July, it was seen that these were caused mainly by non-stomatal factors. Determination of chlorophyll fluorescence showed that, from changes in Fv/Fm, *H. ammodendron* and *C. mongolicum* had lowest Fv/Fm values in a day, which then stopped reducing and started to rise. However, the timing of this reduction and rise, and the degree of change differed for different species. A comparative analysis of the relationship between Pn and Fv/Fm showed that the two lowest values always occurred at the same time. The minimum value of Fv/Fm in *C. mongolicum* was 0.79, while it was 0.73 in *H. ammodendron*. Thus, photoinhibition of *H. ammodendron* is serious. In order to further explore the diurnal changes in chlorophyll fluorescence under different environmental regimes, we compared three such regimes: (1) an arid environment (AE) where plants were exposed without precipitation, and with daytime air RH ranging from 9 to 35% – this situation is considered normal for the temperate arid desert region (Su et al. 2007); (2) moist atmospheric conditions during post-rain (PR), which occurs frequently when rainfall exceeds 8 mm, and with a daytime RH ranging from

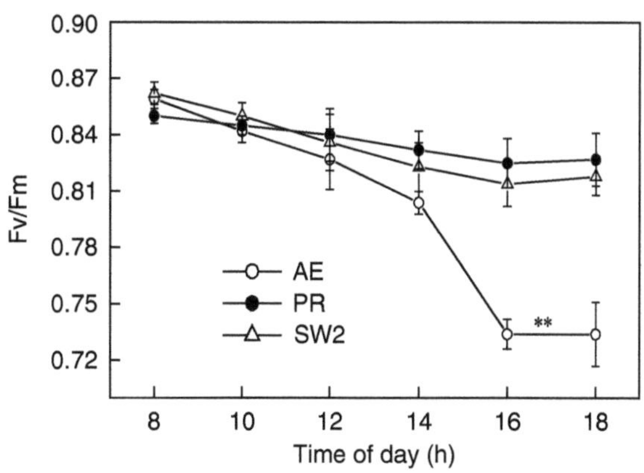

Fig. 12.6 Diurnal changes in maximum photochemical efficiency of PS II (Fv/Fm) of *Haloxylon ammodendron* under three environmental regimes. *AE* Arid environment, *PR* moist atmospheric conditions during post-rain, *SW2* on day 2 after watering. *Asterisks* indicates significant differences ($P < 0.01$) for the AE compared to PR and SW2 between 14:00 and 18:00 hours. *Bars* SE

20 to 70%; (3) soil water supplemented (SW) by the addition of 100 l water at a depth of 25 cm – on day 2 (SW2) after watering, soil volumetric moisture content in the soil layer (20–60 cm) increased by 236%.

It can be seen from the photochemical efficiency of PS II that the Fv/Fm ratio for *H. ammodendron* was lowest between 14:00 and 16:00 hours under AE. Fv/Fm under AE was 0.86 at 08:00 hours and declined to 0.73 at 16:00 hours (Fig. 12.6). Statistical analyses show that there was no significant difference in Fv/Fm during the period 08:00–12:00 hours among the three environmental regimes, but there were significant differences ($P < 0.01$) between the AE and PR or SW2 treatments during the period 14:00–18:00 hours. Minimum values of Fv/Fm recorded under SW2 and PR conditions were similar; these values exceeded 0.81 (Fig. 12.6). In general, healthy plants have an Fv/Fm value of 0.8–0.9 (Odasz-Albrigtsen et al. 2000).

12.6 Changes of Photosynthetic and Physiological Parameters under CO_2 Enrichment

Studies of the responses of C_4 desert species to light intensity and CO_2 have shown that, in the current environment (CO_2 concentration of 360 μmol mol^{-1}), the light compensation point (LCP) and light saturation point (LSP) of *H. ammodendron* are 79 μmol m^{-2} s^{-1} and 1,660 μmol m^{-2} s^{-1}, respectively; the CO_2 compensation point (CCP) is 2 μmol mol^{-1}. For *C. mongolicum*, LCP, LSP and CCP values are 76 μmol m^{-2} s^{-1}, 1,756 μmol m^{-2} s^{-1} and 4 μmol mol^{-1}, respectively. For *S. arbuscula*, the corresponding values are 220 μmol m^{-2} s^{-1}, 1,820 μmol m^{-2} s^{-1} and 9 μmol mol^{-1}. For *S. collina*, LCP and CCP are 152 μmol m^{-2} s^{-1} and 7 μmol mol^{-1}, respectively. Apparent quantum yield (AQY) of *H. ammodendron* and *C. mongolicum* were 0.044 and 0.055 mol mol^{-1}, respectively. AQY for *S. arbuscula* and *S. collina* were 0.045 and 0.085 mol mol^{-1}, respectively. Dark respiration rates of *S. arbuscula* and *S. collina* were 10.2 and 14.3 μmol CO_2 m^{-2} s^{-1}, respectively.

When CO_2 was enriched to 450 μmol mol^{-1} from the current CO_2 level, photosynthesis in *H. ammodendron* was improved (Fig. 12.7a), but the LCP remained fairly constant (Table 12.2); when CO_2 was enriched to 650 μmol mol^{-1}, the photosynthetic rate was reduced, LCP rose, LSP was lowered (Table 12.2), the efficiency for solar energy utilisation (ESEU) was lowered, and the maximum photosynthetic rate was also reduced. Maximum photosynthetic rate refers to the photosynthetic intensity at which the plants reach LSP. Maximum photosynthetic rate and AQY can be used to explain CO_2 consumption and transmission in plant tissues, and the changes in storage capacity of carbohydrate compounds of different plant species or the same species under different CO_2 concentration; they can also reflect changes in content and activity of photosynthetic enzymes. Furthermore, maximum photosynthetic rate and AQY are also related to CO_2 transfer and concentrating mechanisms (Jiang and Lin 1997). When CO_2 was enriched to 450 μmol mol^{-1}, the type of Pn/PFD curve of *C. mongolicum* became open (Fig. 12.7b),

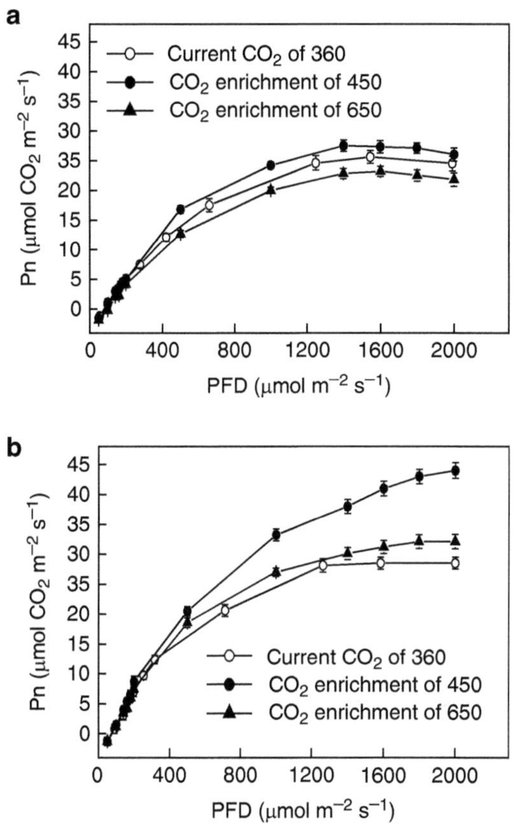

Fig. 12.7 Changes in net photosynthetic rate (Pn) of *Haloxylon ammodendron* (**a**) and *Calligonum mongolicum* (**b**) with different photon flux density (PFD) under different CO_2 concentrations

Table 12.2 Photosynthetic and physiological parameters of *Haloxylon ammodendron* and *Calligonum mongolicum* under different CO_2 concentrations

Species	CO_2 concentration (μmol mol^{-1})	Light compensation point (μmol m^{-2} s^{-1})	Light saturation point (μmol m^{-2} s^{-1})	Apparent quantum yield (mol mol^{-1})	Maximum photosynthetic rate (μmol m^{-2} s^{-1})
H. ammodendron	360	79***	1,660***	0.044***	27.2***
	450	80***	1,600***	0.044***	27.6***
	650	97***	1,533***	0.043***	22.2***
C. mongolicum	360	76***	1,756***	0.057***	30.6***
	450	73***	–	0.064***	–
	650	76***	1,843***	0.057***	31.3***

** $P \leq 0.01$, *** $P \leq 0.001$

LCP was reduced (Table 12.2), photosynthetic duration was elongated, and ESEU significantly enhanced. However, when CO_2 was enriched to 650 μmol mol^{-1}, the photosynthetic rate was reduced almost to that of the current CO_2 level, while LCP and AQY became equal (Table 12.2).

Larcher (1995) reported that the LSP of C_4 plants was higher than 1,500 μmol m^{-2} s^{-1}, and that of C_3 plants was 1,000–1,500 μmol $m^{-2}s^{-1}$. Our results as reported above are consistent with those of Larcher. *H. ammodendron* and *C. mongolicum* all had high Pn, with a maximum of more than 30 μmol CO_2 m^{-2} s^{-1} in the diurnal course. The rise in atmospheric CO_2 concentration has the most direct and most rapid influence on the photosynthesis of plants. Plants with different carbon assimilation pathways (C_3, C_4 or CAM) differ in their response to rising CO_2 concentrations. It has been generally considered that, in the presence of doubled CO_2 concentration, enhancement of the photosynthetic rate of C_4 plants was lower than 10%, or even could not be increased (Cure and Acock 1996). Our study has shown that when CO_2 was enriched to 450 μmol mol^{-1}, the photosynthetic rate of the C_4 plant *H. ammodendron* increased, but when CO_2 was enriched to 650 μmol mol^{-1}, its photosynthetic rate was reduced (Fig. 12.7a, b). This is consistent with the results of an experiment on the response of the herbal plant *Panicum maximum* within Biosphere 2 (Jiang and Lin 1997). When CO_2 was enriched to 450 μmol mol^{-1}, the photosynthetic rate of the C_4 plant *C. mongolicum* was apparently raised and no light saturation point occurred, but when CO_2 was enriched to 650 μmol mol^{-1}, the photosynthetic rate was reduced nearly to the level seen with current CO_2 levels.

Acknowledgements We are grateful for the financial support of the National Natural Sciences Foundation of China (40771005, 30870382) and the key project of the Chinese Academy of Sciences (KZCX2-XB2–04–01, KZCX2-XB2–09–03). This study was carried out in Linze Inland River Basin Research Station.

References

Bender MM (1971) Variations in the $^{13}C/^{12}C$ ratios of plants in relation to the pathway of photosynthetic carbon dioxide fixation. Phytochemistry 10:1239–1244

Carolin RC, Jacobs SWL, Vesk M (1975) Leaf structure in Chenopodiaceae. Bot Jahrb Syst Pflanzengesch Pflanzengeogr 95:226–255

Cure JD, Acock B (1996) Crop response to carbon dioxide doubling: a literature survey. Agric For Meteorol 38:127–145

Demmig-Adams B, Adams WW (1992) Photoprotection and other response of plants to high light stress. Annu Rev Plant Physiol Plant Mol Biol 43:599–622

Farquhar GD (1983) On the nature of carbon isotope discrimination in C_4 species. Aust J Plant Physiol 10:205–226

Farquhar GD, Sharkey TD (1982) Stomatal conductance and photosynthesis. Annu Rev Plant Physiol 33:317–345

Farquhar GD, O'Leary MH, Berry JA (1982) On the relationship between carbon isotope discrimination and the intercellular carbon dioxide concentration in leaves. Aust J Plant Physiol 9:121–137

Farquhar GD, Hubick KT, Condon AG, Richards RA (1989) Carbon isotope fractionation and plant water use efficiency. In: Rundel PW, Ehleringer JR, Nagy KA (eds) Stable isotope in ecological research. Springer, Berlin, pp 21–40

Hatch MD (2002) C_4 photosynthesis: discovery and resolution. Photosyn Res 73:251–256

Hattersley PW (1982) $\delta^{13}C$ values of C_4 types in grasses. Aust J Plant Physiol 9:139–154

Jia HS, Lu CM (2003) Effects of abscisic acid on photoinhibition in maize plants. Plant Sci 165:1403–1410

Jiang GM, Lin GH (1997) Changes in photosynthetic capacity of several species experienced in high CO_2 concentrations in Biosphere 2. Chin Sci Bull 42:434–437

Johnson GN, Young AJ, Scholes JD, Horton P (1993) The dissipation of excess excitation energy in British plant species. Plant Cell Environ 16:673–679

Kloeppel BD, Gower ST, Treichel IW, Kharuk S (1998) Foliar carbon isotope discrimination in *Larix* species and sympatric evergreen conifers: a global comparison. Oecologia 114:153–159

Körner C, Farquhar GD, Wong SC (1991) Carbon isotope discrimination by plants follows latitudinal and altitudinal trends. Oecologia 88:30–40

Krause GH, Weis E (1991) Chlorophyll fluorescence and photosynthesis: the basics. Annu Rev Plant Physiol Plant Mol Biol 42:313–349

Lajtha K, Michener RH (1994) Stable isotopes in ecology and environmental science. Blackwell, London, pp 1–5

Larcher W (1995) Physiological plant ecology: ecophysiology and stress physiology of functional groups. Springer, Berlin, pp 94–111

Marshall JD, Zhang P (1994) Carbon isotope discrimination and water-use efficiency in native plants of the North-central Rockies. Ecology 75:1887–1895

Odasz-Albrigtsen AM, Tømmervik H, Murphy P (2000) Decreased photosynthetic efficiency in plant species exposed to multiple airborne pollutants along the Russian-Norwegian border. Can J Bot 78:1021–1033

Oquist G, Chow WS, Anderson JH (1992) Photoinhibition of photosynthesis represents a mechanism for the long-term regulation of photosystem II. Planta 186:450–460

Pyankov VI, Black CC Jr, Artyusheva EG, Voznesenskaya EV, Ku MSB, Edwards GE (1999) Features of photosynthesis in *Haloxylon* species of Chenopodiaceae that are dominant plants in central Asian deserts. Plant Cell Physiol 40:125–134

Schreiber U, Bilger W, Neubauer C (1994) Chlorophyll fluorescence as a nonintrusive indicator for rapid assessment of in vivo photosynthesis. In: Schulze ED, Caldwell MM (eds) Ecophysiology of photosynthesis. Springer, Berlin, pp 49–70

Schulze ED, Hall AE (1982) Stomatal responses, water loss and CO_2 assimilation rates of plants in contrasting environments. In: Lange OL, Nobel P S, Osmond CB, Ziegler H (eds) Encyclopedia of plant physiology, new series vol 12 B. Springer, Berlin, pp 181–224

Smith BN, Epstein S (1971) Two categories of $^{13}C/^{12}C$ ratios for higher plants. Plant Physiol 47:380–384

Su PX (2007) Geographical distribution of C_4 woody plants in relation to climate in the desert regions of China. Postdoctoral Research Dissertation, Cold and Arid Regions environmental and Engineering Research Institute, Chinese Academy of Sciences, Lanzhou, pp 56–59

Su PX, Cheng GD, Yan QD (2007) Photosynthetic regulation of C_4 desert plant *Haloxylon ammodendron* under drought stress. Plant Growth Regul 51:139–147

Su PX, Chen HS, Li QS (2003) Characteristics of $\delta^{13}C$ values of desert plants and their water use efficiency indicated by $\delta^{13}C$ values in the desert of central Hexi Corridor region. J Glaciol Geocryol 25:597–602

Su PX, Yan QD, Chen HS (2005) $\delta^{13}C$ values and water use efficiency of the leaves and assimilating shoots of desert plants. Acta Bot Boreal Occident Sin 25:727–732

Su PX, Yan QD (2008) Stable carbon isotope variation in plants and their indicating significances along the inland Heihe River basin of northwestern China. Acta Ecol Sin 28:1616–1624

Sun ZJ, Livingston NJ, Guy RD, Ethier GJ (1996) Stable carbon isotopes as indicators of increased water use efficiency and productivity in white spruce (*Picea glauca* Voss) seedlings. Plant Cell Environ 19:887–894

Voznesenskaya EV, Franceschi VR, Kiirats O, Freitag H, Edwards GE (2001) Kranz anatomy is not essential for terrestrial C_4 plant photosynthesis. Nature 414:543–546

Winter K, Smith JAC (1996) Crassulacean acid metabolism. Springer, Berlin, pp 2–10

Chapter 13
Polyamines and Plant Adaptation to Saline Environments

Vladimir V. Kuznetsov and Nina I. Shevyakova

Abstract Polyamines are universal organic polycations implicated in a wide array of fundamental processes in plants, ranging from signalling, genome expression, and plant growth and development, to plant adaptation to abiotic stresses. Stress-induced accumulation of polyamines often correlates with improvements in plant tolerance. Polyamines can protect nucleic acids and proteins and modulate the functions of macromolecules under extreme environments. Polyamines are also regulators of expression of genes encoding stress proteins. They possess antioxidant properties. Taken together, these recent findings have promoted intense efforts to characterise in detail the mechanisms of regulation of polyamine homeostasis, and to elucidate realisation of their multifaceted role in plants under stress. However, the molecular mechanisms underlying polyamine participation in plant adaptation to stress remain incompletely understood. In order to better understand the role of polyamines in plant adaptation, we focus on data concerning gene expression obtained by molecular biology methods using natural salt-tolerant species (halophytes) and also mutant and transgenic plants manifesting a high tolerance to salinity. The restriction of plant growth and productivity caused by salinity is especially acute in arid and semi-arid regions. In these regions, the influence of salt stress is aggravated by the additional action of other xerothermic factors, in particular drought and high temperature. In this chapter, particular emphasis will be paid to the possible role of polyamines in ameliorating the detrimental effects of salinity on plants during adaptation processes.

V.V. Kuznetsov (✉) and N.I. Shevyakova
Timiryazev Institute of Plant Physiology, Russian Academy of Sciences, Botanicheskaya str 35, 127276 Moscow, Russia
e-mail: vlkuzn@ippras.ru

K.G. Ramawat (ed.), *Desert Plants*,
DOI 10.1007/978-3-642-02550-1_13,

13.1 Introduction

Salinity is a major environmental constraint to crop productivity in arid and semi-arid regions of the world. An excess of salts in the soil adversely affects almost all aspects of plant growth and development. Soil salinity imposes ion toxicity, osmotic stress, nutrient (N, Ca, K, P, Fe, Zn) deficiency, and oxidative stress on the plant and thus limits plant productivity. Nearly 20% of the world's irrigated lands are affected by salinity (Anonymous 2004; Sairam et al. 2006). Plants are classified as glycophytes or halophytes according to their capacity to grow on a high-salt medium. Flowers and Colmer (2008) defined halophytes as plants capable of completing their life cycle at a salt concentration of at least 200 mM NaCl, i.e. under conditions similar to those that might be encountered in the natural environment. This definition, based on the completion of the life cycle, allows a separation of so-called "natural halophytes" and plants tolerating salt but normally not living under saline conditions. Among all terrestrial plants, halophytes comprise only 2% (Dajic 2006). Most plants are glycophytes (non-halophytes) and cannot tolerate salt stress. Development of salinity-tolerant crops is essential if we are to sustain agricultural production. To achieve salt tolerance, plants must either prevent or alleviate the damage associated with salt stress, and be able to tolerate a high salt environment. Plant adaptation to various abiotic stresses is one of most complex adaptive processes, involving numerous changes, including the increased expression of many stress-related genes responsible for osmotic adjustment, ion compartmentation, accumulation of compatible solutes, expression of antioxidant enzymes, and suppression of energy-consuming pathways. Transcriptome analysis of the salt-stressed glycophyte plant *Arabidopsis* using a GeneChip microarray showed that salinity affected approximately 13% of the genome: about 424 genes were up-regulated in the roots, and about 278 genes were up-regulated in the leaves (Kreps et al. 2002). Importantly, many of the affected genes, which included transcription factors and kinases, could function in salt stress-responsive regulatory circuits (Seki et al. 2002).

Much effort has been devoted toward understanding the mechanisms of plant salt tolerance with the eventual goal of improving the performance of crop plants in saline soils. Natural halophytic plants, which grow rapidly at high inland salinity, have long been used in an attempt to isolate genes inducing mechanisms of adaptation to salinity. However, an understanding of the combined nature of adaptive traits to salinity is necessary in order to determine the major criteria of salt tolerance and lead to a clear understanding of the physiological, biochemical, and molecular basics of natural halophytic plants.

The use of model halophytic species, such as *Mesembryanthemum crystallinum* and *Thellungiella halophila*, to enhance further research, provides a useful system with which to solve this problem. Such plants make it possible to test the view that halophytes have similar salt tolerance genes to glycophytes, and thus similar regulatory networks to those found in glycophytes, but with different thresholds (Zhu 2000). It should be added, however, that the response of plants to different stress conditions varies but different kinds of stresses often lead to identical or similar responses.

In this context, it is important to note that metabolism of polyamines (PAs) – low-molecular-weight organic polyfunctional polycations – can be a promising marker of plant resistance to various kinds of stresses. Analysis of the effect of abiotic stress on PA metabolism suggests that these nitrogenous "compatible" compounds accumulate in plants under several abiotic stress stimuli, including salinity (Bouchereau et al. 1999; Kuznetsov et al. 2006a).

13.2 What are Polyamines?

Polyamines [putrescine^{+2} (Put), spermidine^{+3} (Spd), spermine^{+4} (Spm), and cadaverine^{+2} (Cad)] are present in all compartments of the plant cell (Bouchereau et al. 1999; Kaur-Sawhney et al. 2003). Like hormones, PAs displaying high biological activity are involved in the processes of replication and gene expression, membrane stabilisation, enzyme activity modulation, cell division and elongation, plant growth and development (Galston et al. 1997; Walden et al. 1997). The concentrations of PAs in the plant (10^{-9}–10^{-5} M) are much higher than those of endogenous phytohormones (10^{-13}– 10^{-6} M). The total PA concentration and the ratios between individual PAs vary markedly depending on plant species, organ, and tissue, and developmental stage.

13.2.1 Biosynthesis

The pathways of PA biosynthesis in higher plants are regulated by a limited number of key enzymes (Tiburcio et al. 1997). Most of the genes for the enzymes involved are cloned; and their regulation is actively studied (Kakkar and Sawhney 2002).

Put is produced directly from ornithine by ornithine decarboxylase (ODC, EC 4.1.1.17) or indirectly from arginine by arginine decarboxylase (ADC, EC 4.1.1.19) via agmatine (Agm). Conversion of Agm to Put requires two distinct enzymes: agmatine iminohydrolase (EC 3.5.3.12) and *N*-carbamoylputrescine amidohydrolase (EC 3.5.1) (Fig. 13.1).

ADC has been shown to be localised in essentially all organs (flowers, stems, seeds, leaves, and roots; Bortolotti et al. 2004; Paschalidis and Roubelakis-Angelakis 2005a, 2005b). It was demonstrated that ADC protein was present in two different compartments – chloroplasts in the leaves and nuclei in the roots – which may be related to specific functions of ADC in different cell types (Borrell et al. 1995; Bortolotti et al. 2004).

Like in animals, plant ODC is localised in the nucleus (Hanfrey et al. 2001; Kakkar and Sawhney 2002). The ODC pathway functions more actively in plants at

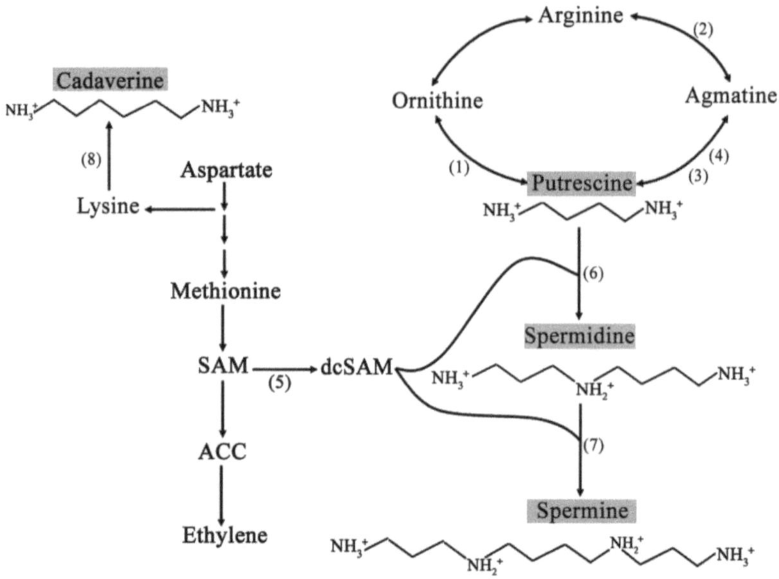

Fig. 13.1 Pathways of biosynthesis of main plant polyamines [PAs: putrescine (Put), spermidine (Spd), spermine (Spm) and cadaverine (Cad)]. Enzymes involved: *1* ornithine decarboxylase (ODC), *2* arginine decarboxylase ADC, *3* agmatine iminohydrolase, *4* N-carbamoylputrescine amidohydrolase, *5* SAM decarboxylase (SAMC), *6* spermidine synthase (SPDS), *7* spermine synthase (SPMS), *8* lysine decarboxylase (LDC). *ACC* 1-aminocyclopropane-1-carboxylic acid, *dcSAM* decarboxylated S-adenosyl-methionine. The genes involved are *ADC1*, *ADC2*, *ODC*, *SPDS1*, *SPDS2*, *ACL5*, and *SPMS*

early stages of development. This means that ODC dominates PA biosynthesis during periods of active cell division.

Spd and Spm are synthesised by successive attachment of aminopropyl, first to Put and then to Spd (Fig. 13.1). These reactions are catalysed by aminopropyl-transferases, Spd synthase (SPDS, EC 2.5.1.16), and Spm synthase (SPMS, EC 2.5.1.22). Aminopropyl is formed due to decarboxylation of S-adenosylmethionine (SAM) by SAM decarboxylase (SAMDC, EC 4.1.1.50), which has a very short half-life (from 5 to 60 min) and is a rate-limiting enzyme of Spd and Spm biosynthesis. SPDS and SAMDC are known to be localised in the cytoplasm (Bouchereau et al. 1999; Kaur-Sawhney et al. 2003). SAM is produced from methionine and ATP by SAM synthetase (SAMS, EC 2.5.1.6). It is important to note that SAM is not only the substrate for SAMDC, providing aminopropyl for PA synthesis, but also a basic donor of methyl groups for numerous reactions of transmethylation and a principal negative regulator of threonine and methionine biosyntheses. *Arabidopsis* contains two different genes encoding SPDS (*SPDS1* and *SPDS2*) and SPMS [*SPMS* and *ACAULIS5* (*ACL5*)]. It has been demonstrated that exogenous abscisic acid (ABA) up-regulated *SPMS* but not *ACL5* expression (Hanzawa et al. 2000; Urano et al. 2003). Thus, four enzymes are involved directly or indirectly in the formation of Spd and Spm.

Cad is a diamine, which are found in plants relatively infrequently; it is derived from lysine, a byproduct of the aspartate pathway of methionine synthesis (Shevyakova and Kir'yan 1995). Its synthesis from lysine is catalysed by lysine decarboxylase (LDC, EC 4.1.1.18) via pyridoxal phosphate-dependent decarboxylation (Fig. 13.1). As distinct from Put, Spd, and Spm, which are synthesised on thylakoid membranes (Borell et al. 1995), Cad is produced in the chloroplast stroma (Herminghaus et al. 1991).

The enzymes of PA biosynthesis are controlled at the transcriptional, translational, and posttranslational levels. It is believed that all enzymes of PA biosynthesis, including SAMDC, are synthesised initially as inactive proteins, which are subjected to posttranslational processing with the formation of mature enzymes.

13.2.2 Catabolism

Polyamine degradation in plants is catalysed by two oxidative enzymes: copper-containing diamine oxidase (DAO, EC 1.4.3.6) and flavoprotein-dependent polyamine oxidase (PAO, EC 1.5.3.3; Rea et al. 2004; Cona et al. 2006). DAO is localised in the cytoplasm and cell walls, where it provides the hydrogen peroxide required for suberinisation and lignification, which confers firmness to the cell walls (Cona et al. 2003). DAO catalyses oxidation of primary amino groups in many biogenic amines, including mono-, di-, and polyamines, with Put and Cad as most preferable substrates. As a result of DAO-catalysed oxidation of Put, Spd, and Spm, amino aldehydes, hydrogen peroxide, and ammonia are produced. PAO catalyses oxidation of secondary amino groups of Spd and Spm with the formation of H_2O_2, 1,3-diaminopropane (1,3-DAP), Δ^1-pyrroline and the corresponding aldehyde. Barley (*Hordeum vulgare*) PAO has been reported to localise in the vacuole (Cervelli et al. 2004). *Arabidopsis thaliana* PAO could also be localised in the cytoplasm (Tavladoraki et al. 2006). A recent paper (Moschou et al. 2008b) described a novel PAO (PAO3) located in *A. thaliana* peroxisomes; in mammals, it can convert Spm to Spd, and Spd to Put. This work demonstrated a new role for plant peroxisomal PAO in PA oxidation and back-conversion that could be of physiological importance in the case of abiotic stress.

13.2.3 Intracellular and Inter-Organ Transport

In order to understand the physiological role of PAs or any other biologically active compounds, their capacity for inter-organ transport should be considered. The presence of large amounts of PAs in the xylem sap and phloem exudates, which was first detected by Friedman et al. (1986) and then in our experiments (Shevyakova et al. 2001; Kuznetsov et al. 2002), is a good argument for inter-organ PA transport. Initial studies were directed towards the kinetics of Put transport.

DiTomaso et al. (1992) presented comprehensive data on the kinetics of Put transport, its subcellular distribution, and excretion, which were obtained on intact maize seedling roots. According to the results, roots absorbed 0.05 and 1.0 mM Put linearly for 30–40 min. Initially, Put penetrated the root apoplast, followed by its transport across the plasma membrane. Such reports suggest that a portion of the exogenously applied Put is metabolised in maize root cell walls by DAO, but the bulk of the Put is transported across the plasmalemma by a carrier-mediated process, similar to that in animal systems. It was also shown that Put accumulated in root-cell vacuoles, which served as PA storage depots. From the vacuole, Put could be transported back across the tonoplast and plasma membrane into the apoplast of the cortex and epidermis cells.

13.2.4 Polyamine Conjugates

Much attention is paid to the formation of PA conjugates with hydroxycinnamic acids (acid-soluble fraction). In this case, PAs produce amide bonds using CoA esters for activation of carboxylic groups with the help of enzymes known as transferases (Bagni and Tassoni 2001; Martin-Tanguy 2001). PA conjugates with phenolic acids combine the properties of both parent compounds and display great chemical, metabolic, and functional diversity (Edreva et al. 2007). Soluble conjugates are important for the control of intracellular PA concentrations; for their interaction with compounds of the cell wall, especially hemicelluloses and lignin; and are a vehicle for PA transport (Havelange et al. 1996).

The formation of PA conjugates with proteins (acid-insoluble conjugates) is based on posttranslational covalent binding of the primary amino groups of PAs to the γ-carboxamide group of protein endoglutamine residues. This is catalysed by Ca^{2+}-dependent and Ca^{2+}-independent transglutaminases (TGase, KΦ. 2.3.2.13; Dondini et al. 2003). Some derivatives also confer additional positive charges to the protein: monoderivatives can interact via their free amino groups with other molecules by various types of binding (Dondini et al. 2003). Such protein modifications, together with the involvement of PA conjugates, could be important for stabilisation of molecular complexes in the thylakoid membranes of osmotically stressed leaves (Legocka and Zaichert 1999).

13.2.5 Components of Signalling Systems

A series of studies performed by Italian researchers (Tassoni et al. 1998) was aimed at determining the general principles of PA-specific binding to plasma membrane proteins. This binding can fulfill a dual role: (1) an early event in PA signal recognition, or (2) binding to a specific transporter. PAs are thought to mediate phytohormone signaling, i.e. they fulfill the role of second messengers (Galston et al. 1997; Kakkar and Sawhney 2002). Experimental evidence for this PA function was first obtained in animal cells (Koenig et al. 1983). Later, in work with cultured animal cells, it was shown that, along with control of the intracellular Ca^{2+} level, PAs were involved in hormonal signal transduction via their binding to G-proteins, which activate hormone recognition by the receptor.

The role of PAs as second messengers in plants was demonstrated in a series of studies by Messiaen and van Cutsem (1999). These authors focussed on the presence of PAs in the cell walls, where they produced complexes with acidic polysaccharides (pectins); these complexes were considered earlier as one of the factors involved in the control of pH, thus affecting cell expansion, or in the control of methylesterase activity in the cell walls. It was also known from some studies that pectin fragments (α-1,4-oligogalacturonides) were capable of modulation of various morphological and physiological processes in the cell walls, and at the level of the whole plant. However, PA involvement in the maintenance of plant growth and development, and their interaction with phytohormones was not studied properly at the molecular level. Some reports about the more complex character of interaction between some phytohormones and PAs appeared (Rakova and Romanov 2005). These latter authors speculated that the physiological role of PA-induced inhibition of cytokinin effects could be involved in the compensatory regulation of intracellular cytokinin content if their concentration became excessive.

13.3 Polyamines and Protection of Plants against Salt Stress

In order to properly understand the role of PAs during plant development under normal and stress conditions, it is important to analyse experimental data obtained with halophytes. We also utilise data from transgenic and mutant plants displaying changed PA metabolism, and from the use of specific inhibitors. Major changes in PA metabolism occur in response to various abiotic stress conditions (Bouchereau et al. 1999; Kuznetsov et al. 2006a). Stress-induced PA accumulation, and the protective function of PAs against various stresses play important roles in ameliorating plant responses to abiotic stresses, in particular salinity.

13.3.1 Regulation of Endogenous Plants Polyamine Content under Stress

Stress-induced changes in PA metabolism and their role in plant responses to abiotic stressors remain one of the key problems in the field of plant adaptive strategies. A large body of information has appeared in the literature concerning stress-dependent PA accumulation under osmotic shock (Bagga et al. 1997), drought (Erdei et al. 2005; Yang et al. 2007) and chilling (Shen et al. 2000).

In the case of salt stress, an increase in PA levels was reported for several glycophytic crops, including rice (*Oryza sativa*), mung bean (*Vigna radiata*), maize (*Zea mays*), barley (*Hordeum vulgare*), and sorgum (*Sorghum bicolor*).

The principal enzymes of PA biosynthesis are under complex metabolic, developmental, and environmental control. It was shown that, in transgenic rice plants overexpressing ADC, the activity of SAMDC – the rate-limiting enzyme of Spd and Spm biosyntheses – was enhanced by Put and suppressed by Spd (Roy and Wu 2002). In one rice cultivars, Spd and Spm accumulation was related to enhanced expression of ADC (Chattopadhyay et al. 1997), whereas in salt-tolerant rice and tomato plants, accumulation of these PAs was correlated with enhanced expression of SAMDC (Krishnamurthy and Bhagwat 1989). It should be noted that the induced expression of SAMDC resulting in Spd and Spm accumulation is often considered to be more important for plant defence against stress than Put accumulation. The instability of SAMDC (half-life 5–60 min) permits rapid changes in the amount of this enzyme in the cell and, consequently, in the PA level in response to new conditions.

It is of interest that, in detached oat leaves subjected to osmotic stress, processing of the inactive ADC precursor was inhibited by Spm treatment (Borrell et al. 1996). The authors believe that, when Spm is absent, transcription of the *ADC* gene is enhanced, and inactive ADC precursor is synthesised, which is processed into active ADC. In the presence of exogenous Spm, the level of mRNA increased but the number of active enzyme molecules decreased and Put accumulation was blocked.

In some plants, *Arabidopsis* for example, Put can be synthesised only by ADC (Hanfrey et al. 2001). This plant, like some Brassicaceae plants, contains two genes for ADC (*ADC1* and *ADC2*; Galloway et al. 1998). Using the methods of molecular biology, it was shown that mechanical injury to *Arabidopsis* leaves or their treatment with jasmonic acid resulted in the activation of only a single gene for Put synthesis: the *ADC2* gene (Perez-Amador et al. 2002). This was accompanied by a transient increase in the content of Put but not of Spd or Spm. It is of interest that the sequences of the ADC1 and ADC2 proteins are 80% identical, differing only in their terminal fragments. It might be that these proteins differ in their localisation and fulfill different biological functions (Perez-Amador et al. 2002).

PA catabolism is an efficient regulator of the free PA level in the cell; its products fulfill an important physiological role under both normal and stress conditions (Martin-Tanguy 2001; Paschalidis and Roubelakis-Angelakis 2005b).

Spontaneous cyclisation of amino aldehydes derived from Put results in the formation of 1-pyrroline, which is converted into γ-aminobutyric acid (GABA) by pyrroline dehydrogenase; GABA is a potential modulator of many physiological processes. Most of the GABA accumulated in stressed plants (Turano and Kramer 1993) appears to be due to glutamate oxidation. However, pulse-chase experiments with 1,4-^{14}C-Put introduced into the roots of the halophyte *Limonium tataricum* demonstrated rapid metabolisation of this diamine into GABA via the DAO pathway (Duhazé et al. 2002). Moreover, the incorporation of ^{14}C into GABA was detected after the introduction of labeled Spd into the roots, which indicated the possibility of Put formation via PAO-catalysed degradation of Spd, and a further conversion of diamine into GABA. As distinct from Put, amino aldehyde, an initial product of oxidation of Cad terminal amino groups, is further converted into 1-piperedine, a precursor of alkaloids.

Investigations into PA catabolism have focussed mainly on changes in their levels and spectra, leaving their biological significance still to be determined. Paschalidis and Roubelakis-Angelakis (2005b) revealed that the specific activities of the enzymes of PA catabolism increased basipetally in the leaf central, basal, petiolar, and internodal regions throughout development. These results permit the supposition that, in stressed plants, developmental changes in PA catabolism are enhanced because DAO and PAO expression and H_2O_2 production occur in cells destined to undergo lignification (Cona et al. 2003).

Stress-induced accumulation of PAs often correlates with improvements in plant tolerance (Bouchereau et al. 1999). Exogenous application of PAs could improve plant survival under saline conditions (Chattopadhyay et al. 2002; Roy and Wu 2002; Wang et al. 2003; Zhao and Qin 2004; Zhu et al. 2006; Liu and Moriguchi 2006), chilling stress (Chen and Murata 2002; Kim et al. 2002), osmotic and acidic stresses (Fujihara and Yoneyama 1993), radiation-induced oxidative stress (Von Detsch et al. 2005), and so on. If PA synthesis is blocked and cell growth is stopped, the provision of exogenous PA can restore growth. All these results indicate that exogenous PAs are involved in the stress responses and tolerance of various glycophytic plants.

13.3.2 Possible Mechanisms of the Protective Actions of Polyamines in Plants under Salt Stress

13.3.2.1 Polyamine-Mediated Regulation of K^+/Na^+ Homeostasis

K^+ deficiency was possibly the first reported stress condition to result in an increase in the Put level in barley (Richard and Coleman 1952). Loss of K^+ is one of the detrimental consequences of salt stress, and mitigation of this loss correlates strongly with salt tolerance in glycophytic plants (Carden et al. 2003; Chen et al. 2005). Since K^+ is an activator of a large number of enzymes, the consequences of K^+ loss to cell metabolism are rather drastic. Like glycophytes, halophytes also require K^+ for the same metabolic functions (Yeo and Flowers 1986; Flowers and Colmer 2008). All halophytes display a common need to regulate cellular Na^+, Cl^- and K^+ concentrations as they adjust to the external water potential under salinity (Flowers and Colmer 2008). Some halophytes (*Suaeda maritima*) grown in non-salinised solution accumulate high endogenous K^+ concentrations (Yeo and Flowers 1986). However, K^+ could not stimulate the growth of *S. maritima* and other succulent halophytes (*Atriplex nummularia*, *A. inflata*). Moreover, their growth was inhibited by high KCl concentrations in culture solution in the range of 30–200 mM and above (Ramos et al. 2004). According to Flowers and Colmer (2008), halophytes, such as *S. maritima*, were able to retain Na^+ but not K^+ in vacuoles.

This view was supported by the findings that exogenous application of PAs could improve plant performances by decreasing shoot Na^+ but increasing shoot K^+ levels under saline conditions (Chattopadhyay et al. 2002; Zhao and Qin 2004; Zhu et al. 2006). Moreover, after salt shock, the concentrations of free PAs increased in roots and leaves of the halophyte *M. crystallinum* (Shevyakova et al. 2006a, 2006b; Kuznetsov et al. 2007), and this could play a role in the regulation of K^+ flux (Flowers and Colmer 2008). Direct effects of PAs (Put, Spd, and Spm) on the two principle vacuolar ion channels were studied on membrane patches isolated from vacuoles of the red beet taproot (Dobrovinskaya et al. 1999).

Investigations into ion channel activities by applying patch-clamp techniques to *Hordeum vulgare* seedlings established that PAs (Put, Spd, and Spm) blocked inward Na^+ and K^+ currents in root epidermal and cortical cells (Zhao et al. 2007). In the root xylem parenchyma, inward K^+ currents were blocked, while outward K^+ currents were enhanced by extracellular Spd. At the whole-plant level, root K^+ content, as well as root and shoot Na^+ levels, was decreased significantly by exogenous Spd. Thus, it was found that exogenous application of PAs could decrease Na^+ accumulation in roots and maintain a high K^+ level in shoots by affecting Na^+ and K^+ currents in specific root cells (Zhao et al. 2007). In another study (Shabala et al. 2007), it was shown that externally applied PAs were also efficient in preventing NaCl-induced K^+ efflux from the leaf mesophyll of the *Pisum sativum* mutant. This effect can be attributed to the inhibition of nonselective cation channels in the mesophyll. The authors suggested that elevated levels of cellular PAs may modulate the activity of plasma membrane ion channels, improving ionic relations and helping in plant adaptation to salinity. These data suggest one possible explanation for the physiological role of PA accumulation in plants in response to salt stress.

Recent advances in investigations of K^+/Na^+ homeostasis in the model halophyte *Th. halophila* showed that, over a wide Na^+ concentration range, these plants were able to maintain stable concentrations of K^+ (Volkov et al. 2004; Vera-Estrella et al. 2005). Measurements of K^+ in various tissues showed that K^+ to Na^+ ratios, as a measure of the plant ability to discriminate between the two ions, might reflect its salinity tolerance (Flowers and Colmer 2008). The observed results demonstrated enhanced expression of specific mechanisms of Na^+ transport. It is also not possible to rule out PA participation in the regulation of the K^+/Na^+ homeostasis described above.

13.3.2.2 Antioxidant Role of Free and Conjugated Polyamines

The effects of oxidative stress are among the most deleterious of any of the environmental stresses on plants, and are characterised by the accumulation of harmful reactive oxygen species (ROS) in tissues: $\bullet O_2^-$, H_2O_2, and $HO\bullet$. These toxic ROS are capable of causing oxidative damage to proteins, nucleic acids, lipids, and other molecules (Apel and Hirt 2004). ROS are produced in various cellular compartments, including chloroplasts, mitochondria, peroxisomes, glyoxysomes, cell walls, plasma membrane, and apoplasts. Plants have evolved several antioxidant strategies. ROS-scavenging enzymes are superoxide dismutase (SOD), catalase (Cat), glutathione reductase (GR), and various peroxidases. Non-enzymatic antioxidants include various plant pigments (carotenoids, tocopherols) and phenolic acids. Numerous reports on stress-induced accumulation of free and conjugated PAs in various plant species have appeared. Since the 1990s, PA participation in oxyradical detoxification was studied predominantly in relation to ozone $(O_3)^-$ pollution (Bouchereau et al. 1999). When exogenous PAs were fed to tomato and tobacco plants, there was a significant suppression of O_3-induced leaf injury (Ormrod and Beckerson 1986). In O_3-treated barley leaves, ADC activity increased. When α-difluormethylarginine (DFMA) was applied to the leaves, the rise in ADC activity was prevented, and injury to ozone exposure was considerably enhanced (Rowland-Bamford et al. 1989). These results suggest that PAs may have a protective role against O_3-induced damage, but the mechanism involved is not clear. One possible mechanism of PA protective action during oxidative stress is activation of expression of genes encoding antioxidant enzymes. In particular, it was shown that Cad induced transcription of the gene for cytoplasmic Cu/Zn-SOD in the halophytic ice common plant (Fig. 13.2; Aronova et al. 2005).

The antioxidant properties of free PAs were first noted by Drolet et al. (1986). Subsequently, Bors et al. (1989) observed that PAs exhibited their most significant antioxidant properties when they form conjugates with phenolic acids. Actually, tobacco leaf injury caused by O_3 was weakened by treatment with exogenous free PAs (Put, Spd, or Spm), which caused a 4- to 6-fold increase in soluble conjugated PAs, especially those associated with the cell wall and membrane fractions (Langebartels et al. 1991). According to their study, PA conjugates with caffeic, cinnamic, and ferulic acids displayed a higher constant of binding to ROS than free PAs, contrasting with earlier notions about the important role of free PAs as radical scavengers (Drolet et al. 1986). It was thus concluded that free PAs could not account for the protection against ozone damage as free radical scavengers.

In contrast, reliable proof of free PA functioning as ROS scavengers was obtained in experiments performed in vitro in a system generating free radicals (Ha et al. 1998; Kuznetsov et al. 2007). As shown by Kuznetsov et al. (2007), when total DNA isolated from leaves of the halophytic plant *M. crystallinum* was

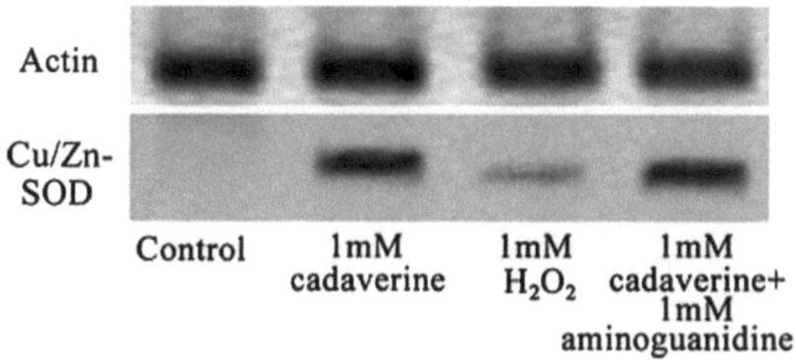

Fig. 13.2 The influence of Cad and H_2O_2 on the expression of Cu/Zn-superoxide dismutase (SOD) gene in roots of *Mesembryanthemum crystallinum* plants

incubated in the presence of the OH•-generating system, practically no DNA was detected, indicating that the OH• attack led to DNA oxidative degradation. H_2O_2 alone did not attack DNA (Fig. 13.4). The addition of Cad or Spm to the OH•-generating system suppressed DNA damage. These PAs inhibited DNA degradation most efficiently at concentrations of 1–5 mM. The active formation of OH• in stressed plants can occur due to simultaneous accumulation of H_2O_2 and Fe^{2+} in the cells.

Recently, it was found that, under saline conditions, iron, in the form of ferritin, accumulated in chloroplasts of *M. crystallinum* leaf parenchymal cells (Paramonova et al. 2007; Fig. 13.3).

Ferritin, a ubiquitous multimeric iron storage protein, plays an important role in the oxidative stress response by protection of chloroplasts. Maximum accumulation of ferritin was found at sites of increased H_2O_2 accumulation and correlated with the highest guaiacol peroxidase activity, demonstrating the role of ferritin in oxidative stress during salinity.

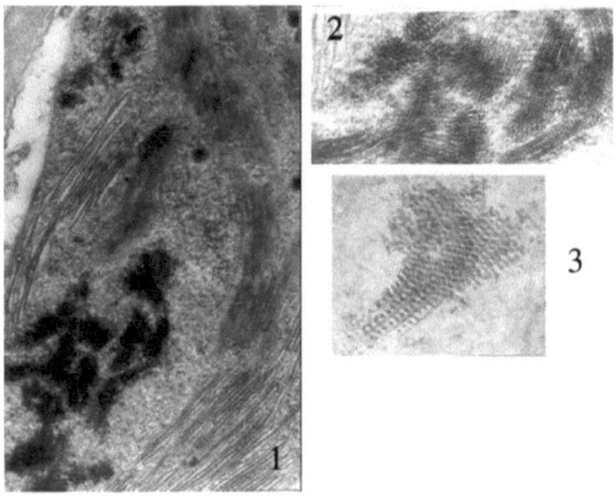

Fig. 13.3 Massive deposits of ferritin accumulate in chloroplasts of *M. crystallinum* plants under salinity. *1* Vesicular parenchyma, *2* enlarged fragment of *1*, *3* crystalloid ferritin

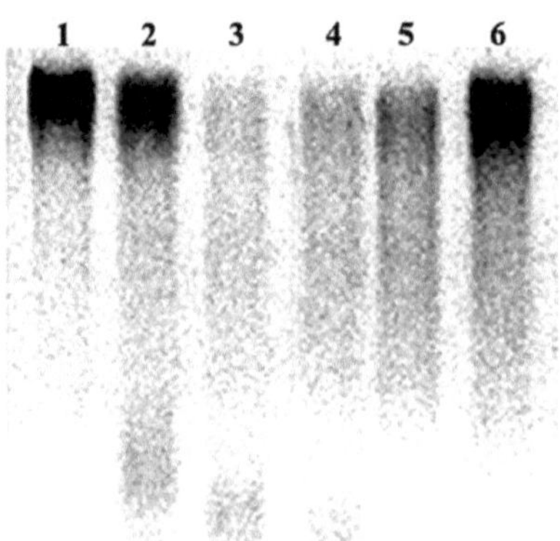

Fig. 13.4 Protection against hydroxyl-induced DNA oxidative damage by Cad. DNA was isolated from *M. crystallinum* plants. To assess damage, 0.2 μg DNA was added to the medium of incubation. Lanes: *1* DNA (control), *2* DNA (+ 10 mM H_2O_2), *3* DNA (+ 10 mM H_2O_2 + 50 μM Fe^{2+}), *4* DNA (+ 10 mM H_2O_2 + 50 μM Fe^{2+} + 0.5 mM Cad); *5* DNA (+ 10 mM H_2O_2 + 50 μM Fe^{2+} + 1.0 mM Cad); *6* DNA (+ 10 mM H_2O_2 + 50 μM Fe^{2+} + 5.0 mM Cad)

Accumulation of free Put and especially Spd was found in *M. crystallinum* leaves under NaCl treatment (Paramonova et al. 2003). The joint treatment of these plants with NaCl and Put increased the formation of suberin plates in the cell wall, activated ionically bound peroxidase, and decreased lipid peroxidation in plastoglobules in chloroplasts. Moreover, the highest number of ferritin deposits was observed in plants affected by NaCl or (NaCl + Put). It appears that the capacity of *M. crystallinum* plants to accumulate ferritin deposits in the chloroplasts in the presence of elevated PA content belongs to the adaptive traits toward oxidative stress.

At the same time, the functional activity of PA conjugates in plant cells is more efficient than those of parent compounds, and conjugates often acquire new properties permitting them to be involved in a wider range of biological processes. Thus, some recent reports contain data about stronger protective antioxidant activity of conjugates than free PAs and phenolic acids (Edreva et al. 2007). This means that plants subjected to abiotic stresses not only accumulate free PAs functioning as scavengers of free radicals but also produce their conjugates, which are more efficient antioxidants.

However, limited information is available concerning the physiological role of PA conjugates in plant responses to various abiotic stresses. A comparative analysis of the content of free PAs, perchloric (PCA)-soluble and PCA-insoluble conjugated polyamines in mature leaves and roots of the halophytic C_3-CAM species, *M. crystallinum*, has been carried out with adult plants exposed to salinity (Shevyakova et al. 2006a). In adult plants, the process of CAM (crassulacean acid metabolism) induction under salinity is linked to oxidative stress and activation of antioxidant defensive responses (Shevyakova et al. 2006a; Ślesak et al. 2003). It was found that, under normal conditions or salinity (400 mM NaCl) grown leaves contained PCA-insoluble (bound) conjugates of Put, Spd, and especially Spm, which showed a tendency for growth with increased duration of salt treatment (from 1.5 to 48 h).

In roots, all forms of PA conjugates (PCA-soluble and -insoluble) were detected. In roots, in contrast to leaves, the formation of PCA-soluble conjugates of all PAs, except Spm, was decreased under long-term salinity. Under these conditions, the content of bound conjugates of Cad was decreased, but those of Put, Spd, and especially Spm were increased. Thus, under saline conditions, during functioning of CAM-type photosynthesis with enhanced ROS formation, *M. crystallinum* could accumulate both free and bound Spm in both leaves and roots, which supports the involvement of this high-molecular weight PA in the development of salt resistance (Shevyakova et al. 2006a, 2006b).

13.3.2.3 Pro-oxidant Role of Polyamines and H_2O_2

Only a few publications have focussed on PA oxidative degradation under the effect of abiotic factors, although, as mentioned above, PAs are a source of H_2O_2 – the most common ROS even under normal conditions. In this sense, the question arises whether H_2O_2 produced in the reactions of PA catabolism contributes much to the damaging effects of abiotic stresses on plant cells, and whether peroxide is involved in adaptation processes.

In order to examine control of the level of endogenous PAs by their oxidative degradation, *Nicotiana tabacum* plants transformed with constructs containing PAO cDNA from *Zea mays* (*MPAO*) and DAO cDNA from *Pisum sativum* (*PcuAO*) were obtained (Rea et al. 2004). These studies showed that both types of transgenic plants (MPAO and PcuAO) produced a large amount of H_2O_2 in the presence of exogenous substrates (Spd and Put). In spite of the fact that both recombinant proteins were actively synthesised in tobacco plants and, like native proteins in wild-type plants, were present in the apoplast, their activities were limited by the low PA content in the intercellular space, which was characteristic of both transgenic and wild-type plants. High activities of DAO and PAO in transgenic plants reduced the level of endogenous free PAs only insignificantly. The amount of H_2O_2 produced in suspension cells derived from transgenic tobacco leaves after addition of 1 mM Spd into the culture medium was sufficient to trigger the apoptosis program.

These studies with transgenic plants proved experimentally for the first time that modulations in the level of endogenous PAs depend somewhat on their oxidative degradation under normal physiological conditions, indicating the existence of compensatory mechanisms maintaining PA homeostasis in cells.

Since both enzymes of PA oxidative degradation are localised mainly in the apoplast and are associated with the cell wall, they are considered as H_2O_2-generating systems required for lignification, suberinisation, and the formation of cross bridges between the components of the cell wall during plant normal growth, and as a defensive factor under unfavorable conditions (Cona et al. 2003).

Until recently, H_2O_2 was often considered only as a toxic metabolite, and as a cause of programmed cell death (Scandalios 1993). In recent years, our notions about peroxide have changed from simply acknowledging its presence in the plant cell to recognition of its signalling function (Neill et al. 2002). Thus, it has been established that generation of H_2O_2, a relatively weak oxidiser and a long-lived molecule capable of diffusion from its site of production to neighbouring cells and tissues, could fulfill a signalling role in plant adaptation (Neill et al. 2002). Plant cells have a rather wide range of peroxide sources: from electron transport chains of chloroplasts and mitochondria to the NADPH-oxidase system of the plasma membrane; however, these sources differ in their efficiency (Vranova et al. 2002). PAs are less studied sources of peroxide. Free and conjugated forms of Spd or Spm are believed to be the most efficient antioxidants, and are considered scavengers of oxyradicals (Drolet et al. 1986; Bors et al. 1989; Ha et al. 1998). The involvement of PAs in oxyradical scavenging is based on the ease of oxygen-dependent autooxidation and enzymatic oxidation of amino groups catalysed by DAO and PAO, and also on the capacity of PA to accumulate under stress conditions. However, a high level of endogenous PAs and plant tolerance to oxidative stress can be based not only on stress-induced but also on the constitutively high PA biosynthesis. In plants resistant against oxidative stresses, in particular to paraquat (methylviologen), whose breakdown results in the formation of $\bullet O_2^-$, a high level of constitutive synthesis of both ADC and ODC was found, and the PA content was two to three times higher than in sensitive cultivars (Ye et al. 1997). In this case, as shown for a resistant *Conyza bonariensis* biotype, plant pretreatment with paraquat did not induce the accumulation of Put and Spd but activated antioxidant enzymes. A similar pattern was observed for a wheat cultivar displaying cross-reactivity to drought and paraquat, and for another species of *Coniza* (*C. canadensis*) resistant to paraquat. Moreover, activities of antioxidant enzymes were high only in resistant biotypes. Treatment of such plants with Put improved further resistance to oxidative stress, but this effect was not observed for sensitive biotypes.

The observed involvement of constitutively high activity of the enzymes of PA biosynthesis in plant defence against oxidative stress motivates studies of PA metabolism and the role of these compounds in naturally tolerant ecological groups of plants, such as halophytes, xerophytes, and heavy-metal accumulators.

The decreased content of conjugated Cad in roots under salinity could be explained by the faster oxidation of free Cad under these conditions (Kuznetsov et al. 2009). The effect of exogenous free Cad on its conjugate formation was analysed in the roots of *M. crystallinum.* It was found that Cad could induce an oxidative burst in the roots of adult plants, as was evident from the sharp decrease in the content of soluble or insoluble Cad conjugates. This unusual effect was associated with increased oxidative degradation of exogenous Cad and intense H_2O_2 production in the roots of adult plants. This negative effect was reversed by exogenous Cad treatment in combination with aminoguanidine (AG), an inhibitor of DAO. One possible reason for the negative effect of exogenous diamine on the formation of conjugated forms in roots was found to be alkalisation of the root apoplast upon Cad addition to nutrient medium and the unusual O_2^- synthase function as a pH-dependent guaiacol peroxidase in the presence of a high content of H_2O_2. Thus, the inhibitory effect of exogenous Cad was induced by enhanced formation of H_2O_2 during oxidative degradation of this diamine by DAO.

This experimental evidence of similar plant response with oxidative burst was obtained by transgenic tobacco plants overexpressing polyamine oxidase (PAO; Moschou et al. 2008a). One would expect that, because antioxidant enzymes (peroxidase, superoxide dismutase, and catalase) were activated, the transgenic plants would be able to cope with the increased H_2O_2 generated by PAO. However, despite the enhanced antioxidant machinery, the further increase in intracellular ROS caused by exogenous H_2O_2 resulted in oxidative stress, which induced programmed cell death (PCD). Thus, at low concentrations, H_2O_2 can participate in the signalling cascade, inducing the expression of various stress-responsive genes, whereas at high concentration, it participates in the induction of the oxidative burst and of PCD syndrome (Moschou et al. 2008a; Kuznetsov et al. 2009).

13.3.2.4 Molecular Mechanisms of Polyamine Protective Activity

The abundance of information concerning stress-dependent PA accumulation raises a question as to their possible biological role in the adaptation process.

The most widely accepted, and indeed experimentally proven, view is that PAs exert their protective action due to their chemical structure, i.e. as polycations. This is determined largely by the shift in their electron density toward nitrogen atoms under physiological pH values, when PAs behave as bases (pK = 9–11). This explains the readiness of PA electrostatic interaction with negatively charged phosphate groups of phospholipids and nucleic acids, and with carboxylic groups of proteins; and also the PA capability of covalent binding with proteins at the stage of their posttranslational modification (Galston et al. 1997; Walden et al. 1997; Bouchereau et al. 1999).

Such defensive properties are ascribed primarily to high-molecular-weight PAs (Spd and Spm; Kakkar and Sawhney 2002) and unusual multipolyamines with longer and often branched molecules, whose efficiency is related directly to the increased number of amino groups in their molecules (Bagga et al. 1997). PA binding to proteins or nucleic acids not only protects them from degradation but also provides the molecule with the most stable conformation under stress conditions. Thus, Spd and Spm retard cell aging, which is accelerated under stress conditions due to suppression of biopolymer-degrading enzymes (DNases, RNases, and proteases) and prevention of chlorophyll breakdown. Exogenous application of Spd stabilised the native structure of thylakoid proteins D1 and D2, cytochromes, and also the key photosynthetic enzyme Rubisco in oat plants subjected to osmotic stress (Tiburcio et al. 1994).

All PAs are capable of binding to A- and B-DNA: in A-DNA, binding occurs mainly to the major groove, whereas in B-DNA, Put and Cad bind to the sugar-phosphate backbone. Spd and Spm, which contain more amino groups, bind to both the sugar-phosphate backbone and the major and minor grooves (Bryson and Greenall 2000). Experiments with B-DNAs differing in the guanine to cytosine ratio showed that high-molecular-weight PAs interacted mainly with phosphate groups and did not affect the native secondary structure of DNA, thus allowing normal transcription of stress-induced genes. This interaction was evidently unspecific, and did not really depend on DNA nucleotide sequence (Deng et al. 2000). PAs could inhibit DNA methylation, permitting expression of specific genes responsible for the synthesis of stress proteins (Ruiz-Herrera et al. 1995). Spm and, to a lesser degree, Spd are capable of shifting the dynamic equilibrium between B- and Z-DNA, and are involved in DNA spiralisation. The protective role of PAs is manifested by their capability of neutralising the action of ROS, which are dangerous for cell structures and accumulate under the effect of various abiotic and biotic stresses (Ha et al. 1998; Aronova et al. 2005; Kuznetsov et al. 2007).

Recently, it was found that PAs could substantially affect the conductivity of ionic channels in plants. Thus, Put, Spd, and Spm blocked fast and slow vacuolar channels, including calcium channels, and the effect was proportional to PA charge ($Spm^{+4} > Spd^{+3} > Put^{+2}$; Dobrovinskaya et al. 1999). The capability of biogenic amines to affect stomatal conductivity under stress conditions was also correlated with their charge. It was shown that this common property of the plant physiological response to stress was based on the PA-induced blockage of potassium channels in the plasma membrane of guard cells, which increased their turgor and, as a consequence, resulted in a decrease in the stomatal aperture. In particular, PAs blocked the potassium channel in the plasma membrane into mesophyll cells harbouring the *KAT1* gene encoding one such channel. It is of interest that, despite the induction of one and the same response by PAs and ABA, the underlying mechanisms are different because ABA inhibits inward potassium channels. PAs also affected stomatal closure after penetrating the cytosol, implying the presence of an intermediate cytoplasmic factor involved in the induction of this response (Alcázar et al. 2006). PA control of ionic channels might be adaptive under stress conditions. Thus, potassium channels are efficient regulators of cell stimulation and a major target for extracellular and intracellular factors. Blocking potassium channels with Spd was shown to be a major impulse permitting adaptation of cell stimulation in response to numerous biological stimuli. It was established, for example, that spinach ADC was associated with LHC of photosystem II (Borrell et al. 1995; Legocka and Zaichert 1999). PAs synthesised in chloroplasts evidently stabilise photosynthetic complexes of thylakoid membranes under stress conditions (Borrell et al. 1995). The regulatory role of PAs manifest in the activation of protein and nucleic acid syntheses has been demonstrated in both prokaryotes and eukaryotes (Bouchereau et al. 1999).

Plant cell metabolism is changed to prevent the damaging consequences of the action of stress-causing agents. This is attained by realisation of two pathways, operating simultaneously or successively, used by living organisms to adapt to extreme factors: (1) induction of the synthesis of new macromolecules with new properties, which allow cell metabolism to function as normal under stress conditions; and (2) optimisation of the intracellular medium to allow functioning of enzyme systems via the accumulation of low-molecular-weight organic compounds with protective and/or osmoregulatory properties. Both pathways of adaptation are directed towards achieving the same ends, namely, providing the organism with energy, reductants, precursors of nucleic acids and proteins, and also ensuring the maintenance of cell regulatory system functions under stress conditions.

Despite great progress in the elucidation of the mechanisms of PA anabolism and catabolism in plant cells, a general scheme of how endogenous levels of PA are controlled under stress conditions has not yet been proposed.

Reviewing all the published data, we can state that stress-causing agents induce a transient accumulation of free PAs in plants during the initial minutes and hours of stress; thereafter, days are necessary to maintain PA homeostasis in the cells at the level required for the development of long-term plant adaptation to stress. The initial changes in PA metabolism in the plant cell can be described as a primary response to rapid disturbances that would be dangerous for plant life: turgor loss and ROS generation. These events activate the signalling cascades that induce a transient increase in Put synthesis in stress-sensitive plant species. In stress-tolerant species, the Spd and Spm levels required for long-term plant adaptation to stress is maintained constitutively by the high activities of the genes encoding the enzymes of their biosynthesis (Kasukabe et al. 2004; Maiale et al. 2004). The level of Put decreases because it is consumed as a precursor in these syntheses. In Cad-containing stress-tolerant plant species, increased levels of Spm and Cad are maintained, compensating for the reduced level of Put in the cells.

13.3.3 Interactions Between Polyamines and Other Signal Molecules

During the first few years after PAs were detected in plant cells, much attention was paid to the involvement of PAs in the initiation of cell division and expansion, in plant morphogenesis, flowering, senescence, and in the interaction between PAs and phytohormones (Galston et al. 1997). It was shown that, in some plants, auxin, gibberellin, and cytokinin stimulated biosynthesis and increased the content of PAs, whereas exogenous PAs affected the level of endogenous phytohormones. In a recent study performed in *Arabidopsis*, it was demonstrated that IAA induced the *ACL5* gene encoding Spm synthase, but ABA or gibberellic acid could not induce this gene (Hanzawa et al. 2000). Inactivation of *ACL5* retarded stem elongation and suppressed cell expansion. It is well known that this mechanism can operate during plant adaptation to extreme conditions, when the retardation of growth processes is required for plant survival (Hasegawa et al. 2000). Despite recent reports describing the involvement of PA in various abiotic stresses, little is known about stress signalling pathways regulating PA metabolism.

Typical examples of multifunctional interactions in plants under stress are those between PAs of the Put family (Spd and Spm) and ABA or ethylene (Eth). Both hormones are considered "stress hormones," although they may fulfill several other functions in the absence of stress.

13.3.3.1 Polyamines and Abscisic Acid

It is well known that the ABA concentration increases under water or salt stress as well as under other abiotic stresses (Christmann et al. 2005). ABA induces the expression of multiple genes involved in defence against water and salt stresses (Wang et al. 2002b; Bartels and Souer 2004). It has also been noted that ABA induces thermo-tolerance in some plants (Gong et al. 2005). Thus, plant adaptation processes are, to a large degree, mediated by ABA.

ABA is synthesised from carotenoids by ABA-synthesising enzymes induced in the root tip cells or parenchyma cells of vascular bundles by drought and salt stresses (Koiwai et al. 2004). ABA synthesised in the roots enters the xylem vessels and from here is transported to the leaves (Sauter et al. 2002). This is very important for plant stress adaptation because the root systems of any plant are the first barrier to immediately counteract salinity, water deficit, and other environmental stresses. The roots are less capable of basic biosyntheses than leaves, but they can synthesise ABA, PAs (Shevyakova et al. 2006b), and various other secondary metabolites, such as phenolic acids, alkaloids, and others.

It was recently established that, in *A. thaliana*, ABA can induce PA biosynthesis, especially under water and salt stresses that elevate the level of free PAs (Kasinathan and Wingler 2004; Urano et al. 2003; Alcázar et al. 2006). Application of exogenous ABA was useful in identifying stress-signalling pathways regulating PA metabolism. *Arabidopsis* contains doubled genes encoding ADC (*ADC1* and *ADC2*), SPDS (*SPDS1* and *SPDS2*), and SPMS (*SPMS* and *ACL5*; Panicot et al. 2002), but has no ODC (Hanfrey et al. 2001).

The possible role of exogenous ABA in the modulation of PA metabolism under water stress was studied in *A. thaliana* (Urano et al. 2003). The application of exogenous ABA induced expression of *ADC2* but not the *ADC1* gene. ABA up-regulated *SPMS* expression (Hanzawa et al. 2000; Urano et al. 2003) but did not affect *ACL5*. In different plant species, S-adenosylmethionine synthase is encoded by three isogenes (*SAM1*, *SAM2*, and *SAM3*), which are expressed differentially in response to ABA treatment, salt stress, and osmotic stress (Espartero et al. 1994).

Alcázar et al. (2006) also showed that ABA modulated PA metabolism at the transcriptional level in response to water deficit by up-regulating *ADC2*, *SPDS1*, and *SPMS* expression. As shown in this latter work, the highest expression of these genes was induced by water stress, with maximum induction after 8–24 h of treatment. *ADC2* and *SPDS1* expression increased 32- and 25-fold, respectively, after 8 h, and *SPMS* was induced 75-fold after 24 h. These increases in expression were not observed in non-stressed well-watered plants. In water-stressed plants, free Put levels increased up to 1.8-fold. The levels of free Spd and Spm did not increase above the constitutive level during 24 h of stress treatment. In contrast, a transient depletion of free Spd and Spm levels occurred during the first hour of water deficit. However, transient depletions in the free forms occurred in parallel to increased levels of the insoluble conjugated acid-soluble fraction. Thus, these conjugated forms of Spd and Spm increased 8.3- and 16-fold, respectively, in the first hour of water deficit. These responses were not observed in non-stressed plants. The involvement of ABA in the transcriptional regulation of PA metabolism under dehydration was confirmed by expression profiling analyses in ABA-deficient (*aba2-3*) and ABA-insensitive (*abi1-1*) mutants. The authors reasoned that levels of free Put were subjected to homeostatic control to maintain its level within a nontoxic range.

13.3.3.2 Polyamines and Ethylene

Ethylene and PAs are regulators of diverse responses manifested during plant growth and development (Wang et al. 2002a). Ethylene evolution is increased drastically in plants by drought, salt stress, abnormal temperature, wounding, and UV-B irradiation (Rakitin et al. 2004). PAs are universal multifunctional regulators of physiological processes during plant growth and development, and are also involved in severe stress responses. However, the functions of PAs and Eth in stressed plants are still obscure (Kuznetsov et al. 2007). At the same time, it is known that the major natural PAs (Put, Spd, and Spm) interact closely with Eth in regulatory processes (Galston et al. 1997; Kaur-Sawhney et al. 2003).

S-Adenosyl-L-methionine (SAM) takes part in the final biosynthetic stages of these PAs, and is also an Eth precursor (Adams and Yang 1979). This might explain the Eth-induced down-regulation of PA synthesis observed by some researchers (Evans and Malmberg 1989). Along with competition for a common precursor, other interactions could arise between PAs and Eth under stress conditions; these are manifested in the mutual inhibition of their biosynthesis, which becomes a basis for the competition between major PAs and Eth (Galston et al. 1997). Such an interaction occupies a special place in the coordination of physiological processes because PAs and Eth often exert opposite effects (Bagni and Tassoni 2001). For example, Spd retards senescence, whereas Eth accelerates it (Galston et al. 1997). It was demonstrated that the facultative halophyte *M. crystallinum* responded to heat shock with a decrease in the content of the Put family, especially Spd, as well as a fast and transient increase in Eth evolution (Kuznetsov et al. 2007).

To understand the functional role of SAM as an intermediate in PA and ethylene biosynthesis, the following points must be taken into account: (1) SAM is used actively in plant cells as a main donor in transmethylation of proteins, nucleic acids, polysaccharides, and fatty acids; and (2) 5-methylthioadosine (MTA), a by-product of SAM degradation during synthesis of Spd, Spm, and ACC, can be recycled by MTA nucleosidase into methionine and further into SAM (Wang et al. 2002a). It should be kept in mind that, in stressed plants, the pool of SAM could increase due to stress-induced accumulation of *S*-adenosylmethionine synthetase (SAMS) transcripts (Espartero et al. 1994), i.e. under stress conditions, SAM homeostasis could be maintained to increase plant adaptive potential.

Some researchers have demonstrated that interaction between PAs and Eth cannot be limited only by their antagonism. Thus, pea seedlings responded to Eth treatment by reduced activity of ADC, increased activity of LDC, and increased content of Cad (Apelbaum et al. 1985). Although their biosynthetic processes are indirectly interconnected, the stimulatory effect of Eth on Cad biosynthesis did not attract attention for long, because Cad is formed in the side branch of the aspartate pathway resulting in biosynthesis of methionine and SAM. In its turn, SAM is required for the formation of ACC, a precursor of Eth.

The facultative halophyte *M. crystallinum* turned out to be a very convenient model for investigating the interaction between Cad and Eth under stress conditions. In this plant, aspartate, a distant precursor of lysine, is one of the main metabolites produced from oxalacetic acid during CO_2 assimilation in CAM-photosynthesis. It was demonstrated that, in the common ice plants, stress-induced Cad accumulation coincided with the developmental stage when plants transited from C_3- to CAM-photosynthesis (Shevyakova et al. 2001; Kuznetsov et al. 2002). In this period, the common ice plants responded to heat shock (HS) by transient Eth evolution and a subsequent interorgan translocation of Cad. Under NaCl salinity, the level of endogenous Eth in plants increased and Cad accumulated in the leaves. To confirm the possible connection between HS-induced Cad translocation from the leaves to the roots and transient Eth evolution, two lines of *A. thaliana* were used as model plants: wild type (Col-0) and a mutant (*ein4*) displaying disturbed Eth perception. It was established that HS-induced inter-organ translocation of Cad, as distinct from Put and Spd, was related to the functioning of the Eth perception system. Eth-dependent Cad formation was demonstrated by accumulation of this diamine in detached leaves of the common ice plant exposed to an Eth atmosphere or incubated in the presence of its precursor, ACC (Kuznetsov et al. 2002; Shevyakova et al. 2004). The phenomenon of Eth-dependent Cad accumulation permitted the study of putative mechanisms of hormonal signal transduction, which were not examined until recently. It was thought that protein phosphorylation was involved in Eth signal transduction, as in other signalling pathways. All inhibitors tested abolished the stimulatory effect of Eth on LDC activity, and this was the first unambiguous proof of the involvement of protein phosphorylation/dephosphorylation in Eth-induced Cad formation in plants.

It is evident from the results obtained that, along with competitive interrelations between major PAs (Spd and Spm) and Eth, which could be manifested under stress conditions, in some cases the interaction between Cad and Eth may be rather evaluated as synergistic (Fig. 13.5). The observation of this phenomenon permits fresh insight into the problem of the compensatory reactions maintaining PA homeostasis required for plant survival under stress conditions. However, Cad

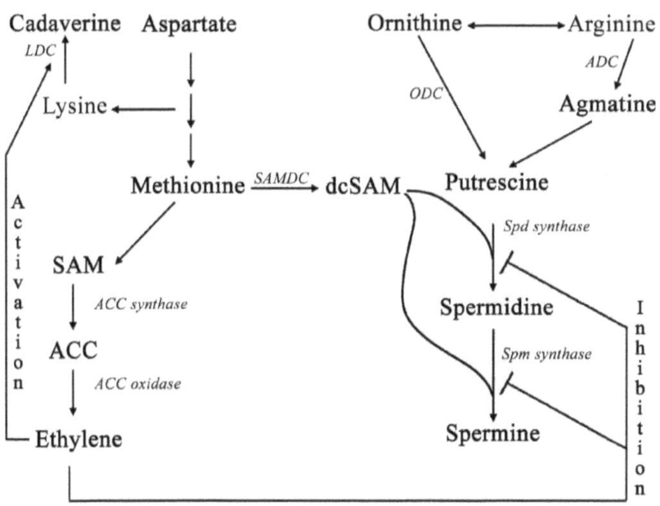

Fig. 13.5 Polyamine biosynthesis in the common ice plant *M. crystallinum* and its possible regulation by ethylene. Cadaverine synthesis is catalysed by LDC, which is induced by ethylene of accumulating in plants under stress. Stress ethylene, on the other hand, can inhibit activity of spermidine synthase and spermine synthase that leads to the reduction of the content of this polyamine in leaves

accumulation and the increased level of endogenous Eth in common ice plants in the period during which CAM-photosynthesis functions did not affect expression of a gene encoding the key enzyme of CAM metabolism, PEPC (Shorina et al. 2005). Cad accumulation occurred at later stages of common ice plant development and was not associated with CAM induction providing for plant adaptation to water deficit. The mechanism by which the common ice plant adapts to salt stress as related to Cad accumulation might be as follows: high Cad concentrations retard cell wall expansion, supplying H_2O_2 for suberinisation and lignification and thus reducing cell wall permeability for salts. This conclusion arises from the analysis of phenotypic responses of common ice plant seedlings to various concentrations of exogenous Cad (Kuznetsov et al. 2007), Cad-dependent induction of Cu/Zn-SOD genes in roots (Aronova et al. 2005; Kuznetsov et al. 2007), and exogenous diamine-caused formation of suberine plates in the cell wall (Paramonova et al. 2003).

13.3.3.3 Polyamines as Components of Stress-Signalling Systems

In contrast to the certain establishment of the role of PA in plant defense against various abiotic and biotic stresses, the involvement of PAs in signal transduction in plants has been detected only in some cases (Messiaen and van Cutsem 1999), whereas in animals and bacteria, it has been reliably shown to be ubiquitous (Bueb et al. 1992).

There is great interest in PAs as modulators of gene expression under oxidative stress. In an exponentially growing culture of *Escherichia coli*, physiological concentrations of Put and Spd significantly increased expression of the *oxyR* and *KatG* genes responsible for defence against oxidative stress (Tkachenko and Nesterova 2003). It was shown that expression of these genes depended on stimulation of PA-induced DNA supercoiling. There are reports indicating that PAs interact with DNA phosphate groups, thus protecting genomic DNA from digestion by DNase, and that they play a crucial role in genomic DNA protection and conformation (D'Agostino et al. 2005). PAs are involved in many cellular processes, including chromatin condensation, maintenance of DNA structure, RNA processing, translation, and protein activation (Childs et al. 2004). The level of PAs in the cell directly affects the range of genes expressed in response both to growth stimulating and to growth inhibiting agents (Lindemose et al. 2005). It has been demonstrated that PAs affect gene expression at the transcriptional level, and this effect is determined most probably by the direct interaction of PAs with DNA and/or transacting protein factors (Wang et al. 2002b; Lindemose et al. 2005). In plants, PAs were found to activate protein phosphorylation and the activities of specific protein kinases (Tassoni et al. 1998).

All these properties of PAs can play a crucial role in ameliorating plant responses to abiotic stresses and PA functioning as components of stress-signalling systems.

PA signalling functions in plant defence against oxidative stress is of great importance. Despite the continued interest in the role of PAs in plants exposed to biotic stresses, only limited information is currently available.

Recently, it was found that 1 mM Cad added to nutrient medium for the halophytic plant *M. crystallinum* for 2 h induced transcription of the gene for the cytoplasmic Cu/Zn-SOD form (Aronova et al. 2005). Addition of an inhibitor of diamine oxidative degradation (1 mM AG) along with Cad to nutrient medium did not reduce the level of mRNA, which indicates that non-oxidised diamine affected transcription of this gene. Root treatment with 1 mM H_2O_2 also increased the level of mRNA, but to a lesser degree. This supports the previously suggested hypothesis (Kuznetsov et al. 2002) that stress-induced Cad accumulation in common ice plants, and its capacity for long-distance transport permitted Cad to play a role as a stress signal that switches on plant defence mechanisms directed, in this case, to the improvement of cell antioxidant activity (Fig. 13.2). In this context, it should be mentioned that ROS generation by the common ice plant increases sharply during the period of CAM-photosynthesis, which, in turn, activates antioxidant systems. On the other hand, enhanced transcription of the genes for antioxidant enzymes did not always correlate with the activity of the corresponding enzymes. A possible cause for this discrepancy might be the direct effects of peroxide on the activities of peroxide-sensitive enzymes. In common ice plants, such an enzyme might be the Cu/Zn-SOD isoform located in the apoplast. We demonstrated that, after treatment of common ice plants with low concentrations of Cad and Spm (below 1 mM), PAs behaved as antioxidants, whereas high PA concentrations manifested pro-oxidant properties due to the active formation of peroxide and increased pH (>7.0) in the apoplast. In this case, PAs facilitated the reverse reaction, with the formation of superoxide radical ($\bullet O_2^-$) from H_2O_2. Superoxide radical produced during a burst of PA oxidative degradation could serve as a signal for enhanced transcription of the corresponding gene. Switching on such a compensatory mechanism is evidently typical for the functioning of defence systems in stress-resistant plant species.

Recent studies provide evidence that Spm is an endogenous inducer and a novel signal transmitter in defence responses against phytopathogens (Takahashi et al. 2004). Some authors implicate PAs – emerging intracellular signalling molecules – as potential physiological regulators of the anti-apoptotic action in animal cells (Kutuzov et al. 2005).

The investigations of Takahashi's group (Takahashi et al. 2004) showed that exogenously applied Spm specifically activated the expression of five hypersensitive response (HR) marker genes [*HSR203J*, *HMGR*, *HSR201*, *HSR515*, and harpin-induced 1 (*NIN1*)] in tobacco leaves. Induction by Spm of these HR marker genes, including *NIN1*, was suppressed by pretreatment with antioxidants, calcium channel blockers, inhibitors of the mitochondrial permeability transition pore openings, and blockers of amine oxidase (AO) or polyamine oxidase (PAO). Free Spm accumulated in the intercellular spaces during tobacco mosaic virus-triggered HR. Under these conditions, Spm is catabolised by AO and/or PAO localised in the apoplast, resulting in the production of H_2O_2 and initiation of Ca^{2+} flux into the cytosol. The down-stream reaction of dysfunction of this organelle is the up-regulation of HR marker genes by pathways separated into at least two branches, namely, independent of and dependent on the activation of two mitogen-activated protein kinases (MAPKs): salicylic acid-induced (SIPK) and wound-induced (WIPK) protein kinases.

In summary, the signalling pathways described above involving Spm and participating in plant responses to abiotic and biotic stresses remain poorly studied.

13.3.4 Some Modern Approaches Toward Elucidation of Polyamine Metabolism and their Protective Functions under Stress

Until recently, a principal tool for studying the biological role of PAs was the use of metabolic chemical inhibitors; most of which were initially used in human cancer chemotherapy to suppress PA accumulation in tumours. The most widely used inhibitors of various enzymes of PA biosynthesis or catabolism are difluoromethylonithine (DFMO) for ODC, α-difluormethylarginine (DFMA) for ADC, methylglyoxalbisguanylhydrazine (MGBG) for SAMDC, cyclohexamine (CHA) for SPMS, and aminoguanidine (AG) for DAO. These inhibitors are used to switch off separate stages of PA biosynthesis (Walden et al. 1997; Bouchereau et al. 1999).

The application of DFMA and DFMO helped to establish that ADC and ODC pathways operate to ensure constitutive Put synthesis under normal conditions, whereas under stress conditions, both ADC and sometimes ODC pathways could be activated further, resulting in Put accumulation. Plants subjected to different type of stresses show a rapid and massive increase in Put levels, which can be inhibited by application of DFMA (Galston et al. 1997). Put reduced cold-induced electrolyte leakage from tomato leaves but DFMO, a biosynthetic PA inhibitor, increased electrolyte leakage from cold-treated leaves. Furthermore, the increased membrane permeability in cold-stressed leaves treated with DFMO was completely abolished by the application of Put to leaves (Kim et al. 2002).

The use of inhibitors has permitted the elucidation of the compensatory reactions accompanying the switching off some PA biosynthesis, which is of importance for understanding the mechanisms of plant-cell homeostasis, especially under stress conditions (Galston et al. 1997). Nevertheless, some limitations of inhibitory analysis should be mentioned, including their possible metabolisation in tissues, differences in the rates of their uptake, insufficient specificity – determined frequently by differences in the localisation of the inhibitor and a target enzyme, injurious effect on membranes, and other drawbacks (Kaur-Sawhney et al. 2003).

13.3.4.1 Tolerance of Transgenic Plants with Changed Polyamine Content

Nowadays, other approaches are available for studying the mechanisms of PA biosynthesis. One promising approach is the production of transgenic plants harbouring the genes encoding enzymes of various PA biosynthetic pathways. A list of genes controlling PA metabolism in plants that have been characterised and cloned can be found in a review by Kakkar and Sawhney (2002).

Since the 1990s, studying transgenic plants has helped to answer some important questions concerning the control of PA metabolism. Firstly, overexpression or negative regulation of key genes for GDC, ADC, and SAMDC allowed control of endogenous Put levels. Overexpression of yeast ODC cDNA in tobacco plants or mouse ODC cDNA in tobacco and carrot plants (Bastola and Minocha 1995) increased the level of Put but did not affect the levels of Spd and Spm as compared to wild-type plants. At the same time, transgenic tobacco leaves expressing human SDC cDNA contained much more Spd and Spm and reduced amounts of Put. SAMDC overexpression in transgenic rice plants was accompanied by Spd accumulation and improved salt tolerance as compared to wild-type plants (Roy and Wu 2002). Over expression of Spd synthase and increased Spd content in *Arabidopsis* improved tolerance to various stresses, such as chilling, freezing, and salinity (Kasukabe et al. 2004).

Under drought stress, transgenic rice plants expressing *Datura stramonium ADC* under the control of the monocot *Ubi-1* promoter produced a much higher level of Put and only slightly higher levels of Spd and Spm, ultimately protecting the plants from drought (Capell et al. 2004).

When antisense SAMDC cDNA was inserted into the potato genome, Spd production was sharply reduced, and transgenic tubers displayed a changed phenotype (Kumar and Minocha 1998). In transgenic tobacco plants transformed with the ADC gene from oat under the control of an inducible promoter (Tet-repressor system), increased levels of this gene transcript, ADC activity, and free Put were observed (Masgrau et al. 1997). Transgenic plants displayed an altered phenotype: necrotic lesions appeared on their leaves, and growth and development were retarded, which was ascribed to a high, toxic level of endogenous Put. On the other hand, transgenic potato harbouring an antisense SAMDC cDNA under the control of the cauliflower mosaic virus (CaMV) 35S promoter displayed an abnormal phenotype (growth retardation, non-flowering plants, leaf chlorosis, etc.) against a background of a decreased SAMDC transcript level, reduced enzyme activity, reduced Put level, but enhanced Eth evolution. All attempts to obtain transgenic plants with a SAMDC construct in the normal orientation were unsuccessful, leading to the supposition that constitutive overexpression of this enzyme might be lethal (Kumar and Minocha 1998). At the same time, in order to elucidate specificity in PA metabolism and developmental regulation controlled by PAs, it is necessary to change the PA level in various tissues by expression of sense and antisense constructs under the control of tissue-specific promoters (Martin-Tanguy 2001). In general, the levels of Spd and Spm in the cells are the least changeable because of the functioning of homeostatic regulation (Kaur-Sawhney et al. 2003), which might be related to the supramolecular organisation of the enzymes involved in their biosyntheses (Bhatnagar et al. 2001). Transgenic tobacco plants with antisense constructs of cDNAs encoding senescence-related 1-aminocyclopropane-1-carboxylic acid (ACC) synthase and ACC oxidase were more tolerant to abiotic stresses than wild-type plants (Wi and Park 2002). This shows a positive correlation between enhanced PA content in plants and their stress tolerance.

The functioning in plants of two alternative pathways of Put biosynthesis does not exclude a dependence of their regulation on mutual intracellular conversions of their substrates (ornithine and arginine) or their availability. Experiments with a cell line of *Populus nigra* x *maximowiczii* transformed with mouse ODC cDNA help to highlight such problems (Bhatnagar et al. 2001). This latter study demonstrated the capacity of plant cells to overexpress a foreign *ODC* gene to maintain a high level of Put by switching on the homeostatic mechanism. The increased activity of ODC induced an increased production of ornithine and its precursor glutamate. Earlier, it had been shown that transgenic animals, which could not tolerate excessive production of Spd and Spm in their cells, excreted the precursor Put, i.e. PA overproduction induced the cell homeostatic response (Halmekytö et al. 1993). In addition, using a transgenic system, it was demonstrated that plant ODC could use ornithine synthesised directly from glutamate rather than ornithine produced from arginine in the urea cycle as a substrate. Thus, use of transgenic plants has helped to decipher the compensatory mechanisms in PA metabolism that could play an important role in the maintenance of PA homeostasis that is required under stress conditions.

In the opinion of Galston and coworkers, some of the discrepancies in PA studies using transgenic plants could depend on various factors: transgene source, position effects, recipient plant system, the plant material used for transformation, promoter type, and other factors (Kaur-Sawhney et al. 2003). The hierarchy of PA accumulation in different transgenic tissues/organs has been studied (Lepri et al. 2001). In general, less metabolically active tissues accumulate higher levels of PAs (Lepri et al. 2001). More significant results concerning the control of transgene expression have been obtained with inducible or tissue-specific promoters (Mehta et al. 2002). Thus, fruit-specific expression of heterologous SAMDC in tomato resulted in the ripening-specific accumulation of Spd and Spm, which improved fruit juice quality (Mehta et al. 2002). In addition to the agronomic interest of these findings, this study presented striking evidence regarding the in vivo involvement of PAs in a particular developmental process.

13.3.4.2 Tolerance of Plant Mutants Displaying Defects in Polyamine Metabolism

One of the genetic approaches towards the investigation of the mechanisms of PA signal perception and transduction in stressed plants is the biochemical and physiological analyses of mutants displaying different phenotypes. To date, several types of plant mutants with induced changes in PA metabolism have been obtained. Among them, there are mutants of tobacco, petunia, tomato, and *Arabidopsis* deficient in PAs and in the genes of their biosynthesis, and mutants resistant to PAs and the inhibitors of their biosynthesis. A review by Kakkar and Sawhney

(2002) comprised a list of such mutants and the characteristics of their phenotypic and biochemical defects.

Mutant tobacco lines resistant to MGBG are of interest for establishing the morphogenic role of PA. These mutants display dwarfism and altered morphology of floral organs; they manifest enhanced SAMDC activity and have an expanded PA pool (Fritze et al. 1995). In petunia, a mutant line with changed flower morphology also showed high levels of endogenous PAs and enhanced ADC activity (Gerats et al. 1988). In the leaves of the non-flowering tobacco *rmb7* mutant, PA conjugates, which are thought to be transported to stem apices toward floral buds to induce flowering, were not found (Martin-Tanguy 2001).

Some types of mutants are useful for studying the role of PAs in stress physiology. Thus, tobacco DFMO-resistant mutant with a high PA concentration was resistant to low pH values inducing an acidic stress in plants (Hiatt and Malmberg 1988). The *flacca*-ABA-deficient tomato mutant is characterised by high ADC and low ODC activities at late developmental stages, which was accompanied by a reduced total level of PAs. Such a mutant is of importance in the study of interactions between ABA and PAs during adaptation to abiotic factors (Kim et al. 2002).

Recently, a group of Japanese researchers has described *Arabidopsis* insertion mutants harbouring T-DNA for two genes of Spd synthase, *SPDS1* and *SPDS2* (Imai et al. 2004a). While each mutant allele showed a normal phenotype, *spds1-1 spds2-1* double-mutant seeds were shrunken and had embryos that were arrested morphologically at the heart-torpedo transition stage. This mutation was lethal. These seeds contained a reduced level of Spd and, in contrast, a high level of Put. These data provide the first genetic evidence indicating the critical role of Spd synthase in plant embryo development. On the basis of these data, we may suppose that a double coding of PA synthesis enzymes in higher plants is essential for plant survival under extreme conditions. At the same time, Imai et al. (2004a, 2004b) showed that, as distinct from Spd, Spm was not necessary for normal development in *Arabidopsis*. Earlier, it had been shown that disruption of the *ACL5* gene, encoding Spm synthase in *Arabidopsis* and required for stem elongation, resulted in a severely dwarfed phenotype (Hanzawa et al. 2000). However, exogenous Spm could not restore normal stem growth. The authors believe that this can be explained by the fact that exogenous Spm did not reach the proper intracellular compartment or did not produce a conjugate required for the manifestation of its action.

13.3.4.3 A New Model Halophytic Species for Elucidating the Molecular Mechanisms of Polyamine Participation in Salt Tolerance

Many halophytes, in particular *M. crystallinum*, have evolved unique regulatory pathways of salt tolerance and polyamine metabolism that are not found in glycophytes (Cushman and Bohnert 2000; Kuznetsov et al. 2006a, 2007, 2009). This yields the prospect of using the ice plant (*M. crystallinum*) and other halophytes to our further understanding of the ameliorative role of PA under salt stress.

Unfortunately, none of these plants are suitable genetic models. Until now, mutant collection of this species have not been produced because the set of genes involved in the control of PA metabolism has not been completely elucidated.

Recently, several research teams have reported the use of a close relative of *Arabidopsis*, *Thellungiella halophila* Mey. (salt cress), as a valuable halophytic model (Inan et al. 2004; Amtmann et al. 2005; Gong et al. 2005). *Th. halophila* has a small genome of approximately twice the size of the *Arabidopsis* genome and a short life cycle; it is characterised by self-pollination and abundant seed production, thus allowing for fast and efficient genetic analysis (Taji et al. 2004; Gong et al. 2005). Also noteworthy is the fact that, although salt cress is classified as a halophyte, it has no specialised morphological features, such as salt glands, bladder cells or succulence, developed by other natural halophytes to tolerate high salinity. However, *Th. halophila* plants contain a large amount of "compatible" solutes, especially proline, in the cytoplasm to balance the osmotic potential of the Na^+ and Cl^- accumulated in the vacuole (Volkov et al. 2004; Vera-Estrella et al. 2005; Kant et al. 2006).

In our study, we first tested the ability of *Th. halophila* to alter its content of free PAs under short-term exposure to salinity (Kuznetsov et al. 2006b; Radyukina et al. 2007a, 2007b). In control plants (without salt treatment) of this halophyte, the constitutive level of free Spd and Spm and the expression of their genes (*SPDS*, *SPMS*) were significantly higher as compared with that of glycophytes, for example *Plantago major* L. (Fig. 13.6), and this remained unchanged after long-term NaCl treatment because their inducible systems are not activated during this time period. In contrast, glycophytes responded to salinity immediately by activation of PA-related genes. Thus, proposals regarding the role of PA in plant adaptation to salt stress could be supported by transgenic *A. thaliana* with enhanced constitutive levels of Spd synthase (Kasukabe et al. 2006).

Furthermore, it was shown that, after 24-h treatment with 100–200 mM NaCl, the contents of Put, Spd and, especially, Spm were increased in leaves of *Thellungiella*, but not in leaves of *Plantago*. However, in *Thellungiella* plants treated with 300 mM NaCl for 72 h, a decrease in Put and Spd was observed. In contrast, Spm increased still further. Spm also accumulated in leaves of another halophyte, *M. crystallinum*, in response to NaCl treatment (Shevyakova et al. 2006b). Using RT-PCR, it was shown that the level of mRNA of the gene encoding Spd synthase did not change upon NaCl treatment of *Thellungiella* plants. Stress-dependent regulation of free Spd content was thus suggested to take place at the posttranscriptional level. In addition, the data obtained support the conclusion of Kaur-Sawhney et al. (2003) that Spd is effectively channelled to SPMS synthase to control the formation of the end product Spm, thereby regulating the synthesis of unusual high-molecular PAs. These data fit with *Thellungiella* transcript and metabolite profile results, leading to the conclusion that the stress-regulated genes in this genome function in activation of osmolyte production, in protein posttranslational modification, and protein redistribution (Gong et al. 2005). Identification of salt tolerance relevant genes from *Thellungiella* could thus provide new insights into plant salt tolerance

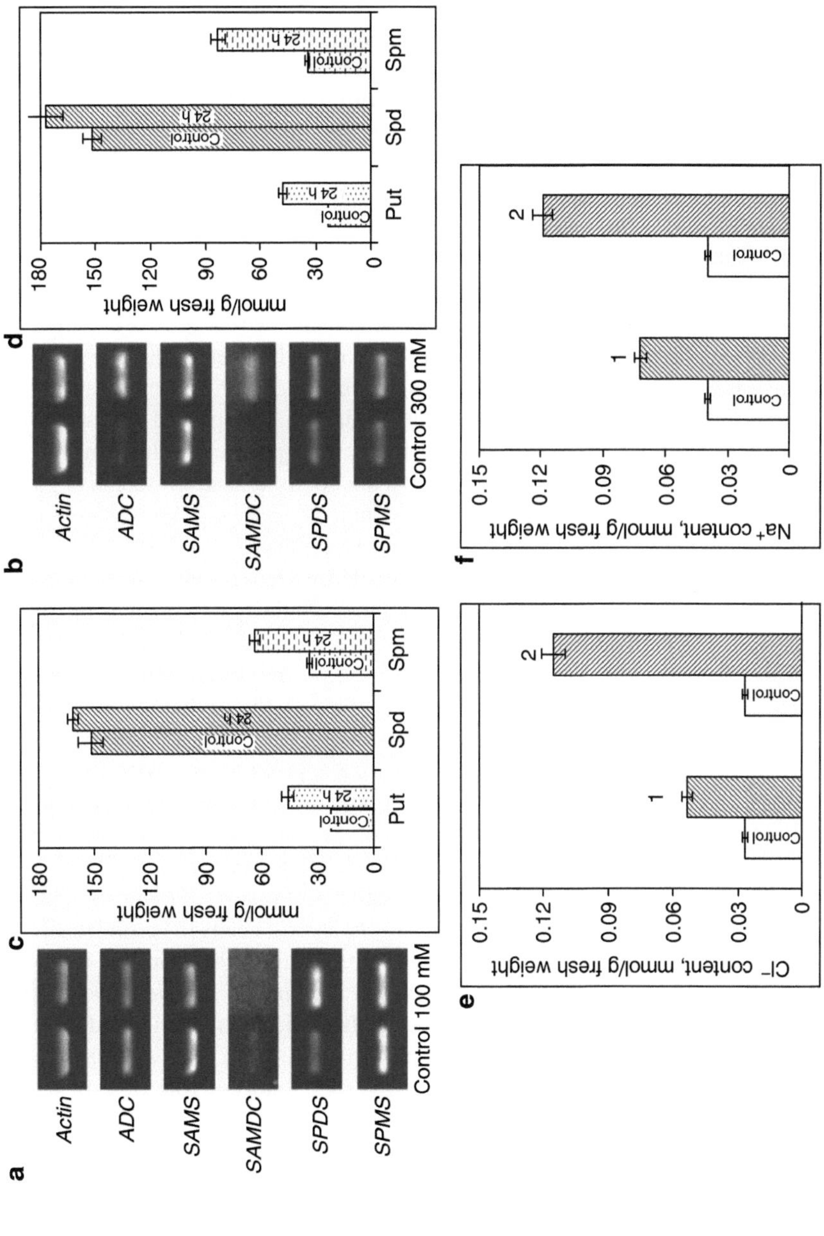

Fig. 13.6 Gene expression pattern (**a**, **b**), PA content (**c**, **d**), Na^+ and Cl^- content (**e**, **f**) in leaves of *Thellungiella. halophila* under 100 mM (**a**, **c**, **e1**, **f1**) and 300 mM NaCl (**b**, **d**, **e2**, **f2**) treatment

13.4 Conclusions

Considerable evidence indicates that PAs are involved in a wide range of plant processes, including adaptation to abiotic stresses. However, their precise role in these specific processes remains to be elucidated. PA biosynthetic pathways are ubiquitous in living organisms but include only a limited number of enzymes. Thus, these pathways represent an excellent model with which to test the hypothesis that PAs are involved in plant protection against stresses. In recent years, many alternative approaches have been developed to manipulate PA metabolism: specific inhibitors, mutants, and transgenic plants. Taken together, some results concerning the elevated levels of Spd and Spm in stress-tolerant plant species suggest that the cellular levels of these PAs are under strict homeostatic regulation due to the supramolecular organisation of some of these enzymes. Application of advanced genomic and proteomic approaches will help elucidate the role of PAs in particular plant processes involved in stress tolerance (Kaur-Sawhney et al. 2003).

Many of the recent advances in understanding the role of PAs in the plant response to abiotic stress have come from studies that employed stress-tolerant plant genotypes. It is thought that halophytes use specific mechanisms in the regulation of PA metabolism under stress that are not used in stress-sensitive plants (glycophytes). Currently, the facultative halophyte *M. crystallinum* is widely used as a model plant to search for such mechanisms (Kuznetsov et al. 2007, 2009). This halophyte is one of the most appropriate models for studying the multifunctional role of PAs in plant stress-response and adaptation. Salinity treatment of this plant induces CAM accompanied by diverse adaptive changes in physiological, biochemical and molecular mechanisms. However, the relatively large genome of this plant species limits its genetic analysis as well as the production of mutants and transgenic plants (Cushman and Bohnert 2000). Salt cress (*Th. halophila*), a close relative of *Arabidopsis* with a genome size approximately twice that of the *Arabidopsis* genome, has recently been reported as an appropriate halophytic model. *Th. halophila* plants display "extreme" tolerance to high salinity (Gong et al. 2005). Many researchers (Inan et al. 2004; Vera-Estrella et al. 2005) have reported the ability of salt cress to survive long-term (for several months) high (up to 500 mM NaCl) salinity stress. However, *Th. halophila* needs to be subjected to a gradual increase in the Na^+ concentration to precondition the plants over a long period (about 1 month). The involvement of PAs in *Thellungiella* adaptation mechanisms can be expected. The analysis of differences in transcript and metabolite profiles supported by microarray results showed that *Thellungiella* induced genes that should provide more penetrating insight into the role of PAs in the adaptation to salt stress. Thus, *Thellungiella* could be a valuable model for the study of abiotic stress tolerance as well as an excellent tool for studying the molecular mechanisms underlying the role of PA in stress tolerance.

An important obstacles to the understanding of the biological role of PAs is that not much is known about their exact cellular and subcellular localisation and their biosynthetic enzymes in plants, especially under stress conditions. There are gaps in the information available on the translocation of free PAs and their interaction with hormones, on their role in gene expression, on the possible role of bound PAs, and on some other aspects of their action in plants experiencing stress. The use of molecular approaches (cloning of genes for PA biosynthetic enzymes in particular), production of transgenic plants, and isolation and characterisation of mutants defective in PA biosynthesis will lead to a better understanding of the role of PA in adaptation of plants to stress conditions. Efforts to improve crop abiotic stress tolerance by cellular and molecular modifications of PA metabolism are in progress. A thorough comparative study of the expression and function of genes involved in PA metabolism in extreme halophytes and xerophytes will eventually help in breeding of stress-tolerant crops.

Acknowledgements The authors are grateful to Professor Nella L. Klachko from the Institute of Plant Physiology (Moscow, Russia) for valuable discussion. This work was partially supported by the Russian Foundation for Basic Research and by the program of the Presidium of RAS (Cell and Molecular Biology).

References

Adams DO, Yang SF (1979) Ethylene biosynthesis identification of 1-aminocyclopropane-1-carboxylic acid as an intermediate in conversion of methionine to ethylene. Proc Natl Acad Sci USA 76:170–174

Alcázar R, Cuevas JC, Patron M, Altabella T, Tiburcio F (2006) Abscisic acid modulates polyamine metabolism under water stress in *Arabidopsis thaliana*. Physiol Plant 128:448–455

Amtmann A, Bohnert HJ, Bressan RA (2005) Abiotic stress and plant genome evolution. Search for new models. Plant Physiol 138:127–130

Anonymous (2004) FAO production yearbook. FAO, Rome

Apel K, Hirt H (2004) Reactive oxygen species: metabolism, oxidative stress and signal transduction. Annu Rev Plant Biol 55:373–379

Apelbaum A, Goldlust A, Isekson I (1985) Control by ethylene of arginine decarboxylase activity in pea seedlings and its implication for hormone regulation of plant growth. Plant Physiol 79:635–647

Aronova EE, Shevyakova NI, Sretsenko LA, Kuznetsov VlV (2005) Cadaverine-induced induction of superoxide dismutase gene expression in *Mesembryanthemum crystallinum* L. Doklady Biol Sci 403:1–3

Bagga S, Rochford J, Klaene Z, Kuehn CD, Phillips GC (1997) Putrescine iminopropyltransferase is responsible for biosynthesis of spermidine, spermine and multiple uncommon polyamines in osmotic stress-tolerant alfalfa. Plant Physiol 114:445–454

Bagni N, Tassoni A (2001) Biosynthesis, oxidation and conjugation of aliphatic polyamines in higher plants. Amino Acids 20:301–317

Bartels D, Souer E (2004) Molecular responses of higher plants to dehydration. In: Hirt H, Shinozaki K (eds) Plant responses to abiotic stress, vol 4. Topics in current genetics. Springer, Berlin, pp 9–38

Bastola DR, Minocha SC (1995) Increased putrescine biosynthesis through transfer on mouse ornithine decarboxylase cDNA in carrot promotes somatic embryogenesis. Plant Physiol 109:63–71

Bhatnagar P, Glasheen BM, Bains SK, Long SL, Minocha R, Walter C, Minocha SC (2001) Transgenic manipulation of the metabolism of polyamines in poplar cells. Plant Physiol 125:2139–2153

Borrel A, Culianez-Macia A, Atabella T, Besford RT, Flores D, Tiburcio AF (1995) Arginine decarboxylase is localized in chloroplasts. Plant Physiol 109:771–776
Borrel A, Besford RT, Atabella T, Masgrau C, Tiburcio AF (1996) Regulation of arginine decarboxylase by spermine in osmotically stressed oat leaves. Plant Physiol 98:105–110
Bors W, Langebartels C, Michel C, Sandermann H (1989) Polyamines as radical scavengers and protectants against ozone damage. Phytochemistry 28:1589–1595
Bortolotti C, Cordeiro A, Alcázar R, Borrell A, Culiañez-Macià FA, Tiburcio AF, Atabella T (2004) Localization of arginine decarboxylase in tobacco plants. Physiol Plant 120:84–92
Bouchereau A, Aziz LF, Martin-Tanguy J (1999) Polyamines and environmental challenges: recent development. Plant Sci 140:103–125
Bryson K, Greenall RJ (2000) Binding sites of the polyamines putrescine, cadaverine, spermidine and spermine on A- and B-DNA located by stimulated annealing. J Biomol Struct Dyn 18:393–412
Bueb L, Da Silva A, Mousli M, Landry Y (1992) Natural polyamines stimulate G-proteins. Biochem J 282:545–550
Capell T, Bassie L, Christou P (2004) Modulation of the polyamine biosynthetic pathway in transgenic rice confers tolerance to drought stress. Proc Natl Acad Sci USA 101:9909–9914
Carden DE, Walker DJ, Flowers TJ, Miller AJ (2003) Single-cell measurements of the contributions of cytosolic NA^+ and K^+ to salt tolerance. Plant Physiol 131:676–683
Cervelli M, Caro O, Penta A, Angelini R, Federico R, Vitale A, Mariottini P (2004) A novel C-terminal sequence from barley polyamine oxidase is a vacuolar sorting signal. Plant J 40:410–418
Chattopadhyay MK, Gupta S, Sengupta DH, Ghosh B (1997) Expression of arginine decarboxylase in seedlings by salinity stress. Plant Mol Biol 34:477–483
Chen TH, Murata N (2002) Enhancement of tolerance of abiotic stress by metabolic engineering of betaines and other compatible solutes. Curr Opin Plant Biol 5:250–257
Chen Z, Newman I, Zhou M, Mendham N, Zhang G, Shabala S (2005) Screening plants for salt tolerance by measuring K^+ flux: a case study for barley. Plant Cell Environ 28:1230–1246
Childs AC, Mehta DJ, Gerner EW (2004) Polyamine-dependent gene expression. Cell Mol Life Sci 60:1394–1406
Christmann A, Hoffmann T, Teplova I, Grill E, Müller A (2005) Generation of active pools of abscisic acid revealed by in vivo imaging of water stressed Arabidopsis. Plant Physiol 137:209–219
Cona A, Cenci F, Cervelli M, Federico R, Mariottini P, Moreno S, Angelini R (2003) Polyamine oxidase, a hydrogen peroxide-producing enzyme, is up-regulated by light and down-regulated by auxin in the outer tissues of the maize mesocotyl. Plant Physiol 131:803–813
Cona A, Rea G, Angelini R, Tavladoraki P (2006) Functions of amine oxidases in plant development and defence. Trends Plant Sci 11:80–88
Cushman JC, Bohnert HJ (2000) Genomic approaches to plant stress tolerance. Curr Opin Plant Biol 3:117–124
Dajic Z (2006) Salt stress. In: Madhava Rao KV, Raghavendra AS, Junardhan Reddy K (eds) Physiology and molecular biology of stress tolerance in plants. Springer, The Netherlands, pp 270–296
D'Agostino L, Massimiliano DP, Luccia AD (2005) Nuclear aggregates of polyamines are supramolecular structures that play a crucial role in genomic DNA protection and conformation. FEBS J 272:3777–3787
Di Tomasso JM, Hart JJ, Kochian LV (1992) Transport kinetics and metabolism of exogenously applied putrescine in roots of intact maize seedlings. Plant Physiol 98:611–620
Deng H, Bloomfield VA, Benevides JM (2000) Structural basis of polyamine–DNA recognition: spermidine and spermine interactions with genomic B-DNAs of different GC content probed by Raman spectroscopy. Nucleic Acids Res 28:3379–3385
Dobrovinskaya OR, Muniz J, Pottosin (1999) Inhibition of vacuolar ion channels by polyamines. J Membr Biol 162:127–140

Dondini L, del Duca S, Doll'Agata L, Bassu R, Gastaldeli M, Della Mea M, di Sandro A, Claparols I, Serafini-Fracassini D (2003) Suborganellar localisation and effect of light on *Heliantus tuberosus* chloroplast transglutaminases and their substrates. Planta 217:84–95
Drolet G, Dumbroff EB, Leggee RL, Thompson JE (1986) Radical scavenging properties of polyamines. Phytochemistry 25:367–371
Duhazé C, Gouzerh G, Gagneul D, Larher F, Bouchereau A (2002) The conversion of spermidine to putrescine and 1,3-diaminopropane in the roots of *Limonium tataricum*. Plant Sci 163:639–646
Erdei E, Lee SJ, Wei Q, Wang LE, Song YS, Bovbjerg D, Berwick M (2005) Reliability of mutagen sensitivity assay: an inter-laboratory comparison. Mutagenesis 21:261–264
Edreva AM, Velikova VB, Tsonov TD (2007) Phenylamides in plants. Russ J Plant Physiol 54:289–302
Espartero J, Pintor-Toro JA, Pardo JN (1994) Differential accumulation of S-adenosylmethionine synthase transcripts in response to salt stress. Plant Mol Biol 25:217–227
Evans PT, Malmberg RL (1989) Do polyamines have roles in plant development? Annu Rev Plant Physiol Plant Mol Biol 40:235–269
Flowers TJ, Colmer TD (2008) Salinity tolerance in halophytes. New Phytol 179:945–963
Friedman R, Altman A, Levin N (1986) Presence and identification of polyamine in xylem and phloem exudates of plants. Plant Physiol 82:1154–1157
Fritze K, Czaja I, Walden R (1995) T-DNA tagging of genes influencing polyamine metabolism – isolation of mutant plant lines and rescue of DNA promoting growth in the presence of polyamine biosynthetic inhibitor. Plant J 7:261–271
Fujihara S, Yoneyama T (1993) Effects of pH and osmotic stress on cellular polyamine contents in the soybean rhizobia *Rhizobium fredii* P220 and *Bradyrhizobium japonicum* A1017. Appl Environ Microbiol 59:1104–1109
Galloway GL, Malmberg RL, Price RA (1998) Phylogenetic utility of the nuclear gene arginine decarboxylase: an example from Brassicaceae. Mol Biol Evol 15:1312–1320
Galston AW, Kaur-Sawhney R, Atabella T, Tiburcio AF (1997) Plant polyamines in reproductive activity and response to abiotic stress. Bot Acta 110:197–207
Gerats AGM, Kaye C, Collins C, Malmberg RL (1988) Polyamine level in Petunia genotypes with normal and abnormal floral morphologies. Plant Physiol 86:390–393
Gong Q, Li P, Ma S, Rupassara SI, Bohnert HJ (2005) Salinity stress adaptation competence in the extremophile *Thellungiella halophila* in comparison with its relative *Arabidopsis thaliana*. Plant J 41:1–14
Ha HL, Sirisoma NS, Kuppusamy P, Zweller JL, Woster PM, Casero RA (1998) The natural polyamine spermine functions as a free radical scavenger. Proc Natl Acad Sci USA 95:11140–11145
Halmekytö M, Alhonen L, Alakuijala L, Jänne J (1993) Transgenic mice over-producing putrescine in their tissues do not convert the diamine into higher polyamines. Biochem J 291:505–508
Hanfrey C, Sommer S, Mayer MJ, Burtin D, Michael AJ (2001) Arabidopsis polyamine biosynthesis: absence of ornithine decarboxylase and the mechanism of arginine decarboxylase activity. Plant J 27:551–560
Hanzawa Y, Takahashi T, Michael AJ, Burtin D, Long D, Pineiro M, Coupland G, Komeda Y (2000) *ACAULIS5*, an Arabidopsis gene required for stem elongation, encodes a spermine synthase. EMBO J 19:4248–4256
Hasegawa PM, Bressan RA, Zhu J-K, Bohnert HJ (2000) Plant cellular and molecular responses to high salinity. Annu Rev Plant Physiol Mol Biol 51:463–497
Havelange A, Lejeune P, Bernier A, Kaur-Sawhney R, Galston AW (1996) Putrescine export from leaves in relation to floral transition in *Sinapis alba*. Physiol Plant 96:59–65
Herminghaus S, Schreirer PH, McCarthy JEG, Landsmann J, Botterma J, Berlin J (1991) Expression of bacterial lysine decarboxylase gene and transport of the protein into chloroplasts of transgenic tobacco. Plant Mol Biol 17:475–486

Hiatt AC, Malmberg RL (1988) Utilization of putrescine in tobacco cell lines resistant to inhibitors of polyamine synthesis. Plant Physiol 86:441–446

Imai A, Akiyama T, Kato T, Sato S, Tabata S, Yamamoto KT, Takahashi T (2004a) Spermine is not essential for survival of *Arabidopsis*. FEBS Lett 556:148–152

Imai A, Matsuyama T, Hanzawa Y, Akiyama T, Tamaoki M, Saji H, Shirano Y, Kato T, Hayashi H, Shibata D, Tabata S, Komeda Y, Takahashi T (2004b) Spermidine synthase genes are essential for survival of *Arabidopsis*. Plant Physiol 135:1565–1573

Inan G, Zhang Q, Li P, Wang Z, Cao Z., Zhang H, Zhang C, Quist TM, Goodwin SM, Zhu J, Shi H, Damsz B, Charbaji T, Gong Q, Ma S, Fredricksen M, Galbraith DW, Jenks MA, Rhodes D, Hasegawa PM, Bohnert HJ, Joly RJ, Bressan RA, Zhu JK (2004) Salt cress. A halophyte and cryophyte Arabidopsis relative model system and its applicability to molecular genetic analysis of growth and development of extremophiles. Plant Physiol 135:1718–1737

Kakkar RK, Sawhney VK (2002) Polyamine research in plants – a changing perspective. Physiol Plant 116:281–292

Kant S, Kant P, Raven E, Barak S (2006) Evidence that differential gene expression between the halophyte, *Thellungiella halophyla*, and *Arabidopsis thaliana* is responsible for higher levels of the compatible osmolyte proline and tight control of Na^+ uptake in *T. halophyla*. Plant Cell Environ 29:1220–1234

Kasinathan V, Wingler A (2004) Effect of reduced arginine decarboxylase activity on salt tolerance and on polyamine formation during salt stress in *Arabidopsis thaliana*. Physiol Plant 121:101–107

Kasukabe Y, He L, Nada K, Misawa S, Ihara I, Tachibana S (2004) Overexpression of spermidine synthase enhances tolerance to multiple environmental stresses and up-regulates the expression of various stress-regulated genes in transgenic *Arabidopsis thaliana*. Plant Cell Physiol 45:712–722

Kasukabe Y, He L, Watakabe Y, Otani M, Shimada T, Tachibana S (2006) Improvement of environment stress tolerance of sweet potato by introduction of genes for spermidine synthase. Plant Biotechnol 23:75–83

Kaur-Sawhney R, Tiburcio AF, Atabella T, Galston AW (2003) Polyamines in plants: an overview. J Cell Mol Biol 2:1–12

Kim TE, Kim S-K, Han TJ, Lee JS, Chang SC (2002) ABA and polyamines act independently in primary leaves of cold-stressed tomato (*Licopersicon esculentum*). Physiol Plant 115:370–376

Koenig H, Goldstone A, Lu CY (1983) Polyamines regulate calcium fluxes in a rapid plasma membrane occurrence. Nature 305:530–534

Koiwai H, Nakaminami K, Seo M, Mitsuhasi W, Toyomasu T, Koshiba T (2004) Tissue-specific localization of an abscisic acid biosynthesis enzyme, AAO#, in *Arabidopsis*. Plant Physiol 134:1697–1707

Kreps JA, Wu Y, Chang HS, Zhu T, Wang X, Harper JF (2002) Transcriptome changes for *Arabidopsis* in response to salt, osmotic, and cold stress. Plant Physiol 130:2129–2141

Krishnamurthy R, Bhagwat KA (1989) Polyamines as modulators of salt tolerance in rice cultivars. Plant Physiol 91:500–504

Kumar A, Minocha SC (1998) Transgenic manipulation of polyamine metabolism. In: Lindsey K (ed) Transgenic research in plants. Harwood, London, pp 189–199

Kutuzov MA, Andreeva AV, Voyno-Yasennetskaya TA (2005) Regulation of apoptosis signal-regulating kinase 1 (ASK1) by polyamine levels via protein phosphatase 5. J Biol Chem 280:25388–25395

Kuznetsov VlV, Rakitin VYu, Sadomov NG, Dam DV, Stetsenko LA, Shevyakova NI (2002) Do polyamines participate in the long-distance translocation of stress signals in plants? Russ J Plant Physiol 49:136–147

Kuznetsov VlV, Radyukina NL, Shevyakova NI (2006a) Polyamine and stress: biological role, metabolism, and regulation. Russ J Plant Physiol 53:583–604

Kuznetsov VlV, Rakitin VYu, Radyukina NL, Ivanov Vyu, Kartashov AV, Shevyakova NI (2006b) Stress-accelerated spermine production in leaves of *Thellungiella halophyla* is not controlled at level of expression of *SPDS* gene. Abstract P08002 in American Society of Plant Biologists; http://abstracts.aspb.org/pb2006/public/P08/P08002.html

Kuznetsov VlV, Shorina M, Aronova E, Stetsenko L, Rakitin V, Shevyakova N (2007) NaCl- and ethylene-dependent cadaverine accumulation and its possible protective role in the adaptation of the common ice plant to salt stress. Plant Sci 172:363–370

Kuznetsov VlV, Stetsenko LA, Shevyakova NI (2009) Exogenous cadaverine induces oxidative burst and reduces cadaverine conjugate content in the common ice plant. J Plant Physiol 166:40–51

Langebartels C, Kerner KJ, Leonardi S, Schraudner M, Trost M, Heiller W, Sanderman H (1991) Biochemical plant response to ozone. Differential induction of polyamine and ethylene biosynthesis in tobacco. Plant Physiol 9:882–887

Legocka J, Zaichert J (1999) Role of spermidine in the stabilization of apoprotein of the light-harvesting chlorophyll a/b-protein complex of photosystem 11 during leaf senescence process. Acta Physiol Plant 21:127–137

Lepri O, Bassie L, Safwat G, Thu-Hang P, Trung-Nghia P, Hölttä E, Christou P, Capell T (2001) Over-expression of human ornithine decarboxylase cDNA in transgenic rice plants alters the polyamine pool in a tissue-specific manner. Mol Gen Genet 266:303–312

Lindemose S, Nielson PE, Møllegaard NE (2005) Polyamines preferentially interact with bent adenine tracts in double-stranded DNA. Nucleic Acids Res 33:1790–1803

Liu J-H, Moriguchi T (2006) ADC pathway plays an important in salt stress response of apple in vitro callus. Plant Genomic Chin 124:1315–1325

Maiale S, Sanchez DH, Guirado A, Vidal A, Ruiz OA (2004) Spermine accumulation under salt stress. J Plant Physiol 161:35–42

Martin-Tanguy J (2001) Metabolism and function of polyamines in plants: recent development (new approaches). Plant Growth Regul 34:135–148

Masgrau C, Altabella T, Farras R, Flores D, Thompson AJ, Besford RT, Tiburcio AF (1997) Inducible overexpression of oat arginine decarboxylase in transgenic tobacco plants. Plant J 11:465–473

Mehta RA, Cassol T, Li N, Ali N, Handa AK, Mattoo AK (2002) Engineered polyamine accumulation in tomato enhances phytonutrient content, juice quality and vine life. Nat Biotechnol 20:613–618

Messiaen J, van Cutsem P (1999) Polyamines and pectins. Modulation of pectic-signal transduction. Planta 208:247–250

Moschou PN, Dellis I, Paschalidis K, Robelakis-Angelakis KA (2008a) Transgenic tobacco plants over-expressing polyamine oxidase are not able to cope with oxidative burst generated by abiotic factors. Physiol Plant 133:140–156

Moschou PN, Sanmartin M, Andriopoulou AH, Rojo E, Sanchez-Serrano LJ, Roubelakis-Angelakis KA (2008b) Bridging the gap between plant and mammalian polyamine catabolism: a novel peroxisomal polyamine oxidase responsible for a full back-conversion pathway in Arabidopsis. Plant Physiol 147:1845–1857

Neill SJ, Desikan R, Hancock JT (2002) Hydrogen peroxide signalling. Curr Opin Plant Biol 5:388–395

Ormrod DP, Beckerson DW (1986) Polyamines as antiozonats for tomato. Hortic Sci 21: 1070–1071

Panicot M, Minguet EG, Ferrando A, Alcázar R, Blázquez MA, Carbonell J, Atabella T, Koncz C, Tiburcio AF (2002) A polyamine metabolon involving aminopropyl transferase complexes in *Arabidopsis*. Plant Cell 14:2539–2551

Paramonova NV, Shevyakova NI, Shorina MV, Stetsenko LA, Rakitin VYu (2003) The effect of putrescine on the apoplast ultrastructure in the leaf mesophyll of *Mesembryanthemum* crystallinum under salinity stress. Russ J Plant Physiol 50:661–674

Paramonova NV, Shevyakova NI, Kuznetsov VlV (2007) Ultrastructural of ferritin in the leaves of *Mesembryanthemum crystallinum* under stress conditions. Russ J Plant Physiol 54:244–256

Paschalidis KA, Roubelakis-Angelakis KA (2005a) Spatial and temporal distribution of polyamine levels and polyamine anabolism in different organs/tissues of tobacco plants. Correlations with age, cell division/expansion, and differentiation. Plant Physiol 138:142–152

Paschalidis KA, Roubelakis-Angelakis KA (2005b) Sites and regulation of polyamine catabolism in the tobacco plant. Correlations with cell division/expansion, cell cycle progression, and vascular development. Plant Physiol 138:2174–2184

Perez-Amador MA, Leon J, Green PJ, Carbonell J (2002) Induction of the arginine decarboxylase *ADC2* gene provides evidence for the involvement of polyamines in the wound response in *Arabidopsis*. Plant Physiol 130:1454–1463

Radyukina NL, Ivanov YuV, Kartashov AV, Shevyakova NI, Rakitin VYu, Khryanin VN, Kuznetsov VlV (2007a) Inducible and constitutive mechanisms of salt stress resistance in *Geum urbanum* L. Russ J Plant Physiol 54:692–698

Radyukina NL, Kartashov AV, Ivanov YuV, Shevyakova NI and Kuznetsov VlV (2007b) Functioning of defence systems in halophytes and glycophytes under progressing salinity. Russ J Plant Physiol 54:806–815

Rakitin VYu, Prudnikova ON, Rakitina TYa, Vlasov PV, Karyagin VV (2004) UV-B induced evolution, accumulation of ABA and putrescine in *Arabidopsis thaliana* plants. Abstract Botanikertagung, Braunschweig, 5–10 September, Proc Dtsch Bot Ges Verein Angew Bot pp 420

Rakova NU, Romanov GA (2005) Polyamines suppress manifestation cytokinine primary effects. Russ J Plant Physiol 52:50–57

Ramos J, Lopez MJ, Benlloch M (2004) Effect of NaCl and KCl salts on the growth and solute accumulation of the halophyte *Atriplex nummularia*. Plant Soil 259:163–168

Rea G, de Concetta PM, Tavazza R, Biondi S, Gobbi V, Ferrante P, De Gara L, Federico R, Angelini R, Tavladoraki P (2004) Ectopic expression of maize polyamine oxidase and pea copper amine oxidase in the cell wall of tobacco plants. Plant Physiol 134:1414–1426

Richard FJ, Coleman RG (1952) Occurrence of putrescine in potassium-deficient barley. Nature 170:460

Rowland-Bamford AJ, Barland AM, Lea PJ, Mansfield TA (1989) The role of arginine decarboxylase in moduling the sensitivity of barley to ozone. Environ Pollut 61:93–99

Roy M, Wu R (2002) Overexpression of S-adenosylmethionine decarboxylase gene in rice increases polyamine level and enhances sodium chloride stress tolerance. Plant Sci 163: 987–992

Ruiz-Herrera J, Ruiz-Medrano R, Dominguez A (1995) Selective inhibition of cytosine-DNA methylases by polyamines. FEBS Lett 357:192–196

Sairam RK, Tyagi A, Chinnusamy V (2006) Salinity tolerance: cellular mechanisms and gene regulation. In: Huang B (ed) Plant–environment interactions. Taylor and Francis, New York, pp 121–309

Sauter A, Dietz K-J, Hartung W (2002) A possible physiological role of abscisic acid conjugates in root-to-shoot signalling. Plant Cell Environ 25:223–22

Scandalios JG (1993) Oxygen stress and superoxide dismutases. Plant Physiol 101:7–12

Seki M, Narusaka M, Ishida J (2002) Monitoring the expression profiles of 7000 Arabidopsis genes under drought, cold and high-salinity stresses using a full-length cDNA microarray. Plant J 31:279–292

Shabala S, Cuin TA, Pottosin I (2007) Polyamines prevent NaCl-induced K^+ efflux from pea mesophyll by blocking non-selective cation channels. FEBS Lett 581:1993–1999

Shen W, Nada K, Tachibana S (2000) Involvement of polyamines in the chilling tolerance of cucumber cultivars. Plant Physiol 124:431–439

Shevyakova NI, Kir'yan IG (1995) Osobennosti regulyatsii biosinteza metionina v soleustoichivykh kletkakh *Nicotiana sylvestris* L. Fiziol Rast 42:94–99

Shevyakova NI, Rakitin VYu, Duong DB, Sadomov NG, Kuznetsov VLV (2001) Heat shock-induced cadaverine accumulation and translocation throughout the plant. Plant Sci 161: 1125–1133

Shevyakova NI, Shorina MV, Rakitin VYu, Stetsenko LA, Kuznetsov VlV (2004) Ethylene-induced production of cadaverine is mediated by protein phosphorylation and dephosphorylation. Dokl Biol Sci 395:283–285

Shevyakova NI, Rakitin VYu, Stetsenko LA, Aronova EE, Kuznetsov VlV (2006a) Oxidative stress and fluctuations of free and conjugated polyamines in the halophyte *Mesembryanthemum crystallinum* L. under NaCl salinity. Plant Growth Regul 50:69–78

Shevyakova NI, Shorina MV, Rakitin VYU, Kuznetsov VLV (2006b) Stress-dependent accumulation of spermidine and spermine in the halophyte *Mesembryanthemum crystallinum* L. under salinity conditions. Russ J Plant Physiol 53:739–745

Shorina MV, Ragulin VV, Kuznetsov VlV, Shevyakova NI (2005) Does cadaverine and ethylene involved in CAM-type photosynthesis induction in the common ice plant? Dokl Biol Sci 400:115–120

Ślesak J, Karpinska B, Surówka E, Miszalski Z, Karpinski S (2003) Redox changes in the chloroplast and hydrogen peroxide are essential for regulation of C_3-CAM transition and photooxidative stress responses in the facultative CAM plant *Mesembryanthemum crystallinum* L. Plant Cell Physiol 44:573–581

Takahashi Y, Uehara Y, Berberich T, Ito A, Saitoh H, Miyazaki A (2004) A subset of hypersensitive response marker genes, including HSR203J, is the downstream target of a spermine signal transduction pathway in tobacco. Plant J40:586–595

Taji T, Motoaki S, Masakazu S, Tetsuya S, Masatomoto K, Ishiyama K, Narusaka Y, Narusaka M, Zhu J-K, Shinozaki K (2004) Comparative genomics in salt tolerance between *Arabidopsis* and *Arabidopsis*-related halophyte salt stress using *Arabidopsis* microarray. Plant Physiol 135:1697–1709

Tassoni A, Antognoni F, Battistini ML, Sanvido OA, Bagni N (1998) Characterization of spermidine binding to solubilized plasma membrane proteins from Zucchini hypocotyls. Plant Physiol 117:971–977

Tavladoraki P, Rossi MN, Saccuti G, Perez-Amador MA (2006) Heterologous expression and biochemical characterization of a polyamine oxidase from Arabidopsis involved in polyamine back conversion. Plant Physiol 149:15–1532

Tiburcio AF, Besford RT, Capell T, Borell A, Testillano PS, Risueûo MC (1994) Mechanisms of polyamines action during senescence responses induced by osmotic stress. J Exp Bot 45:1789–1800

Tkachenko AG, Nesterova LY (2003) Polyamines as modulators of gene expression under oxidative stress in *Escherichia coli*. Biochemistry 68:850–856

Turano FJ, Kramer GF (1993) Effect of metabolic intermediates on the accumulation of polyamines in detached soybean leaves. Phytochemistry 34:959–968

Urano K, Yoshiba Y, Nanjo T, Igarashi Y, Seki M, Sekiguchi F, Yamaguchi-Shinozaki K, Shinozaki K (2003) Characterization of *Arabidopsis* genes involved in biosynthesis of polyamines in abiotic stress responses and developmental stages. Plant Cell Environ 26:1917–1926

Vera-Estrella R, Barkla BJ, García-Ramírez L, Pantoja O (2005) Salt stress in *Thellungiella halophila* activates Na^+ transport mechanisms required for salinity tolerance. Plant Physiol 139:1507–1517

Volkov V, Wang B, Dominy PJ, Fricke W, Amtman A (2004) *Thellungiella halophyla*, a salt-tolerant relative of *Arabidopsis thaliana*, between potassium and sodium. Plant Cell Environ 27:1–14

Von Detsch AW, Mitchell CD, Williams CE, Dutt K, Silvestrov NA, Klement BJ, Abukhalaf JK, van Dentsch DA (2005) Polyamines protect against radiation-induced oxidative stress. Gravit Space Biol Bull 18:109–110

Vranova E, Atichartpongkus S, Villarroll R, Van Montagu M, Inzé D, Van Camp W (2002) Comprehensive analysis of gene expression in *Nicotiana tabacum* leaves acclimated to oxidative stress. Proc Natl Acad Sci USA 99:10870–10875

Walden R, Cordeiro A, Tiburcio F (1997) Polyamines: small molecules triggering pathways in plant growth and development. Plant Physiol 113:1009–1013

Wang KL-C, Li H, Ecker JR (2002a) Ethylene biosynthesis and signaling networks. Plant Cell S131–S151

Wang W, Vinocur B, Altman A (2003) Plant responses to drought, salinity and extreme temperatures: towards genetic engineering for stress tolerance. Planta 218:1–14

Wang Y, Devereux W, Stewart TM, Casero RA (2002b) Polyamine-modulated factor 1 binds to the human homologue of the 7a subunit of the *Arabidopsis* COP9 signalosome: implications in gene expression. Biochem J 366:79–86

Wi SJ, Park KY (2002) Antisense expression of carnation cDNA encoding ACC synthase or ACC oxidase enhances polyamine content and abiotic stress tolerance in transgenic tobacco plants. Transgenic Res Newslett 13:209–220

Yang J, Zhang J, Liu K, Wang Z, Liu L (2007) Involvement of polyamines in the drought resistance of rice. J Exp Bot 58:1545–1555

Ye B, Müller H, Zhang J, Cressel J (1997) Constitutively elevated level of putrescine and putrescine-generation enzymes, correlated with oxidant stress resistance in *Coniza bonariensis* and wheat. Plant Physiol 115:1443–1451

Yeo AR, Flowers TJ (1986) Ion transport in *Suaeda maritima*: its relation to growth and implications for the pathway of radial transport of ions across the root. J Exp Bot 37:143–159

Zhao F, Chun-Peng S, He J, Zhu H (2007) Polyamines improve K^+/N^+ homeostasis in barley seedling by regulating root ion channel activities. Plant Physiol 145:1061–1072

Zhao FG, Qin P (2004) Protective effects of exogenous polyamines on root tonoplast function against salt stress in barley seedlings. Plant Growth Regul 42:97–103

Zhu J-K (2000) Genetic analysis of plant salt tolerance using *Arabidopsis*. Plant Physiol 124:941–948

Zhu H, Ding GH, Fang K, Zhao FG, Qin P (2006) New perspective on the mechanism of alleviating salt stress by spermidine in barley seedlings. Plant Growth Regul 49:147–156

Chapter 14
Ecology of Inland Saline Plants

Pawan K. Kasera and Sher Mohammed

Abstract This chapter describes the ecology and adaptive strategies of inland halophytes growing in natural saline areas, with special reference to classification, metabolic products, soil–water relationships, the role of proline in their survival, etc. Studies on eight saline plants, viz. *Aeluropus lagopoides* (Poaceae), *Cressa cretica* (Convolvulaceae), *Salsola baryosma* (Chenopodiaceae), *Sesuvium sesuvioides* (Aizoaceae), *Sporobolus helvolus* (Poaceae), *Suaeda fruticosa* (Chenopodiaceae), *Trianthema triquetra* (Aizoaceae), and *Zygophyllum simplex* (Zygophyllaceae), selected from different Indian desert sites, is described. Among these species, *S. fruticosa* is distributed widely and is highly salt tolerant. The major cations and anions in saline areas of the Indian arid zone are Na^{+}, K^{+} and Ca^{++}, and Cl^{-}, respectively. The inhibition of seed germination of saline plants is controlled by both osmotic and ionic factors. Soil salinity increases especially during dry periods, which results in accumulation of more ions in leaves. The metabolic products of different plant species respond differently to higher salinity. Free proline accumulation during unfavourable conditions increases with increasing salinity, which helps in the survival of saline plants.

14.1 Introduction

Salinity problems are of great concern in arid and semi-arid regions, where soil salt contents are high and precipitation is insufficient for their leaching. In these regions, planting salt-tolerant species, particularly N_2-fixing species, is the most useful

P.K. Kasera (✉)
Laboratory of Plant Ecology, Department of Botany, Jai Narain Vyas University, Jodhpur, 342 033, India
e-mail: jnvupkkasera@yahoo.co.in

S. Mohammed
Department of Botany, Government Lohia (PG) College, Churu, 331 001, India
e-mail: smdchuru@gmail.com

K.G. Ramawat (ed.), *Desert Plants*,
DOI 10.1007/978-3-642-02550-1_14,

approach to rehabilitating salt-affected degraded lands. There are a large number of plant species that are regarded as salt-tolerant, the most competitive being those that are able to become established, grow to maturity and survive until they are able to reproduce (Heidari-Sharifabad and Mirzaie-Nodoushan 2006).

Salt-affected soil is defined as soil that has been adversely affected to the extent that it is no longer suitable for normal plant growth due to presence of excess soluble salts. Such soils include both saline and sodic soils. The loss of plant productivity due to excess salinity is a worldwide problem (Evangelou 1994; Sen et al. 2002). In salt-affected soils, high electrolyte content or extreme pH conditions limit the development of the majority of plants, and such soils serve only as habitat for species that can survive or tolerate the unfavourable conditions caused by the salinity or alkalinity. Salt-affected soils can be characterised as soils formed under the dominant influence of different salts in their solid or liquid phases, which then have a decisive influence on the development, characteristics, physical, chemical and biological properties, and, eventually, the fertility of the soil. In saline soils, the high salt concentration in the solid or liquid phase results in high osmotic pressure, hindering the normal development of plants, i.e. the stress factor is the salinity, with all its disadvantageous consequences to plant life (Szabolcs 1994).

Agricultural production worldwide is greatly affected by a number of environmental hazards, one of the most important being salinity associated with aridity. Salinity is a problem not only in India, but also throughout the world. Saline lands are not only distributed in desert and semi-desert regions, but also occur frequently in fertile alluvial plains, rivers, valleys, and coastal regions close to densely populated areas and irrigation systems. Salinisation can occur in areas with arid and semi-arid climates or in coastal regions, where salts are transported by groundwater and precipitation. The direct effects of salts on plant growth can be divided into three broad categories: (1) a reduction in the osmotic potential of the soil solution, which reduces the amount of water available to plants; (2) a deterioration in the physical structure of the soil, which decreases permeability to water and gases; and (3) specific ion toxicity (Dudley 1994). Halophytes thrive under varying soil salinity conditions and may be irrigated with brackish water or with a certain percentage of seawater without any major ill effects on growth and reproduction (Parida and Das 2005).

The vegetation of saline habitats is designated as "halophytic", as opposed to the vegetation of non-saline habitats, which is sometimes referred to as glycophytic. Phenologically, halophytic plants may be succulent or xeromorphic, having small or grass-like leaves and often also salt-secreting glands. In their saline environments, halophytes are exposed not only to salt stress, but the root system may also be exposed to osmotic water and low oxygen pressure (Palfi and Juhasz 1970). The climate of the Thar desert is one of the most important factors influencing salinity. Various climatic factors, such as low rainfall, temperature, wind direction, velocity, etc., play a key role in the formation of saline–alkali soils. Rainfall has a greater effect than temperature in determining soil salinity. In areas receiving low rainfall and high temperature, evaporation takes place, leaving behind accumulated soluble salts. High wind velocity also plays a part in evaporation of surface water as well as transporting some of the salt from salt-affected areas to adjoining areas (Poljakoff-Mayber and Gale 1975; Sen 1990; Sen and Mohammed 1994; Sen et al. 2002). The aridity caused by low rainfall and extreme high or low temperatures are very often compounded by high salinity in the soil, resulting in severe problems in absorption of water, thus curtailing productivity (Mohan Ram and Gupta 1997).

Salinity plays an important role in vegetation cover. An excess of salts in the soil inhibits plant growth in various ways. The Thar – the Indian desert – includes the arid and semi-arid portions of western India and part of Pakistan. Rajasthan State alone owns the maximum arid area, which represents 62% of the total arid area of India and 56% of the total area of the State. The salt basins of the Indian arid zone are of the inland type, and differ greatly from other saline areas in vegetation make up; they support a relatively small number of plant species, namely those capable of tolerating a high degree of salinity. The saline areas of the Indian desert can be divided into two categories: (1) salt lakes such as the Sambhar, Didwana, and Kuchaman, located in the eastern part of Rajasthan; and (2) salt basins such as Pachpadra, Thob, Sanwarla, Luni, Bap, Kaparda, Taal Chhapar, Pokran, Lunkaransar, Lanela, Sakhi, Khajuwalla, etc., located in the western half of Rajasthan (Fig. 14.1).

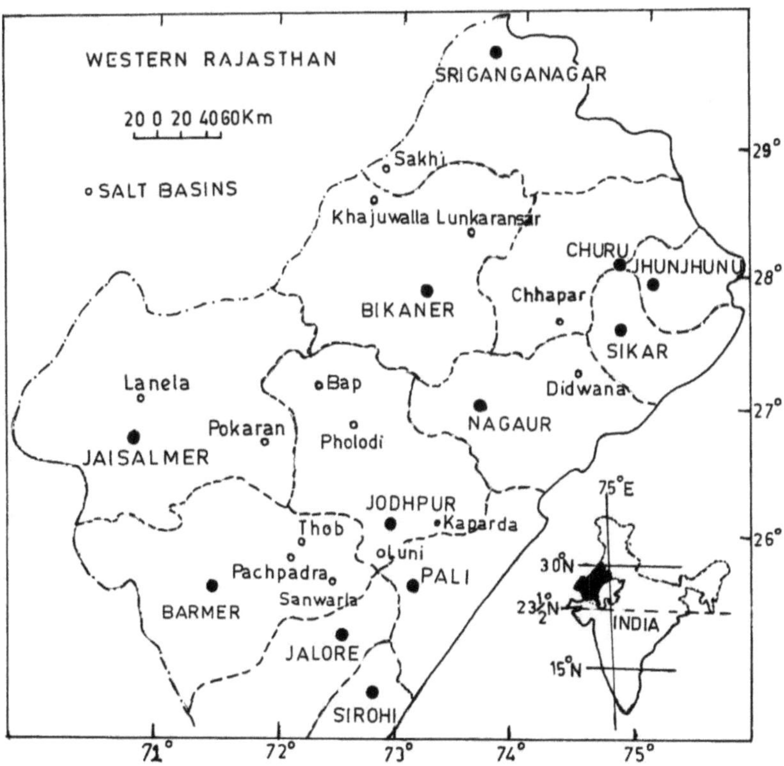

Fig. 14.1 Some important saline areas of western Rajasthan, India (after Mohammed 1988)

Halophytes are defined as plants that grow and complete their entire life-cycle in saline habitats. Halophytic species that live under conditions of high salinity and exhibit succulence might resort to other physiological adaptations to overcome the adverse saline environment in the soil. Their extreme tolerance to salinity is related to their ability to maintain a high salt concentration within the cells. In desert areas salinity is often very prominent, caused by the input of sodium chloride (NaCl) and other salts over long periods of time combined with lack of drainage. On such saline soils, Chenopodiaceae very often dominates typical plant associations that develop along salt gradients (Waisel 1972; Breckle 1986).

14.2 Classification of Inland Saline Vegetation in Western Rajasthan

The saline areas of the Indian arid zone have a mosaic of successive zones represented by characteristic taxa of limited distribution. These zones differ from one another in soil characteristics. Each zone supports a distinct plant grouping with its own characteristic species. The precise determination of the effect of salinity, and the sites where salinity may affect plants are not easily assessed. First, since both salt combination and salt concentration differ from one habitat to another, the term "salinity" usually has a loose meaning. In certain cases, it is not the absolute amount of a certain ion that may affect plants, but rather the composition and total concentration of salts. Certain plant species may be found in sites where the NaCl concentration is beyond their theoretical tolerance but where high concentrations of calcium, potassium, or sulfate are found as supplementary ions. These ions moderate the toxic effects of sodium and chloride, thus enabling plants to exist. In additions, contact between the salt and the plant may involve different tissues of the plant.

Sen and Rajpurohit (1978) classified the vegetation of some Indian salt basins. Later, Rajpurohit (1980) surveyed some of the salt basins and reported a total of 122 plant species, including 10 true halophytes, 48 facultative halophytes, and 64 glycophytes. According to the new classification of Mohammed and Sen (1994), plant species in this region can be divided into four groups:

1. True halophytes: plants that grow in extreme saline conditions (above 1.5% NaCl) and resist high salinity (Fig. 14.2), including *Suaeda fruticosa* (Figs. 14.3, 14.4), *Salsola baryosma* (Fig. 14.5), *Cressa cretica*, *Zygophyllum simplex* (Fig. 14.6), *Haloxylon recurvum* (Fig. 14.7), *Sporobolus helvolus* (Fig. 14.8), *Heliotropium curassavicum* (Fig. 14.9), and *Aeluropus lagopoides*.
2. Facultative halophytes: plants that grow in saline to non-saline conditions, or mainly saline, but cannot tolerate high salinity (0.5% NaCl level), such as *Sesuvium sesuvioides* (Fig. 14.10), *Trianthema triquetra* (Fig. 14.11), *Tamarix* spp. (Fig. 14.12), *Chloris virgata* (Fig. 14.13), *Eleusine compressa*, *Dipcadi erythraeum*, and *Portulaca oleracea*.

Fig. 14.2 View of salt lake at Didwana

Fig. 14.3 Field view of *Suaeda fruticosa* at Pachpadra salt basin

Fig. 14.4–14.14 Halophytes growing under different natural field conditions
Fig. 14.4 *Suaeda fruticosa* (Pachpadra)

Fig. 14.5 *Salsola baryosma* (Jodhpur)

Fig. 14.6 *Zygophyllum simplex* (Pachpadra)

Fig. 14.7 *Haloxylon recurvum* (Jodhpur)

Fig. 14.8 *Sporobolus helvolus* (Pachpadra)

Fig. 14.9 *Heliotropium curassavicum* (Didwana)

3. Transitional halophytes: plants that grow only at the transition of saline and non-saline areas, including *Cassia italica*, *Haloxylon salicornicum*, *Fagonia cretica*, *Cyperus* spp., *Dactyloctenium aegyptium*, *Salvadora persica* (Fig. 14.14), *Tragus racemosus*, and *Boerhavia diffusa*.

Fig. 14.10 *Sesuvium sesuvioides* (Pachpadra)

Fig. 14.11 *Trianthema triquetra* (Pachpadra)

4. Glycophytes: mainly non-saline plants, but growing in saline areas only for a short duration when salinity levels are reduced, including *Brachiaria ramosa*, *Digera alternifolia*, *Eragrostis ciliaris*, *Heliotropium marifolium*, *Convolvulus microphyllus*, *Dicoma tomentosa*, *Farsetia hamiltonii*, *Gisekia pharnacioides*, *Polygala chinensis*, *Calotropis procera*, *Aerva persica*, *Oldenlandia aspera*, *Prosopis juliflora*, *Acacia jacquemonti*, etc.

These species were classified on the basis of (1) their distribution in the salt basins, and (2) their basic requirements, especially for salt and water (Mohammed 1988; Sen 1990; Mohammed and Sen 1994).

Fig. 14.12 *Tamarix troupii* (Luni)

Fig. 14.13 *Chloris virgata* (Jodhpur)

The ecological limits of the distribution of plant communities depend upon the presence of soluble salts in the water or soil. The water of the habitat is the dominant ecological factor determining the distribution of species. Salinity is probably the major factor determining the characteristics of the habitat, and salts affect natural selection more than any other factor (Waisel 1972). Thus, the distribution of a halophytic community appears to be limited by salinity and the depth of the water table as well as by the competitive ability of members of the neighbouring community in the halosere (Reed 1947; Sen and Mohammed 1994; Sen et al. 1997).

Fig. 14.14 *Salvadora persica* (Luni)

14.3 Adaptive Strategies of Inland Halophytes

Plant adaptations dealing with water conservation have particular meaning in dry environments where water stress is either permanent or temporary and severely limits plant growth. Plants living in such environments get adapted by increased drought tolerance and water use efficiency. There are a number of modifications to plant structures and processes as a consequence of drought stress. These include sensitivity of stomatal response, osmotic adjustment, smaller cell volume, reduced leaf area, increased leaf thickness, hairy leaves, and increased root:shoot ratio, as well as various changes in enzyme and hormone production and activity (Pugnaire et al. 1994).

In the world where most of the water is saline, halophytic plants are examples of effective adaptation to increased salinity. Halophytic plants grow in a wide range of environmental conditions. Some grow in high moisture saline areas such as mangrove swamps; others exist in fluctuating moisture conditions in tidal zones; and still others live in inland saline areas in arid climate. In all these cases, in order to take up water halophytes must adjust their tissue water potential to a level that is lower than that of the soil water potential in the habitat where they are growing. Without sufficient moisture, halophytes can become stunted and reproduction becomes very limited (Weber 1995).

Halophytes are adapted to survive in a range of saline environments. In any environment, halophytes require water for growth and development. The key mechanism used by halophytes to obtain sufficient water for growth and development is osmotic adjustment. Halophytes take up ions to increase the osmotic level in their tissues, which permits moisture to move from the soil into the tissues. On the other hand, excess salt ions can have a toxic effect on plant cells. Some of the mechanisms used by halophytes to counter the potential toxic effect of high concentrations of ions involve exclusion of salts by the roots, dilution of ions through succulence, synthesis of organic osmotic compounds that can reduce the need for salt ions, and compartmentalisation of excess salt ions into tissues, organs, or cell vacuoles. Halophytes need to obtain sufficient ions to maintain growth, while avoiding a water deficit or an excess of ions. The net result is that halophytes grow successfully in saline environments whereas glycophytes cannot (Khan and Ungar 1995).

Sen and Rajpurohit (1982) and Sen et al. (2002) classified halophytes into two categories on the basis of salt accumulation and secretion:

1. Salt accumulating halophytes: plants growing under saline conditions maintain a high concentration of osmotically active substances in order to compete successfully with the water-retaining capacity of the surrounding medium. The increase in osmotic potential of cell sap associated with osmotic adjustment of plants to salinity is accomplished by accumulation of ions. Increased succulence has a diluting effect on the toxic ion content of cells, thus enabling plants to withstand the effects of larger quantities of salts. Examples include: *Suaeda fruticosa*, *Salsola baryosma*, *Zygophyllum simplex*, *Trianthema triquetra*, *Haloxylon recurvum*, *Heliotropium curassavicum*, etc.
2. Salt secreting halophytes: secretion of salt by salt glands is a mechanism for regulating mineral content in many non-succulent halophytic species. The salt glands act as desalination devices in halophytes and thus appear to maintain the salt balance in the leaves by secretion of excess salts, e.g. *Aeluropus lagopoides*, *Cressa cretica*, *Sporobolus helvolus*, *Chloris virgata*, *Tamarix* spp., *Atriplex* spp., etc.

A saline environment induces morphological, physiological, and phenological adaptations in plants (Waisel 1972). In saline areas, the effect of salinity on growth and salinity tolerance is temporary in glycophytes and permanent in halophytes; both try to adapt themselves accordingly. Plants that are well adapted to saline soils from various survival aspects are able to grow only in these habitats.

The resistance of plants to salinity leads to several adaptations. Most of these plants avoid salinity, some evade it, and a few others tolerate it. Most plants avoid salinity by: (1) limiting reproduction, growth, and germination during specific periods of the year; (2) limiting uptake of salt; and (3) allowing roots to penetrate into non-saline soils. Evasion of salt can be achieved through the accumulation of salts into certain specific cells and trichomes or by secretion of excess salts through special mechanised salt-secreting glands (Waisel 1972; Sen et al. 1997).

The leaves of saline plants play an important role under physiological drought conditions, developing certain xeromorphic characteristics such as succulence, reduction in surface area, thick cuticle or waxy layers on the epidermis, a hair cover, and salt glands. The hairs on leaves can accumulate salts and act as salt concentration sites although their salt storage capacity is limited (Ungar 1995). A thick cuticle and a cover of waxy layers, such as is present in *S. fruticosa*, *S. baryosma*, and *H. recurvum*, characterises xerosucculents. Cuticle and waxy layers have also been reported on the leaf surfaces of *C. cretica*, *A. lagopoides*,

S. helvolus, and *C. virgata*. Halophytes such as *C. cretica*, *C. virgata*, *S. helvolus*, and *A. lagopoides* show an additional mode of adaptation to their habitat. The leaves and stems of these plants remain covered with hairs (trichomes), giving the plant a greyish appearance. Their effectiveness in reducing the loss of water is small, but they are able to protect the leaf surface against dust.

Halophytic cells need to have high osmotic pressure and at the same time prevent excess ions from inhibiting enzymatic processes. If excess ions are stored in the vacuole, then metabolic activity can carry on in the cytoplasm where the ion content is lower. The lower salt concentration prevents organelles such as chloroplasts from being damaged by excess ions (Khan and Ungar 1995). When a change in metabolism results in a change in the ability to resist stress conditions, anthocyanin may develop in the leaves or stems of plants. Some halophytes of the Indian region, such as *S. fruticosa*, *S. baryosma*, *T. triquetra*, and *Z. simplex*, exhibit this characteristic under osmotic stress. Thus, the development of anthocyanin is a well-known characteristic of plants exposed to drought, osmotic drought, or physiological drought.

The existence of a plant species in its natural habitat depends on its ability to reproduce under those ecological conditions. The presence of excess salt in the soil is one of the critical factors that adversely affects seed germination under such conditions, thereby preventing plant species from saline environments from becoming established successfully (Mohammed 1988; Mohammed and Sen 1990a; Ungar 1995).

Haloxylon recurvum and *H. salicornicum* are two halophytes characteristic of the Indian Thar desert. Seeds of these two species have an extremely fast seed germination time – commonly occurring within an hour – as first reported for any Indian plant species by Sharma and Sen (1989). This rapid germination serves to illustrate the adaptive strategy of these plants, as water with reduced NaCl content in soil during the rainy season is available only for a short period of time because evaporation of moisture under bright sunlight and heat results in increasing the salt content by capillary movement.

Heliotropium curassavicum is one of the rare herbaceous species of the family Boraginaceae, which has been found to flourish in inland salines of Rajasthan. This plant is a perennial herb with spreading habit. The leaves are succulent and glabrous in nature. When growing away from moist areas, the plants prefer to grow in shade and show vigorous growth. The fully grown plants appear characteristically bluish-green. Since the plants grow in a gregarious fashion, they form patches on the ground. The characteristic feature of this species is that, upon the onset of dry conditions, more and more salt is accumulated in the fleshy leaves. These salt saturated leaves dry up to maintain a balanced osmotic level in the plant. New leaves sprout, and this process continues during the whole lifespan of the plant (Sen et al. 2002).

Rajput (1992) reported that with an increase in the duration of leaching in *Atriplex* species, the amount of ions leached from the leaves also increased, with the maximum value being observed after 72 h. Maximum values of leacheable Na^+ and Cl^- were observed in young leaves of *A. halimus* and *A. nummularia*, respectively, during the summer, while a minimum Na^{+} values were recorded in young leaves of *A. argentina* during the rainy season, probably due to the leaching of salts from leaves by rainwater. The young leaves and stems of *Atriplex* spp. contained quite high levels of sodium and chloride ions in comparison to mature parts, due to their fleshiness and the presence of more salt bladders in the epidermis of tender plant parts.

As mentioned above, the presence of excess salt in the soil is one of the critical factors adversely affecting seed germination under such conditions, thereby preventing plant species from inhabiting saline environments successfully. Halophytes show a reduction in germination when subjected to salinities above 1% NaCl, and increasing salt concentrations also delay germination (Chapman 1974). Information regarding the germination behaviour of Indian halophytes has been provided by various workers from the Ecology Laboratory, Department of Botany, Jai Narain Vyas University, such as Rajpurohit (1980), Sen and Rajpurohit (1982), Jhamb (1984), Jhamb and Sen (1984), Mohammed (1988), Sharma (1991), Rajput (1992), Thomas (1992), etc. Mohammed (1988) and Mohammed and Sen (1990a) collected seeds of halophytes from four different sites (Pachpadra, site-I; Didwana, site-II; Jodhpur, site-III; and Luni, site-IV) and studied the effect of various salts on seed germination behaviour. Various salts that are common in saline soils, viz. NaCl, Na_2SO_4, $MgSO_4$, KCl, and $CaCl_2$ were selected. Single salt solutions at concentrations of 100, 1,000, 5,000 and 10,000 mg l^{-1} were used. The aim of the investigation was to understand how individuals from a plant species from different localities behave with a particular salt, and whether the inhibition of germination is due to the osmotic or toxic effect of different ions. In order to make a distinction between osmotic and ionic effects, or the combined effect of these two factors on seed germination inhibition, those seeds that remained ungerminated in the saline medium at 10,000 mg l^{-1} concentration were transferred to distilled water individually to determine the additional germination per treatment. The results revealed that the germination percentage varied with different salt solutions (Mohammed and Sen 1990a, 1991). Higher concentrations of all the salts affected germination directly, thus reducing germination percentage. Seeds of *C. cretica* collected only from site-I showed dormancy, as no germination was observed in the control; no salt solution improved germination in this case. Maximum (40%) germination was recorded with 100 mg l^{-1} Na_2SO_4. Seeds of *Z. simplex* possessed extreme dormancy, because no germination was observed control. However, Mohammed and Sen (1992a) observed that one-year-old seeds of *Z. simplex* showed 50% germination when pretreated with gibberellic acid (GA_3) for 48 h. Khan and Ungar (1997a) reported that growth regulator treatments increased germination to over 80% in non-saline conditions in *Z. simplex*.

After 10 days of salt treatments, ungerminated seeds from the 10,000 mg l^{-1} concentration of the five salt solutions were transferred individually to distilled water. Mohammed and Sen (1990a) discovered that germination inhibition in saline media is due to osmotic stress or specific ion toxicity, because the germination percentage increased when seeds were transferred from salt solution to distilled water (Table 14.1). Variations in temperature appear to play an important role in the recovery of germination of halophytes from salt stress when seeds are

Table 14.1 Additional mean germination percentage of some halophytes observed after transfer of ungerminated seeds from 10,000 mg l^{-1} concentration of each salt solution to distilled water after 10 days (after Mohammed and Sen 1990a, 1991)

Species	Salt solutions					
	Site	NaCl	Na_2SO_4	$MgSO_4$	$CaCl_2$	KCl
Salsola baryosma	I	36	16	23	10	40
	III	53	13	23	20	40
Sesuvium sesuvioides	I	20	10	20	70	40
	II	10	13	16	20	50
	III	13	16	10	30	10
	IV	16	20	13	60	10
Suaeda fruticosa	I	36	26	16	20	20
	II	40	26	30	16	26
	III	40	30	10	30	26
Trianthema triquetra	I	16	30	20	13	30
	II	10	10	13	10	10
	III	13	13	10	10	10

transferred to distilled water (Khan and Ungar 1997b). Therefore, Mohammed and Sen (1990a) proved that higher concentrations of salts retarded germination because of osmotic effects, as the process of seed germination speeded up after their transfer to non-saline medium. This may be of significance under natural conditions, especially for inland desert salines, because seeds that could not germinate under extreme salinity stress may have evolved a mechanism to germinate rapidly when the salt stress is relieved. Although NaCl is the major salt in most salt-affected soils, other salts also present in the soil play a combined role in the salt tolerance of a species at the time of germination.

A significant characteristic of halophyte seeds, which distinguishes them from glycophytes, is their ability to maintain seed viability for extended periods of time during exposure to hypersaline conditions and then to initiate germination when the salinity stress is reduced (Chapman 1974; Ungar 1982). The enforced dormancy response of halophyte seeds to saline conditions is of selective advantage to plants growing in highly saline habitats. These seeds can withstand high salinity stress and provide a viable seed bank for recruitment of new individuals. However, seed germination is limited to periods when soil salinity levels are within the species tolerance limits (Ungar 1982).

Rajpurohit and Sen (1977) concluded that, under field conditions, the highest germination percentage in *C. cretica*, *S. fruticosa*, *S. baryosma*, *S. sesuvioides*, and *T. triquetra* can be achieved after rain that is heavy enough to leach out the salt from the close environment of the seeds. Several authors have shown that the increase in salinity leads to dormancy of seeds in halophytes and glycophytes (Rajpurohit and Sen 1977; Jhamb 1984; Mohammed 1988; Mohammed and Sen 1990a; Ungar 1995). Germination was inhibited or severely reduced at salinity levels above 250 mM NaCl level in *Atriplex lentiformis* (Malcolm et al. 2003). Huang et al. (2003) reported that germination of *Haloxylon ammodendron* seeds decreased with salinity, and was substantially inhibited at 1.2 mol l^{-1} NaCl; however, maximum values were seen in the non-saline control. Mohammed and Sen (1990a) thus observed that inhibition of germination of saline plants is due to both osmotic and ionic factors.

14.4 Proline Accumulation under Salt Stress

The phenomenon of free proline accumulation in plants exposed to diverse environmental stresses has considerable ecophysiological significance. Water stress produces numerous metabolic irregularities in plants (Levitt 1980). The increased proline concentration in water-stressed plants is due either to the inhibition of protein oxidation or to the breakdown of precursor proteins (Barnetts and Naylor 1966). Heidari-Sharifabad and Mirzaie-Nodoushan (2006) documented that proline accumulation is a good indicator of salinity stress. A significant increase in proline was observed under NaCl salinity stress in *Sorghum* spp. (Thakur and Sharma 2005).

Mohammed and Sen (1987) examined 65 plant species in the Indian desert for proline content; of these, 54 showed the presence of proline. Furthermore, they concluded that some of well-adapted desert plants do not accumulate proline at all. Mohammed and Sen (1990b) and Sen and Mohammed (1992) observed that proline accumulation in desert plants is not governed by the environment but rather by some innate factors.

Data on the presence of proline accumulation in the leaves of halophytes in the Indian desert revealed that plants growing in saline areas accumulated maximum proline during winter, followed by summer, with a minimum in the rainy season (Table 14.2). All plant species at site-I (Pachpadra salt basin) accumulated more proline compared to those at sites-II (Didwana salt lake) and -III (Jodhpur non-saline), which may be due to the high salinity of this former habitat. Since site-I is more saline than the latter two, it can be hypothesised that salt stress caused more accumulation of proline. Perhaps free proline content play an essential role in plant survival.

Table 14.2 Seasonal variations in proline (μg g^{-1} fresh weight) content in some halophytes growing at sites I–III (after Sen et al. 2002). R Rainy, W winter, S summer

Species	Pachpadra (site-I)			Didwana (site-II)			Jodhpur (site-III)		
	R	W	S	R	W	S	R	W	S
Aeluropus lagopoides	6.2	26.6	–[a]	24.6	–	–	–	–	–
Cressa cretica	0.5	108.0	3.8	4.8	–	–	–	–	–
Salsola baryosma	0.1	7.1	5.7	4.3	–	–	0.1	7.0	2.7
Sesuvium sesuvioides	2.0	–	–	4.3	–	–	1.8	–	–
Sporobolus helvolus	5.3	100.0	7.9	6.1	8.7	6.2	–	–	–
Suaeda fruticosa	1.4	11.9	7.0	2.7	9.9	6.6	1.9	5.8	5.0
Trianthema triquetra	3.4	74.7	–	3.7	14.3	–	5.2	15.6	8.2
Zygophyllum simplex	5.7	109.9	–	–	–	–	–	–	–

[a]Plant not seen

14.5 Soil–Plant Analyses

All desert plants and most saline plants of the Indian arid zone are totally dependent upon the availability of water in the rainy season; water controls seed germination, seedling growth, and plant survival. Rainfall leaches salts down the soil profile as far as the groundwater, with a compensating upward movement as a result of capillary action (Jackson et al. 1956). Decreases in soil moisture and the intensity of evaporation lead to increases in soil salinity.

Ion analyses of the soils of saline and non-saline areas of the Indian arid zone revealed that Na^+ and Ca^{2+} were among the major cations, and Cl^- among anions at site-I; while at sites-II and -III, Na^+ and K^+ were found as the major cations, and Cl^- among the anions. All ions in soils, such as Na^+, K^+, Ca^{2+} and Cl^-, were maximum at site-I, followed by site-II, and minimum at site-III (Table 14.3). Seasonal variations in the salt content of the soil depend on soil moisture, depth of groundwater, soil texture, habitat vegetation, the occurrence of a downward flow of water due to rainfall, flow of groundwater towards the surface in dry periods, and loss of moisture to the atmosphere (Waisel 1972; Rajpurohit and Sen 1979). Jackson et al. (1956) described the seasonal downward and upward movements of salts in the soil profile. A considerable seasonal variation in soil salinity was also observed. Higher salinity levels in the dry period have been observed as a result of the evaporative mechanism (Sharma and Tongway 1973).

Halophytes absorb salts continuously from their surrounding medium. *S. baryosma*, *S. sesuvioides*, *S. fruticosa*, *T. triquetra*, and *Z. simplex* continue to accumulate salt in their tissues. *A. lagopoides*, *C. cretica*, and *S. helvolus* secrete excess salt through the entire shoot. Fine streaks of white salt are seen on the stem and leaves throughout. On the basis of ion analyses, Na^+ and K^+ were among the major cations and Cl^- among the anions absorbed in large quantities by these halophytes (Table 14.4). Considering the habit as well as the Cl^- content of individual halophytes, it is concluded that: (1) the amount of Cl^- absorbed by the leaves of *Z. simplex* (11–18%) and *S. fruticosa* (13–19%) was nearly equal, and much higher as compared with other species; (2) the internal Cl^- content of ion-accumulating species was higher as compared to ion-secreting grass species (*A. lagopoides* and *S. helvolus*); and (3) among ion-accumulating species, *S. baryosma* (6–11%), *S. sesuvioides* (3–7%), and *T. triquetra* (3–8%) accumulated much less Cl^-. Since Cl^- is the dominant ion present in the medium at both saline sites, it can be

Table 14.3 Range of ionic contents in soil at sites I–III (after Sen and Mohammed 1994)

Site	Depth (cm)	Ion (mg 100 g^{-1} dry soil)			
		Na^+	K^+	Ca^{++}	Cl^-
I	0–5	187–4,125	39–71	93–262	0.4–4
	20–25	269–1,812	16–33	14–166	0.4–0.9
II	0–5	55–1,280	2–35	0.6–6	0.04–0.06
	20–25	105–603	15–62	0.7–3	0.1–1.2
III	0–5	41–219	2–87	0.7–5	0.03–0.1
	20–25	35–176	2–76	0.7–19	0.03–0.1

Table 14.4 Range of ionic content (mg g^{-1}) accumulated by leaves of some halophytes growing at sites I–III (after Sen and Mohammed 1994)

Species	I				II				III			
	Na^+	K^+	Ca^{++}	Cl^-	Na^+	K^+	Ca^{++}	Cl^-	Na^+	K^+	Ca^{++}	Cl^-
Aeluropus lagopoides	55–100	5–14	1–22	5.8–6.3	10	7	10	4	–[a]	–	–	–
Cressa cretica	20–110	1–31	3–30	4–10	42–49	9–14	19–21	4–8	–	–	–	–
Salsola baryosma	44–291	27–78	4–43	6–11	–	–	–	–	37–241	8–49	2–50	1.4–2.9
Sesuvium sesuvioides	16–100	14–36	5–37	3–7	47–98	9–40	11–50	2–4	72–106	9–21	2–20	2–3
Sporobolus helvolus	7–70	4–37	1–89	3–5	22–28	4–12	5–7	1–4	–	–	–	–
Suaeda fruticosa	43–313	10–45	3–55	13–19	45–119	7–63	5–26	7–21	48–315	8–20	1–37	7–15
Trianthema triquetra	29–115	3–33	2–28	3–8	27–112	7–34	3–87	1–11	22–225	9–24	1–30	1.9–2.2
Zygophyllum simplex	42–105	8–15	5–52	11–18	–	–	–	–	–	–	–	–

[a]Plant not seen

concluded that *S. fruticosa*, *T. triquetra*, *S. baryosma*, *C. cretica*, and *Z. simplex* are well suited to these habitats, and thus they are the most salt tolerant species (Mohammed and Sen 1992b, c).

Ion uptake by plants was dependent largely upon their availability in the soil. When ion concentrations fluctuated in the soil by upward or downward movements, their uptake by plants was also affected. Like soil salinity, higher amounts of elements were observed in plants during dry periods in the Indian arid zone.

14.6 Metabolic Behaviour

Salinity is known to affect almost all aspects of plant metabolism. The leaves of plants subjected to water stress often show a decrease in starch, which is usually followed by an increase in sugar content (Levitt 1980; Mohammed 1988; Mohammed and Sen 1992b, c). Plant species from site-II (Didwana), which is less saline as compared to site-I (Pachpadra), showed maximum sugar content during the summer season, when plant-water stress was higher than in winter or in the rainy seasons. Plant species at site-I, which is extremely saline, exhibited higher sugar content during the rainy season, followed by winter, with lowest values in summer (Table 14.5). These observations of varying sugar content may be due to higher salinities at particular sites. Heidari-Sharifabad and Mirzaie-Nodoushan (2006) reported that soluble sugars increased as a result of salinity, which may act as an osmotic adjustment or osmotic conservation factor in *Salsola* species. They further stated that accumulation of soluble sugars acts as an osmotic adjustment factor to maintain turgor, or is related to stabilising cell membranes and proteins and also may be due to further transformation of starch to sugars or a lower consumption of carbohydrates by the tissues.

Metabolism of halophytes is affected by a general increase in salinity as well as by the types of ions present. Salt-tolerant species are able to withstand the

Table 14.5 Seasonal variations in total sugars (mg g^{-1}; dry weight) and crude protein (%; dry weight) contents in some halophytes growing at sites-I and -III (source: Sen and Mohammed 1994). R Rainy, W winter, S summer

Species	Total sugars						Crude protein					
	Site-I			Site-III			Site-I			Site-III		
	R	W	S	R	W	S	R	W	S	R	W	S
Aeluropus lagopoides	34	–[a]	–	–	–	–	12	–	–	–	–	–
Cressa cretica	39	11	17	–	–	–	21	19	17	–	–	–
Salsola baryosma	33	16	18	5	10	18	20	16	15	36	16	22
Sesuvium sesuvioides	22	9	–	16	17	–	20	16	–	19	9	–
Sporobolus helvolus	29	8	14	–	–	–	18	12	8	–	–	–
Suaeda fruticosa	39	35	18	5	14	25	29	25	19	28	25	17
Trianthema triquetra	10	4	–	5	15	19	19	17	–	21	15	12
Zygophyllum simplex	18	25	–	–	–	–	18	17	–	–	–	–

[a]Plant not seen

metabolic disturbances that occur as a result of high salt content. Such changes in ionic content and ionic composition of plant cells induce changes in the activity of certain metabolic systems. The nutritive pattern of plants is very important when fodder values and productivity are taken into consideration. Protein is the most important constituent of cells from both structural and functional points of view. Under conditions of extreme salinity, proteins are precipitated. The protein content of various plant tissues declined under drought or saline conditions because of increased proteolysis and decreased protein synthesis (Waisel 1972). Doddema et al. (1986) observed that soluble protein decreased under saline conditions, being maximal in the rainy season when salinity was less in *Arthrocnemum fruticosum*. The protein content in saline plants at both saline and non-saline sites was maximum during the rainy season, when plant-water status was higher than in winter or summer seasons (Table 14.5).

14.7 Conclusions

It is generally observed that an excess of salts in the soil inhibits plant growth. Halophytes survive under conditions of high salinity, exhibit succulence, and can resort to other physiological adaptations to overcome the adverse saline environment in the soil. The ionic and toxic effects of various salts, especially those of NaCl, play a major role in halophytism. Most halophytes avoid salinity, some evade it, and a few others tolerate it. The accumulation of proline in plants is correlated with the extent of the water stress in the plant. In halophytes, a positive correlation is seen between proline content and the amount of Na^+ and Cl^- in the cell sap. Salt stress induces accumulation of more proline in halophytes, and perhaps plays an essential role in their survival. Ion uptake by plants is dependent largely upon their availability in the soil. Higher levels of minerals in halophytes have been observed during dry periods. Although NaCl is the major salt present in most salt-affected soils, other salts, such as $MgCl_2$, $MgSO_4$, Na_2SO_4, etc., are also present, which play a combined role in the salt tolerance of a species at the time of seed germination. The maximum seed germination in halophytes has been reported during the rainy season due to leaching of salts in deeper soil layers through rainfall. Due to leaching action, there is a decrease in soil salinity. Salinity also affects almost all aspects of plant metabolism. Maximum values of carbohydrate and crude protein in halophytes have been observed during rainy seasons, when plant-water status is higher than in winter or summer seasons.

Most halophytic species investigated so far seem to prefer saline conditions. Salinity plays an important role in the existence and distribution of plants because plant species respond differently to soil salinity. Plant species growing in an area may provide useful information regarding the degree of salinisation and consequent soil deterioration. Such information may be helpful in the more effective planning of practical reclamation of saline wastelands.

Acknowledgements The authors express their gratitude to their esteemed teacher Professor Dr. David N. Sen, a pioneer worker in the Indian desert since 1963, for his help in the preparation and editing of the manuscript. A number of research scholars, namely, Dr. K.S. Rajpurohit, Dr. (Mrs.) R.B. Jhamb, Dr. T.P. Sharma, Dr. (Mrs.) Pramila Rajput, Dr. T.P. Thomas, among others, have worked on the biology of saline plants in our laboratory, and their studies have been profusely included in this chapter. We are also highly thankful to the Professor and Head of the Department of Botany, J.N.V. University, Jodhpur for providing the necessary facilities.

References

Barnetts NM, Naylor AW (1966) Amino acids and protein metabolism in Bermuda grass during water stress. Plant Physiol 41:1222–1230

Breckle SW (1986) Studies on halophytes from Iran and Afghanistan. III. Ecology of halophytes along salt-gradients. Proc R Bot Soc 89B:203–215

Chapman VJ (1974) Salt marshes and salt desert of world. Cramer, Bremerhaven, Germany.

Doddema H, Saadeddin R, Mahasneh A (1986) Effects of seasonal changes of soil salinity and soil nitrogen on the N-metabolism of the halophyte *Arthrocnemum fruticosum* (L.) Moq. Plant Soil 92:279–293

Dudley LM (1994) Salinity in the soil environment. In: Pessarakli M (ed) Handbook of plant and crop stress. Dekker, New York, pp 13–30

Evangelou VP (1994) Influence of sodium on soils of humid regions. In: Pessarakli M (ed) Handbook of plant and crop stress. Dekker, New York, pp 31–62

Heidari-Sharifabad H, Mirzaie-Nodoushan H (2006) Salinity-induced growth and some metabolic changes in three *Salsola* species. J Arid Environ 67:715–720

Huang Z, Zhang X, Zheng G, Gutterman Y (2003) Influence of light, temperature, salinity and storage on seed germination of *Haloxylon ammodendron*. J Arid Environ 55:453–464

Jackson EA, Blackburn G, Clarke ARP (1956) Seasonal changes in soil salinity at Trintinara, South Australia. Aust J Agric Res 7:20–24

Jhamb RB (1984) Biology of halophytes. PhD Thesis, University of Jodhpur, Jodhpur, India

Jhamb RB, Sen DN (1984) Seed germination behaviour of halophytes in Indian desert. I. *Suaeda fruticosa* (Linn.) Forsk. Curr Sci 53:100–101

Khan MA, Ungar IA (eds) (1995) Biology of salt tolerant plants. University of Karachi, Pakistan

Khan MA, Ungar IA (1997a) Alleviation of seed dormancy in the desert forb *Zygophyllum simplex* L. from Pakistan. Ann Bot 80:395–400

Khan MA, Ungar IA (1997b) Effects of light, salinity and thermoperiod on the seed germination of halophytes. Can J Bot 75:835–841

Levitt J (1980) Responses of plants to environmental stresses, vol 2, Academic, New York

Malcolm CV, Lindley VA, O'Leary JW, Runciman HV, Barrett-Lennard EG (2003) Halophyte and glycophyte salt tolerance at germination and the establishment of halophyte shrubs in saline environment. Plant Soil 253:171–185

Mohammed S (1988) Comparative studies of saline and nonsaline vegetation in Indian arid zone. PhD Thesis, University of Jodhpur, Jodhpur, India

Mohammed S, Sen DN (1987) Proline accumulation in arid zone plants. J Arid Environ 13: 231–236

Mohammed S, Sen DN (1990a) Germination behaviour of some halophytes in Indian desert. Indian J Exp Biol 28:545–549

Mohammed S, Sen DN (1990b) Environmental changes and proline content in some desert plants. J Arid Environ 19:241–243

Mohammed S, Sen DN (1992a) Effect of GA_3 and different nitrates on seed germination of *Zygophyllum simplex* (Linn.), an inland halophyte of Indian desert. Proc Nat Acad Sci India 62 (B) III 393–397

Mohammed S, Sen DN (1992b) Ecophysiological studies on saltwort in Indian desert. Ann Arid Zone 31:115–118

Mohammed S, Sen DN (1992c) Metabolic and mineral responses of *Salsola baryosma* (Roem. et Schult.) Dandy: a halophyte of Indian inland salines. J Indian Bot Soc 71:161–163

Mohammed S, Sen DN (1994) Vegetation patterns in saline areas of Indian arid zone. Bull Life Sci 4:1–8

Mohan Ram HY, Gupta P (1997) Plant life under extreme environments. Curr Sci 72:306–315

Palfi G, Juhasz J (1970) Increase of free proline level in the water deficient leaves as a reaction to saline or cold root media. Acta Agron Acad Sci Hung 19:278–287

Parida AK, Das AB (2005) Salt tolerance and salinity effects on plants: a review. Ecotoxicol Environ Saf 60:324–349

Poljakoff-Mayber A, Gale J (1975) Plants in saline environments. Springer, Berlin

Pugnaire FI, Endolz LZ, Pardos J (1994) Constraints by water stress on plant growth. In: Pessarakli M (ed) Handbook of plant and crop stress, Dekker, New York, pp 247–259

Rajpurohit KS (1980) Soil salinity and its role on phytogeography of western Rajasthan. PhD Thesis, University of Jodhpur, Jodhpur, India

Rajpurohit KS, Sen DN (1977) Soil salinity and seed germination under water stress. Trans ISDT UCDS 2:106–110

Rajpurohit KS, Sen DN (1979) Seasonal variation in chloride percentage of plants and soil of Pachpadra salt basin in Indian desert. Indian J Bot 2:17–23

Rajput P (1992) Ecological studies on introduced species of *Atriplex* in Indian desert. PhD Thesis, JNV University, Jodhpur, India

Reed JF (1947) The relation of the *Spartinetum glabrae* near Beaufort, North Carolina, to certain edaphic factors. Am Midl Nat 38:605–614

Sen DN (1990) Ecology of saline areas of Rajasthan and exploitation of saline ecosystem for increased productivity. DOEn Final Technical Progress Report, University of Jodhpur, Jodhpur, India

Sen DN, Mohammed S (1992) Proline accumulation in some halophytes in Indian desert. In: Prasad BN, Ghirmire GPS, Agrawal VP (eds) Role of biotechnology in agriculture. Oxford & IBH, New Delhi, pp 129–137

Sen DN, Mohammed S (1994) General aspects of salinity and the biology of saline plants. In: Pessarakli M (ed) Handbook of plant and crop stress. Dekker, New York, pp 125–145

Sen DN, Mohammed S, Kasera PK (1997) Biology of plants in saline environment. In: Grover IS, Thukral AK (eds) Environment and development. Scientific Publishers, Jodhpur, pp 117–126

Sen DN, Rajpurohit KS (1978) Plant distribution in relation to salinity in Indian desert. In: Proceedings of Second International Congress of Ecology, Jerusalem, Abstract, p 340

Sen DN, Rajpurohit KS (1982) (eds) Contributions to the ecology of halophytes. Junk, The Hague

Sen DN, Mohammed S, Kasera PK (2002) Biology and physiology of saline plants. In: Pessarakli M (ed) Handbook of plant and crop physiology, 2nd edn. Dekker, New York, pp 563–581

Sharma ML, Tongway DJ (1973) Plant induced soil salinity patterns in two salt bush (*Atriplex* spp.) communities. J Range Manage 26:121–125

Sharma TP (1991) Ecology and biology of saline ecosystem in Indian desert. PhD Thesis, University of Jodhpur, Jodhpur, India

Sharma TP, Sen DN (1989) A new report on abnormally fast germinating seeds of *Haloxylon* spp. – an ecological adaptation to saline habitat. Curr Sci 58:382–385

Szabolcs I (1994) Soil and salinization. In: Pessarakli M (ed) Handbook of plant and crop stress. Dekker, New York, pp 3–11

Thakur M, Sharma AD (2005) Salt-stress induced proline accumulation in germinating embryos: evidence suggesting a role of proline in seed germination. J Arid Environ 62:517–523

Thomas TP (1992) Ecology of some halophytes in Indian desert with special reference to introduction of *Atriplex* spp. PhD Thesis, JNV University, Jodhpur, India

Ungar IA (1982) Germination ecology of halophytes. In: Sen DN, Rajpurohit KS (eds) Contributions to the ecology of halophytes. Junk, The Hague, pp 143–154

Ungar IA (1995) Seed bank ecology of halophytes. In: Khan MA, Ungar IA (eds) Biology of salt tolerant plants. University of Karachi, Pakistan, pp 65–79

Waisel Y (1972) Biology of halophytes. Academic, New York

Weber DJ (1995) Mechanisms and reactions of halophytes to water and salt stress. In: Khan MA, Ungar IA (eds) Biology of salt tolerant plants. University of Karachi, Pakistan, pp 170–180

Chapter 15
Ecophysiology of *Prosopis* Species From the Arid Lands of Argentina: What Do We Know About Adaptation to Stressful Environments?

Pablo E. Villagra, Alejandra Vilela, Carla Giordano, and Juan A. Alvarez

Abstract The expansion of the *Prosopis* genus from the sub-humid Chaco towards colder and drier zones such as Monte, Prepuna and Patagonia biogeographical regions would have implied the acquisition and/or adjustment of morphological and physiological adaptations to stressful environments. In this chapter, we discuss the phenological, morphological and physiological features of seven *Prosopis* species native to Argentinean arid regions that allow them to avoid or tolerate water stress, salinity, and other environmental stress factors in arid lands. Some of these adaptations appear to be spread over the genus and should confer the capability to deal with the most common stressful factor of arid lands (i.e. water availability); however, other morphological or physiological adaptations appear to be specific to each species, and should be the cause of niche differentiation among species and the occupation of particular environments within arid lands (e.g. sand dunes, saline environments). Finally, we discuss some consequences of these adaptations for the management of *Prosopis* species. The inter- and intra-specific variability observed in their adaptation to stressful factors suggest that some *Prosopis* species may be a good option to be used in the restoration of degraded areas or in afforestation projects with productive objectives.

P.E. Villagra (✉) and J.A. Alvarez
Departamento de Dendrocronología e Historia Ambiental, Instituto Argentino de Nivología, Glaciología y Ciencias Ambientales. Centro Científico Tecnológico CONICET Mendoza, CC. 330–5500, Mendoza, Argentina
e-mail: villagra@mendoza-conicet.gov.ar

P.E. Villagra
Facultad de Ciencias Agrarias, Universidad Nacional de Cuyo, Argentina

A. Vilela
Museo Egidio Feruglio-CONICET, Av. Fontana 140, 9100 Trelew, Chubut, Argentina

C. Giordano
Instituto Argentino de Investigaciones de las Zonas Áridas. CONICET, Mendoza, Argentina

K.G. Ramawat (ed.), *Desert Plants*,
DOI 10.1007/978-3-642-02550-1_15,

15.1 Introduction

The genus *Prosopis* L. (Fabaceae, Mimosoideae) includes 45 species distributed mainly in arid and semi-arid lands of America, Southwestern Asia and Northern Africa. Most *Prosopis* species are distributed in South America (Burkart 1976; Fagg and Stewart 1994). The genus includes trees, shrubs and, rarely, sub-shrubs. The great morphological diversity of South American species suggests that the main centre of irradiation of *Prosopis* is located in the Argentinean-Paraguayan Chaco (Burkart 1976; Burkart and Simpson 1977; Roig 1993b). This region occupies the warm plains of Northern Argentina and Paraguay; the mean annual rainfall ranges from 550 to 1,300 mm/year and the mean annual temperature from 20 to 23°C (Cabrera 1976). The expansion of the genus towards colder and drier zones, such as the Monte, Prepuna and Patagonia biogeographical regions (mean annual rainfall between 50 and 400 mm; Fig. 15.1; Cabrera 1976), implies the acquisition and/or adjustment of morphological and physiological features that allow plants to avoid or tolerate resource limitation and the harsh environmental conditions characteristic of these regions. These adaptations include the change of growth habit from tree to shrub form, leaf size reduction, anatomical modifications, increased thorniness and several physiological mechanisms related to stress avoidance or tolerance (Burkart 1976; Burkart and Simpson 1977; Roig 1993b).

In arid zones, such as Monte, Prepuna and Patagonia, water availability is the most important factor limiting seedling establishment, consequently shaping ecosystem dynamics and structure. Edaphic factors can also control plant establishment and growth, mainly by modification of water availability (Noy-Meir 1973). However, other limiting factors could act during the short periods in which water is sufficiently available; e.g. soil texture can act as a mechanical factor that controls root penetration (Brar and Palazzo 1995), salinity can affect survival through the toxic effects of certain ions (Munns 2002), and the spatial and temporal variability in the distribution of nutrients limit their availability to short pulses and particular microhabitats (Aguiar and Sala 1999; Austin et al. 2004).

The variety of adaptations to severe environmental conditions, and the ecological and economic importance of shrubs and trees for rural dwellers confer on *Prosopis* species the potential to be used in the restoration of degraded areas and to become the biological axis of new production systems (Fagg and Stewart 1994; Roig 1993a). In this chapter, we will give an overview of the physiological data available for *Prosopis* species native to the arid zones of Argentina, mainly the Monte and Patagonia Deserts. We will analyse the phenological, morphological and physiological features that allow these perennials to avoid or tolerate factors that produce plant stress. Finally, we will discuss the consequences of these adaptations for the management of *Prosopis* species.

We will focus the analysis on three arboreal species (*Prosopis flexuosa*, *P. chilensis* and *P. ferox*) and four shrubby species (*Prosopis argentina*, *P. alpataco*, *P. strombulifera* and *P. denudans*; Figs. 15.1, Figs. 15.2). This group of species includes adaptations to very different environments. *Prosopis flexuosa* grows in

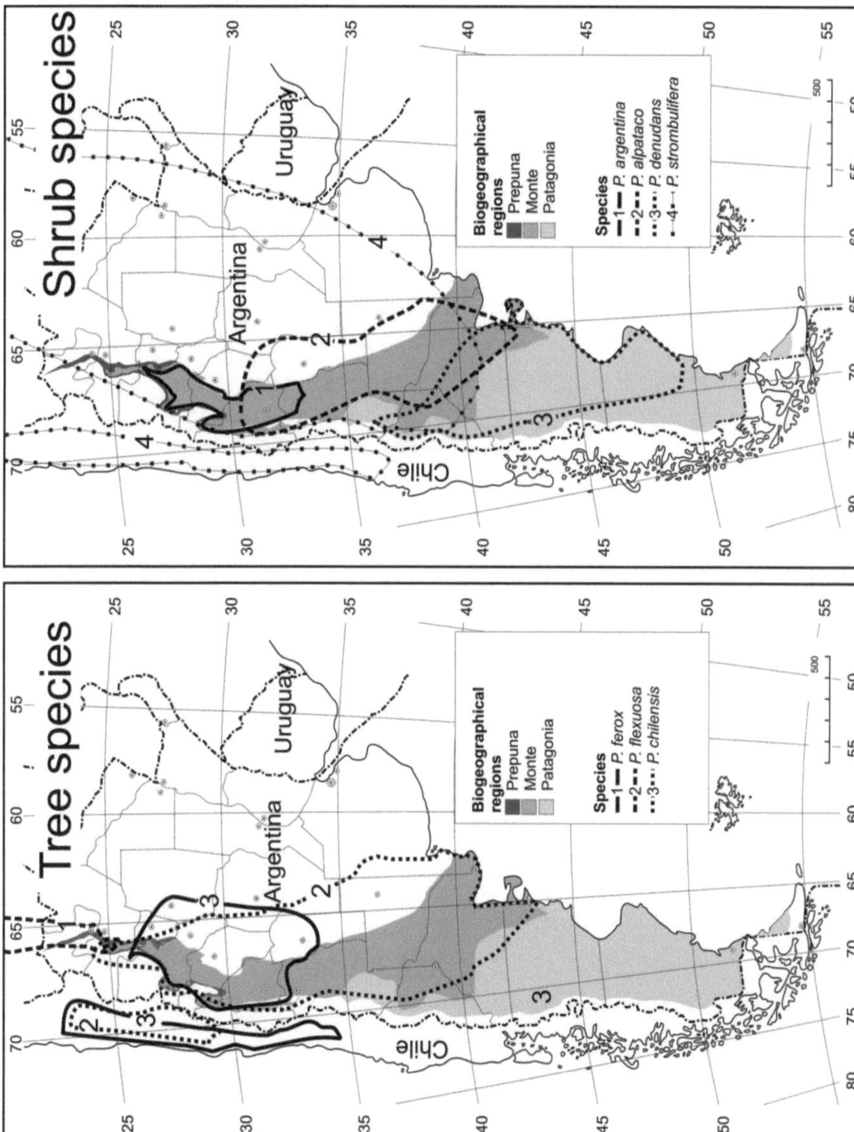

Fig. 15.1 Biogeographical regions of Argentinean arid lands and distribution of the seven *Prosopis* species studied in this review (Cabrera 1972; Roig 1993b)

Fig. 15.2 a–g Growth habit of the studied *Prosopis* species and the main environments they occupy. **a** *Prosopis flexuosa* tree occupying interdune valleys from the Northern Monte (Fiambalá, Catamarca Province); **b** *P. ferox* tree from a dry intermountain valley from the Prepuna region at

basin floors and lower alluvial fans, and *P. chilensis* along washes and rivers of the Monte desert. Both species also grow in the mesic plains of the Chaco (Morello 1958; Rundel et al. 2007). *Prosopis ferox* occupies the dry intermountain valleys at elevations between 2,500 and 4,000 m in the Prepuna region (Morales et al. 2001). Among shrubs, *P. argentina* and *P. alpataco* are sympatric in Northern and Central Monte, but *P. argentina* occupies sand dunes environments while *P. alpataco* is a dominant species in clayish and saline soils where sporadic flooding occurs (Villagra and Roig 2002). *Prosopis strombulifera* is a sub-shrub species distributed in saline areas of central-western Argentina (Reinoso et al. 2004), and *P. denudans* is an extreme xerophyte endemic in Patagonia, adapted to the desert steppe with very cold winters. This last species reaches nearly 48°S latitude, which marks the southernmost limits of the genus (Figs. 15.1, Figs. 15.2).

15.2 Germination and Early Seedling Growth

A thick and impermeable coat provides *Prosopis* seeds with a physical dormancy allowing the temporal and spatial regulation of germination, which is likely to be an advantage in unpredictable environments (Catalán and Balzarini 1992; Villagra 1995). This coat, in addition to the nutritious mesocarp, constitutes an adaptation to endozoic dispersal (Campos and Ojeda 1997; Peinetti et al. 1993), which liberates seeds from the indehiscent pod (Ortega Baes et al. 2002), and increases germination rate through chemical scarification (Campos and Ojeda 1997), and seed longevity through predation avoidance (Villagra et al. 2002). For cultivation purposes, a number of scarification techniques have been developed. These methods include mechanical and chemical scarification, and hot water treatments (Ffolliot and Thames 1983; Ortega Baes et al. 2002; Vilela and Ravetta 2001; Villagra 1995).

The fast germination process (2–4 days for radical emergence, and 8–10 days for leaf appearance; Cony and Trione 1996; Vilela and Ravetta 2001; Villagra 1995) and the high rate of root development (Salih 1998) allow an efficient use of water during the short period of high availability. High inter- and intra-specific variability has been observed in the cardinal temperatures for germination (Cony and Trione 1996; Villagra 1995). The optimum germination temperature for *P. chilensis* ranges from 25 to 40°C, while *P. flexuosa* has lower requirements (20–25°C). The shrubby species *P. argentina* and *P. alpataco* show an optimum germination temperature of 35°C (Villagra 1995). The high germination temperatures would suggest a thermal adjustment to the season of high water availability in the Monte desert (rainfalls

3,000 m altitude (Quebrada de Humahuaca, Jujuy); **c** *P. chilensis* individual in a dry wash from the Northern Monte (Talampaya Canyon, La Rioja); **d** *P. alpataco* shrubland in a clayish environment of the Central Monte (Asunción, Mendoza); **e** *P. argentina* shrub occupying a sand dune in the Central Monte (Tucunuco, San Juan); **f** *P. denudans* shrublands in a Patagonian plateau (Chubut); **g** *P. strombulifera* individuals colonising a clayish and saline environment (Mendoza). (Photo credits: **a**, **c**, **d**, **e**, **g** P.E. Villagra; **b** M.S. Morales; **f** A. Vilela) concentrated in summer). Furthermore, there is evidence that minimum germination temperatures are related to the geographical distribution of each species: species from northern areas with higher mean annual temperatures, such as *Prosopis chilensis* and *P. argentina*, show higher minimum germination temperatures than species from southern areas with lower mean temperatures, such as *P. flexuosa* and *P. alpataco* (Cony and Trione 1996; Villagra 1995).

Germination is affected by salinity and water stress at water potential values lower than –1.4 MPa. However, a high variability within- and among-species has been observed (Cony and Trione 1998). Tolerance to drought and saline conditions during germination appears to be adaptive to native soil characteristics since those species, or provenance within species, coming from harsh environments are less affected by soil moisture or salinity than those from milder ones. *Prosopis alpataco* and *P. strombulifera*, dominant species in saline environments, show higher salinity tolerance than *P. argentina*, which is dominant in sand dunes (Sosa et al. 2005; Villagra 1997). Among trees, *Prosopis flexuosa* shows a higher germination capacity than *P. chilensis* under both saline and water stress (Cony and Trione 1998). In addition, Sosa et al. (2005) reported that the effect of salinity on germination of *P. strombulifera* is influenced strongly by the nature of the ions in the salt solutions and their interactions.

Soil type also affects germination as well as seedling biomass accumulation, rate of leaf appearance, shoot:root ratio and CO_2 uptake in *P. chilensis* and *P. flexuosa* (Vilela and Ravetta 2001). The effect of soil texture and its interaction with salinity and water availability on seedling emergence and early growth have been suggested as the cause of exclusion of *P. alpataco* from sandy soils and *P. argentina* from clayish soils (Villagra and Cavagnaro 2000, 2005, 2006). In addition, the ability of *P. alpataco* to occupy areas sporadically flooded could be related to its germination capacity under hypoxic conditions (Villagra 1998).

The interaction of environmental factors affecting germination and seedling establishment in *Prosopis* has been reported only sparsely. However, initial evidence suggests that the final effect of each factor on germination will depend on the interaction with other relevant factors. For example, Villagra (1997) showed an increasing detrimental effect of salinity with increasing temperature on germination of *P. argentina* and *P. alpataco*, and Vilela et al. (2003) reported a detrimental effect of nutrient addition under water limitations on seedling biomass accumulation.

15.3 Growth Patterns

Growth studies in adult individuals have been limited to the arboreal species of the genus. The presence of defined annual growth rings allows for the determination of growth rate and the ecological factors that control it (Castro 1994; Villalba 1985; Villalba and Boninsegna 1989; Villalba et al. 2000). The growth rate of *P. flexuosa*, distributed widely in the Monte desert, is highly variable among- and within-populations (Alvarez 2008; Villagra et al. 2005a, 2005b). The mean growth rate

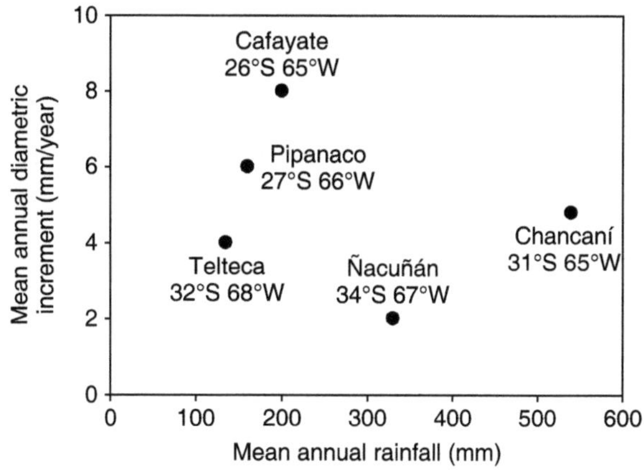

Fig. 15.3 Diametric growth of *P. flexuosa* trees in localities with different rainfall regime (Villagra et al. 2005a)

decreases from north to southern Monte, between 25°S and 36°S (Fig. 15.3; Villagra et al. 2005a). This latitudinal gradient in growth rate could be related to the length of the growing season, which is longer in northern than in southern areas. In addition, *P. flexuosa* populations also differ in growth habit: individuals growing in northern populations are predominantly erect, and are only single-stemmed; while in southern populations, individuals are characteristically decumbent and multi-stemmed (Villagra et al. 2005a).

Cony (1996) reported a high genetic variability, both among and within populations, and a high heritability of this variability. He found that populations from Northern Monte have higher growth potential than southern ones when grown in a common garden. However, Vilela et al (2003) observed that tree-type and shrub-type individuals responded in a similar way in biomass partition, C:N ratio, energy partition and CO_2 uptake. These findings suggest a complex set of genetic and environmental factors that determines the growth differences among populations and growth forms.

The productivity of *P. flexuosa* is also affected by the growth habit of individuals. Alvarez (2008) generated a dendrochronological model to compare the productivity of single-stemmed with multi-stemmed individuals, and found that the annual increments in dry weight during the first years of growth are higher in multi- than single-stemmed individuals; however, multi-stemmed individuals decrease their growth rates rapidly after 60 years of age, while single-stemmed individuals constantly increase their growth rates, at least during their first 100 years. This variability in growth pattern could be attributed to differences in the photosynthetic capacity throughout an individual's life, and could imply interference among crowns of different stems of multi-stemmed individuals that should be more strongly expressed in old individuals than in young ones (Alvarez 2008; Duff et al. 1994).

In *P. ferox,* dendrochronological techniques allowed detection of a strong relationship between radial growth chronologies and climatic variables in the subtropical desert from the high valleys of the North-western Argentina. In this species, growth is correlated positively to summer rainfalls and negatively to summer mean temperatures, suggesting the strong control of growth by water availability (Morales et al. 2001).

15.4 Phenology and Fruit Production

Prosopis species resprout in the Monte Desert at the beginning of spring (average temperature 16°C) and stay in leaf until autumn. Initiation of leaf production and cambium activity appears to be rather independent of rainfall (Mooney et al. 1977; Villalba 1985). Most *Prosopis* species produce abundant flowers at a predictable time of year, since they bloom independently of yearly rainfall fluctuation (Simpson et al. 1977), responding rather to photoperiod and to the length of the growing season (Solbrig and Cantino 1975). Despite the predictability of the blooming of *Prosopis* species, great variability in fruit production has been observed (Ffolliot and Thames 1983; Salvo et al. 1988). Dalmasso and Anconetani (1993) reported seed production of *P. flexuosa* in the Central Monte desert to vary between 80,000 and 800,000 seeds/ha in different years. Resource limitation, the time of rainfall, frost and wind occurrence, and soil water content have been suggested as physical causes of different abortion rates (Cariaga et al. 2005; Solbrig and Cantino 1975; Villagra 2000). The absence of an adequate pollinator system, and predation by bruchids (Coleoptera) might be among the ecological causes (Simpson et al. 1977; Smith and Ueckert 1974; Toro et al. 1993).

Surveillance of native stands of *Prosopis alpataco* and *P. denudans* indicated that only between 0.05 and 0.25% of the flowers buds initiated fruits, and 20–45% of initiated fruits reached full size (Cariaga et al. 2005). Similar patterns have been observed in *P. flexuosa* and *P. chilensis* (Mooney et al. 1977). Flower mortality includes either inflorescence abortion (common in *P. denudans*) and/or intra-inflorescence flower abortion (common in *P. alpataco*; Cariaga et al. 2005). Inflorescence abortion could result from competition for carbohydrate reserves among reproductive structures within a branch, while flower abortion has been explained in other species as a strategy to avoid the energetic and mechanical costs of maturing several offspring at one site (Stephenson 1981) or to avoid direct herbivory by insects (Siemens et al. 1994). The abortion of fruits as a physiological response to drought (Pawsey 1960) seems the most likely explanation to these observations in the Patagonian steppe.

In years of abundant fructification of *P. denudans*, competition for resources among fruits produces pods with a lower proportion of pericarp. However, the number of seeds per fruit, the seed mass, and the fruit chemical composition (protein and total non-structural carbohydrate contents) was not affected by fruit yield (Agüero et al. 2006).

15.5 Mechanisms to Avoid or Tolerate Water Stress

15.5.1 Water Sources and Soil Exploration

Prosopis trees and shrubs typically develop an extensive dimorphic root system with a vertical tap root several metres long, and surface roots that extend laterally far beyond the canopy area, which facilitates access to both deep and surface soil water (Gile et al. 1997; Morello 1958; Philips 1963; Snyder and Williams 2003). In *Prosopis* tree-species from different deserts of the world, access to underground water results in relatively high and seasonally stable predawn water potentials (Nilsen et al. 1983, 1984; Snyder and Williams 2003), higher transpiration and photosynthesis than shallow-rooted perennials (Nilsen et al. 1984), and higher net primary productivity than that predicted by precipitation (Sharifi et al. 1982). Studies of water relations in *Prosopis* species of the arid zones of Argentina are still incipient; however, there is some evidence of phreatophytism in the Monte desert.

Morello (1958) proposed that, in the Monte, *Prosopis*-tree species cannot grow independently from an accessible phreatic layer in areas where the annual rainfall is lower than 350 mm. This idea was supported by several lines of indirect evidence, such as the development of *Prosopis* woodlands in low lands or river margins where subterranean water did not exceed 20 m depth; the observation of *Prosopis* roots at great depths (up to 17 m); and the elevation of the phreatic layer after tree removal. Anatomical studies of *P. flexuosa* wood structure from localities with contrasting mean annual precipitation revealed marked differences in the mean area of pores between localities; this could be correlated to the exploitation of different sources of water (Villalba and Boninsegna 1989). The inner rings of wood sections from the drier site presented a smaller vessel area than the outer rings, which might reflect water stress during the first years of growth and higher water availability in subsequent years, probably due to the reach of a stable water table accessible at this site. This pattern was not observed in wood sections of the same species growing in a more mesic site, in which vessel area was distributed uniformly among tree rings (Villalba and Boninsegna 1989). Dendroecologial studies of *P. flexuosa* trees from a range of localities with different rainfall regimes within the Monte phytogeographical province demonstrated that the diametric structure, wood productivity of the population, and tree annual diameter increment did not correlate with mean annual precipitation, suggesting alternative controls on growth and regeneration of this species (Fig. 15.3; Villagra et al. 2005a). In addition, physiological studies on *P. flexuosa* trees from Ñacuñán, Mendoza (mean annual rainfall = 350 mm), demonstrated that they were able to maintain relatively stable pre-dawn water potentials and high turgor pressures throughout the growing season independently of rainfall dynamics, suggesting access to stable deep water reservoirs (Cavagnaro and Passera 1993).

The first direct evidence of groundwater use was recently found in *P. flexuosa* trees and *P. alpataco* shrubs from Telteca, Mendoza, by means of analysis of isotope composition of xylem water (Jobbágy et al. 2008). In this work, the authors demonstrated that *P. flexuosa* trees that grow in valleys where the water table is 7–13 m depth, absorbed underground water during drought, and ground and rainfall water in the rainy period. *Prosopis flexuosa* trees from adjacent slopes of 20 m-dunes did not tap groundwater, and used rainfall water throughout the growing season. *Prosopis alpataco* from valleys used groundwater exclusively both in drought and rainy periods. These results suggest that *P. flexuosa* trees behave as facultative phreatophytes in this extreme arid environment, and that their dependence on rainfall water might vary among landscape units. These results also suggest that different *Prosopis* species co-existing in valleys and dune slopes might display alternative strategies of soil exploration and water use, and might be affected differently by fluctuations in the rainfall regime and water table level.

The fact that *P. flexuosa* is able to survive and grow without tapping the water table in the dune slopes in regions with low precipitation (< 350 mm) suggests that this species might display some degree of morphological and physiological plasticity to exploit different water reservoirs and to tolerate drought. Preliminary results from recent studies on the root system of *P. flexuosa* adult trees indicate that trees growing in sand dunes positioned their coarse surface-roots deeper in the soil profile than did trees from the adjacent valley (Table 15.1; J.C. Guevara and C. Giordano, unpublished data), which might grant access to rainfall water that drained to subsurface soil layers in the dunes and accumulated at a depth of around 1–3.5 m (Jobbágy et al. 2008). Additionally, surface roots presented a higher number of primary branches per unit length in the dune than in the valley, which might enhance soil exploration and rainfall water absorption in trees that do not access the water table (Table 15.1).

15.5.2 Morpho-Physiological Adaptations to Tolerate Water Deficits

Despite using the phreatic layer as the main mechanism to avoid water stress, *Prosopis* species show several other morpho-physiogical adaptations to tolerate

Table 15.1 Phenotypic plasticity of coarse surface roots of *Prosopis flexuosa* trees with different access to the water table in the hyper-arid extreme of the Monte desert

Surface root trait[a]	Landscape unit		*P*-value
	Dune slope[b]	Valley[c]	
Horizontal pathway depth (cm)	26.30 (± 3.86)	55.70 (± 7.40)	0.014*
Horizontal roots that turned vertically downward (% of excavated roots)	70%	40%	nd
Branching[d] (number m^{-1})	11.05 (± 3.4)	7.18 (± 2.51)	0.067

*Statistical analysis: two tailed *t*-test, *n*=10
[a]Measurements performed in Telteca (32° 20′S; 68° 00′W, Mendoza, Argentina) in 2007–2008
[b]Trees did not access the water table
[c]Trees did access the water table (Jobbágy et al. 2008)
[d]First order branches

water deficit. Besides, it has been observed that species occupying different environments show different degrees of tolerance, suggesting that adaptations to drought may play a central role in niche differentiation between species (Villagra 1998). *Prosopis argentina*, a species dominant in sand dunes with no access to a stable phreatic layer, shows a higher degree of xeromorphism than *P. alpataco*, a species dominant in clayish soils with the phreatic layer near the surface. *Prosopis argentina* presents low foliar area, thick leaf cuticle, dense leaf pubescence, green photosynthetic stems and high number of small grouped vessels in the wood – all features associated with the hydraulic safety of the conducting system – while the opposite features where observed in *P. alpataco* (Vilela 1996; Villagra and Roig Juñent 1997). As a consequence, *P. argentina* was less affected by drought than *P. alpataco* in a greenhouse experiment during the 1st year of growth (Villagra and Cavagnaro 2006).

Most species of the genus shows petiolar glands. These extra-floral nectaries secrete nectar during the hottest time of the day, which could be considered a mechanism to avoid desiccation (Vilela and Palacios 1997). At the same time, nyctinastic movements reduce the surface of the leaves exposed to high solar radiation. Leaf folding protects against heat shock and water stress. *Prosopis chilensis* shows a decrease in CO_2 assimilation at the time of day with the highest evaporative demand (high temperature and low relative humidity). This decrease in CO_2 assimilation was associated with leaf and foliole movements, and not with stomatal closure, which resulted in a reduced leaf surface but not in reduced capacity of leaf refreshing (Ortiz et al. 1995). The stomatal control at low water potentials, leaf folding to reduce transpiring area (Table 15.2; C. Giordano, unpublished data), and osmotic regulation (Cavagnaro and Passera 1993) displayed by *P. flexuosa* during drought, suggest physiological adaptation of this species to tolerate water stress. Similar responses to water deficit were observed in *Prosopis* species from other South and North American ecosystems and appear to be common over the genus (Ansley et al. 1992; Nilsen et al. 1983; Ortiz et al. 1995; Wan and Sosebee 1991).

Table 15.2 Water status, leaf conductance and leaf folding of *P. flexuosa* trees with different access to the water table in the hyper-arid extreme of the Monte desert

Measurement[a]	Landscape unit		*P*-value
	Dune slope[b]	Valley[c]	
$\Psi_{Predawn}$ (MPa)	–4 (± 0.24)	–2 (± 0.22)	0.0002***
Ψ_{Midday} (MPa)	–4.3 (± 0.20)	–3.9 (± 0.18)	0.1236
Leaf conductance (mmol m^{-2} s^{-1})[d]	66.12 (± 8.75)	194.6 (± 15.68)	<0.0001***
Adaxial angle between foliolules (degrees)[e]	20.6 (± 8.0)	96.3 (± 18.2)	0.0028**

** $P < 0.05$, *** $P < 0.001$; statistical analysis: two tailed t-test, n=11

[a]Measurements performed in Telteca (32° 20′S; 68° 00′W, Mendoza, Argentina) on December 2007

[b]Trees did not access the water table

[c]Trees did access the water table (Jobbágy et al. 2008)

[d]Measured at 0800 hours (maximal conductance registered)

[e]Measured at midday (minimal angle registered)

15.6 Salinity Tolerance

Profuse evidence of the high inter- and intra-specific variability in salt tolerance has been reported for *Prosopis* species. Furthermore, salt tolerance is also variable among different ontogenetic stages of an individual (Cony 1998; Villagra et al. 2004). In this sense, the salt tolerance of *P. flexuosa* and *P. chilensis* increase throughout ontogeny: i.e. the effect of salinity on seed germination is higher than on the growth of 30-day seedlings, and on the survival of adult plants (Catalán et al. 1994; Cazebbone et al. 1999; Cony 1998). This increasing tolerance throughout ontogeny suggests that the effect of salinity is critical during early seedling establishment.

Differences in salt tolerance among species can be observed in different life stages and in different physiological processes. The mortality of *P. flexuosa* seedlings is lower than that of *P. chilensis* in saline cultures with salt concentrations approximating that of sea water (3.3% NaCl; Rhodes and Felker 1987). In a greenhouse experiment, a stronger decrease in biomass accumulation in *P. argentina*

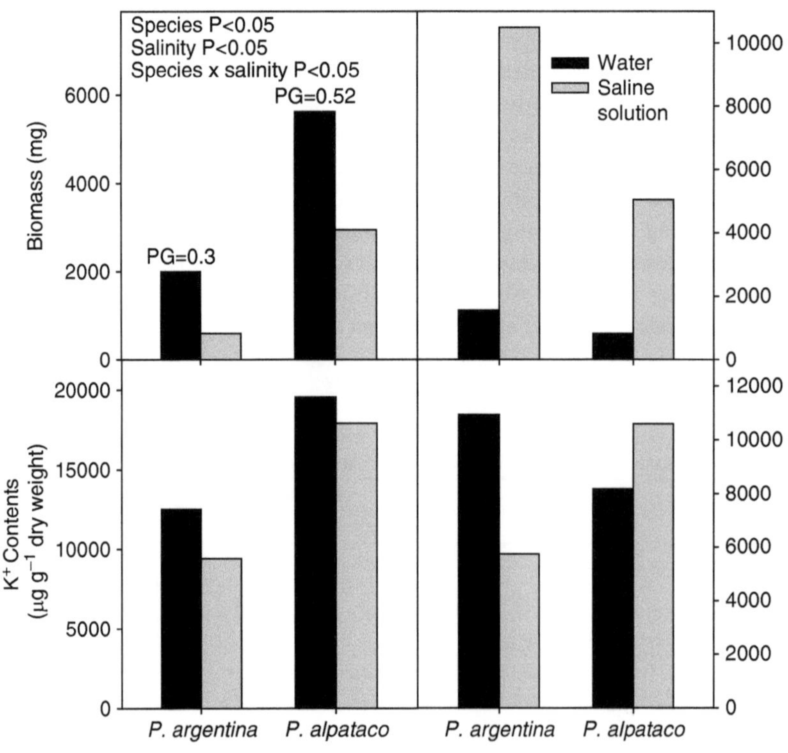

Fig. 15.4 Effects of salinity on biomass accumulation and ion content of 1-year-old plants of *P. argentina* and *P. alpataco*. PG Proportional growth, the ratio between the biomass reached in each saline treatment and the respective control (redrawn from Villagra and Cavagnaro 2005)

than *P. alpataco* was observed under saline conditions (Fig. 15.4; Villagra and Cavagnaro 2005). The high tolerance of *P. alpataco* to salinity is attributed to its ability to regulate and control absorption and transport of ions. This species maintains low contents of Na^+ and high K^+ and Ca^{++} in leaves, thus avoiding the toxic effects of salinity (Fig. 15.4). In contrast, the exclusion of *P. argentina* from saline environments appears to be caused by salt toxicity, despite its tolerance to the osmotic effects of salts (Villagra and Cavagnaro 2005, 2006). These findings agree with other authors that have postulated that the physiological responses to salt stress are different in xerophytes and halophytes species: while xerophytes are capable of counteracting the osmotic effect of salts, halophytes are capable of counteracting both the osmotic and toxic effect of salts (Zhao et al. 2003; Zhao and Harris 1992).

Studies on *P. strombulifera*, a species considered highly salt-tolerant, provide understanding of some of the mechanisms used to avoid salt stress. Sosa et al. (2002) found that low concentrations of NaCl stimulate growth of *P. strombulifera*, while low concentrations of SO_4Na_2 did not provoke any growth stimulation. Moreover, high concentrations of SO_4Na_2 reduce growth to a higher extent than high concentrations of NaCl. Two aspects stand out from these experiments: (1) the initial growth stimulation by salt in *P. strombulifera* differs from the response of other *Prosopis* species, suggesting that this species is an euhalophyte; and (2) the response to salt content during early growth depends on the type of salts in the soil solution. From an anatomical point of view, Reinoso et al. (2004) found that, in the absence of secretory organs, salinity induces changes in roots, hypocotyls, young stems, and leaflets of *P. strombulifera* that are related to metabolic adaptations, such as the early development of an endodermal barrier for ion exclusion, to allow survival in high salinity.

15.7 Other Factors Affecting *Prosopis* Growth

Whereas water availability, salinity, and temperature are the most conspicuous and well-known factors controlling the growth of arid land species, knowledge relating to the effects of other factors on *Prosopis* species growth is scarce. However, other factors, such as nutrients, act during the short periods of water availability, and can affect the resource utilisation of this species (Austin et al. 2004).

Seedlings of *Prosopis* coming from environments with diverse resource availabilities are affected distinctively by resource supply. For example, seedlings of *Prosopis* trees coming from mesic sites are able to respond to water and nitrogen supplies, modifying their growth rate, biomass partition and C:N ratio; while seedlings of *Prosopis* shrubs, coming from low-resource environments are not able to increase utilisation of these resources in a way that results in greater biomass accumulation or a shift in the pattern of resource allocation (Vilela et al. 2003).

The symbiosis of *P. chilensis* seedlings with bacteria of the genus *Rhizobium* increased the N supply for the growing plant, as neither C nor N content differed between *Rhizobium* inoculated and N-fertilised seedlings (Aiazzi et al. 1996).

In this species, nitrogen supply favours aerial growth, and nitrogen restriction promoted dry weight partitioning to roots and the retaining of nutrients, especially phosphorous, in roots (Aiazzi et al. 1996; Imo and Timmer 1992).

Irradiation is not a limiting factor in arid lands; however, it effects on plant development appear to be important in the spatial patterning of seedling establishment. Shading decreases biomass accumulation and increases shoot:root ratio in *P. chilensis* and *P. flexuosa* seedlings. In addition, shading increases protein content in leaves and reduces total non-structural carbohydrates in roots, the C:N ratio, and stored carbohydrates in these species. These changes could reduce the probability of survival under stressful conditions (Vilela and Ravetta 2000). On the other hand, high levels of photosynthetically active radiation (PAR) and UV-B radiation produced photochemical damage in *Prosopis chilensis*, as well as high concentrations of antioxidative β-carotene and stable chlorophyll–protein complexes (Lehner et al. 2001).

15.8 Implications for *Prosopis* Use and Management

Prosopis trees and shrubs have been proposed as candidate species for the management of native populations and cultivation in arid lands (Fagg and Stewart 1994; Ravetta and Soriano 1998; Vilela et al. 2009). Traditional uses include the provision of shade, food, forage, firewood, timber, and charcoal. Recently, other services have been considered, including gum production (Vilela and Ravetta 2005), and the restoration of degraded areas (Cony 1995).

Variability in population structure and growth pattern of *P. flexuosa* woodlands along the latitudinal gradient of its area of distribution requires different management approaches in order to attain sustainability. The northern and most productive woodlands could be used for lumber production, while the southern populations could sustain only a combination of local firewood production and other activities such as extensive grazing by livestock (Villagra et al. 2005a). The application of forestry practices to improve the growth habit, and the quality and production of wood have been suggested for the management of *Prosopis* woodlands around the world. Pruning is the most commonly suggested practice (Elfadl and Luukkanen 2003; Pasiecznik et al. 2001; Patch and Felker 1997a, 1997b). The high proportion of multi-stem individuals described for the Central Monte, and the fast decrease in growth rate observed in multi-stemmed individuals after 60 years of age suggest that elimination of some stems could improve the growth rate of the remaining stems. Recent studies report a significant increment of crown growth immediately after pruning, though stem diameter and plant height remain unaffected (Fig. 15.5; Alvarez 2008). Long-term studies are needed to make a more conclusive evaluation of the advantages of this practice.

In highly degraded arid ecosystems, the variety of adaptations shown by different species of the genus, and the variability observed within species, suggest that

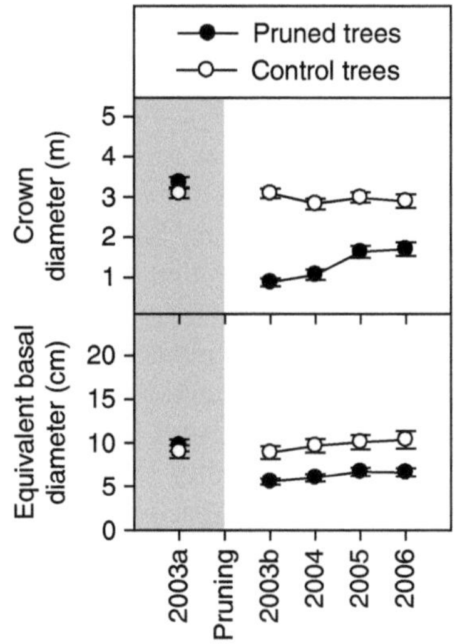

Fig. 15.5 Effects of pruning on the crown diameter and basal diameter of *P. flexuosa* growing in the Central Monte (Mendoza) (Alvarez 2008)

species of this genus may be a good option for use in restoration and afforestation projects. Cony (1995) proposed that the use of multi-purpose trees, such as *P. chilensis* and *P. flexuosa*, be considered in afforestation programmes. Several studies on the genetic variability of these species suggest that the high variability – and high heritability – observed in different physiological and growth features, confer on these species the potential to be included in genetic improvement programmes for the reafforestation of the arid lands of Argentina (Cony 1996, 1998).

Fruits of *Prosopis* spp. have provided a food source for humans and domestic cattle in rural communities of the Monte Desert since ancient times (Burkart 1952). Fruits are an industrial source of seed oil with major components of palmitic, oleic and linoleic acids (Madrinán Polo et al. 1976), and the pod mesocarp is a source of flour and gum (7–10% DW; Kalman 2000), both of which are suitable for human consumption. *Prosopis* flour has a taste and aroma in the same general class as coffee/cacao/cinnamon/mocha (Felker and Guevara 2003). According to cost estimations for milled flour produced in Argentina, a major proportion (56%) of the cost depends on the proportion of mesocarp of the fruit, suggesting that the cost of producing flour would be lower using pods with a higher pericarp:seed ratio. As reproductive output affects the weight of pods and the pericarp:seed ratio, during years of heavy fructification, pod thinning could be one option to decrease industrial costs. The chemical quality of the flour would not be affected by this practice (Agüero et al. 2006). However, the variability in seed production observed as a result of flower and fruit abortion suggests that the utilisation of native populations of *Prosopis denudans* and *P. alpataco* for fruit production in the Patagonian steppe and *P. flexuosa* in the Monte Desert, could not guarantee a consistent supply for the food industry (Cariaga et al. 2005; Mooney et al. 1977).

15.9 Concluding Remarks

The evidence reported in the numerous studies reviewed here is consistent with the idea that the ability of *Prosopis* species to become dominant components of arid ecosystems from Argentina is related to the acquisition of morpho-physiological adaptations to the stressful environment. Some of these adaptations appear to be spread over the genus and should confer the capability to deal with the most common stressful factor of arid lands (i.e. water availability); however, other morphological or physiological adaptations appear to be specific to each species and should be the cause of niche differentiation among species and the occupation of particular environments within arid lands (e.g. sand dunes, saline environments).

The inter- and intra-specific variability observed in their adaptation to stressful factors suggest that some *Prosopis* species may be good plants to use in the restoration of degraded areas or in afforestation projects with productive objectives. However, new studies on the adaptive features and plasticity of each species, on their performance under different environmental conditions, and on their genetic variability should be carried out in order to select and improve target genotypes for different project objectives and to propose management technologies to improve system productivity.

References

Agüero PR, Cariaga RE, Ferrari M, Ravetta DA, Vilela AE (2006) The impact of reproductive output on *Prosopis denudans* pod partition and chemical composition: implications for economic utilization. Econ Bot 60:292–295

Aguiar MR, Sala OE (1999) Patch structure, dynamics and implications for the functioning of arid ecosystems. Trends Ecol Evol 14:273–277

Aiazzi M, Argüello J, Abril A (1996) Nodulated and non-nodulated *Prosopis chilensis* (Mol) St. seedlings. economy of carbon and nitrogen. Forest Ecol Manage 89:25–29

Alvarez JA (2008) Bases ecológicas para el manejo sustentable del bosque de algarrobos (*Prosopis flexuosa* D.C.) en el noreste de Mendoza. Argentina. Universidad Nacional del Comahue, Bariloche (Argentina)

Ansley RJ, Jacoby PW, Meadors CH, Lawrence BK (1992) Soil and leaf water relations of differentially moisture-stressed honey mesquite (*Prosopis glandulosa* Torr.). J Arid Environ 22:147–159

Austin AT, Yahdjian L, Stark JM, Belnap J, Porporato A, Norton U, Ravetta DA, Schaeffer SM (2004) Water pulses and biogeochemical cycles in arid and semiarid ecosystems. Oecologia 141:221–235

Brar GS, Palazzo AJ (1995) Shoot and root development of tall and hard fescues in two different soils. J Environ Qual 24:771–781

Burkart A (1952) Las Leguminosas Argentinas Silvestres y Cultivadas. Acme, Buenos Aires

Burkart A (1976) A monograph of the genus *Prosopis* (Leguminosae subfam. Mimosoideae). J Arn Arbor 57:219–249; 450–455

Burkart A, Simpson BB (1977) The genus *Prosopis* and annotated key to the species of the world. In: Simpson BB (ed) Mesquite. Its biology in two desert scrub ecosystems. US/IBP Synthesis Series 4. Hutchinson & Ross, Dowden, pp 201–216

Cabrera AL (1976) Regiones Fitogeográficas Argentinas. In: Kugler WF (ed) Enciclopedia Argentina de Agricultura y Jardinería. Editorial ACME, Buenos Aires, pp 85

Campos CM, Ojeda RA (1997) Dispersal and germination of *Prosopis flexuosa* (Fabaceae) seeds by desert mammals in Argentina. J Arid Environ 35:707–714

Cariaga RE, Aguero PR, Ravett DA, Vilela AE (2005) Differences in production and mortality of reproductive structures in two *Prosopis* L. (Mimosaceae) shrub species from Patagonia, Argentina. J Arid Environ 63:696–705

Castro MA (1994) Maderas Argentinas de *Prosopis*. Atlas Anatómico. Secretaría General de la Presidencia de la Nación, República Argentina

Catalán LA, Balzarini M (1992) Improved laboratory germination conditions for several arboreal *Prosopis* species: *P. chilensis*, *P. flexuosa*, *P. nigra*, *P. alba*, *P. caldenia* and *P. affinis*. Seed Sci Technol 20:293–298

Catalán L, Balzarini M, Taleisnik E, Sereno R, Karlin U (1994) Effects of salinity on germination and seedling growth of *Prosopis flexuosa* (D.C.). Forest Ecol Manage 63:347–357

Cavagnaro JB, Passera CB (1993) Relaciones hídricas de *Prosopis flexuosa* (algarrobo dulce) en el Monte, Argentina. In: IADIZA (ed) Contribuciones Mendocinas a la Quinta Reunión de Regional para América Latina y el Caribe de la Red de Forestación del CIID. Conservación y Mejoramiento de Especies del Género *Prosopis*., Mendoza, Argentina, pp 73–78

Cazebbone C, Vega AI, Varela DA, Cardemil LA (1999) Salinity effects on germinations and growth of *Prosopis chilensis*. Rev Chil Hist Nat 72:83–91

Cony M (1995) Reforestación racional de zonas áridas y semiáridas con árboles multipropósitos. Interciencia 20:249–253

Cony MA (1996) Genetic variability in *Prosopis flexuosa* D. C., a native tree of the Monte phytogeographic province, Argentina. For Ecol Manage 87:41–49

Cony MA (1998) Importancia de los estudios fisiológicos, ecofisiológicos y genéticos sobre las especies del género *Prosopis* para la recuperación de ecosistemas áridos degradados de América Latina. In: Cuba Salerno AB, Silva Peralta A, Cornejo Flores C (eds) Bosques secos y desertificación. INRENA – Proyecto Algarrobo, Lambayeque, Perú, pp 255–276

Cony MA, Trione SO (1996) Germination with respect to temperature of two Argentinian *Prosopis* species. J Arid Environ 33:225–236

Cony MA, Trione SO (1998) Inter- and intraspecific variability in *Prosopis flexuosa* and *P. chilensis*: seed germination under salt and moisture stress. J Arid Environ 40:307–317

Dalmasso AD, Anconetani J (1993) Productividad de frutos de *Prosopis flexuosa* (Leguminosae), algarrobo dulce, en Bermejo, San Juan. Multequina 2:173–181

Duff AB, Meyer JM, Pollock C, Felker P (1994) Biomass production and diameter growth of nine half-sib families of mesquite (*Prosopis glandulosa* var. *glandulosa*) and a fast growing *Prosopis alba* half-sib family grown in Texas. For Ecol Manage 67:257–266

Elfadl MA, Luukkanen O (2003) Effect of pruning on *Prosopis juliflora*: considerations for tropical dryland agroforestry. J Arid Environ 53:441–455

Fagg CW, Stewart JL (1994) The value of *Acacia* and *Prosopis* in arid and semi-arid environments. J Arid Environ 27:3–25

Felker P, Guevara JC (2003) Potential of commercial hardwood forestry plantations in arid lands – an economic analyses of *Prosopis* lumber production in Argentina and the United States. For Ecol Manage 186:271–286

Ffolliot PF, Thames JL (1983) Recolección, manipuleo, almacenaje y pre-tratamiento de la semillas de *Prosopis* en América Latina. FAO, Tucson, Arizona

Gile LH, Gibbens RP, Lenz JM (1997) The near-biquitous pedogenic world of mesquite roots in an arid basin floor. J Arid Environ 35:39–58

Imo M, Timmer VR (1992) Growth, nutrient allocation and water relations of mesquite (*Prosopis chilensis*) seedlings at differing fertilization schedules. For Ecol Manage 55:279–294

Jobbágy EG, Nosetto MD, Villagra PE, Jackson RB (2008) Isotopos estables como trazadores de las fuentes de agua de bosques de algarrobo en un desierto arenoso. XXI Congreso Argentino de Ciencias del Suelo, Potrero de los Funes, San Luis

Kalman DM (2000) Variabilidad interespecífica e interpoblacional en el contenido de goma en frutos de *Prosopis* L. (Mimosaceae) y su relación con el ataque de insectos. Universidad de Buenos Aires, Buenos Aires

Lehner G, Delatorre J, Lütz C, Cardemil L (2001) Field studies on the photosynthesis of two desert Chilean plants: *Prosopis chilensis* and *Prosopis tamarugo*. J Photochem Photobiol B: Biol 64:36–44

Madrinán Polo C, Hunziker J, Cattaneo P (1976) Aceite de semilla de especies de *Prosopis* y *Prosopidastrum* (Leguminosas). An Asoc Quim Argent 64:127–138

Mooney HA, Simpson BB, Solbrig OT (1977) Phenology, morphology, physiology. In: Simpson BB (ed) Mesquite. Its biology in two desert scrub ecosystems. U.S./ibp synthesis series 4. Hutchinson & Ross, Dowden, pp 26–43

Morales MS, Villalba R, Grau HR, Villagra PE, Boninsegna JA, Ripalta A, Paolini L (2001) Potencialidad de *Prosopis ferox* Griseb (Leguminosae, subfamilia: Mimosoideae) para estudios dendrocronológicos en desiertos subtropicales de alta montaña. Rev Chil Hist Nat 74: 889–896

Morello J (1958) La Provincia Fitogeográfica del Monte. Opera Lilloana 2:5–115

Munns R (2002) Comparative physiology of salt and water stress. Plant Cell Environ 25:239–250

Nilsen ET, Sharifi MR, Rundel PW, Jarrell WM, Virginia RA (1983) Diurnal and seasonal water relations of the desert phreatophyte *Prosopis glandulosa* (Honey mesquite) in the Sonoran Desert of California. Ecology 64:1381–1393

Nilsen ET, Sharifi MR, Rundel PW (1984) Comparative water relations of phreatophytes in the Sonoran Desert of California. Ecology 65:767–776

Noy-Meir I (1973) Desert ecosystems: environment and producers. Annu Rev Ecol Syst 4:25–51

Ortega Baes P, de Viana M, Sühring S (2002) Germination in *Prosopis ferox* seeds: effects of mechanical, chemical and biological scarificators. J Arid Environ 50:185–189

Ortiz C, Bravo L, Pinto M, Cardemil L (1995) Physiological and molecular responses of *Prosopis chilensis* under field and simulation conditions. Phytochemistry 40:1375–1382

Pasiecznik NM, Felker P, Harris PJC, Harsh LN, Cruz G, Tewari JC, Cadoret K, Maldonado LJ (2001) The *Prosopis juliflora – Prosopis pallida* complex: a monograph. HDRA, Coventry

Patch NL, Felker P (1997a) Influence of silvicultural treatments on growth of mature mesquite (*Prosopis glandulosa var. glandulosa*) nine years after initiation. Forest Ecol Manage 94: 37–46

Patch NL, Felker P (1997b) Silvicultural treatments for sapling mesquite (*Prosopis glandulosa var. glandulosa*) to optimize timber production and minimize seedling encroachment. Forest Ecol Manage 96:231–240

Pawsey CK (1960) Cone production reduced, apparently by drought, in the southeast (of South Australia). Aust For 24:74–75

Peinetti R, Pereyra M, Kin A, Sosa A (1993) Effects of cattle ingestion on viability and germination rate of caldén (*Prosopis caldenia*) seeds. J Range Manage 46:483–486

Philips WS (1963) Depth of roots in soil. Ecology 44:424

Ravetta DA, Soriano A (1998) Alternatives for the development of new industrial crops for Patagonia. Ecol Aust 8:297–307

Reinoso H, Sosa L, Ramírez L, Luna V (2004) Salt-induced changes in the vegetative anatomy of *Prosopis strombulifera* (Leguminosae). Can J Bot 82:618–628

Rhodes D, Felker P (1987) Mass screening *Prosopis* (mesquite) seedlings for growth at sea water salinity. For Ecol Manage 24:169–176

Roig FA (1993a) Aportes a la etnobotánica del género *Prosopis*. In: IADIZA (ed) Contribuciones Mendocinas a la Quinta Reunión de Regional para América Latina y el Caribe de la Red de Forestación del CIID. Conservación y Mejoramiento de Especies del Género *Prosopis*., Mendoza, Argentina, pp 99–119

Roig FA (1993b) Informe Nacional para la Selección de Germoplasma en Especies del Género *Prosopis* de la República Argentina. In: IADIZA (ed) Contribuciones Mendocinas a la Quinta Reunión de Regional para América Latina y el Caribe de la Red de Forestación del CIID.

Conservación y Mejoramiento de Especies del Género *Prosopis*. IADIZA-CRICYT-CIID, Mendoza, Argentina, pp 1–36

Rundel P, Villagra PE, Dillon MO, Roig-Juñent SA, Debandi G (2007) Arid and semi-arid ecosystems. In: Veblen TT, Young K, Orme A (eds) The physical geography of South America. Oxford University Press, Oxford, pp 158–183

Salih AA (1998) Root and shoot growth of *Prosopis chilensis* in response to soil impedance and soil matric potential. J Arid Environ 40:43–52

Salvo B, Botti C, Pinto M (1988) Flower induction and differentiation in *Prosopis chilensis* (Mol.) Stuntz and their relationship with alternate fruit bearing. In: Habitat MA (ed) The current state of knowledge of *Prosopis juliflora*. FAO, Rome, pp 269–275

Sharifi MR, Nilsen ET, Rundel PW (1982) Biomass and net primary production of *Prosopis glandulosa* (Fabaceae) in the Sonoran Desert of California. Am J Bot 69:760–767

Siemens DH, Ralston BE, Johnson CD (1994) Alternative seed defense mechanism in a palo verde (Fabaceae) hybrid zone: effects on bruchid beetle abundance. Ecol Entomol 19: 381–390

Simpson BB, Neff JL, Moldenke AR (1977) *Prosopis* flowers as a resource. In: Simpson BB (ed) Mesquite. Its biology in two desert scrub ecosystems. US/IBP Synthesis Series 4. Hutchinson & Ross, Dowden, pp 84–105

Smith LL, Ueckert DN (1974) Influence of insects on mesquite seed production. J Range Manage 27:61–65

Snyder KA, Williams DG (2003) Defoliation alters water uptake by deep and shallow roots of *Prosopis velutina* (Velvet Mesquite). Funct Ecol 17:363–374

Solbrig OT, Cantino PD (1975) Reproductive adaptations in *Prosopis* (Leguminosae, Mimosoideae). J Ann Arbor 56:185–210

Sosa L, Reinoso H, Frutos M, Luna V (2002) Efecto de distintas sales de Na y K en soluciones monosalinas y bisalinas isoosmóticas sobre la germinación de *Prosopis strombulifera*. XI Reunión Latinoamericana de Fisiología Vegetal, Punta del Este (Uruguay)

Sosa L, Llanes A, Reinoso H, Reginato M, Luna V (2005) Osmotic and specific ion effects on the germination of *Prosopis strombulifera*. Ann Bot 96:261–267

Stephenson AG (1981) Flower and fruit abortion: proximate causes and ultimate functions. Annu Rev Ecol Syst 12:253–279

Toro H, Chiappa E, Covarrubias R, Villaseñor R (1993) Interrelaciones de polinización en zonas áridas de Chile. Acta Entomol Chil 18:20–29

Vilela A (1996) Morfología y anatomía foliar de especies sudamericanas del género *Prosopis* (Leguminosae–Mimosoideae): un enfoque adaptativo. Universidad de Buenos Aires

Vilela AE, Palacios RA (1997) Distribución de nectarios extraflorales en especies sudamericanas del género *Prosopis*. Bol Soc Argent Bot 32:163–170

Vilela AE, Ravetta DA (2000) The effect of radiation on seedling growth and physiology in four species of *Prosopis* L. (Mimosaceae). J Arid Environ 44:415–423

Vilela AE, Ravetta DA (2001) The effect of seed scarification and soil-media on germination growth, storage, and survival of seedlings of five species of *Prosopis* L. (Mimosaceae). J Arid Environ 48:171–184

Vilela AE, Ravetta DA (2005) Gum exudation in South-American species of *Prosopis* L. (Mimosaceae). J Arid Environ 60:389–395

Vilela AE, Rennella MJ, Ravetta DA (2003) Responses of tree-type and shrub type *Prosopis* (Mimosaceae) taxa to water and nitrogen availabilities. J Arid Environ 186:327–337

Vilela A, Bolkovic ML, Carmanchahi P, Cony M, de Lamo D, Wassner D (2009) Past, present and potential uses of native flora and wildlife of the Monte Desert. J Arid Environ 73:238–243

Villagra PE (1995) Temperature effects on germination of *Prosopis argentina* and *P. alpataco* (Fabaceae, Mimosoideae). Seed Sci Technol 23:639–646

Villagra PE (1997) Germination of *Prosopis argentina* and *P. alpataco* seeds under saline conditions. J Arid Environ 37:261–267

Villagra PE (1998) Comparación del comportamiento fitosociológico y ecofisiológico de *Prosopis argentina* y *P. alpataco* (Fabaceae, Mimosoideae). Programa de Postgrado en Biología. Universidad Nacional de Cuyo, Mendoza, pp 130

Villagra PE (2000) Aspectos ecológicos de los algarrobales argentinos. Multequina 9:21–36

Villagra PE, Cavagnaro JB (2000) Effects of clayish and sandy soils on the growth of *Prosopis argentina* and *P. alpataco* seedlings. Ecol Austr 10:111–119

Villagra PE, Cavagnaro JB (2005) Effects of salinity–soil type interactions on the establishment, growth and water relations of *Prosopis argentina* and *P. alpataco* seedlings. Implications for their ecological success. Aust Ecol 30:325–335

Villagra PE, Cavagnaro JB (2006) Water stress effects on the seedling growth of *Prosopis argentina* and *Prosopis alpataco*. J Arid Environ 64:390–400

Villagra PE, Roig FA (2002) Distribución geográfica y fitosociología de *Prosopis argentina* y *P. alpataco* (Fabaceae, Mimosoidea). Bol Soc Argent Bot 37:99–106

Villagra PE, Roig Juñent FA (1997) Wood structure of *Prosopis alpataco* and *P. argentina* growing under different edaphic conditions. IAWA J 18:37–51

Villagra PE, Marone L, Cony MA (2002) Mechanism affecting the fate of *Prosopis flexuosa* seeds during secondary dispersal en the Monte desert. Aust Ecol 27:416–421

Villagra PE, Cony MA, Mantován NG, Rossi BE, González Loyarte MM, Villalba R, Marone L (2004) Ecología y Manejo de los algarrobales de la Provincia Fitogeográfica del Monte. In: Arturi MF, Frangi JL, Goya JF (eds) Ecología y Manejo de Bosques Nativos de Argentina. Editorial Universidad Nacional de La Plata

Villagra PE, Boninsegna JA, Alvarez JA, Cony M, Cesca E, Villalba R (2005a) Dendroecology of *Prosopis flexuosa* woodlands in the Monte desert: implications for their management. Dendrochronologia 22:209–213

Villagra PE, Villalba R, Boninsegna JA (2005b) Structure and growth rate of *Prosopis flexuosa* woodlands in two contrasting environments of the central Monte desert. J Arid Environ 60:187–199

Villalba R (1985) Xylem structure and cambial activity in *Prosopis flexuosa* D.C. IAWA Bulletin n.s. 6:119–130

Villalba R, Boninsegna JA (1989) Dendrochronological studies on *Prosopis flexuosa* D.C. IAWA Bulletin 10:155–160

Villalba R, Villagra PE, Boninsegna JA, Morales MS, Moyano V (2000) Dendroecología y dendroclimatología con especies del género *Prosopis*. Multequina 9(2):1–18

Wan C, Sosebee RE (1991) Water relations and transpiration of honey mesquite on 2 sites in West Texas. J Range Manage 44:156–160

Zhao K, Hai F, San Z, Jie S (2003) Study on the salt and drought tolerance of *Suaeda salsa* and *Kalanchoe claigremontiana* under iso-osmotic salt and water stress. Plant Sci 165:837–844

Zhao KF, Harris PJC (1992) The effects of iso-osmotic salt and water stresses on the growth of halophytes and non-halophytes. J Plant Physiol 139:761–763

Chapter 16
Plant Growth Inhibitors From Mesquite (*Prosopis juliflora*)

Hiroshi Nakano

Abstract This chapter describes investigations into the allelopathy of mesquite (*Prosopis juliflora*) leaves and the isolation of allelochemicals from aqueous exudates and leachates from the leaves. The plant growth inhibitors isolated from the exudates of mesquite leaves were L-tryptophan (**14**), syringin (**15**), and (–)-lariciresinol (**16**). L-Tryptophan (**14**) was present in higher content in the exudates of mesquite leaves, and inhibited plant growth more than syringin (**15**) and (–)-lariciresinol (**16**). To confirm whether L-tryptophan (**14**) was allelopathic, we measured its content in leachates from mesquite leaves and assayed the effect of the leachates on the root growth of lettuce (*Lactuca sativa*) and barnyard grass (*Echinochloa cru-galli*). L-Tryptophan (**14**) was responsible for most of the root growth suppression by the leachates. These results suggest that L-tryptophan (**14**) leaching from leaves is a major contributor to the allelopathic effect of mesquite. Other plant growth inhibitory alkaloids and their structure–activity relationships were also investigated. After isolation from the methanol extracts of mesquite leaves, we examined the plant growth inhibitory activities of $3''''$-oxojuliflorine ($3''''$-oxojuliprosopine; **4**), secojuliprosopinal (**13**), a 1:1 mixture of $3'$-oxo- (**9**) and 3-oxo-juliprosine (**10**), juliprosine (**8**), and juliflorine (juliprosopine; **3**). The compound with the highest activity appeared to be juliprosine (**3**), followed by a 1:1 mixture of $3'$-oxo- (**9**) and 3-oxo-juliprosine (**10**), juliflorine (juliprosopine; **3**), $3''''$-oxojuliflorine ($3''''$-oxojuliprosopine; **4**), and secojuliprosopinal (**13**). These results suggest that the active sites within the chemical structure of alkaloids from mesquite leaves are the functional groups at C-3 and C-$3'$ of the piperidine and indolizine skeletons.

H. Nakano
National Agricultural Research Center for Kyushu Okinawa Region, Chikugo, Fukuoka, 833-0041, Japan
e-mail: nakanohr@affrc.go.jp

K.G. Ramawat (ed.), *Desert Plants*,
DOI 10.1007/978-3-642-02550-1_16, © Springer-Verlag Berlin Heidelberg 2010

16.1 Introduction

Allelopathy has been defined as any direct or indirect beneficial or detrimental effect of one plant on another (including microorganisms) via release of chemical compounds to the environment (Molisch 1937). Mesquite (*Prosopis juliflora*, Leguminosae; Fig. 16.1) is widespread in India, Pakistan, Saudi Arabia, Brazil, Mexico, and the United Arab Emirates (Sankhla et al. 1965; Nasir and Ali 1972; Al-Humaid and Warrag 1998; Tabosa et al. 2000; Almaraz-Abarca et al. 2007; El-Keblawy and Al-Rawai 2007). Mesquite inhibits the germination or growth of many plant species growing in its vicinity through the release of allelochemicals in exudates from leaves, roots, or fruits (Sankhla et al. 1965; Pandit et al. 1995;

Fig. 16.1 Mesquite (*Prosopis juliflora*)

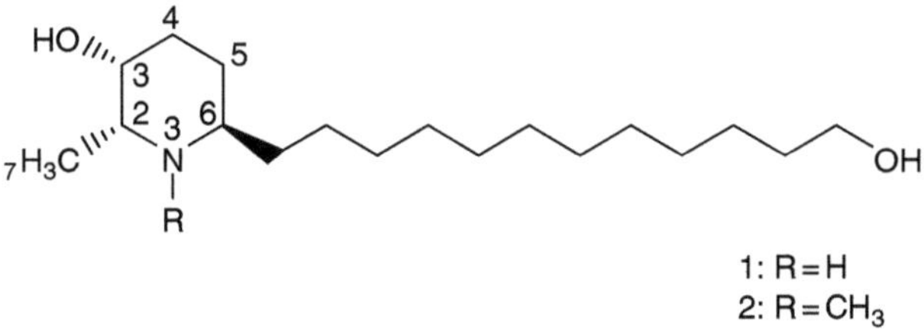

Fig. 16.2 Chemical structures of julifloridine (**1**) and N-methyljulifloridine (**2**)

Al-Humaid and Warrag 1998). The plant growth inhibitors involved in the allelopathic effect of mesquite have been isolated and identified for the first time by our group and are summarised in this chapter.

There are many bioactive alkaloids (Toyooka 2001). Julifloridine (**1**), which contains a piperidine ring, and its N-methyl derivative N-methyljulifloridine (**2**), were isolated from the extracts of mesquite leaves (Fig. 16.2; Ahmad et al. 1978; Ahmad and Qazi 1983; Ahmad and Sultana 1990). Juliflorine (juliprosopine; **3**) which has two piperidine rings and a hexahydroindolizine ring, and its stereoisomers juliflorícine (**5**, **6**) and juliflorinine (**7**) were isolated from the extracts of leaves of both mesquite and *Prosopis glandulosa* (Fig. 16.3; Ahmad and Mohammad 1979; Ott-Longoni et al. 1980; Ahmad and Qazi 1985; V.U. Ahmad et al. 1989). Juliflorine (juliprosopine; **3**) possesses antidermatophytic, antibacterial, and DNA-binding activities (Khursheed et al. 1986; Ahmad et al. 1986; A. Ahmad et al. 1989; Tapia et al. 2000), and juliflorícine (**5**) has antidermatophytic and antibacterial activities (Aqeel et al. 1989). Juliprosine (**8**), which contains two piperidine rings and a dihydroindolizine ring, and its stereoisomer isojuliprosine (**11**), were isolated from the extracts of mesquite leaves (Fig. 16.4; Dätwyler et al. 1981; V.U. Ahmad et al. 1989). Juliprosine (**8**) was found to possess DNA-binding activity (Tapia et al. 2000). Juliprosinene (**12**), which has two piperidine rings and an indolizine ring, was isolated from the extracts of mesquite leaves (Fig. 16.5; V.U. Ahmad et al. 1989). Despite the isolation of these alkaloids from mesquite, no associated plant growth inhibitory activity had previously been identified before the work reported herein.

This chapter describes a series of investigations designed to define the allelopathy of mesquite leaves and to isolate the allelochemicals from aqueous exudates and leachates, and plant growth inhibitors from the methanol extracts of mesquite leaves.

16.2 Allelopathy of Mesquite Leaves in the Laboratory

Because natural mesquite communities frequently contain no other plant species (Sankhla et al. 1965; Pandit et al. 1995; Al-Humaid and Warrag 1998), we investigated the allelopathy of mesquite leaves in the laboratory (Nakano et al. 2001).

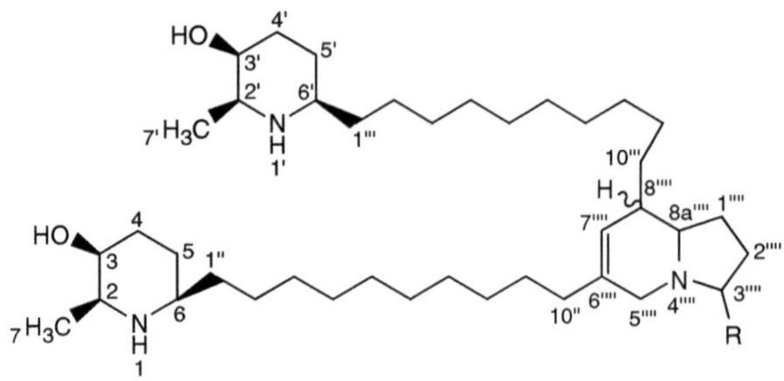

3: R = H
4: R = = O

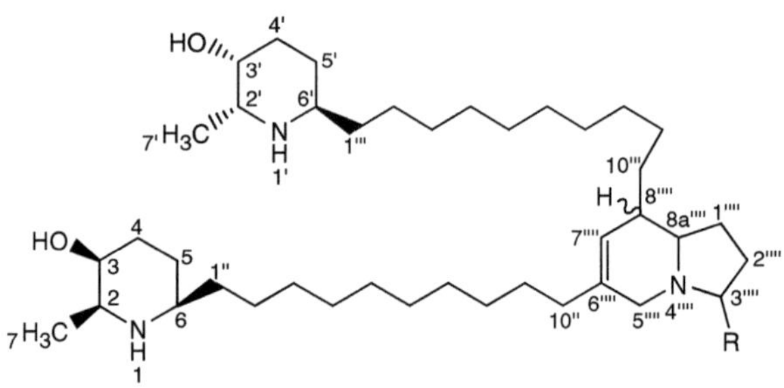

5

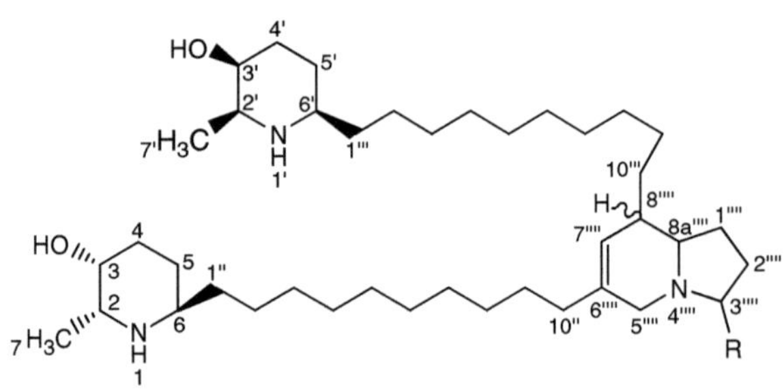

6

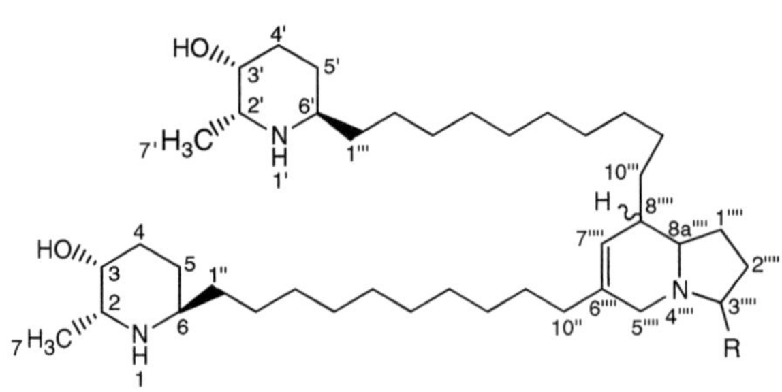

7

Fig. 16.3 Chemical structures of juliflorine (juliprosopine; **3**), 3''''-oxojuliflorine (3''''-oxojuliprosopine; **4**), julifloricine (**5**, **6**), and juliflorinine (**7**)

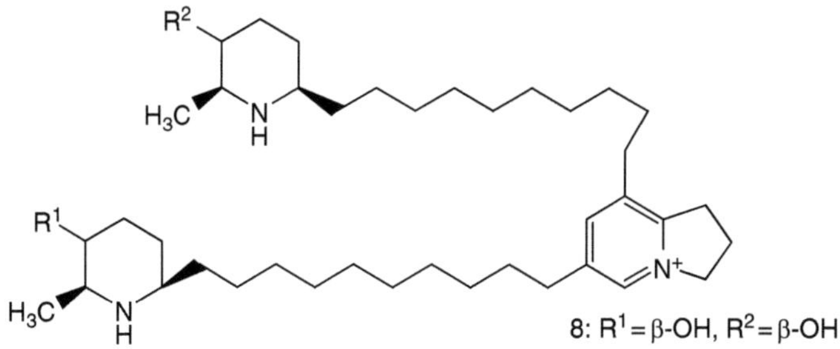

8: R^1 = β-OH, R^2 = β-OH
9: R^1 = =O, R^2 = β-OH
10: R^1 = β-OH, R^2 = =O

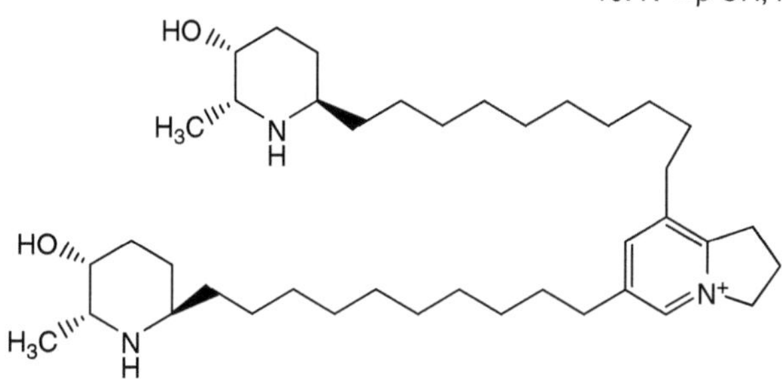

11

Fig. 16.4 Chemical structures of juliprosine (**8**), 3′-oxo- juliprosine (**9**), 3-oxo-juliprosine (**10**), and isojuliprosine (**11**)

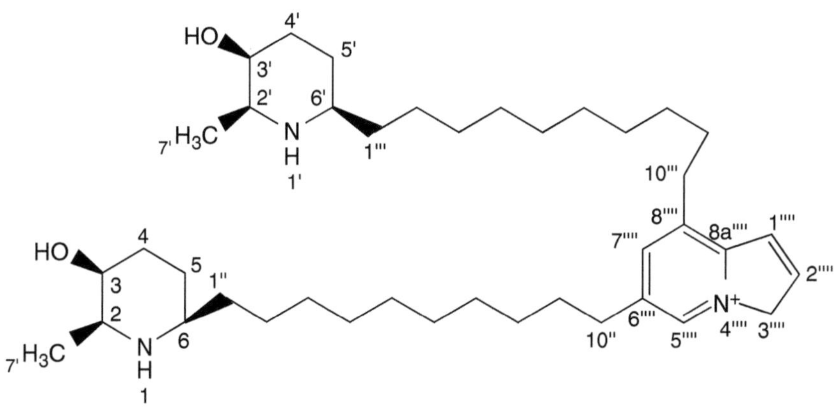

12

Fig. 16.5 Chemical structures of juliprosinene (**12**)

Freeze-dried leaves (200 mg) were placed in the centre of a dish (9 cm diameter × 4.5 cm deep) containing 200 ml 0.5% agar culture medium. Lettuce (*Lactuca sativa*) seeds were placed on the culture medium at 0, 10, 20, and 30 mm from the dried leaves and kept in the dark at 25°C (Fig. 16.7). After 3 days, seedling root and shoot

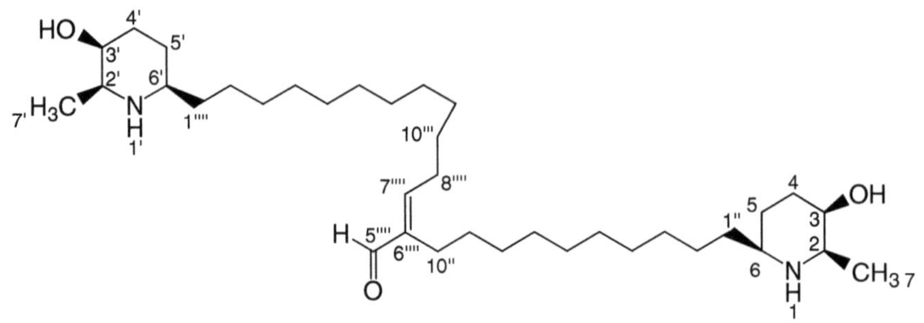

Fig. 16.6 Chemical structure of secojuliprosopinal (**13**)

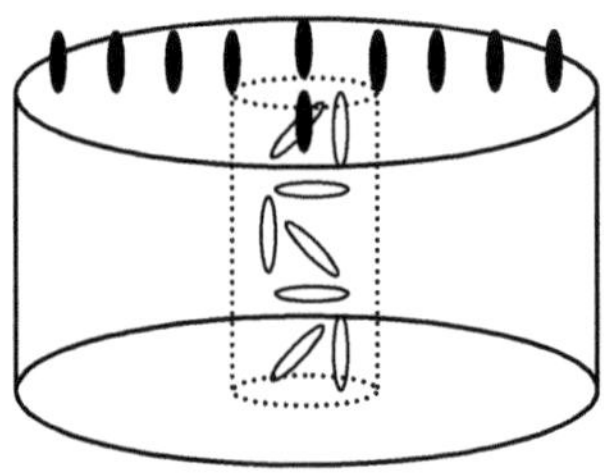

Fig. 16.7 A schematic presentation of the method used to measure the effect of mesquite leaves on the root and shoot growth of lettuce seedlings

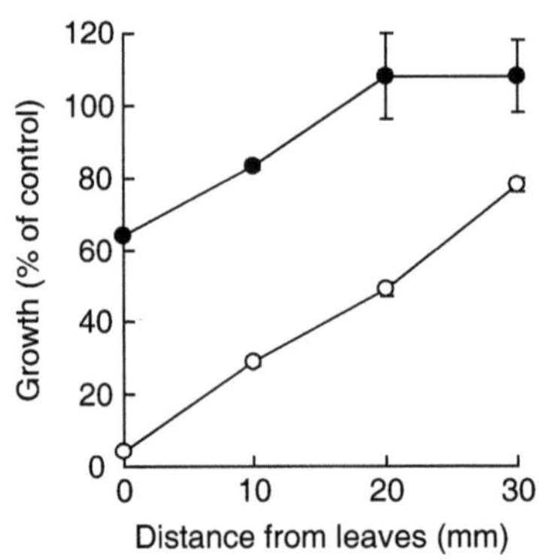

Fig. 16.8 Effect of mesquite leaves on the root (○) and shoot (•) growth of lettuce seedlings. Means ± SE of results from three replicates of four plants

lengths were measured. Growth was inhibited, especially that of roots (Fig. 16.8). The inhibitory activity was more pronounced as the distance from the leaves was reduced. These results suggest that allelochemicals inhibiting the root growth of lettuce were exuded from the mesquite leaves into the agar culture medium.

16.3 Plant Growth Inhibitors from the Exudates of Mesquite Leaves

L-Tryptophan (**14**), syringin (**15**), and (–)-lariciresinol (**16**) were isolated from the aqueous exudates of mesquite leaves (Fig. 16.9) as plant growth inhibitors, as determined by lettuce and barnyard grass root assays (Nakano et al. 2001, 2002).

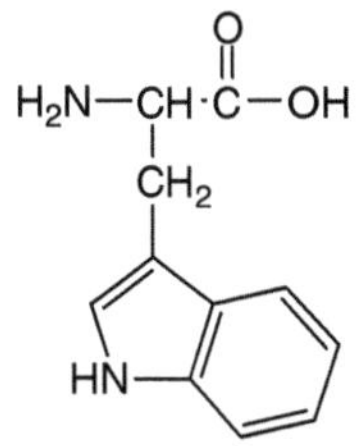

14

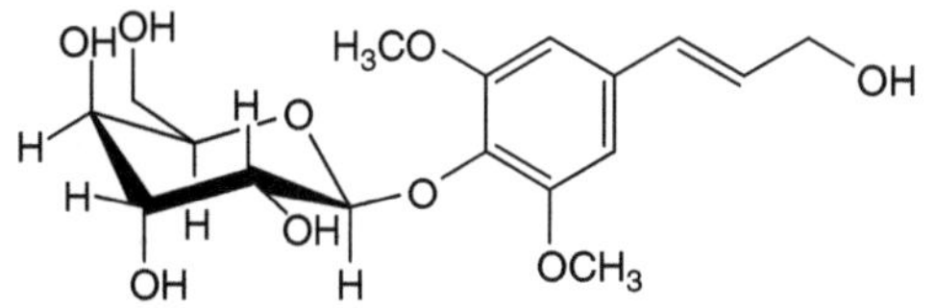

15

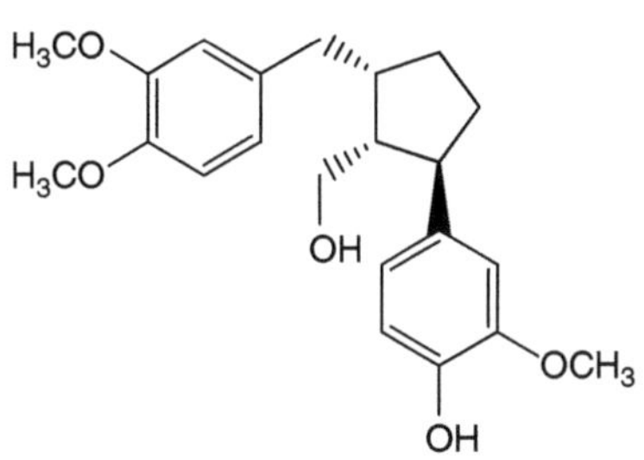

16

Fig. 16.9 Chemical structures of L-tryptophan (**14**), syringin (**15**), and (–)-lariciresinol (**16**)

The chemical structures of these compounds were elucidated from spectroscopic analyses. L-Tryptophan (**14**) was isolated from oat (*Avena sativa*) and wheat (*Triticum aestivum*) and investigated for its activity as an allelochemical (Kato-Noguchi 1994a, b; Nakano et al. 2006; Nakano 2007a, 2007b). Syringin (**15**) was isolated from *Syringa vulgaris* and *Foniculum vulgare* and found to act as a hypotensive agent and antioxidant (Ahmad and Aftab 1995; Nakayama et al. 1996). (–)-Lariciresinol (**16**) was isolated from *Dirca occidentalis* and *Wikstroemia elliptica* and proven to have efficacy as an anticancer agent (Badawi et al. 1983; Duh et al. 1986). However, neither syringin (**15**) nor (–)-lariciresinol (**16**) has previously been reported to act as a plant growth inhibitor. To evaluate whether these compounds have allelopathic potential, we measured their contents in the exudates of freeze-dried mesquite leaves and examined the effect of the exudates on the root and shoot growth of lettuce and barnyard grass (Nakano et al. 2001, 2002). Ten seeds per dish of either lettuce or barnyard grass were placed on a filter paper moistened with 500 μl sample solution made from an extract of agar culture medium containing the exudates from freeze-dried mesquite leaves (1 mg leaf equivalent; Fig. 16.7) or from different concentrations of authentic solutions of these compounds in a 3.3 cm Petri dish. The dishes containing the seeds were kept for 3 days at 25°C in the dark. The L-tryptophan (**14**) content in the extract was estimated to be 4.8 μM by HPLC analysis. The extract inhibited lettuce roots by

Table 16.1 Plant growth inhibitory activity of compounds isolated from exudates of mesquite leaves as measured by root growth tests with lettuce and barnyard grass

Test plant	Plant growth inhibitor	I_{50} (μM)
Lettuce	L-Tryptophan (**14**)	50
	Syringin (**15**)	>1,000
	(–)-Lariciresinol (**16**)	500
Barnyard grass	L-Tryptophan (**14**)	10
	Syringin (**15**)	50
	(–)-Lariciresinol (**16**)	100

44% relative to the control. By comparison, 4.8 μM authentic L-tryptophan (**14**) inhibited lettuce roots by 30%. L-Tryptophan (**14**) was present in higher content in the extract and inhibited plant growth more than syringin (**15**) and (–)-lariciresinol (**16**; Table 16.1). These results suggest that L-tryptophan (**14**) plays an important allelopathic role, and that syringin (**15**) and (–)-lariciresinol (**16**) contribute in supplementary roles.

16.4 Leaching of L-Tryptophan from the Leaves of Mesquite Plants

In general, allelochemicals can be released from plants to the environment in one of the following ways (Rice 1984): (1) by volatilisation from plants (e.g. camphor and 1,8-cineole from *Salvia* species; Muller 1965), (2) as exudates from roots (e.g. scopoletin from oat; Martin 1957), (3) by leaching from plants (e.g. chlorogenic, *p*-coumarylquinic, and gentisic acids from *Eucalyptus globulus*; del Moral and Muller 1969), or (4) by decomposition of residues (e.g. salts of acetic, propionic, and butyric acids from wheat; Tang and Waiss 1978). To confirm whether leachates from mesquite plants contain enough L-tryptophan (**14**) to be allelopathic, we studied the L-tryptophan (**14**) content in leachates, and the effect of the leachate on the root growth of barnyard grass (Nakano et al. 2003). To reproduce leaching by fog, mist, and rain, we sprayed 500 ml distilled water from above at 10 ml min^{-1} onto the leaves of 1-year-old mesquite plants that had been grown in a greenhouse for 1 year (Fig. 16.10). Water passing through the leaves of the mesquite plants was collected in four dishes (9 cm diameter × 4.5 cm deep) placed under the branches. Ten seeds each of lettuce and barnyard grass were placed on a filter paper moistened with 500 μl of a solution made from a concentrate of the leachates from mesquite leaves (30 ml equivalent) or different concentrations of authentic L-tryptophan (**14**) solution in a 3.3-cm Petri dish and kept for 3 days at 25°C in the dark. The L-tryptophan (**14**) content in the extract was estimated to be 17.9 μM by HPLC. The extract inhibited both lettuce and barnyard grass roots by 70%. Authentic L-tryptophan (**14**) at 17.9 μM inhibited lettuce roots by 40% and barnyard grass roots by 65%. These results suggest that L-tryptophan (**14**) leaching from leaves is a major contributor to the allelopathy of mesquite leaves.

Fig. 16.10 Leaching of L-tryptophan from the leaves of mesquite plants by sprayed water

16.5 Plant Growth Inhibitory Alkaloids in the Extracts of Mesquite Leaves

3''''-Oxojuliflorine (3''''-oxojuliprosopine; **4**), secojuliprosopinal (**13**; Fig. 16.6), a 1:1 mixture of 3'-oxo- (**9**) and 3-oxo-juliprosine (**10**), juliprosine (**8**), and juliflorine (juliprosopine; **3**) were isolated from the methanol extracts of mesquite leaves and investigated for their activity as plant growth inhibitors (Figs. 16.3, 16.4, and 16.6; Nakano et al. 2004a, b). 3''''-Oxojuliflorine (3''''-oxojuliprosopine; **4**) contains two piperidine rings and a hexahydroindolizine ring. Secojuliprosopinal (**13**) contains two piperidine rings and an indolizine ring. 3'-Oxo- (**9**) and 3-oxo-juliprosine (**10**) contain two piperidine rings and a dihydroindolizine ring. These structures were elucidated from spectroscopic analyses. The structure–activity relationships of these alkaloids were assessed by examining the inhibition of root and shoot growth of the monocots barnyard grass (*Echinochloa cru-galli*), rice (*Oryza sativa*), and timothy (*Phleum pratense*), and the dicots amaranth (*Amaranthus viridis*), lettuce, and cress (*Lepidium sativum*). All of these alkaloids generally inhibited growth of seedlings of both monocots and dicots. All alkaloids inhibited root growth more than shoot growth of all plant species used, except juliflorine (juliprosopine; **3**), which had lower activity against root growth of rice. The most active compound appeared to be juliprosine (**8**), followed by a 1:1 mixture of 3'-oxo- (**9**) and 3-oxo-juliprosine (**10**), and juliflorine (juliprosopine; **3**). The activity of juliprosine (**8**), which contains 2-methyl piperidine bearing hydroxyl groups at C-3 and C-3', was higher than that of 3'-oxo- and 3-oxo-2-methylpiperidine. The activities of 3'-oxo- (**9**) and 3-oxo-juliprosine (**10**) and juliprosine (**8**), which all contain a dihydroindolizine ring, were greater than that of juliflorine (juliprosopine; **3**), which contains a hexahydroindolizine ring, whereas the activity of 3''''-oxojuliflorine (3''''-oxojuliprosopine; **4**), which contains a hexahydroindolizine ring, was weaker. Secojuliprosopinal (**13**), which has no indolizine ring,

Table 16.2 Plant growth inhibitory activity of alkaloids from extracts of mesquite leaves on root growth

Test plant	Alkaloid	I_{50} (μM)	
		Root	Shoot
Dicotyledons			
Amaranth	3″-Oxojuliflorine (3″-oxojuliprosopine; **4**)	210	400
	Secojuliprosopinal (**13**)	280	270
	A (1:1) mixture of 3′-oxo- (**9**) and 3-oxo-juliprosine (**10**)	94	350
	Juliprosine (**8**)	29	320
	Julifloiine (juliprosopine; **3**)	140	230
Lettuce	3″-Oxojuliflorine (3″-oxojuliprosopine; **4**)	300	>1,000
	Secojuliprosopinal (**13**)	300	1,000
	A (1:1) mixture of 3′-oxo- (**9**) and 3-oxo-juliprosine (**10**)	120	540
	Juliprosine (**8**)	27	69
	Julifloiine (juliprosopine; **3**)	240	600
Cress	3″-Oxojuliflorine (3″-oxojuliprosopine; **4**)	400	>1,000
	Secojuliprosopinal (**13**)	600	>1,000
	A (1:1) mixture of 3′-oxo- (**9**) and 3-oxo-juliprosine (**10**)	100	240
	Juliprosine (**8**)	80	>1,000
	Julifloiine (juliprosopine; **3**)	430	760
Monocotyledons			
Barnyard grass	3″-Oxojuliflorine (3″-oxojuliprosopine; **4**)	170	370
	Secojuliprosopinal (**13**)	200	590
	A (1:1) mixture of 3′-oxo- (**9**) and 3-oxo-juliprosine (**10**)	82	230
	Juliprosine (8)	17	10
	Julifloiine (juliprosopine; **3**)	140	280
Rice	3″-Oxojuliflorine (3″-oxojuliprosopine; **4**)	290	>1,000
	Secojuliprosopinal (**13**)	420	430
	A (1:1) mixture of 3′-oxo- (**9**) and 3-oxo-juliprosine (**10**)	47	240
	Juliprosine (**8**)	17	130
	Julifloiine (juliprosopine; **3**)	140	30
Timothy	3″-Oxojuliflorine (3″-oxojuliprosopine; **4**)	170	240
	Secojuliprosopinal (**13**)	290	310
	A (1:1) mixture of 3′-oxo- (**9**) and 3-oxo-juliprosine (**10**)	19	190
	Juliprosine (**8**)	12	20
	Julifloiine (juliprosopine; **3**)	120	220

was very weak. Thus, the active sites in these alkaloids are the functional groups at C-3 and C-3′ of the piperidine and indolizine skeletons.

References

Ahmad A, Khan KA, Ahmad VU, Qazi S (1986) Antibacterial activity of juliflorine isolated from *Prosopis juliflora*. Planta Med 4:285–288

Ahmad A, Khursheed AK, Sabiha Q, Viqaruddin A (1989) Antifungal activity of some hydro-soluble *Prosopis juliflora* alkaloids. Fitoterapia 60:86–89

Ahmad M, Aftab K (1995) Hypotensive action of syringin from *Syringa vulgaris*. Phytother Res 9:452–454

Ahmad VU, Mohammad ZG (1979) Studies on the structure of juliflorine. J Chem Soc Pak 1:137–138

Ahmad VU, Qazi S (1983) The absolute configuration of julifloridine. Z Naturforsch 38:660

Ahmad VU, Qazi S (1985) Studies on the structure of juliflorícine. J Chem Soc Pak 7:347–350

Ahmad VU, Sultana A (1990) A new alkaloid from *Prosopis juliflora*. Sci Pham 58: 409–411

Ahmad VU, Basha A, Haque W (1978) New alkaloids from *Prosopis juliflora* DC. Z Naturforsch 33:347–348

Ahmad VU, Sultana A, Qazi S (1989) Alkaloids from the leaves of *Prosopis juliflora*. J Nat Prod 52:497–501

Al-Humaid AI, Warrag MOA (1998) Allelopathic effects of mesquite (*Prosopis juliflora*) foliage on seed germination and seedling growth of bermudagrass (*Cynodon dactylon*). J Arid Environ 38:237–243

Almaraz-Abarca N, Campos MDG, Avila-Reyes JA, Naranjo-Jimenez N, Corral JH, Gonzalez-Valdez LS (2007) Antioxidant activity of polyphenolic extract of monofloral honeybee-collected pollen from mesquite (*Prosopis juliflora*, Leguminosae). J Food Compos Anal 20:119–124

Aqeel A, Khursheed AK, Viqaruddin A, Sabiha Q (1989) Antimicrobial activity of julifloricine isolated from *Prosopis juliflora*. Arzneim-Forsch/Drug Res 39:652–655

Badawi MM, Handa SS, Kinghorn AD, Cordell GA, Farnsworth NR (1983) Plant anticancer agents. XXVII: antileukemic and cytotoxic constituents of *Dirca occidentalis* (Thymelaeaceae). J Pharm Sci 72:1285–1287

Dätwyler P, Ott-Longoni R, Schöpp E, Hesse M (1981) Über Juliprosin, ein weiteres Alkaloid aus *Prosopis juliflora* A. DC. Helv Chim Acta 64:1959–1963

Del Moral R, Muller CH (1969) Fog drip: a mechanism of toxin transport from *Eucalyptus globulus*. Bull Torrey Bot Club 96:467–475

Duh CY, Phoebe CH, Pezzuto JM, Kinghorn AD, Farnsworth NR (1986) Plant anticancer agents. XLII: Cytotoxic constituents from *Wikstroemia elliptica*. J Nat Prod 49:704–705

El-Keblawy A, Al-Rawai A (2007) Impacts of the invasive exotic *Prosopis juliflora* (Sw.) D.C. on the native flora and soils of the UAE. Plant Ecol 190:23–35

Kato-Noguchi H, Kosemura S, Yamamura S, Mizutani J, Hasegawa K (1994a) Allelopathy of oat. I. Assessment of allelopathic potential of extract of oat shoots and identification of an allelochemical. J Chem Ecol 20:315–319

Kato-Noguchi H, Mizutani J, Hasegawa K (1994b) Allelopathy of oat. II. Allelochemical effect of L-tryptophan and its concentration in oat root exudates. J Chem Ecol 20:309–314

Khursheed AK, Arshad HF, Viqaruddin A, Sabiha Q, Sheikh AR, Tahir SH (1986) In vitro studies of antidermatophytic activity of juliflorine and its screening as carcinogen in salmonella/ microsome test system. Arzneim-Forsch/Drug Res 36:17–19

Martin P (1957) Die Abgabe von organischen Verbindungen insbesondere von Scopoletin aus den Keimwurzeln des Hafers. Z Bot 45:475–506

Molisch H (1937) Der Einfluss einer Pflanze auf die andere Allelopathie. Fischer, Jena

Muller CH (1965) Inhibitory terpenes volatilized from *Salvia* shrubs. Bull Torrey Bot Club 92:38–45

Nakano H (2007a) Identification of L-tryptophan as an allelochemical in wheat bran extract. Allelopathy J 19:461–468

Nakano H (2007b) Effects of wheat bran extracts on seedling growth of plants. Allelopathy J 19:487–494

Nakano H, Fujii Y, Suzuki T, Yamada K, Kosemura S, Yamamura S, Suzuki T, Hasegawa K (2001) A growth-inhibitory substance exuded from freeze-dried mesquite (*Prosopis juliflora* (Sw.) DC.) leaves. Plant Growth Regul 33:165–168

Nakano H, Fujii Y, Yamada K, Kosemura S, Yamamura S, Hasegawa K, Suzuki T (2002) Isolation and identification of plant growth inhibitors as candidate(s) for allelopathic substance(s), from aqueous leachate from mesquite (*Prosopis juliflora* (Sw.) DC.) leaves. Plant Growth Regul 37:113–117

Nakano H, Nakajima E, Fujii Y, Yamada K, Shigemori H, Hasegawa K (2003) Leaching of the allelopathic substance, L-tryptophan from the foliage of mesquite (*Prosopis juliflora* (Sw.) DC.) plants by water spraying. Plant Growth Regul 40:49–52

Nakano H, Nakajima E, Hiradate S, Fujii Y, Yamada K, Shigemori H, Hasegawa K (2004a) Growth inhibitory alkaloids from mesquite (*Prosopis juliflora* (Sw.) DC.) leaves. Phytochemistry 65:587–591

Nakano H, Nakajima E, Fujii Y, Shigemori H, Hasegawa K (2004b) Structure–activity relationships of alkaloids from mesquite (*Prosopis juliflora* (Sw.) DC.). Plant Growth Regul 44:207–210

Nakano H, Morita S, Shigemori H, Hasegawa K (2006) Plant growth inhibitory compounds from aqueous leachate of wheat straw. Plant Growth Regul 48:215–219

Nakayama R, Kikuzaki H, Nakatani N, Horiuchi H (1996) Antioxidative action of constituents from fennel seeds. J Home Econ Jpn 47:1193–1199

Nasir E, Ali SI (1972) Flora of West Pakistan. Fakhri, Karachi

Ott-Longoni R, Viswanathan N, Hesse M (1980) Die Konstitution des Alkaloides Juliprosopin aus *Prosopis juliflora* A. DC. Helv Chim Acta 63:2119–2129

Pandit BR, Mahesh KR, Kotiwar OS (1995) Effect of *Prosopis juliflora* (Sw) DC. extracts on root and shoot growth of bajra seedlings. Geobios 22:145–148

Rice EL (1984) Allelopathy, 2nd edn. Academic, London

Sankhla N, Baxi MD, Chatterji UN (1965) Eco-physiological studies on arid zone plants. I. Phytotoxic effects of aqueous extract of mesquite, *Prosopis juliflora* DC. Curr Sci 21:612–614

Tabosa IM, Souza JCD, Graca DL, Barbosa-Filho JM, Almeida RN, Riet-Correa F (2000) Neuronal vacuolation of the trigeminal nuclei in goats caused by ingestion of *Prosopis juliflora* Pods (Mesquite Beans). Vet Human Toxicol 42:155–158

Tang CS, Waiss AC (1978) Short-chain fatty acids as growth inhibitors in decomposing wheat straw. J Chem Ecol 4:225–232

Tapia A, Feresin GE, Bustos D, Astudillo L, Theoduloz C, Schmeda-Hirschmann G (2000) Biologically active alkaloids and a free radical scavenger from *Prosopis* species. J Ethnopharmacol 71:241–246

Toyooka N (2001) Synthesis and its application the synthesis of biologically active natural products of new and versatile chiral building blocks. Yakugaku Zasshi 121:467–479

Part D
Biotechnological Studies

Chapter 17
Genetic Variation in the Tunisian Date Palm (*Phoenix dactylifera* L.)

Soumaya Rhouma, Salwa Zehdi-Azouzi, Sonia Dakhlaoui-Dkhil, Ali Ould Mohamed Salem, Ahmed Othmani, Emira Cherif, Mohamed Marrakchi, and Mokhtar Trifi

Abstract The date palm – one of the oldest domesticated fruit crops – is the tree most adapted to growing in desert areas. It has always been looked upon as a key source of stability, survival and evolution of the oasis agro-system as it constitutes the basic features of the ecological pyramid in desert regions. Tunisian date palm germplasm is characterised by high genetic diversity, with more than 250 varieties identified. However, this patrimony is seriously menaced by severe genetic erosion due to different biotic and abiotic factors. In Tunisia, as well as in North African countries, dates are cultivated for fruit production, and all parts of the tree are used for many other artisanal and/or industrial purposes. Recent efforts have focussed on the development of phenotypic, biochemical and DNA-based markers useful in characterising the genetic diversity of date palm populations and to establish the relationships between different cultivars. This chapter reviews current efforts made towards developing such selection markers for Tunisian date palm cultivars for use in breeding programmes.

17.1 Introduction

17.1.1 History

The date palm is recorded in ancient history extending over an area from the Indus Valley (now Pakistan) to Mesopotamia (now Iraq), the Nile Valley, Southern Persia, the Eastern Mediterranean and the Horn of Africa. References to the date

S. Rhouma, S. Zehdi-Azouzi, S. Dakhlaoui-Dkhil, A. Ould Mohamed Salem, A. Othmani, E. Cherif, M. Marrakchi, and M. Trifi (✉)
Laboratory of Molecular Genetics, Immunology and Biotechnology, Tunis Faculty of Sciences Campus University, 2092 El Manar Tunis, Tunisia
e-mails: mokhtar.t@fst.rnu.tn; trifimokhtar@yahoo.fr

K.G. Ramawat (ed.), *Desert Plants*,
DOI 10.1007/978-3-642-02550-1_17, © Springer-Verlag Berlin Heidelberg 2010

in the Nile Valley and Tigris/Euphrates valleys suggest that it has been under cultivation for at least the last 5,000 years. Such a wide distribution implies that *Phoenix dactylifera* ($2n = 36$) is well adapted to quite extensive geographic, soil and climatic conditions. Another species, *Phoenix sylvestris* (sugar date palm or toddy palm) still occurs in the wild throughout northern India; its sap is used to produce a crude sugar.

Phoenix dactilyfera most likely grew wild as a natural hybrid of *P. sylvestris* in the Indus Valley, where it was appreciated as a wild fruit and probably cultivated as early as the sixth millennium BC; there have been finds of date palm seeds in association with human settlement from 5000 BC onwards. The oldest radiocarbon dated discovery of date seeds was on Dalma island, part of the Abu Dhabi Island group. Two seeds were found in 1998, the oldest was 5110 BC and the other 4670 BC. As there was no evidence of cultivation of date palms in the region at that time, it is probable that these seeds came from traders (Zohary and Spiegel-Roy 1975; Wrigley 1995).

17.1.2 Botanical Profile

Belonging to the Angiosperms-Monocotyledones, Palmaceae is a family of about 200 genera and 1,500 species (Dowson 1982). The genus *Phoenix* (Coryphoideae Phoeniceae) contains a dozen species, all native to the tropical or subtropical regions of Africa or Southern Asia, including *Phoenix dactylifera* L. (Munier 1973). According to Dransfield and Uhl (1986), the date palm is classified as follows:

- Group: Spadiciflora
- Order: Palmea
- Family: Palmaceae
- Sub-family: Coryphyoideae
- Tribe: Phoeniceae
- Genus: *Phoenix*
- Species: *dacttylifera* L.

Twelve species of the genus *Phoenix*, along with their geographical distribution, were first listed (Table 17.1) by Chevalier (1952). Besides date palm, five of these species bear edible fruit (*P. atlantica* Chev., *P. reclinata* Jacq., *P. farinifera* Roxb., *P. humilis* Royle., and *P. acaulis* Roxb.).

The species *Phoenix dactilyfera* has about 19 known relatives, including *Phoenix canariensis* (Canary Island palm), *P. reclinata* (Senegal date palm) and *P. sylvestris* (Indian sugar date palm). All are members of the plant family, Arecaceae. The scientific name was derived from 'Phoenix', the legendary bird of ancient Greece. The Phoenicians dyed cloth a purple colour using dye from the Murex shellfish; this colour was also called Phoenix, possibly because it had such great appeal and value. The same colour was noted on the fruit of the date, hence the

Table 17.1 Distribution of different species of the genus *Phoenix*

Species	Common name	Distribution
Phoenix dactylifera L.	Date palm	Mediterranean countries, Africa and part of Asia; introduced in North America and Australia
P. atlantica A. Chev.		Occidental Africa and Canary Islands
P. canariensis chabeaud.	Canary Palm	Canary Islands and Cape Verde
P. reclinata Jacq.	Dwarf Palm	Tropical Africa (Senegal and Uganda) and Yemen (Asia)
P. sylvestris Roxb.	Wild date palm or sugar palm	India and Pakistan
P. hanceana Naudin.		Meridional China and Thailand
P. robelinic O'Brein.		Sri Lanka, Toukin, Annam, Laos and Thailand
P. farinifera Roxb.	Pigmy palm	India, Ceylon and Annam
P. rupicola T. Anders.	Rocky date palm	India
P. acaulis Roxb.	Dwarf palm	Bangladesh and India
P. paludosa Roxb.	Hental or Juliana palm	Bangladesh, Tenasherim, Andaman, Nikobaren and Thailand

date palm genus became *Phoenix*. The specific name *dactylifera* came from the shape of the fruit, 'dactylos' being the ancient Greek word for 'finger'. Date palms are dioecious; i.e. the male and female parts are on separate plants. The date palm is the tallest of the *Phoenix* species, growing to 30 m in some places. The trunk, in cultivation, is surrounded from the ground upward in a spiral pattern of leaf bases. The leaves are large, 4–5 m, alternate, sheathing in dense terminal rosettes, and pinnately lobed. The ends of leaf fronds are needle sharp to help protect the growth tip from grazing animals.

The fruit is also the largest of the species, with a few varieties reaching up to 100 mm x 40 mm in size. *P. dactylifera* is now found in tropical and sub-tropical regions all over the world as well as in temperate and arid regions of the United States, Australia, southern Spain and the Mediterranean coast of Africa and West Asia. The fruit is a 'drupe', with a single seed in each date. Fruit is borne on clusters often weighing 10 kg or more. A fully productive palm can support up to ten clusters carrying as much as 100 kg fruit. From the time of pollination, the fruit takes 200 days to reach the fully ripened stage.

17.1.3 Production Levels and Locations

The FAO estimate that worldwide production of dates peaked in 1996 at 4,492,000 tons. The world largest producer is Iran at 765,000 Mt followed by Egypt (680,000 Mt), Saudi Arabia (597,000 Mt), Iraq (550,000 Mt), Pakistan (533,000 Mt), Algeria (361,000 Mt) then the United Arab Emirates (UAE) at 240,000 Mt. Other significant producers are Libya, Morocco, Sudan, Tunisia, China, Oman, Yemen, Qatar,

Bahrain, the United States and Jordan. In UAE, the Abu Dhabi Emirate is by far the largest producer of dates. There is said to be approximately 16 million date palms in the Emirate, with approximately 4 million in the remaining six Emirates. Turkey and Spain produce small quantities; Spain is of particular interest because, after its introduction by the Moors, the Spanish took the date subsequently to Mexico from where it migrated to California. Major commercial plantings in California were however, imported some time later direct from Saudi Arabia (Hodel and Johnson 2007).

17.1.4 Propagation

While dates grow readily from seed, the quality of the resultant plants and reliability of the crop is too 'hit and miss'; therefore, the most common method of reproduction is the planting of suckers. There have been further refinements in propagation methods including, over the last 20 years, production of 'tissue culture' dates in laboratories in various parts of the world.

17.1.5 Date Varieties

There are more than 600 varieties, including cultivars, grown worldwide, and different countries have their favourites. In UAE the preferred fruit is Khalas but others such as Zaghloul, Khuneizi, Hilali, Howaiz, Naghal and Jaberi Fardh have their followings. These dates have different colours, flavours, sweetness, acidity and textures. A popular imported variety, mainly from Morocco, is Majool, which is a very large fruit. All major date-producing countries have their own cultivars and favoured varieties, such as Amir Hajj and Ashrashi from Iraq; Saidy and Hayany from Egypt, Deglet Nour and Thoory from Algeria, and Ruzeiz, Bukeira, Nebut, Seif and Barhi from Oman.

17.2 Date Palms in Tunisia

In Tunisia (Fig. 17.1), as well as in North African countries and several other tropical countries located in the Middle East and the Arabic peninsula, oasis cultures consist of date-palm groves and constitute one of the main factors determining social, environmental and economic stability in these areas. Date palm groves also most important factor in the establishment of favourable conditions in the oasian agro-system. On a commercial scale, Tunisia is one of the main date palm producing countries in the world. In addition, date palms constitute the principal financial resources and food sources of oasis cultivators, and contribute

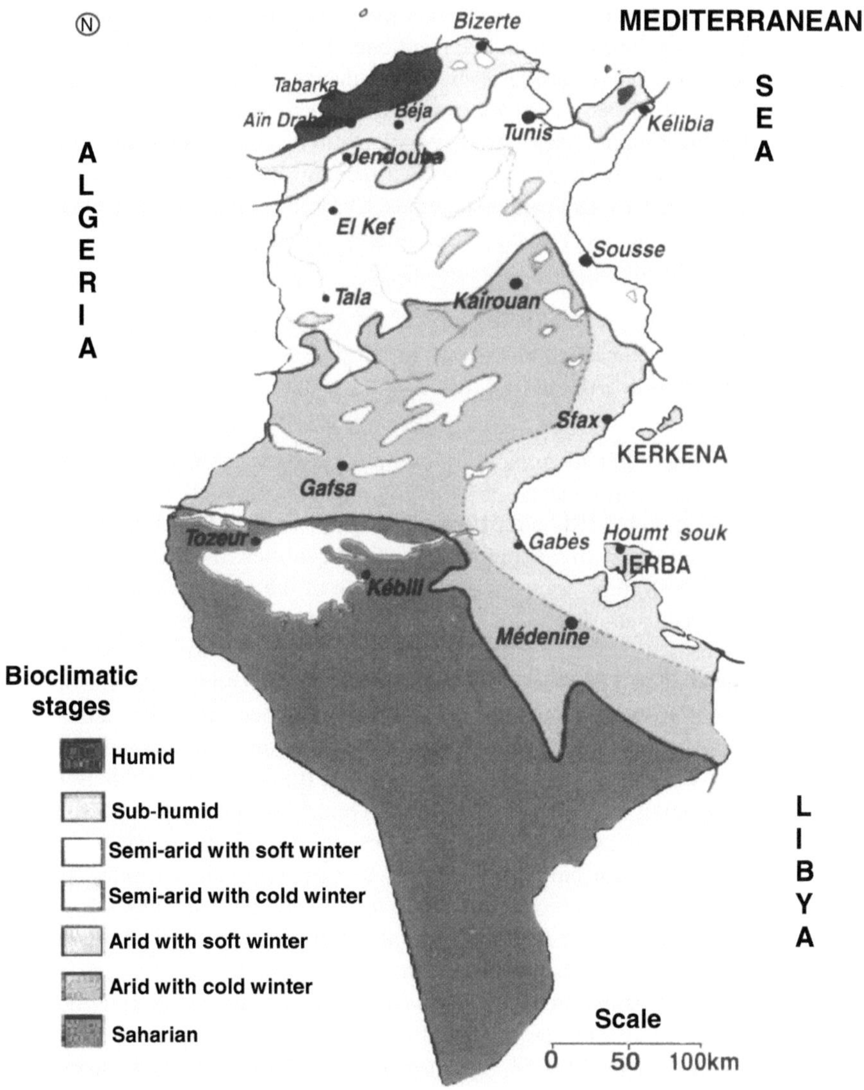

Fig. 17.1 Bioclimatic map of Tunisia showing the main Tunisian date palm oases (from Zehdi-Azouzi 2005)

to the development of adjacent culture of fruit trees, saffron, forage plants and vegetables. Indeed, more than 10% of Tunisians depend on date palm culture for their livelihood (Haddouch 1996).

Well-adapted ecotypes empirically selected by farmers for their attractive fruit qualities are cultivated. As a consequence, the local germplasm is composed of a variable scale of selected ecotypes of locally named cultivars, and exhibits a great diversity. For instance, more than 220 and 800 cultivars have been reported in

Morocco and in Algeria, respectively (Benkhalifa 1996), and 250 have been identified in Tunisia (Rhouma 2005). It should be mentioned here that the date palm is propagated clonally via offshoots produced by the mother tree. However, this method of propagation is relatively slow since the mother tree produces a very limited number of offshoots – too low a number to establish new plantations. To overcome this difficulty, plantlets are generated on a large-scale via in vitro culture methods, which has allowed the expansion of modern date palm plantations. Characterisation of genetic variation in this crop has become imperative to check fidelity of date palm trees produced from offshoots and/or the in vitro-derived plantlets.

17.3 Genetic Variation in Date Palm

In Tunisia, this important subtropical fruit crop is currently in danger due to severe genetic erosion as a consequence of the predominance of the elite cultivar Deglet Nour in modern cultures (Rhouma 1994). This tendency has led to the disappearance of many cultivars with medium and low fruit qualities. It is therefore imperative to elaborate a strategy aimed at the evaluation of the genetic diversity and the preservation of Tunisian date palm germplasm. Many studies have addressed this issue, and describe the use of either morphological traits or isozyme markers to identify Tunisian date palm varieties (Reynes et al. 1994; Rhouma 1994; Bouabidi et al. 1996; Ould Mohamed Salem et al. 2001). Among these methods, those based on morphological traits are of some benefit in the evaluation of date palm genetic resources (Mohamed et al. 1983; Reynes et al. 1994; Bouabidi et al. 1996; Belguedj 2002; El Houmaizi et al. 2006; Rhouma 2005; Ould Mohamed Salem et al. 2007). Therefore it has been assumed that criteria related to either vegetative or fruit parameters are useful for cultivar characterisation, phenotypic diversity analysis and the exploration phylogenic relationships among date-palm ecotypes.

Evaluation of phenotypic diversity is a logical first step in the elaboration of a program to improve germplasm management and utilisation of any crop. However, most morphological traits are highly influenced by environmental conditions or vary with the developmental stage of the plant, and isozymes are limiting due to low levels of polymorphism (Fig. 17.2). Consequently, DNA-based techniques have been developed, and have proved effective in assessing genetic diversity because they access an almost unlimited source of potential markers to uncover differences at the molecular level.

Microsatellites or simple sequence repeats (SSRs) consist of variable numbers of tandemly repeated units, each of 1–6 bp, and represent a class of repetitive DNA commonly found in eukaryotic genomes (Tautz and Renz 1984). They are characterised by their great abundance (Roder et al. 1995), high variability, and extensive distribution throughout different genomes (Roder et al. 1998). Microsatellites are typically multi-allelic loci since more than five alleles per locus are commonly observed in plant populations (Senior and Heun 1998). In addition, automated polymerase chain reaction (PCR)-based techniques, which enable high

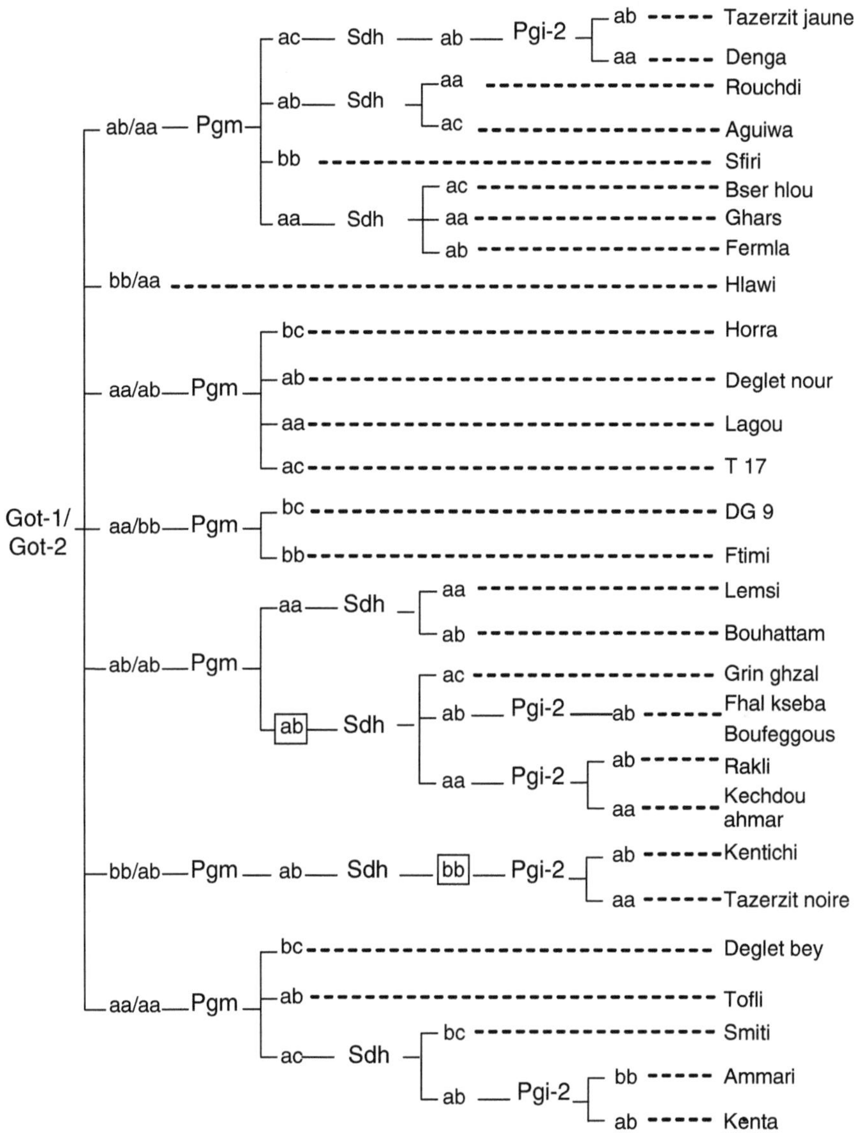

Fig. 17.2 Diagram illustrating the discrimination of 29 Tunisian date-palm ecotypes based on isozyme markers (from Ould Mohamed Salem 2001)

throughput data collection and good analytical resolution at a low cost, have been developed for microsatellites (Kresovich et al. 1995). Data based on molecular markers such as restriction fragment length polymorphisms (RFLPs), rapid amplification of polymorphic DNA (RAPDs), inter simple sequence repeats (ISSRs), amplified fragment length polymorphisms (AFLPs), random amplified microsatellite polymorphisms (RAMPOs) and SSRs have been used to characterise

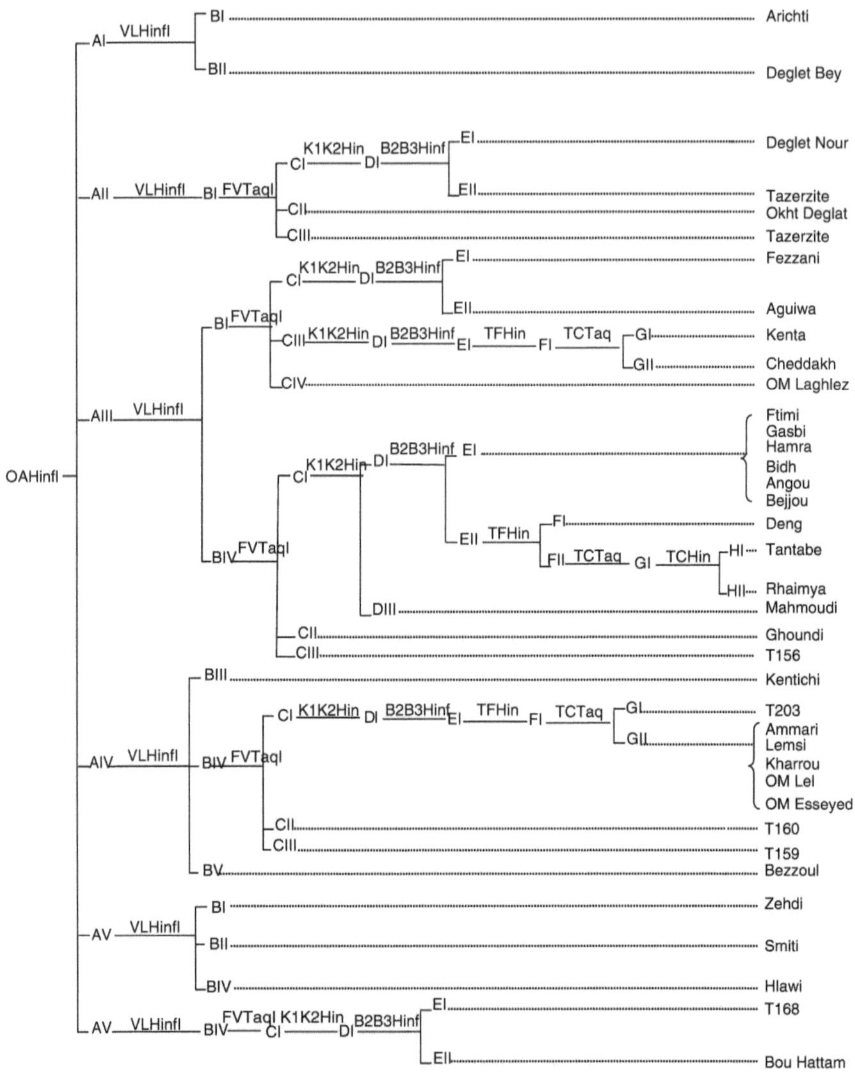

Fig. 17.3 Diagram illustrating the discrimination of 38 Tunisian date-palm ecotypes based on cleaved amplified polymorphic sequences (CAPS) chloroplast (ct)DNA (from Sakka et al. 2004)

(Figs. 17.2–17.4) date palm genotypes (Cornicquel and Mercier 1994, 1997; Sedra et al. 1998; Ben Abdallah et al. 2000; Trifi et al. 2000; Trifi 2001; Zehdi et al. 2002; Sakka et al. 2004; Al Khalifah and Askari 2003; El Assar et al. 2005; Hussein et al. 2005; Rhouma et al. 2008). It is evident from these data that microsatellite DNA (Fig. 17.4), ISSRs, RAMPO and cleaved amplified polymorphic sequences (CAPS; Fig. 17.3) provided clear polymorphism as compared to isoenzyme patterns (Fig. 17.2). A large number of SSR alleles have been revealed, with a mean 7.14

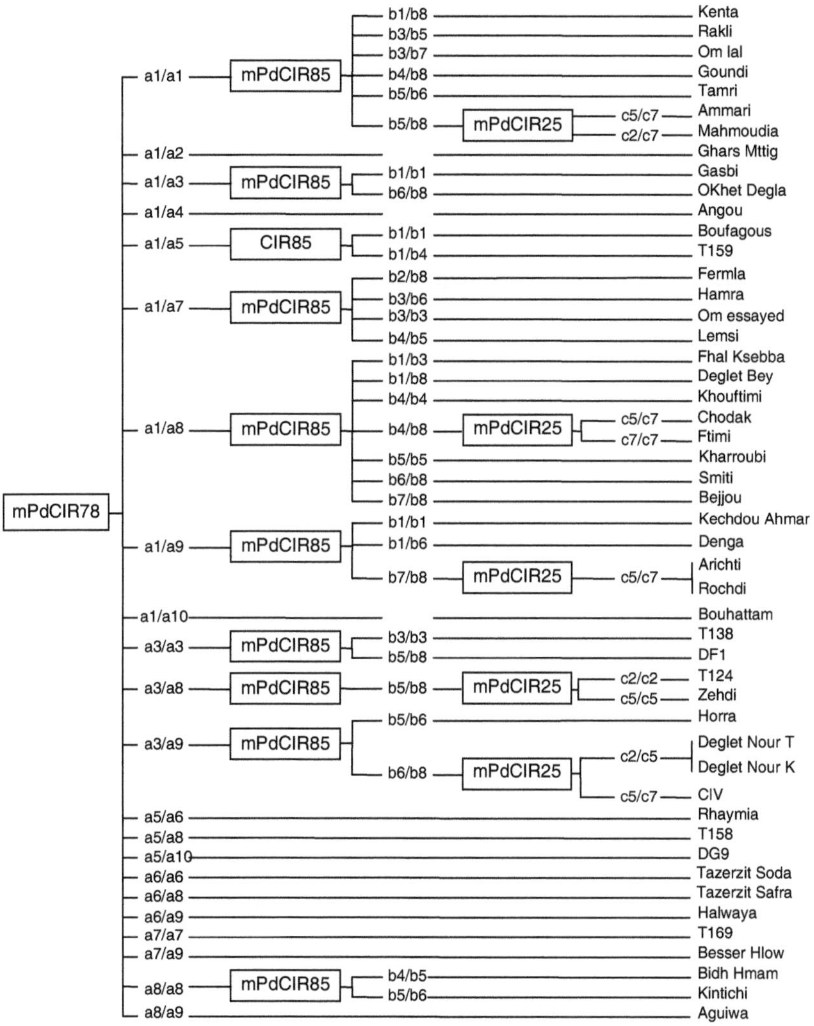

Fig. 17.4 Diagram illustrating the discrimination of 49 Tunisian date-palm ecotypes based on three microsatellite loci. Alleles significance (size in base pairs): *a1* 138, *a2* 142, *a3* 144, *a4* 148, *a5* 153, *a6* 157, *a7* 159, *a8* 165, *a9* 171, *a10* 173, *b1* 175, *b2* 181, *b3* 183, *b4* 185, *b5* 187, *b6* 189, *b7* 195, *b8* 197, *c1* 219, *c2* 231, *c3* 233, *c4* 236, *c5* 246, *c6* 248, *c7* 250 (from Zehdi et al. 2004)

per locus, which has permitted detection of a high degree of genetic variability in date palm. The scored values of diversity are higher at the intra-group level than at the inter-group level (Zehdi et al. 2004). In comparison to vegetative and fruiting characters for distinguishing cultivars, SSR alleles were used successfully to discriminate molecularly between a nearly unlimited number of date palms cultivars. Compared to studies reported in date palms, the scored percentage of resolution is higher than that observed using isoenzymes (Booij et al. 1995;

Ould Mohamed Salem et al. 2001) and plastid DNA haplotypes (Sakka 2003). These studies have permitted the identification of markers suitable for identification of date palm varieties. However, the search for other markers continues in order to obtain a deeper comprehension of the genetic organisation in Tunisian date palm germplasm.

17.4 Molecular Diversity and Development of an Identification Key

The survey of genetic variation at the DNA level and establishment of precise fingerprints has become an important task in plant breeding programs and in germplasm management. For this purpose, studies reported on Tunisian date palms describe the use of isozymes (Ould Mohamed Salem 2001; Bennaceur et al, 1991; Booij et al. 1995), plastid DNA (Sakka et al. 2004) and nuclear DNA markers (Zehdi et al. 2004). In fact, all these latter authors, working independently, have clustered cultivars from different oases based on geographical origin and/or the sex of the trees. They concluded that these plants had a common genetic basis at the DNA level in spite of their phenotypic divergence. Importantly, this is in agreement with the unique domestication origin of this crop. When other molecular markers such as SSRs were used, the data showed large genetic variation in different defined groups. This is well exemplified by SSRs in the Tozeur oasis, which showed a significant deficiency in heterozygocity. However, the remaining two groups (Gabès and Kebili) showed no deviation from Hardy-Weinberg equilibrium (HWE). This result can be explained by the stronger selection pressure operating in the Tozeur oasis compared to in the Kebili and Gabès oases (Zehdi et al. 2004). In fact, scored values of diversity are higher at the intra-group level than at the inter-group level. Similar results have been reported in Moroccan, Algerian and Tunisian datepalm cultivars using isozyme markers (Torres and Tisserat 1980; Ould Mohamed Salem et al. 2001). These results are also comparable to those reported in other long-lived cultivated species such as olive (Ouazzani et al. 1995) and fig (Salhi-Hannachi et al. 2005). Taken together, our present data and the available prior isozyme information (Ould Mohamed Salem et al. 2001) suggest that genetic diversity in Tunisian date palms is high. This could be attributed to the dioecious nature of this crop. Date palms in the Tozeur oasis and male groups showed a significant deficit of heterozygocity. Using RAPD and ISSR markers, similar results have been reported in date palms (Sedra et al. 1998; Trifi et al. 2000; Zehdi et al. 2002). These authors have suggested a common genetic basis among date palm genotypes in spite of the distinctiveness of their morphometric parameters, particularly those related to fruit traits. Hence, our data suggest the existence of one ancestral date palm population, and are in agreement with the unique Mesopotamian domestication origin of this crop (Wrigley 1995).

17.5 Biotechnology in Tunisian Date Palms

As noted above, the date palm is propagated clonally via offshoots produced by the mother plant. This conventional propagation mode has several drawbacks: it is time consuming and produces only a low number of shoots. For instance, several genotypes do not produce offshoots and rooting percentage is low in such shoots. Moreover, a period of 5–7 years is required to verify the fidelity of the sucker-derived plants (Nixon and Carpenter 1978). To overcome these drawbacks, in vitro methods have been developed to provide an alternative strategy aimed at the mass propagation of date palm plants (El Hadrami 1995; Sharma et al. 1986). The in vitro propagation of endangered Tunisian date palm elite and/or fruity cultivars (i.e. Deglet nour, Deglet bey, Boufaggous, Gondi) through organogenesis and somatic embryogenesis has been successfully achieved (Drira and Benbadis 1985; Othmani et al. 2009a, b; Fig. 17.5). Assessment of certification of the plant tissue culture-derived

Fig.17.5a–g Induction of somatic embryogenesis and plant regeneration from leaf explants of date palm cv. Deglet Nour. **a** Embryogenic callus within proembryogenic globular structures obtained after 6-month culture period on Murashige-Skoog (MS) medium including 1 mg l^{-1} 2,4-D (M_2). **b** Direct embryogenesis at the basic part of a juvenile leaf cultured on MS medium including 1 mg l^{-1} 2,4-D for 6 months of culture. **c** Initiation of differentiation of embryogenic callus 1 month after transfer to MS medium supplemented with 0.1 mg l^{-1} 2,4-D. **d** Matured somatic embryos obtained after 10 weeks of transfer of differentiated embryogenic callus on MS medium lacking 2,4-D. **e** Hardened plantlets with full radicle and shoot obtained after 3 months of transfer to half-strength MS liquid medium supplemented with 1 mg l^{-1} indole-3-butyric acid (IBA). **f** Potted plantlets 3 months after transfer to a greenhouse. **g** Two-year-old plants after transfer to free-living conditions. *Bars* **a**, **c** 10 mm; **b** 5 mm; **d** 20 mm; **e** 15 mm; **f** 100 mm; **g** 300 mm (from Othmani et al. 2009a)

progeny has, however, never been reported. Among the available molecular biology methods, we investigated RAPD (Williams et al. 1990) and AFLP (Vos et al. 1995) methods, and found them to be reliable, fast and inexpensive procedures with which to identify clones and to assess somaclonal variation in this crop. The RAPD and AFLP banding profiles (op cit.) of the derived progenies were very similar to those of the mother plant (Rhouma 2008), suggesting that no variation had cropped in. In fact, the number as well the sizes of the generated bands are similar in these profiles. This result strongly supports the true-to-type nature of the in vitro-derived date palm plantlets. Moreover, 2,4-D did not induce somaclonal variation in this crop. Similar results have been reported in plantlets regenerated via embryogenic suspension cultures, in comparison to the mother tree of the date palm cv. Deglet nour (Fki et al. 2003). In fact, using flow cytophotometry analysis, these authors examined the ploidy level in these plantlets and revealed the levels to be identical in the mother tree and its in vitro progeny. Moreover, among 100 microsatellite alleles, a difference of only one allele size was registered in only one plantlet over 150 samples studied (Zehdi-Azouzi 2005). Similar results from somatic embryogenesis have also been reported in date palm (Cohen et al. 2004; Sharma et al. 1986) and in other crops such as Norway spruce (Heinze and Schmidt 1995), conifers (Tautorus et al. 1991), *Hevea brasiliensis* (Michaux-Ferrière et al. 1992), and cereals (Vasil 1995).

17.6 Conclusions

Date palm – one of the oldest domesticated fruit crops – is the tree most adapted to growing in desert areas. It has always been looked upon as a key source of stability, survival and evolution in oasis agro-systems as it constitutes the basic features of the ecological pyramid in desert regions. Ecotypes are well suited to various usages, as locally known cultivars. All over the world, date palm germplasms are characterised by the presence of a large number of cultivars (Djerbi 1985; Rizvi and Davis 1983; Ben Khalifa 1996; Sedra 1996). In early periods, germplasm was characterised using classical morphometric and vegetative criteria. Despite their usefulness of the latter methods in the establishment of phenotypic divergence in the cultivars studied, in order to set up a catalogue of the most important date palm cultivars, both in North African and other producing countries, analytic fruit parameters and isozyme markers were used. However, taking advantage of the large panel of DNA-based markers developed in the last two decades, investigations have focussed on identifying DNA markers suitable for fingerprinting of date palm cultivars and/or for varietal identification, as well as for the survey of the genetic organisation of this crop. Data have clearly shown that among the procedures currently available, PCR-based methods are the best suited to surveying genetic variation and to precisely characterise cultivars in this crop. Moreover, among these markers, microsatellites (SSRs) are ideal for the unambiguous differentiation between cultivars via multilocus genotyping.

In addition, taking into account the mode of propagation, preservation of date palm biodiversity has been made possible through in vitro multiplication methods by providing attractive strategies for mass propagation of fruity and/or endangered cultivars. As a result, the in vitro mass propagation of several Tunisian date palm elite and/or fruity endangered cultivars has been successfully achieved (e.g. Deglet nour, Deglet bey, Boufaggous, Gondi). Furthermore, PCR-based methods have demonstrated that somaclonal variation did not occur even after cultivation of date palm cells on medium containing 2,4-D, which is known to induce such variations by chromosome aberrations.

It is obviously necessary to extend the use of in vitro methods to many other local and/or introduced cultivars in order to enhance date palm cultivation and to improve biodiversity in local germplasms. In addition, based on SSR evidence, it will be easily possible to fingerprint any cultivar from anywhere in the world.

Acknowledgements This work was partially supported by grants from the IPGRI "CRRAO Degache, and the Tunisian 'Ministère de l'Enseignement Supérieur, de la Recherche Scientifique et de la technologie' "project Lab B02" This paper is dedicated to Professor Mohamed Marrakchi, Director of the Lab B02 and promoter of this work, who succumbed in April 2008 during the investigation and evaluation process of his laboratory research activities.

References

Al Khalifah NS, Askari E (2003) Molecular phylogeny of date palm (*Phoenix dactylifera* L.) cultivars from Saudi Arabia. Theor Appl Genet 107:1266–1270

Belguedj M (2002) Les ressources génétiques du palmier dattier, caractéristiques des cultivars de dattiers dans les palmeraies du Sud-Est Algérien. Institut National de la Recherche Agronomique d'Algérie, pp 289

Ben Abdallah A, Stiti K, Lepoivre P, Du Jardin P (2000) Identification de cultivars de palmier dattier (*Phoenix dactylifera* L.) par l'amplification aléatoire d'ADN (RAPD). Cah Agric 9:103–10

Ben Khalifa A (1996) Diversity of date palm (*Phoenix dactylifera* L.). Options Méditerranéennes 28:160

Bennaceur M, Lanaud C, Chevalier MH, Bounaga N (1991) Genetic diversity of the date palm (*Phoenix dactylifera* L.) from Algeria revealed by enzyme markers. Plant Breed 107:56–69

Booij L, Montfort S, Ferry M (1995) Characterization of thirteen date palm (*Phoenix dactylifera* L.) cultivars by enzyme electrophoresis using the Phast-System. J Plant Physiol 145:62–66

Bouabidi H, Reynes M, Rouissi MB (1996) Critères de caractérisation des fruits de quelques cultivars de palmier dattier (*Phoenix dactylifera* L.) du sud tunisien. Ann Inst Rech Agric Tunisie 69:73–87

Chevalier A (1952) Recherches sur les Phoenix africains. Rev Int Bot Appl Agric Trop 1952: 355–356

Cohen R, Korchinsky R, Tripler E (2004) Flower abnormalities cause abnormal fruit setting in tissue culture propagated date palm (*Phoenix dactylifera* L.). J Hortic Sci Biotechnol 79:1007–1013

Cornicquel B, Mercier L (1994) Date palm (*Phoenix dactylifera* L.) cultivar identification by RFLP and RAPD. J Plant Sci 101:163–172

Cornicquel B, Mercier L (1997) Identification of date palm (*Phoenix dactylifera* L.) cultivars by RFLP: partial characterization of a cDNA probe that contains a sequence encoding a zinc finger motif. Int J Plant Sci 158:152–156

Djerbi M (1985) Précis de Phoéniciculture. FAO, Rome, Italy
Dowson VHW (1982) Date production and protection with special reference to North Africa and the Near East. FAO Technical Bulletin 35:294
Dransfield J, Uhl NW (1986) An outline of a classification of palms. Principes 30:3–11
Drira N, Benbadis A (1985) Multiplication végétative du palmier dattier (*Phoenix dactylifera* L.) par réversion, en culture in vitro d'ébauches florales de pieds femelles adultes. J Plant Physiol 119:227–235
El Assar A, Krueger RR, Devanand PS, Chao CCT (2005) Genetic analysis of Egyptian date (*Phoenix dactylifera* L.) accessions using AFLP markers. Genet Resour Crop Evol 52:601–607
El Hadrami I (1995) L'embryogenèse somatique chez *Phoenix dactylifera* L.: quelques facteurs limitants et marqueurs biochimiques. PhD Thesis, Université Cadi Ayyad, Marakech, Morocco
El Houmaizi MA, Devanand PS, Fang J, Chao CT (2006) Confirmation of 'Medjool' date as a landrace variety through genetic analysis of 'Medjool' accessions in Morocco. J Am Soc Hortic Sci 131:403–407
Fki L, Masmoudi R, Drira N, Rival A (2003) An optimised protocol for plant regeneration from embryogenic suspension cultures of date palm, *Phoenix dactylifera* L., cv. Deglet Nour. Plant Cell Rep 21:517–524
Haddouch M (1996) Situation actuelle et perspectives de développement du palmier dattier au Maroc. Cah Options Méditerranéennes 28:63–79
Heinze B, Schmidt J (1995) Monitoring genetic fidelity vs. somaclonal variation in Norway spruce (*Picea abies*) somatic embryogenesis by RAPD analysis. Euphytica 85:341–345
Hodel DR, Johnson DV (2007) Imported and American varieties of dates in the United States. University of California. Agriculture and Natural Resources, pp 112
Hussein A, Adawy S, Ismail EME, El-Itriby A (2005) Molecular characterization of some Egyptian date palm germplasm using RAPD and ISSR markers. Arab J Biotechnol 8:83–98
Kresovich S, Szewc-McFadden AK, Bliek SM, McFerson JR (1995) Abundance and characterization of simple sequence repeats (SSRs) isolated from a size-fractionated genomic library of *Brassica napus* L. (rapeseed). Theor Appl Genet 91:206–211
Michaux-Ferrière N, Grout H, Carron MP (1992) Origin and ontogenesis of somatic embryos in *Hevea brasiliensis* (Euphorbiaceae). Am J Bot 79:174–180
Mohamed S, Shabana HR, Mawlod BA (1983) evaluation and identification of Iraqi date: fruit characteristics of fifty cultivars. Date Palm J 2:27–56
Munier P (1973) Le palmier dattier. Maisonneuve et Larose, Paris
Nixon RW, Carpenter B (1978) Growing date in the United States. US Dep Bull 207, pp 63
Othmani A, Bayoudh C, Drira N, Marrakchi M, Trifi M (2009a) Somatic embryogenesis and plant regeneration in date palm *Phœnix dactylifera* L., cv. Boufeggous is significantly improved by fine chopping and partial desiccation of embryogenic callus. Plant Cell Tissue Organ Cult 97:71–79
Othmani A, Bayoudh C, Drira N, Marrakchi M, Trifi M (2009b) Regeneration and molecular analysis of date palm (*Phœnix dactylifera* L.) plantlets using RAPD markers. Afr J Biotechnol 8:813–820
Ouazzani N, Lumaret R, Villemar P (1995) Apport du polymorphisme alloenzymatique à l'identification variétale de l'olivier (*Oleae europeae* L.). Agronomie 15:31–37
Ould Mohamed Salem A (2001) Contribution à l'évaluation des resources phytogénétiques du palmier dattier (*Phoenix dactylifera* L.) par analyse du polymorphisme isoenzymatique. PhD Thesis, Faculté des Sciences de Tunis, Université Tunis-El Manar
Ould Mohamed Salem A, Trifi M, Rhouma A, Marrakchi M (2001) Genetic inheritance of four enzymes in date-palm (*Phoenix dactylifera* L.). Genet Resour Crop Evol 48:361–368
Ould Mohamed Salem A, Rhouma S, Zehdi S, Marrakchi M, Trifi M (2007) Morphological variability of Mauritanian date-palm (*Phoenix dactylifera* L.) cultivars as revealed by vegetative traits. Acta Bot Croatica 67:81–90
Reynes M, Bouabidi H, Piombo G, Risterrucci AM (1994) Charactérisation des principales variétés des dattes cultivées dans la région de Tozeur. Fruits 49:289–298

Rhouma A (1994) Le palmier dattier en Tunisie. Le patrimoine génétique Vol I. Arabesques Editions et Créations, Tunis, Tunisie

Rhouma A (2005) Le palmier dattier en Tunisie. Le patrimoine génétique Vol II. IPGRI, Rome, Italy

Rhouma S (2008) Analyse de la diversité génétique chez le palmier dattier (*Phoenix dactylifera* L.) étude transcriptomique de la maladie des feuilles cassantes. PhD Thesis, Faculté des Sciences de Tunis, Université Tunis-El Manar

Rhouma S, Zehdi-Azouzi S, Ould Mohamed Salem A, Rhouma A, Marrakchi M, Trifi M (2008) Genetic diversity analysis in Tunisian date palm (*Phoenix dactylifera* L.) accessions by means of Random Amplified Microsatellite Polymorphisms (RAMPO). Sci Hortic 117:53–57

Rizvi M, Davis J (1983) Structural features of date market in Sind Pakistan. Date Palm J 2:103–122

Roder MS, Plaschke J, Konig SU, Börner A, Sorrells ME, Tanksley SD, Ganal MW (1995) Abundance, variability and chromosomal location of microsatellites in wheat. Mol Gen Genet 246:327–333

Roder MS, Korzun V, Wendehake K, Plaschke J, Tixier MH, Leroy P, Ganal MW (1998) A microsatellite map of wheat. Genetics 149:2007–2023

Sakka H (2003) Analyse de la diversité génétique chez le palmier dattier (*Phoenix dactylifera* L.): polymorphisme de l'AND chloroplastique. PhD Thesis, Universite?Tunis, El Manar, Tunis, Tunisie

Sakka H, Zehdi S, Ould Mohamed Salem A, Rhouma A, Marrakchi M, Trifi M (2004) Tunisian date-palm (*Phoenix dactylifera* L.) genotypes identification mediated by plastid PCR/RFLP based DNA. J Genet Breeding 57:259–264

Salhi-Hannachi A, Chatti K, Mars M, Marrakchi M, Trifi M (2005) Comparative analysis of genetic diversity in two collections figs cultivars based on random amplified polymorphic DNA and Inter Simple Sequence Repeats fingerprints. Genet Res Crop Evol 52:563–573

Sedra My H (1996) La palmeraie marocaine, caractéristique et potentialités. Options Méditerranéennes A 28:163

Sedra My H, Lashermes P, Trouslot P, Combes MC, Hamon S (1998) Identification and genetic diversity analysis of date palm (*Phoenix dactylifera* L.) varieties from Morocco using RAPD markers. Euphytica 103:75–82

Senior ML, Heun M (1998) Mapping maize microsatellites and polymerase chain reaction confirmation of the targeted repeats using a CT primer. Genome 36:884–889

Sharma DR, Deepak S, Chowdury JB (1986) Regeneration of plantlets from somatic tissues of the date palm. Indian J Exp Bot 24:763–766

Tautorus TE, Fowke LC, Dunstan DI (1991) Somatic embryogenesis in conifers. Can J For Res 69:1873–1899

Tautz D, Renz M. (1984) Simple sequences are ubiquitous repetitive components of eukaryotic genomes. Nucleic Acids Res 12: 4127–4138

Torres AM, Tisserat B (1980) Leaf isozymes as genetic markers in date palms. Am J Bot 67:162–167

Trifi M (2001) Polymorphisme moléculaire de variétés Tunisiennes de palmier dattier (*Phoenix dactylifera* L.: relation avec la résistance au bayoud. PhD Thesis, Faculté des Sciences de Tunis, Université Tunis-El Manar

Trifi M, Rhouma A, Marrakchi M (2000) Phylogenetic relationships in Tunisian date palm (*Phoenix dactylifera* L.) germplasm collection using DNA amplification fingerprinting. Agronomie 20:1–7

Vasil IK (1995) Cellular molecular genetic improvement of cereals. In: Terzi M, Cell R, Falavigna A (eds) Current issues in plant molecular and cellular biology. Kluwer, Dordrecht, pp 5–18

Vos P, Hogers R, Bleeker M, Reijans M, van de Lee T, Hornes M, Frijters A, Pot J, Peleman J, Kuiper M, Zabeau M (1995) AFLP: a new technique for DNA fingerprinting. Nucleic Acids Res 23:4407–4414

Williams JGK, Kubelik AR, Livak KJ, Rafalski JA (1990) DNA polymorphisms amplified by arbitrary primers are useful as genetic markers. Nucleic Acids Res 18:6531–6535

Wrigley G (1995) Date-palm (*Phoenix dactylifera* L.). In: Smartt J, Simmonds NW (eds) The evolution of crop plants, 2nd edn. Longman, Essex, pp 399–403

Zehdi S, Trifi M, Ould Mohamed Salem A, Rhouma A, Marrakchi M (2002) Survey of inter simple sequence repeat (ISSR) in Tunisian date-palms (*Phoenix dactylifera* L.). J Genet Breed 56:77–83

Zehdi S, Trifi M, Billotte N, Marrakchi M, Pintaud JC (2004) Genetic diversity of Tunisian date palms (*Phoenix dactylifera* L.) revealed by nuclear microsatellite polymorphism. Hereditas 141:278–287

Zehdi-Azouzi S (2005) Analyse moléculaire de la diversité génétique chez le palmier dattier (*Phoenix dactylifera* L.). PhD Thesis, Faculté des Sciences de Tunis, Université Tunis-El Manar

Zohary D, Spiegel-Roy P (1975) Beginning of fruit growing in the Old World. Science 187:319–327

Chapter 18
Biology and Multiplication of *Prosopis* species Grown in the Thar Desert

R. Raj Bhansali

Abstract *Prosopis* (Mesquite) trees are members of the *Fabaceae* family. They are the 'backbone' plants of many xeriscape environments including the Thar Desert because of their ability to fix nitrogen and tolerate aridity. Leguminous tree products are economically important sources of food, fodder, firewood, timber and soil fertility enrichment. Improvement in quality attributes through selection, modification and mass production of germplasm is desirable. Sexual and asexual methods have been developed and used for multiplication of *P. cineraria* and *P. juliflora*. Propagation through seeds is the most common practice for raising quality trait seedlings for new plantations in arid areas. Vegetative methods for propagation, such as cuttings, suckers, air layering and tissue cultures are available but more efforts towards their refinement are still required, particularly with regards nursery and laboratory techniques, before commercial cultivation. Multiplication of germplasm through plant tissue culture methods is one of the applications of biotechnology via which elite trees can be mass produced rapidly. Various problems are encountered with tissue culture of *P. cineraria* and *P. juliflora* therefore this technique is still at the laboratory stage and requires more research effort. Thus, macro propagation must be utilised for genetic improvement and large-scale production of desired plants. This chapter also describes aspects of the eco-biology and economic importance with regard to the development of arid and semi-arid regions.

18.1 Introduction

Survival and growth of mankind has always depended upon the use of natural resources. Over many decades, habitat destruction and unsustainable utilisation of natural resources of many species, including plants, has led to these coming under

R. Raj Bhansali
Division of Plant Sciences and Biotechnology, Central Arid Zone Research Institute, Jodhpur 342003, India
e-mail: anjalirikhab@gmali.com

K.G. Ramawat (ed.), *Desert Plants*,
DOI 10.1007/978-3-642-02550-1_18,

threat, and their existence is now in the hands of human beings. Arid lands of the world face serious problems of scarcity of vegetation, water, food, fodder, and fuel-wood as well as harsh environmental conditions. Vegetation in dry lands represents one of the most threatened ecosystems. Arid ecosystems have been converted extensively into desert or grasslands due to various abiotic and biotic factors. This is now estimated to affect 3.4 million ha of the word's drylands, constituting 25% of the word's surface land area, which harbours populations of 500 million people and many more animals. Due to the ever-expanding population the people of these drylands belong to some of the world's poorest communities, with problems of food security, inadequate access to drinking water, and poor infrastructure facilities. Such poor economic conditions prevail in the drylands of Southeast Asia (Felker and Moss 1996). Among deserts, the Thar is the most densely populated (61 persons/km^2) desert in the world (Khoshoo and Subrahmanyam 1989). The Indian desert (22°–32° N and 68°–76° E), covering an area of 0.32 million km^2, represents several specific desert characteristics. The climate and soil conditions are characterised by high temperature, low and erratic rainfall, high evaporation, low organic matter in the soil, and strong desiccating winds. Due to obvious pressure on land, marginal lands are cultivated, and available vegetation cover is overexploited, leading to desertification (Paroda 1979; Arya et al. 1995). Fortunately, most of the natural plants of these areas are multipurpose and, if properly selected, conserved and multiplied, then some of these issues can be resolved to some extent. Some indigenous as well as some exotic plants are well adapted to arid conditions and consume less water for their sustenance. The Indian desert is highly fragile with poor primary producers and large liabilities, i.e. the consumers who severely impede 'ecological regeneration' and 'desertification control' efforts.

The population as well as diversity of plant communities declines rapidly with increasing aridity. The National Research Centre Institute for Arid Horticulture, Bikaner (CIAH), Central Arid Zone Research Institute (CAZRI), Jodhpur and Arid Zone Forestry Research Institute (AFRI), Jodhpur are working to collect germ-plasm, and to improve and propagate fruit and forest trees found in hot arid zones. Mesquite trees are members of the genus *Prosopis* of the family *Fabaceae* (legume or bean). Woody trees are vital to arid environments because of their ecofriendly and multipurpose nature, and the fact that they are well able to tolerate drought situations. The centres mentioned above have large number of accessions of *Khejri* (*Prosopis cineraria*) with wide variability in important characters (Pareek 1998). *Prosopis* and *Acacia* are the principal genera in these regions, and both have great biological diversity and ecological plasticity. These are used worldwide in arid regions to improve the local economy. These tree species are biologically diverse and multipurpose, and are well adapted to stress as a result of multiple interbreeding species. The genus *Prosopis* has a significant advantage over *Acacia* and *Casuarina* (Felker and Moss 1996). *P. cineraria* has great historical precedence for providing food for humans and domestic livestock. It is in this context that there is an urgent need to understand the biology of this species, and to develop methods for rapid multiplication of these multipurpose trees for extensive plantation in Indian arid zones (Raj Bhansali and Jindal 2000; Nandwani et al. 2005). This chapter illustrates some aspects of the biology, and methods of sexual and asexual multiplication, of two *Prosopis* species, i.e. *P. cineraria* and *P. juliflora.*

18.2 Origins and Distribution

Prosopis species (Mesquite: meaning "bark that tans") are very well adapted to xerophytic environments. The nitrogen-fixing genus *Prosopis* (ranging from 1 m-tall shrubs to 18 m-tall trees) is native to North and South America, Africa and Asia. Most species of *Prosopis* are diploid ($2n = 28$) except *P. juliflora* in which tetraploid plants are also known. In the past, *Prosopis* spp. played an important role in the traditional agriculture practiced by indigenous populations in Latin America, Africa and Asia. Burkart (1976) described 44 species of *Prosopis* distributed in tropics and subtropics of both hemispheres. Argentina is considered to be the largest centre of diversity for *Prosopis.* Details of botanical description, distribution and biology have been published by Burkart (1952, 1976) and Leakey and Last (1980). In its natural range, *P. cineraria* grows very slowly, especially during its early growth. Realising the slow-growing nature of *P. cineraria*, another *Prosopis* species, *P. juliflora* was introduced into India in 1857 from Latin America. This latter species fulfilled its potential in terms of adaptability and growth in these new environmental conditions immediately after its introduction. The arid and semi-arid parts of the country were the most preferred habitats and this species has spread into other parts of the Indian sub-continent. Depending on climatic and edaphic conditions, the density of species varies from 5 to 100 trees/ha. In 1962–1966, other *Prosopis* species, such as *P. chilensis*, *P. tamarugo*, *P. alba* and *P. pubescens*, were introduced into Rajasthan. Later, *P. alba*, *P. chilensis*, *P. glandulosa*, *P. hassleri*, *P. nigra*, and *P. pallida* were introduced into the saline-sodic soils of Uttar Pradesh (1980) and Karnal, Haryana (1985). In 1991, under a joint Indo-US research project (PL-480), several *Prosopis* species (*P. alba*, *P. chilensis*, *P. flexuosa*, *P. nigra*, a *Prosopis* hybrid and Peruvian *Prosopis* sp.) were introduced for experimental studies but were not extensively cultivated due to their being less competitive (Anonymous 1980).

18.3 *Prosopis cineraria*

P. cineraria is one of the most important leguminous trees of the Indian desert (Fig. 18.1) but is now present only in protected land areas. This highly esteemed tree of dry forest ecosystems is valued as a renewable source of energy and biomass (Leakey and Last 1980). It is used as food (vegetable, dry fruit), fodder and as a rich fuel. It enriches soil-fixing nitrogen and provides canopy cover to the cereal and legume crops of the region. Above all, *P. cineraria* combats drought and counters

Fig. 18.1 *Prosopis cineraria* (*Khejri*) – a most valued leguminous tree of the Indian Thar desert

Fig. 18.2 Lopped *P. cineraria* for production of fodder during the winter season

salinity and alkalinity. It tolerates both high and low temperatures. *Prosopis* are among the trees that have a very high rate of CO_2 fixation. These trees are known as the backbone of the rural economy of arid zones due to their wide utility (Arya et al. 1995).

P. cineraria can grow on a variety of soils but deep sandy loam soil with availability of moisture is optimum. The tree is so hardy that it can survive very well in a dry climate with annual rains ranging between 100 and 500 mm and high summer temperatures varying from 45°C to 48°C. In dryland areas, the tree is lopped heavily during the winter season for fodder use (Fig. 18.2). The leaves of *Khejri* are highly palatable, and the tannin content decreases with the maturity of the leaves. A mature tree yields about 60 kg leaf fodder on full lopping. The percentage of digestibility of leaves lopped in June and July is 41%, whereas it is 48% in the September–October period. The crude protein content ranges from 9.0 to 13.0%. In addition to leaf fodder, the green pods are commonly used as vegetables and ripe pods as animal feed. An adult tree produces 2–5 kg dry pods per year. However, limited fruits are produced in lopped trees. *P. cineraria* also ameliorates soil fertility. The tree yields high economic returns when used as part of an agroforestry system (Fig. 18.3). Field crops like clusterbean, cowpea, mothbean, mungbean and pearl millet grow very well in association of *P. cineraria*. It is essential to select individual superior fast growing trees for their desirable

Fig. 18.3 An excellent agroforestry tree, *P. cineraria* grows well with field crops

Fig. 18.4 Local Bishnoi people worshipping and hugging (*inset*) *P. cineraria* trees, which are vital to local culture; many shrines to local gods reside beneath such trees

characters, viz., leaf protein, absence of protease inhibitors, thorniness/thornless, fast-growing foliage and good timber quality. The tree shows moderate-to-high salinity resistance. However, these plants are frequently attacked by a variety of insects as well as root-rotting and wood-rotting fungi.

18.3.1 Socio-Cultural Values

P. cineraria forms an integral part of desert life and local culture. The population will protect trees of *P. cineraria* with their lives, and they are loved and worshipped by the various communities living in Thar Desert areas. The Bishnois sect Guru Jambheshwar (1451–1485 AD) directed them to love and worship trees and animals. Bishnoi people are farmers with stock and agriculture. Killing animals and the felling of green trees were banned by him. This was one of his 29 principles – "*Jeev daya palni, runkh lilo nahi dhave*", which means to protect trees and animals. Thus, trees and animals are considered sacred by the Bishnois. They are strong lovers of "*Khejri*" (*P. cineraria*) trees and animals such as deer and antelope (blue bulls, black bucks, chinkaras and chowsinghas). Many shrines to local gods are generally beneath the *Khejri* tree. Lord Rama and pandvas also worshipped the tree, which is known as "*Samipuja*". *Khejri* is worshipped by the local and Bishnoi people, as the tree is vital for their future (Fig. 18.4). Similarly, many aboriginal peoples worship the tree for good omens and take blessings before taking part in important social and cultural events in desert areas. For this reason, the tree is also known as "*Kalptaru*", which indicates how people should live in harmony with trees, animals and the environment of the desert.

In 1730 AD, Amrita Devi, a "Khejri" lover, protested against the king's men who were attempting to cut green trees as this was prohibited according to Bishnoi principles. She considered it an act of insult to her religious faith and was prepared to give her life to save the green trees. At that point she spoke these words: *Sar santey rookh rahe to bhi sasto jaan* (If a tree is saved even at the cost of one's head, it's worth it). Having said these words she offered her head. She hugged the *Khejri* tree and paid for this defiance with her life. The axes that had been brought to cut the trees, severed the heads of her and her three young daughters Asu, Ratni and Bhagu. Maharaja Abhay Singh, the ruler of the kingdom of Jodhpur, in the Marwar region wanted to fell green *Khejri* trees from Bishnoi villages as he needed a lot of greenery to burn to make lime for the construction of the Maharaja's new palace. This news spread like wildfire. The Bishnois gathered to decide on the next course of action, since the supreme sacrifice by Amrita Devi and her daughters had not satisfied the royal party, and the felling of green trees continued. It was decided that for every green tree to be cut, one Bishnoi volunteer would sacrifice his/her life. In the beginning, old people voluntarily started holding the trees to be cut in an embrace (as in the twentieth century Chipko Movement). Despite many valiant old persons sacrificing themselves, the *Hakim* (the royal party's leader), Girdhar Das Bhandari, taunted the Bishnois that they were offering unwanted old persons. Soon, young men and women (including the recently married) and children were sacrificing themselves in a similar manner. There was pandemonium. The kings's army was badly shaken by the fact that their mission unfulfilled. They informed the Maharaja of these events. As soon as he learned of it, he ordered the felling of trees to be stopped. By that time, 363 Bishnois, young and old, men and women, married and unmarried, rich and poor, had already become martyrs. Later, followers made a Cenotaph of the Bishnoi martyrs at Khejarli (named after "*Khejri*" trees) for protecting trees. Honouring the sacrifices and courage of the Bishnoi community, Maharaja Abhay Singh apologised for the mistake committed by his officials and issued a royal decree, engraved on a copper plate, ordering the following:

- All cutting of green trees and hunting of animals within the revenue boundaries of Bishnoi villages was strictly prohibited.
- It was also ordered that if any individual violated this order by mistake, he would be prosecuted by the state and a severe penalty imposed.
- Even members of the ruling family would not shoot animals in or near Bishnoi villages.

18.3.2 Tree Description and Characters

P. cineraria is a small-to-moderate sized evergreen thorny tree (up to 7 m tall), often with a straight bole to a height of 2 m and a round crown, with slender

Fig. 18.5 Spike-like racemes with yellow or creamy white flowers develop during March to May after the new flush of leaves

Fig. 18.6 Formation of long narrow and cylindrical pods after fertilisation

branches armed with conical thorns and with light bluish-green foliage (Fig. 18.1). It forms an open crown and has thick, rough grey bark with deep fissures. Leaves are alternate, bipinnately compound with 1–3 pairs of pinnae. Each pinnae has 7–14 pairs of leaflets, 4–15 mm long and 2–4 mm broad. The thorns are straight, with a conical base, and are distributed sparsely along the length of the stem. They first become visible when the seedlings are 6–8 weeks old. At this time, the leaflets are dark green with a thin casting of light shade. The plants produces new flush leaves before summer. The flowers are small in size, and yellow or creamy white in colour; they appear from March to May after the new flush of leaves. The yellow-green flowers (0.6 cm) are borne on 5–23 cm spike-like racemes (Fig. 18.5). The pods are formed soon after fertilisation and grow rapidly in size, attaining full size in about 2 months (Fig. 18.6). Up to 25 dull brown seeds, 0.3–0.8 cm long, are contained in each of the light yellow pods, which are long (8–19 cm), narrow (0.4–0.7 cm), and cylindrical (Fig. 18.7).

P. cineraria improves soil fertility and is an important constituent of the vegetation system. It is well adapted to arid conditions and stands up well to the adverse vagaries of climate and browsing by animals. Camels and goats readily browse it. In areas open to goat browsing, the young plants assume a cauliflower-shaped bushy appearance. The trees demand strong light, and dense shade kills the seedlings.

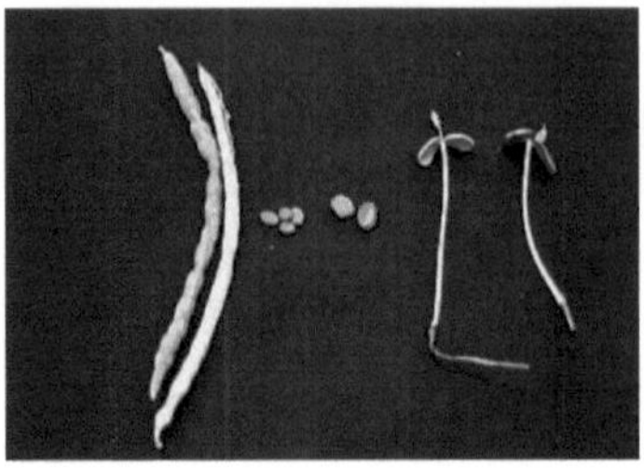

Fig. 18.7 Brown seeds produced in matured pods

Fig. 18.8 Well developed strong root system to provide sustenance in arid conditions

18.3.3 Abiotic Resistance

P. cineraria is drought resistant and tolerates dry and arid condition to such an extent that it is described as an aridity-loving tree. It is the only indigenous tree species that has the potential to withstand well the rigorous and challenging conditions of the desert. The root system of *P. cineraria* is long and well developed (Fig. 18.8). Growth above the ground is slow but below the ground the roots penetrate deeper and deeper to find sub-soil water. Very deep roots help in securing a firm footing and in obtaining moisture supplies from deep soil layers. Taproot penetration up to 35 m depth has been reported. The tree is able to endure the hottest winds and the driest seasons, and remains alive when other plants would succumb. It is even frost resistant. Because of its economic value, these trees are left standing in arable land, and farmers regulate its population by adopting suitable agroforestry management. As stated above, *P. cineraria* is a multipurpose tree, possessing great vitality and rapid growth in its natural habitat in arid and semi-arid zones, and possesses a considerable reproductive ability from coppice shoots as it coppices profusely. Maximum yields of fodder are obtained when the trees are pollard on a 3-year-rotation. Villagers traditionally lop their trees in winter and store the sun-dried leaves for dry season fodder. The trees reach 3–5 m high in 5–6 years with an average diameter of 6 cm. Annual firewood yields of up to 2.9 Mt^3/ha have been reported (Anonymous 1980). A moderate-sized tree may yield 45 kg dry leaf fodder per year. *P. cineraria* wood is reported to have a high calorific value and to provide high quality fuel wood. The lopped branches are good as fencing material.

18.3.4 Biotic Stresses

18.3.4.1 Diseases

Little information is available about insect pests and diseases affecting *P. cineraria* (Raj Bhansali 2008). Recently, heavy mortality of standing aged trees has been recorded due to attack from root-rotting (*Ganoderma lucidum*; Fig. 18.9) and wood-rotting fungi (*Fomes* sp. and *Phillinus* sp.), especially in areas above 250 mm annual rainfall (Figs. 18.10, 18.11; Raj Bhansali and Jindal 1997; Raj Bhansali 2008). *Ganoderma* fungus kills the tree by clogging vascular bundles engaged in conducting water and nutrients. Fungal fruiting bodies develop from the infected stem bases and roots.

Fig. 18.9 Attack of butt and root rotting fungi (*Ganoderma lucidum*) near base of tree and root zone

Fig. 18.10 Fruiting bodies of *Fomes* sp. develop on above-ground portion of tree trunk

Fig. 18.11 Large basidiocarps of *Phillinus* sp. develop on trees during the rainy season, causing wood rot

18.3.4.2 Insect Pests

The most important pest attacking the tree is gallfly, which causes galls on the branches, rachis and pinnae of the leaves. Parihar (1993) described the insect complex associated with *P. cineraria* tree in the western part of the province of Rajasthan, India. Among subterranean insects affecting *P. cineraria*, termites form the main group. Considerable injury to newly planted seedlings and old trees, particularly during periods of drought, is inflicted by the termite species *Odontotermes obesus*, *Microtermes mycophagus* and *Microtermes obesi* (Roonwal 1975). The flowers of *P. cineraria* are entomophilous and depend on insects for seed setting but also have to suffer the consequences of insect pests causing gall formation. Flowering occurs during the driest months of the year. Gall forming insects occur most commonly in arid areas, which reduce vegetative growth and seed formation in *P. cineraria*. Verma (1985) also reported that the leaf beetle *Clytria succimcta* feeds on leaves and branches of *P. cineraria*. Chaffer beetles (*Holotrachia* spp.) are the insects occurring most frequently in trees in arid areas. Four distinct types of gall are present in western Rajasthan (Parihar 1994; Sharma et al. 1995):

1. Galls on branches are solid, hard, woody brown structures (Fig. 18.12), caused by a chalcid, *Pediobopsis* sp. Such galls, which are formed throughout the growing season, measure 11.2–45.2 mm in length and 11–42 mm in breadth, and each produces an oval larval chamber in the centre that opens externally through a small pore in the periphery through which the adults ultimately escape. The chalcids are parasitised by *Eurytoma* sp.
2. Galls on rachis of leaflets are globose, indehiscent and hard, and measure 3.4–10.2 mm in length and 2.0–3.4 mm in breadth. Such galls are prevalent throughout the growing season. The causal insect was found to be a cecidomyid, *Contarinia prosopidis*. Larval development takes place in the gall cavity, where they suck sap from the tissues.
3. Galls on leaflets are of variable size and develop on leaflets mostly during the rainy season (Fig. 18.13). The causal organism is *Eriophyes prosopidis*, a mite. The mites reproduce parthenogenetically, giving rise to large number of galls on the leaflets.

Fig. 18.12 Woody brown galls on twigs of leafy branches

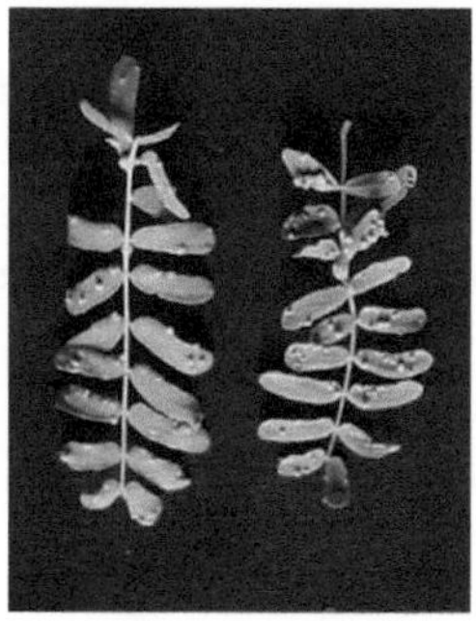

Fig. 18.13 Galls on leaflets caused by a mite (*Eriophyes prosopidis*)

Fig. 18.14 Transformation of florets into galls during flowering season

4. Galls on inflorescences are oval and globose, and masses of such galls are found developing on florets during the flowering season (April–May; Fig. 18.14). Galls have various shapes, sizes and weight. Four types of microlepidopteran insects (*Assura albicostalis*, *Anarsia triaenota*, *Eucosma lioplintha* and *Ascalenia* sp.) have been identified. Larval development takes place in the galls while they are attached to the inflorescence. Older galls with fully developed pupae became detached from the inflorescence and the moths emerge from fallen galls.

18.3.5 Propagation

18.3.5.1 Seeds

Multiplication of *P. cineraria* trees is based on seedlings originating from seeds. Being leguminous seeds, they possess a hard seed coat and, in some cases, a hard endocarp fruit wall. Therefore, seeds require pretreatment to enhance germination success by up to 70–80%. Most *P. cineraria* produce abundant seeds (1 kg/plant) with the exception of lopped trees. Seeds (25,000/kg) remain viable for decades in dry storage and establish well, with 80–90% germination (Mahoney 1990; Manga and Sen 1995). Soaking seeds in tepid water for 24 h is recommended as a pre-germination treatment. The round end of the seed may also be scarified by scratching or nicking with a file or knife.

18.3.5.2 Cuttings

P. cineraria is a cross-pollinated, slow-growing tree therefore the progenies do not resemble their parents. Asexual propagation is the only method to propagate germplasm of tested, elite trees with high growth rate and biomass production. Vegetative propagation allows clonal and mass multiplication of material selected for best adaptation, fruiting, spineless and spiny characters. *P. cineraria* is difficult to propagate by cuttings, although treatment with rooting hormones has proved successful in India. Propagation by root suckers and by air layering has been reported. Recent attention has also been given to micropropagation of this species, but it appears that in vitro propagation is more difficult with *P. cineraria* than with many other *Prosopis* species for several reasons: the tree is also considered slower growing than other *Prosopis*; the tree coppices readily (Anonymous 1980); different characters have to be selected separately; and Stem cuttings of *P. cineraria* are difficult to root. Solanki et al. (1984) described preliminary results of vegetative propagation of *P. cineraria* by air layering. In nature, the plant sprouts at cut and injured portions of stems and roots. Such sprouted shoots/roots can be used for vegetative propagation.

18.3.5.3 Suckers and Grafting

Vegetative propagation is possible by rooting of stem cuttings in the suckers, and from coppice shoots in *P. cineraria*, *P. juliflora* and *P. alba* (Leakey and Last 1980; Felker and Clark 1981). Although the percentage of plantlets obtained through these methods was not very high, it demonstrates the possible application of this method (Fig. 18.15). Considerable improvement in percentage rooting of *P. alba* stem cuttings was obtained by application of fertiliser to the stock plants (De Souza and Felker 1986; Green et al. 1990). Rootstocks of wild seedlings of *P. cineraria* and scion taken from selected mother plants can be grafted through patch budding (Pareek and Purohit 2002). Pareek et al. (2006) have also reported successful grafts among four *Prosopis* species, viz., *P. cineraria* (88.9) and *P. juliflora* (88.9), *P. alba* (72.2) and *P. nigra* (11.1). Patch budding *P. nigra* scion on *P. juliflora* rootstock gave 83.3% successful grafts, suggesting that combinations of *P. cineraria* with any of the other three species of section *Algarobia* are incompatible. Elite *khejri* trees can be multiplied in the nursery by patch budding. This technique can also be used for top-working wild seedling trees.

Fig. 18.15 Emergence of leafy shoots from a cut main stem

18.3.5.4 Tissue Culture

Tissue culture methods have great potential for the mass multiplication of plants using the minimum quantity of explant tissues. Some of the advantages of this technique are that heterozygous materials may be perpetuated easily and quickly without much alteration, dormancy problems are eliminated, and juvenile stages reduced. Nowadays, biotechnology has proved invaluable in solving many intrinsic problems pertaining to agriculture, industry, health and the environment that have direct relevance to sustainable development of desert lands. Since the beginning of domestication and cultivation of plants, human beings have searched for techniques that help to produce the maximum number of individuals from the minimum number / quantity of propagules. Plant tissue culture provides an excellent system with which to study the factors involved in vegetative multiplication and the possibilities of obtaining improved plants by genetic engineering. It also provides a means of perpetuating clones that do not produce viable seeds or that do not produce seeds at all. Tissue culture or micropropagation of plants involves a sequence of steps, each of which requires a specific set of conditions. Three distinct steps are usually involved: (1) establishment of aseptic culture, (2) multiplication of the propagule, and (3) preparation and establishment of the propagule for independent existence by hardening and acclimatisation. These steps involve the use of different chemicals, media and management of light, humidity, temperature, etc., depending on the plant. Vegetative multiplication through tissue culture has emerged as a major applied aspect of plant tissue culture research (Murashige 1974; Bajaj 1986; Arya and Shekhawat 1986; Raj Bhansali 2001, 2003). Currently, a large number of research groups are working on plant biotechnology for the mass regeneration of *Prosopis* species through plant tissue culture. In the near future, this could be an alternative route towards afforestation of arid lands. Besides being slow growing, *Prosopis* species exhibit a wide range of characters due to their heterozygous nature. After careful germplasm selection, an improvement programme for pest and disease resistance could be initiated by conventional breeding methods, or in combination with plant tissue, cell and protoplast culture techniques for the development of clones of elite trees.

Juvenile Tissues

Various parts of young seedling have been used in tissue culture studies. Goyal and Arya (1981) reported culture of hypocotyl segments (0.8–1.0 cm) from 7–10 day old in vitro grown seedlings on Murashige and Skoog (MS) medium containing kinetin and naphthaleneacetic acid (NAA)/indole acetic acid (IAA). Shoot bud differentiation was recorded in these explants. These shoots developed roots on basal White's medium containing indole butyric acid (IBA; 3.0 mg/l). However, this technique is not applicable for multiplication of mature selected trees. Further, Goyal and Arya (1981, 1984) employed lateral buds from leafy shoots of a 7-year-old *P. cineraria* strain (CAZRI elite selection) for initiation of tissue culture. Multiple shoot differentiation was achieved from such explants on MS medium containing 3.0 mg/1 IAA + 0.05 mg/1 kinetin. The developed shoots produced roots on White's basal medium containing IBA (3.0 mg/1) and kinetin (0.05 mg/1). Cotyledon, epicotyl, hypocotyl and apical bud explants excised from seedlings were suitable for multiple axillary shoot regeneration in *P. cineraria* (Fig. 18.16; Puri et al. 1992; Nandwani 1990; Nandwani and Ramawat 1993; Raj Bhansali 2003). A maximum of 11 shoots was obtained per explant obtained from 2-week-old seedlings on MS medium supplemented with 0.5 mg/l IAA and 2–5 mg/l benzylaminopurine (BAP)/kinetin (Fig. 18.17; Raj Bhansali 2003). More healthy shoots were produced from brown, spiny and woody explants than from green and young stem explants. Regeneration has not been successful from callus. Explants of different genotypes varied in their morphogenetic response in vitro and produced a variable number of shoots on the same medium. Regenerated shoots were rooted on MS medium containing 5 mg/l IBA. Plantlets transferred to pots survived well under field conditions (Nandwani and Ramawat 1993).

Mature Tissues

Nodal Explants

In vitro multiplication of shoots from mature aged trees is highly desirable for mass propagation of selected trees. Several workers have reported methods for the rapid propagation of *P. cineraria* through axillary branching. Single node segments from actively growing branches of elite tree have been cultured on MS basal medium containing 3.0 mg/l each of 2-naphthoxyacetic acid (NOA) and NAA (Shekhawat et al. 1993; Raj Bhansali 2001). Such segments produced several axillary shoot

Fig. 18.16 Induction of shoots from the embryonal axis of germinated seeds and callus from cotyledons on Murashige-Skoog (MS) medium containing 3.0 mg/l indole acetic acid (IAA) + 0.05 mg/l kinetin

Fig. 18.17 Development of large number of shoots from 2-week-old seedlings on MS medium supplemented with 0.5 mg/l IAA and 2–5 mg/l benzylaminopurine (BAP)/ kinetin

buds within 7–10 days when 2-cm-long shoots were transferred to modified MS medium containing 3.0 mg NOA/l (Kackar et al. 1991, 1992). Around 80% of shoots showed rooting to a length of about 8 cm within 30 days. Such shoots were then further cut into five to seven segments and planted individually on MS + NOA supplemented medium. Each segment produced a plantlet. By following this procedure, five- to seven-fold multiplications of plants could be achieved within a month; 30% of these plants survived after transplantation. However, these methods still need to be refined before they can be utilised on a large scale. In another experiment, young shoot tips of *P. cineraria* were sterilised and cultured in MS medium with or without 5.0 mg NAA/l. This medium induced rooting, and plants survived when transferred to a soil + nutrient medium (Raj Bhansali 2001).

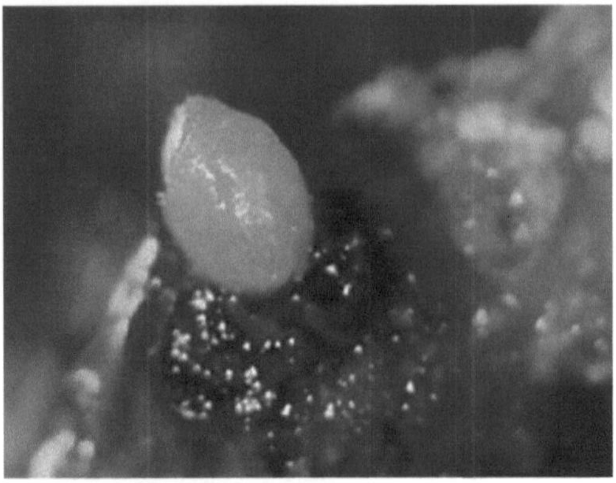

Fig. 18.18 Nodal bud break from on high-cytokinin-containing medium

Fig. 18.19 Proliferation of bud into multiple propagules on MS medium containing 2 mg/l BAP/kinetin

Bud break from nodal explants of *Prosopis* can be obtained on diverse media (Fig. 18.18; Goyal and Arya 1984; Jordan et al. 1987; Wainright and England 1987; Yao et al. 1989; Nandwani 1990). MS medium containing high cytokinin (Nandwani 1990) or auxin (Goyal and Arya 1984) content, or a high concentration of both cytokinin and auxin (Batchelor et al. 1989) supports shoot bud proliferation from stem explants of *P. cineraria* (Fig. 18.19). These findings suggest that the response is not controlled precisely by the concentration and quality of exogenous growth regulators. Perhaps, requirements are quite different from the levels provided exogenously, but subsequent changes in metabolic level and endogenous levels of growth regulators promote shoot formation. This conclusion is supported by the fact that, in most cases, formation of only single shoots has been achieved. Complete bud stimulation in explant is lacking, thus production of adventitious shoots remains suppressed in *P. juliflora* (Nandwani and Ramawat 1991). *P. cineraria* has provided much evidence that explants obtained from different genotypes, as well as explants of the same genotype, differ in their regenerative potential on the same medium. Explants exhibited considerable variation in phenotypic expression in relation to delayed regeneration, size, length and vigour of regenerated shoots and leaf expansion (Nandwani 1990). Thus, the differing results obtained by various workers are due mostly to the different material (genotype) used, making such results very hard to reproduce. Therefore, these studies must report a detailed account of the source of the explant used in the investigation. Some of these species are very recalcitrant to regeneration. Similarly, shoot number could not be enhanced to more than seven in direct adventitious bud formation.

Nandwani and Ramawat (1989, 1992a) have evaluated the precise role of growth hormones, particularly cytokinin, in adventitious shoot formation in various *Prosopis* species. This has resulted in the formation of multiple shoot buds from hypocotyls in successive passage on hormone-free and cytokinin-supplemented media. Embryonic explants isolated from zygotic embryos produced multiple shoots when grown on hormone-free or cytokinin-supplemented media. It was observed that proliferation of shoot buds increased in the second passage. A reduction in the cytokinin concentration in the medium resulted in elongation of shoots from the clump of organogenetic mass. Elongated shoots produced roots on rooting medium containing an auxin. It has also been reported that if both cotyledons are allowed to remain attached with the embryonic axis, multiple shoot formation in the explants is enhanced (Nandwani and Ramawat 1992b, 1993; Raj Bhansali 2001). In the absence of cotyledons, the number of shoot buds remains low. The growth of the explant is delayed but the number of shoots buds is increased by the formation of new adventitious buds. It appears that these buds develop adventitiously at a later stage and form newly proliferating callus cells. In certain cases, apical meristems in axillary positions proliferate to produce a phyllody-like structure.

Root Explants

Root segments from 3- to 4-month-old seedlings have also been used for production of multiple shoots on MS medium containing 3 mg/l NOA (Kackar et al. 1992). Shekhawat et al. (1993) have also used root segments for development of multiple shoots on MS medium containing 0.5 mg/l IBA + 2.5 mg/l BAP. However, all these workers describe methods for propagation of trees that were not more than 7 years old. The main problem is that these methods cannot be applied to multiply field-tested aged trees that have been selected by the users for various purposes. Moreover, selection for any character appearing at ages less than 12–15 years is not possible in *P. cineraria* because of the slow expression of such characters in nature (Shekhawat et al. 1993; Raj Bhansali 2001; Raj Bhansali and Singh 2003). Several intrinsic and extrinsic factors influence the behaviour of explants in culture. The age of the tree, lopping season, harvesting period and sterilisation procedure have more influence than the medium used. Various factors affecting in vitro clonal propagation of *P. cineraria* have been studied in detail (Shekhawat et al. 1993). Genotype, age of tree, nature and size (length and diameter) of explant, season of collection, explant position on medium, plant growth regulators and certain additives (ascorbic and citric acids, adenine sulphate, L-arginine, glutamine and ammonium citrate), incubation conditions, and frequency of subculturing period greatly influenced the in vitro clonal propagation of *P. cineraria* (Shekhawat et al. 1993). The maximum number of 10–12 shoots were induced from the nodal shoot segments from pruned thorny adult trees on MS medium containing 0.1 mg/l IAA + 2.5 mg/l BAP + additives. Differentiated shoots multiplied best on MS medium containing 0.1 mg/l NAA + 1.0 mg/l BAP + additives.

Organogenesis

Callus induction is frequently observed in *P. cineraria* on MS medium containing growth regulators using various explants (Nandwani and Ramawat 1989, 1991). It is possible to induce and maintain callus culture from *P. cineraria* (Nandwani and Ramawat 1989). However, all attempts to regenerate whole plantlets from isolated and subcultured callus have failed. It is interesting to note that Jordan et al. (1987) observed a single embryo/shoot apex-like structure in cultures grown on B5 medium supplemented with 2,4-D. *P. chilensis* also formed differentiating calli. Generation of embryo-like structures has been reported in cell suspensions in a few cultures, but this needs to be confirmed. However, confirmation of somatic embryogenesis/organogenesis has so far not been reported from callus tissues. Greater efforts are urgently needed to overcome this problem, which could open many new avenues for biotechnological research in *Prosopis* improvement programmes. Tissue culture methods should also be attempted to overcome self-incompatibility and to generate plant variability in these species (Jordan 1987, 1988). Efforts were also made to isolate protoplasts from *P. cineraria* (Shekhawat and Kackar 1987). It is mentioned earlier that, once isolated from explants and maintained separately, callus has never regenerated through somatic embryogenesis/organogenesis in *P. cineraria*. However, in a few cases, highly regenerative tissues were found regenerating spontaneously from juvenile explants of *P. tamarugo*; it seems unlikely that such a high regeneration could be obtained without the interface of callus organogenesis. The regenerative mass showed shoot bud development from callus tissues. This represents the first evidence of callus regeneration in any *Prosopis* species. Increased nitrogen content enhanced shoot bud proliferation while adding inhibitors of auxin synthesis and auxin transport enhanced callus formation in regenerative tissues (Nandwani 1990).

Plantlet Formation

Rooting in developed shoots is achieved by pulsing treatment with 100 mg/l IBA for 4 h and then culturing on hormone-free half-strength MS medium. An initial dark incubation for 5 days at high temperature (33±2°C) was found essential for root induction, which was 63% within 2 weeks. The rooted plantlets contained a consistent number of chromosomes ($2n$=28). It has been suggested that the protocol developed could be useful for cloning of mature and tested trees of *P. cineraria* (Shekhawat et al. 1993). A high frequency of shoots (8–12 per explant) were harvested from 10-year-old spineless *P. cineraria* on MS medium containing IAA, BAP (Fig. 18.20) and other additives (Nandwani and Ramawat 1993). Some of these shoots were successfully rooted but rooting in culture is highly unpredictable. Thus, more experiments need to be conducted in order to obtain reproducible

Fig. 18.20 Development of shoots from 10-year-old spineless *P. cineraria* on MS medium containing 1 mg/l IAA and 2 mg/l BAP

protocols that can be followed by anyone working on tissue cultures of *Prosopis* sp. Compared to *P. cineraria*, rooting can be induced with relative ease in other *Prosopis* species such as *P. tamarugo* and *P. juliflora* (Nandwani 1990; Goyal and Arya 1984). The system requires further refinement to increase the frequency and subsequent viability of rooted plants under field conditions. However, in the case of *P. cineraria*, the culture medium, genotype and age of shoots markedly affect the rooting behaviour of in vitro regenerated shoots. Therefore, to develop a reproducible rooting system, all these factors require due consideration. Once shoots are rooted, establishment of these plants does not pose any difficulty except that initial low watering is required.

Cultures of Gall Tissues

Plant tissue culture methods have also been used for in vitro studies of insect-induced plant galls. Stem and rachis galls of *P. cineraria* are caused by a *Chalcid* and *Lobopteromyia prosopidis* (*Contarinia prosopidis*). Galled tissues have been used for induction of callus and differentiation (Kant and Ramani 1990). Laboratory studies to induce gall formation by Cecidomyiid larvae on the leaf rachis of germinated seedlings and callus tissues of *P. cineraria* were carried out; gall induction took place on the leaf rachis of seedlings germinated in vitro. Galls formed in tissue culture were soft and lacked the differential structures normally associated with the gall. Normal and gall callus inoculated with insect larvae showed nodule-like structures with high meristematic activity. This was the first report on in vitro gall formation by a Cecidomyiid larva (Ramani and Kant 1993).

The various methods of propagation of *P. cineraria* are listed in Table 18.1.

18.3.6 Economic Importance

In the Indian sub-continent, only *P. cineraria* is endemic while all others *Prosopis* species have been introduced. These trees are vital components of the traditional

Table 18.1 Various methods of propagation of *Prosopis cineraria*. *IAA* Indole acetic acid, *IBA* indolebutyric acid

Method of propagation	Explant treatment	Response	References
Sexual; seed	Pretreatment of seeds with H_2SO_4; sandy loam soil with compost in ratio of 2:1 in February–March/ September–October	Germination takes place within 15 days; produce transplanting seedling after 6–9 months	Arya et al. 1995; Manga and Sen 1995
Asexual; suckers/ cutting	Root suckers/coppice shoots with IAA/IBA/ NAA singly or in combination	Rooting	Rao 1953; Arya and Tomer 1989; Shekhawat et al. 1993; Arya et al. 1994
Grafting	Scion buds on rootstock seedlings in March	Patch grafted seedlings	Pareek and Purohit 2002
Air layering	Removal of bark then applying 100 ppm IAA/Seradix B3 (IBA) with moss	Rooting	Rawat et al. 1982, 1983; Solanki et al. 1984, 1986
Tissue culture	Stem segment; Cotyledonary and hypocotyl explants from seedlings; nodal buds; leafy buds; culture of insect gall	Organogenesis/ multiple shoots/plantlet/callus/ cell suspension; apical shoot proliferation; multiple shoots; callus	Goyal and Arya 1981, 1984; Shekhawat et al. 1993; Kackar et al. 1991, 1992; Raj Bhansali 2001; Nandwani and Ramawat 1989, 1992b, 1993; Raj Bhansali 2003; Raj Bhansali and Singh 2003; Kant and Ramani 1990; Ramani and Kant 1993

agroforestry system and are grown together with field crops to provide dry season fodder and fuel. Most strains are economically important in that they provide increased fertility to the arid soil.

18.3.6.1 Environmental Protection

P. cineraria has a very deep tap root system (the tap root may penetrate vertically up to 20 m or more; Mahoney 1990) and hence does not generally compete with the associated crops. The improved physical soil conditions as well as the higher availability of nutrients under the *Khejri* canopy explain the better growth of the crops associated with it. Rural communities encourage the growth of *P. cineraria* in their agricultural fields, pastures and village community lands. Because of its extensive root system, it stabilises shifting sand dunes, and is also useful as windbreak shelterbelt and in afforestation of dry areas. *P. cineraria* fixes atmospheric nitrogen through microbial activities, and adds organic matter through leaf litter decomposition, thus rejuvenating poor soils. Since this is the only tree species in these arid regions, it provides much needed shade and shelter to farmers working in the fields as well as to cattle and wildlife during the summer months.

18.3.6.2 Land Stablisation

P. cineraria effectively stabilises sand dunes and can withstand periodic burial (Gates and Brown 1988). Because of its deep taproot, these trees are not believed to compete for moisture or nutrients with crops grown close to the trunk. During the growing season it casts only light shade and is therefore suitable as an agroforestry species. Farmers in and semi-arid regions of India and Pakistan have long believed it to increase soil fertility in crop fields. Yields of sorghum or millet increase when grown under *P. cineraria*, as a result of higher organic matter content, total nitrogen, available phosphorus, soluble calcium, and lower pH (Mann and Shankarnarayan 1980). Other crops traditionally grown amid scattered *Khejri* are maize, wheat, and mustard.

18.3.6.3 Food and Fodder

P. cineraria pods are eaten as a vegetable in the human diet in some areas. In Rajasthan, green pods called *sangri* are boiled and dried (FFN 1991). The dried pods, locally called 'Kho-Kha', are eaten by local inhabitants. The pods also provide a sweet pulp with a pleasant taste. They are also used as famine food and were known even to prehistoric man. Pod yield is nearly 1.4 quintals of pods/ha, with a variation of 10.7% in dry locations. The flowers are valuable for honey production. The bark can be used in leather tanning and yields an edible gum. Bark and flowers are used medicinally (Anonymous 1980). In times of famine, the powdered bark has been mixed with flour and made into cakes (Bhandari 1978).

P. cineraria is the most important top feed species, providing nutritious and highly palatable green as well as dry fodder that is eaten readily by camels, cattle, sheep and goats, constituting a major feed requirement of desert livestock. Locally it is called 'Loong'. Dried pods also form a rich animal feed, which is liked by all livestock. The tree produces leaves during the extremely dry summer months when most other trees are leafless. Leaves contain 13.8% crude protein, 20% crude fibre, and 18% calcium (FFN 1991); thus it is much valued as a fodder tree. The trees are heavily lopped particularly during winter months when no other green fodder is available in dry tracts. There is a popular saying that death will not visit a man, even at the time of a famine, if he has a *P. cineraria*, a goat and a camel, since the three together are said to be what can sustain a man even under the worst drought conditions. Green pods also form a rich animal feed that is made by drying the young boiled pods. The forage yield per tree varies a great deal. On average, the yield of green forage from a full-grown tree is expected to be about 60 kg with complete lopping having only the central leading shoot, 30 kg when the lower two-thirds of the crown is lopped, and 20 kg when the lower one-third of the crown is lopped. The pods have a sweetish pulp and are also used as fodder for livestock.

18.3.6.4 Medicinal Properties

P. cineraria flowers are pounded, mixed with sugar and used during pregnancy as a safeguard against miscarriage. Water-soluble extracts of the residue from the methanol extract of the stem bark exhibits anti-inflammatory properties. *P. cineraria* produces gum, which is obtained during May and June. The bark of the tree is dry, acrid and bitter, with a sharp taste; it is used as a cooling anthelmintic, a tonic, and as a cure for leprosy, dysentery, bronchitis, asthma, leucoderma, piles and tremors of the muscles. Smoke from the leaves is good for eye troubles. The strongly flavoured fruit is dry and hot, is indigestible, causes biliousness, and destroys the nails and the hair. The pod is considered an astringent in Punjab. The bark is used as a remedy for rheumatism, and to treat coughs, colds, and asthma. The plant is recommended for the treatment of snakebite. The bark is prescribed for scorpion string.

18.4 *Prosopis juliflora*

Prosopis juliflora (Sw.) DC is known as mesquite and honey mesquite. It is the most common evergreen tree native to South America, Central America and the Caribbean islands. It has now become established in arid and semi-arid Australia, Africa and Asia (Abdelbari 1986; Alban et al. 2002). The tree is a perennial deciduous thorny shrub or small tree that can grow up to 10 m tall, with a trunk up to 1.2 m in diameter. It is a fast growing biomass producer that can fix nitrogen and tolerate arid conditions and saline soils (Gurumurti et al. 1984). It delivers a variety of valuable goods and services: construction materials, charcoal, soil conservation, and rehabilitation of degraded and saline soils. *P. juliflora* can grow in areas receiving as little as 50 mm rainfall per year. In unmanaged conditions it often colonises disturbed, eroded and over-grazed lands, forming dense impenetrable thickets. Thickets of *P. juliflora* have become established in grazing lands, croplands and along river courses, alarming pastoralists, farmers and conservationists. In many areas, this situation has led to great controversy over the future thrust of its plantation. There is also concern regarding the impact of this tree on the biodiversity of native plants, and on the amount of water in dry land streams. *Prosopis* species have been declared noxious weeds in many countries, including Argentina, Australia, South Africa, India, Pakistan and Sudan. Nevertheless, *P. juliflora* has proved useful in restoring degraded and saline lands, producing a variety of useful products for the local populations. *P. juliflora* has potential as a source of fuel wood, timber, honey and animal forage. In the late 1990s, the Forestry Research Programme of the United Kingdom (UK) Department for International Development supported a project by HDRA (Henry Doubleday Research Association) in the UK, and CAZRI (Central Arid Zone Research Institute), Jodhpur, in India, that was initiated to collate information on the utility of most common *Prosopis* species (Tewari et al. 2000). Three main conclusions were drawn: (1) that *P. juliflora* can be a very valuable resource for dry lands; (2) that efforts to completely eradicate *P. juliflora* are overly expensive and likely to be ineffective; and (3) that *P. juliflora*, when properly managed, can be a very valuable source of commercial products and livelihoods in dry areas. Most importantly, its value as fuel wood, charcoal, timber, as well as in furniture construction, animal feed, human food, and medicinal products has been well documented and is increasingly exploited.

18.4.1 Tree Description and Characters

P. juliflora is predominantly xerophilous, aculeate, and spiny or rarely unarmed (Burkart 1976). The tree usually has a slightly gnarled stem, with a characteristic fibrous-looking bark. Leaves are bipinnate, often with few pairs of opposite pinnae. Petioles have circular, sessile, apical glands. Leaflets are small, numerous, mostly opposite, linear, oblong, fusiform, the same colour on both sides and exhibit pinnate venation. Shoots in most species are dimorphic with long megablasts, flexuous and becoming knotty with age. Brachyblasts or short shoots emerge from multiple axillary buds, from which the cauline spines develop. The flowers are small, actinomorphic, pentamerous and hermaphroditic. The corolla has linear petals. Pollen grains are large or small. The ovary is stipitate, villous or sometimes glabrous. Racemes are spike-like, amentiform, axillary, mostly densiflorous, (Fig. 18.21) but sometimes in globose heads (Burkart 1976). The fruit is a modified, indehiscent, fleshy legume called a 'dropaceous loment' (Burkart 1976). These are linear, straight, falcate, annular to spirally coiled with a fleshy mesocarp. The mesocarp is sugary or fibrous, containing endocarps, divided into one-seeded, coriaceous to bony segments. These are closed, longitudinal or rarely biseriate

Fig. 18.21 Spike-like racemes and inflorescence with axillary and densiflorous florets

and transverse. Seeds are ovoid, compressed, with a fissural line on the faces, known as a pleurogram. They are hard, brown, with a mucilaginous endosperm, typical of the Mimosoideae, surrounding the embryo. Cotyledons are flat, rounded and epigeous in germination.

18.4.2 Pests and Diseases

Many pests inflict damage on *P. juliflora* seedlings and mature trees but without causing economic damage. Nematodes that may be harmful have been recorded from soil under the canopies of *P. juliflora* (Singh 1998; Yousuf and Gaur 1993). Some insects, such as seed feeding beetles, cause widespread damage. Twig girdlers are important pests of several *Prosopis* species (Felker et al. 1983b; Lima and Haji 1993). Control of these insects involves burning of fallen branches to kill any eggs laid on them. Psyllids are another major problem, attacking buds and significantly reducing tree growth, but can be controlled effectively with the application of insecticides (Felker et al. 1983a). Locusts and grasshoppers can cause completely defoliation.

Several wood rot diseases belonging to the group basidiomycetes attack the stem of *P. juliflora*. *Schizophyllum commune* fungus commonly attacks the stem and bark of injured and cut trees (Fig. 18.22). Branches and leaves of *Prosopis* are often found infected with deuteromycetes fungus (Lesney and Felker 1995; Srivastava and Mishra 1998). Some diseases have been observed in young *P. juliflora* plants in India, notably the root rot *Fusarium* spp. and the collar rot *Macrophomina phaseolina* (Srivastava and Mishra 1998). In older *P. juliflora* trees in India, the leaf blights *Colletotrichum capsici*, *C. cymopsicola* and *Ravenelia spicigera*, the twig blight *Diplodia rosopides* and stem canker *Botryodiplodia theobrmae* have been observed (Srivastava and Mishra 1998). Canker and tip-dieback diseases have been found in both field and greenhouse plantings of *Prosopis*. Several fungal species have been found associated with field cankers, most notably *Lasiodiplodia* and *Pestalotiopsis* sp., often with several other associated fungi (including *Alternaria* and *Fusarium*).

Fig. 18.22 Wood rot disease caused by *Schizophyllum commune* (belonging to the group basidiomycetes) attacks the stem and bark

Pestalotiopsis requires high humidity and proved much less damaging compared to *Lasiodiplodia*, which closely mimicked the field symptoms. Benlate was found effective to inhibit these fungal pathogens in fungicide tests in culture.

18.4.3 Propagation by Seeds

18.4.3.1 Seed Collection and Storage

Seeds are collected from trees having desirable characteristics such as high yields of sweet pods, lack of thorns, erect form and free from pests and diseases. *P. juliflora* fruits throughout the year. Pods can be collected from the trees or from the ground. Pods are stored after drying to 10–12% moisture. Pods are stored in rooms and barns, either ventilated to keep pods dry and free from fungal attack, or sealed to prevent insect infestation. Killian (1990) reported reduced viability of seed stored in pods rather than as cleaned seed.

18.4.3.2 Seed Extraction

Seeds with endocarps can be used to propagate *Prosopis*, but germination is poor. Endocarps can more easily be separated after soaking in acid or alkaline solutions. Soaking is much more effective if a small amount of sodium hydroxide (NaOH) is added to the water, followed by rubbing between two folds of a coarse cloth to loosen the endocarps (Saxena and Khan 1974). Pimentel (1982) suggested soaking in a 4% solution of NaOH. Valdivia (1972) found soaking with 5% HCl and rubbing with sand to be effective. Muthana and Arora (1983) recommended shaking in concentrated H_2SO_4 for 5 h to completely destroy the endocarps.

18.4.3.3 Germination Pretreatment

Seed sown in the nursery germinates in a synchronised manner within 7 days. Fresh seeds exhibited higher germination without pretreatment due to their unhardened seed coat (Ffolliot and Thames 1983; Luna 1996). Heat treatment of seed generally involves immersion of seeds in hot water, and is often recommended as a simple and effective pretreatment. However, Pasiecznik et al. (2001) found this method gave germination of only 52% with *P. juliflora*. Sulphuric acid is effective with *P. juliflora* (Muthana 1988; Tewari et al. 1998). Maximum germination rates of over 95% were achieved with *P. juliflora* seed following a 15- or 30-min soak in 97% (concentrated) sulphuric acid, or a 30-min soak in a 60% solution (Pasiecznik et al. 1998). The high rate of seedling emergence from animal dung is common under field conditions, showing that the seeds are not only cleaned but the seed coat is also softened during digestion.

18.4.4 Vegetative Propagation

18.4.4.1 Cutting and Grafting

Various techniques have been employed to root *Prosopis* stem cuttings, including growth chambers, mist or non-mist propagators, and open or sealed nursery bags. Auxins are known to play a significant role in stimulating root initiation in stem cuttings of woody plants. Propagation by stem cuttings is more difficult from mature trees than from juvenile plants. Vegetative propagation of *P. juliflora* by stem cutting was first reported by Kaul (1956), although these cuttings took a long time to sprout. Lima (1990) found that *P. juliflora* is more difficult to root than *P. alba*, *P. chilensis* or *P. pallida*. In contrast, Harris et al. (1996) reported much higher rooting percentages (over 70%). Rooting performance of stem cuttings of three clones of *P. juliflora* in non-mist propagators exhibited significant differences (Wilson et al. 1990), indicating intraspecific as well as interspecific variation in rooting ability. Dick et al. (1991) and Lima (1990) achieved an 80% success rate in rooting cuttings without hormone treatments. Leakey et al. (1990) recommended IBA, while Sandys-Winsch and Harris (1991) used NAA to root *P. juliflora* cuttings. Mixture of IBA and NAA, or IBA and IAA, or IBA, NAA and thiamine are most effective in rooting a variety of *Prosopis* species. Felker (1991) concluded that air layering has serious limitations because of the low success rate, long rooting periods and limited number of suitable branches.

P. juliflora was found to be graft compatible with other species of *Prosopis* (Wojtusik and Felker 1993). Grafted plants survived at rates of up to 100% with healthy stock plants and an experienced grafter. Out of four graft types, cleft grafting was the most successful (Wojtusik and Felker 1993). Scions from the spineless and most erect *P. juliflora* trees were successfully grafted onto *P. alba* rootstock. These scions should have great potential for grafting onto existing weedy *Prosopis* and for grafting onto rootstock previously shown to tolerate salinities of seawater and pH values of 10.3 (Wojtusik et al. 1993).

18.4.4.2 In Vitro Propagation

There are reports of regeneration of complete plants from *P. juliflora* shoot tips (Walton et al. 1990) and nodal explants (Batchelor et al. 1989; Yao et al. 1989; Nandwani and Ramawat 1991). Different species respond differently to the same culture conditions. Walton et al. (1990) reported wide differences in regeneration rates among various *Prosopis* species such as *P. glandulosa* (100%), *P. alba* (94%), *P. juliflora* (74%), *P. chilensis* (67%), *P. cineraria* (9%) and *P. tamarugo* (4%). Ramawat and Nandwani (1991a) found that rooting could be induced in *P. tamarugo* and *P. juliflora* with relative ease compared with *P. cineraria*. Regeneration from juvenile nodal explants from *P. chilensis* was always better than from *P. tamarugo* (Jordan et al. 1987). The age of the mother plant influences the success of regeneration. Arce and Balboa (1991) obtained 80% regeneration from nodal and apical tip of juvenile (1- to 4-month-old) material, compared with 60% from rooted cuttings and none from field collections. Despite these observations, there are reports of regeneration from 10-year-old, field-grown *P. juliflora* by Nandwani and Ramawat (1991). Batchelor et al. (1989) found that the top six nodes of *P. juliflora* and *P. chilensis* were most suitable for micropropagation.

Variations in the medium used have impacts on shoot growth and root initiation (Ramawat and Nandwani 1991a, 1991b). The nitrogen nutrition of *Prosopis* explants is important. Ammonium ions in the medium are inhibitory to shoot production, while glutamine or nitrate proved satisfactory nitrogen sources for short incubations, whereas only glutamine was suitable for longer periods (Tabone et al. 1986). Antioxidant treatments have been used in an attempt to counter the browning of explants and medium, which has been attributed to phenolics (Jordan et al. 1987; Batchelor et al. 1989). Regeneration from a node or shoot tip is easy on NAA-containing rooting medium with a low concentration of kinetin. Nandwani and Ramawat (1991) reported rooting from *P. juliflora* shoots on a medium with NAA or IBA. Arce and Balboa (1991) obtained complete regeneration with NAA and cysteine in another *Prosopis* spp. i.e., *P. chilensis*. Jordan (1987) obtained plant regeneration from nodes of *P. chilensis* with NAA and kinetin, and from *P. alba* with NAA alone. Nandwani and Ramawat (1991) found that a cytokinin was essential for in vitro shoot growth of *P. juliflora*. Batchelor et al. (1989) found high light intensity led to an improvement in *P. juliflora* growth. Multiple shoot formation has been reported from nodal regions of *P. juliflora*.

Nandwani and Ramawat (1991) observed callus formation in *P. juliflora* explants from stems, cotyledons, hypocotyls and inflorescences but, in all cases, callus did not survive upon subculture. Rarely direct organogenesis from hypocotyls, callus production and root regeneration is reported in *P. chilensis* (Batchelor et al. 1989). All of the individual stages of in vitro micropropagation of *Prosopis* have been achieved in short-term experiments. Production programmes have not been established, however, because *Prosopis* species have proved recalcitrant to sustained in vitro culture. After a period in culture, explants tend to become chlorotic and suffer from leaf abscission and shoot necrosis (Yao et al. 1989). Tabone et al. (1986) also found that, after 5–6 weeks in culture, shoots of *P. alba* also became necrotic, and were dead after 8 weeks. Felker (1991) concluded that tissue culture propagation of *Prosopis* has not proven useful in any process related to asexual propagation as new shoots can be produced but these die after about 12 weeks in culture. Despite the fact that many laboratories all over the world have worked on the regeneration of *Prosopis* species, much of this effort has been abandoned because of either recurrent contamination problems or recalcitrance to regeneration. *Prosopis* can be propagated vegetatively in vitro, but unless micropropagation can offer positive advantages over conventional vegetative propagation by stem cuttings, the very much higher labour input, technical level and capital costs are unlikely to make it an attractive alternative.

The various methods of propagation of *P. juniflora* are listed in Table 18.2.

Table 18.2 Various methods of propagation of *Prosopis juliflora*

Method of propagation	Explant treatment	Response	References
Sexual; seed	Soaking in acid (H_2SO_4) or alkaline solutions (NaOH)	Over 95% germination by a 15–30 min soak in 97–60% (H_2SO_4)	Saxena and Khan 1974; Muthana and Arora 1983; Pasiecznik et al. 1998; Muthana 1988; Tewari and Harsh 1998
Asexual; suckers/ cutting	Stem cutting in controlled chambers with auxins singly or in combination	Rooting	Kaul 1956; Harris et al. 1996; Dick et al. 1991; Lima 1990; Leakey et al. 1990; Sandys-Winsch and Harris 1991
Grafting	Graft compatible with other species of *Prosopis*; cleft grafting and patch grafted seedlings	Grafted plants; most successful scions from the spineless and most erect *P. juliflora* trees and grafted onto *P. alba* rootstock	Wojtusik and Felker 1993; Pareek and Purohit 2002
Tissue culture	Shoot tips; nodal and apical tip explants top six nodes of *P. juliflora*	Complete plants; multiple shoot formation and rooting from shoots; micropropagation	Walton et al. 1990; Batchelor et al. 1989, 1990; Yao et al. 1989; Nandwani and Ramawat 1991; Arce and Balboa 1991

18.4.5 Economic Importance

Prosopis plays a leading role in the afforestation of arid lands, and has also been planted widely for firewood. *Prosopis* wood and pods play an important role in local economies in many regions of the world, including both its native range and where introduced. As arid zones are lacking in profitable natural resources, it is essential to promote local products in an attempt at import substitution. Pasiecznik et al. (2001) and Pasiecznik (1999) have provided a comprehensive account of the generic uses of *P. juliflora*. The tree is capable of growing on degraded land under arid conditions. Being a multipurpose tree, *Prosopis* was shown to have fitted in very well into 12 dryland agroforestry systems, controlling soil erosion, stabilising sand dunes, improving soil fertility, reducing soil salinity, providing fuel energy resources, and furnishing construction timber and furniture wood. Bark of *P. juliflora* is rich in tannin, and is used for roofing in Colombia. *P. juliflora* trees have also been used to shelter agricultural crops from wind and to reduce the movement of soil and sand, acting as windbreaks in desert lands where wind velocity is quite high and soil erosion is a serious problem. The wood is used for parquet floors, furniture and turned items, fence posts, pilings, as a substrate for producing single-cell protein, but most of all for fuel.

All these uses will help to increase economic activity in arid zones, promoting investment with a positive impact on livelihood enhancement. *P. juliflora* will enhance the scope for the promotion of goods from arid zones as 'greening the desert'. Import substitution of wood and feed products, which can be produced from stands of *Prosopis*, should be a primary aim of rural development.

18.4.5.1 Fuel

Prosopis firewood and charcoal are traded in all types of arid and semi-arid regions. Such markets evolve locally due to the relative ubiquity of *Prosopis* stands and the very low cost of the raw materials and their product processing. *P. juliflora* produces good quality fuel of high calorific value, which burns well even when freshly cut. It also produces high quality charcoal and its heartwood is strong and durable.

18.4.5.2 Fodder and Food

Mesquite pods are among the earliest known foods of prehistoric man in the New World. The pods are high in protein and sugars and represent a potentially important fodder for livestock, and/or food for humans. The pods and leaves are used as feed and forage for grazing animals, and to supplement human diets, and the plants are also used to promote the production of honey. Nectar from mesquite yields a superior honey. Today, flour products made from the pods are still popular, although only sporadically prepared, mostly by Amerindians. Pods are made into gruels, sometimes fermented to make a mesquite wine. Toasted seeds are added to coffee. However, the pods have been reported to result in facial contortions, impacted rumen and constipation among livestock. These ill effects are sometimes fatal. The leaves contain various chemicals known to affect palatability to livestock, but that also suppress the germination and growth of crops, weeds and other trees (Mwangi and Swallow 2005). The gum, which forms an adhesive mucilage, is used as an emulsifying agent, in confectionary, and in mending pottery. Roots contain 6–7% tannin, which might discourage *Rhizobia*.

18.4.5.3 Folk Medicine

According to Hartwell (1967–1971), the juice is used in folk remedies for cancerous growth "superfluous flesh", catarrh, colds, diarrhoea, dysentery, excrescences, eyes, flu, cold, hoarseness, inflammation, itch, measles, pinkeye, stomachache, sore throat, and wounds (Duke and Wain 1981). Pima Indians drank the hot tea to treat sore throats (Lewis and Elvin-Lewis 1977). Aqueous and alcoholic extracts are markedly antibacterial.

18.5 Conclusion

Around 36%–46% of the Earth's land area is classified as desert. Desert lands around the world presently attract little inward investment due to very low and variable amounts and quality of agricultural and forest products. Many were once fertile but, due to poor management and over-exploitation of natural resouces, are now degraded, desertified or covered with poor vegetation. Traditional agroforestry systems based on experiences gained from the sustainable growth of drought-tolerant and nitrogen-fixing trees may represent a possible long-term solution, leading to sustainable increases in productivity, reclaimation of soils and rehabilitation of rural economies. Thus, both *Prosopis* species described in this chapter are well-suited to desert regions.

The developed nations are reaching the limits of sustainable productivity of their land. Demand on forest land for conversion to agriculture continues unabated, whereas in deserts land is underutilised. Co-ordinated efforts at the regional, national and international level are now required to develop deserts profitable by using new technologies. *Prosopis* trees already play a vital role in the ecology and economy of arid and semi-arid regions. They now form an integral part of several sustainable land use systems that are improving the livelihoods of rural desert populations while preventing further soil degradation and assisting land reclamation. Desertification can be reversed only if economic value can be conferred on arid zones. This involves the production, processing and commercialisation of desert plants. Many have been suggested as 'miracle' plants over the last few decades, but most have failed to be adopted, being relatively intensive crops requiring high levels of investment and relatively better sites and soils. With low and very variable rainfall, returns will be poor unless irrigation is available. *Prosopis*-based agroforestry has been promoted and encouraged for wide adaptation by rural farmers in dry zones.

In deserts, tree planting has been promoted largely for the positive effects on the environment, reasons that are not immediate priorities for land users. Only those trees that produce financial returns have been widely accepted, planted and promoted by farmers. *Prosopis* species are already there to help combat deserts. The highest priorities for future work with *Prosopis* include an urgent need to evaluate valuable germplasm and to continue field-testing to select and improve the best material for specific arid zone sites. However, there is also a great need for the research results obtained so far to be integrated into agricultural and forestry development programmes. Such approaches will provide practical help for commercial plantation of leguminous trees and their products.

Although *P. cineraria* plays a vital role as an agroforestry species in some parts of its natural range, little success has been achieved in planting it elsewhere. Further work is needed to select fast-growing, disease-resistant trees, and to develop rapid clonal propagation methods to establish these in a range of arid conditions under which it might prove useful. *P. cineraria* displays considerable genetic variation, particularly in populations close to the edge of its natural range, which are often threatened by overgrazing. Genetic conservation of this valuable resource is considered a priority. *Prosopis* species are required primarily for biomass production; regeneration through nodal bud culture is an ideal tissue culture approach to produce large number of desired plants. Clonal propagation leading to monoclonal population is not required in such cases. Mixing of selected fast-growing clonally propagated *Prosopis* species followed by planting would be the ideal way from a forestry point of view. However, the present state of knowledge regarding regeneration in *Prosopis* suggests that there is still a wide gap between the laboratory feasibility of micropropagation and the practical application of this technology at field level. Therefore, conventional method of propagation (seeds) must still be used until these barriers are removed. The availability of a highly regenerative system (embryogenesis/organogenesis), allowing promising lines to be followed would be extremely useful for the large-scale multiplication of *P. cineraria* and related *Prosopis* species. The development of desert regions is thus an issue to be managed scientifically.

At present, *P. juliflora* provides more than 70% of the total fuel wood needs of the populations of tropical arid and semi-arid parts of India. Moreover, a sizeable proportion of the cattle, goat and sheep population of these areas feed primarily on the pods and leaves of *P. cineraria* and *P. juliflora*. Thus, eradication of *P. juliflora* is clearly not the answer; indeed, it appears that the deserts can be made profitable by adopting simple management, turning weedy stands into productive agroforestry systems. Thus the introduction and management of various exotic *Prosopis* species can play a significant role in uplifting rural communities, especially in arid and semi-arid tracts of the country. There is also scope for the promotion of goods from arid zones as 'greening the desert'. Import substitution of wood and feed products, which can be produced from *Prosopis*, should be a primary aim of rural development. Undoubtedly both species will have a major impact on man's survival in desert regions.

References

Abdelbari E (1986) The identity of common mesquite *Prosopis* species. Pamphlet No.1, *Prosopis* project supported by IDRC. Khartoum, Sudan

Alban L, Matorel M, Romero J, Grados N, Cruz G, Felker P (2002) Cloning of elite, multipurpose trees of the *Prosopis juliflora/pallida* complex in Piura. Peru Agrofor Syst 54:173–182

Anonymous (1980) Firewood crops, vol. 1. National Academy Press, Washington, DC, pp 150–151

Arce P, Balboa O (1991) Seasonality in rooting of *Prosopis chilensis* cuttings and in vitro micropropagation. For Ecol Manage 40:163–173

Arya HC, Shekhawat NS (1986) Clonal multiplication of tree species in the Thar Desert through tissue culture. For Ecol Manage 16:201–208

Arya S, Tomer R (1989) Study of rooting in juvenile stem cutting of *Prosopis cineraria* (L) MacBride. Jeevanti 7:65–68

Arya S, Tomer R, Toky OP (1994) Effect of plant age and auxin treatment on rooting response in stem cutting of *Prosopis cineraria* (L.) Druce. J Arid Environ 27:99–103

Arya S, Kumar N, Toky OP (1995) *Khejri* (*Prosopis cineraria* L. Druce): its value, research and extension. HDRA-ODA Project, India

Bajaj YPS (1986) Biotechnology of tree improvement for rapid propagation and biomass energy production. In: Bajaj YPS (ed) Biotechnology in agriculture and forestry trees: vol 1. Springer, Heidelberg, pp 1–23

Batchelor CA (1990) In vitro regeneration of tropical leguminous trees and shrubs with particular reference to the genus *Prosopis*. M.Phil/PhD Transfer Report, Department of Biological Sciences, Coventry University, Coventry, UK

Batchelor CA, Yao D, Koechler MJ, Harris PJC (1989) In vitro propagation of *Prosopis* species (*P. chilensis*, *P. cineraria* and *P. juliflora*). Ann Sci For 46:110–112

Bhandari MM (1978) Flora of the Indian Desert. Scientific, Jodhpur, India

Burkart A (1952) Las Leguminosas Argentinas. ACME, Buenos Aires

Burkart A (1976) A monograph of the genus *Prosopis* (Leguminosae sub. Fam. Mimosoideae). (Part 1&2) S J Arnold Arboretum, Harvard University 57:219–249, 450–525

De Souza SM, Felker P (1986) The influence of stock plant fertilization on tissue concentrations of N, P and carbohydrates and the rooting of *Prosopis alba* cuttings. For Ecol Manage 16: 181–190

Dick J McP, East K, Leakey RRB (1991) Influence of propagation environment and cutting length on the rooting of *Prosopis juliflora* (Swartz) DC. Nitrogen Fixing Tree Res Rep 9:114–116

Duke JA, Wain KK (1981) Medicinal plants of the world. Computer index with more than 85,000 entries, vol 3. FAO, Rome, Italy, pp 277–295

Felker P (1991) *Prosopis* in Latin America: genetic conservation and economic development for the 21st century. In: *Prosopis* germplasm workshop, 3–6 December 1991, Mendoza, Argentina, organised by the International Development Research Centre, Ottawa, Canada

Felker P, Clark PR (1981) Rooting of mesquite (*Prosopis*) cuttings. J Range Manag 34:446–448

Felker P, Moss J (1996) *Prosopis*: semiarid fuel wood and forage tree. Building consensus for the disenfranchised – a workshop, 13–15 March, 1996, United States National Academy of Science, Washington DC

Felker P, Cannell GH, Clark PR, Osborn JF, Nash P (1983a) Biomass production of *Prosopis* species (mesquite), leucaena, and other leguminous trees grown under heat drought stress. For Sci 29:592–606

Felker P, Cannell GH, Clark PR, Osborn JF, Nash P (1983b) Effects of irrigation on biomass production of 32 *Prosopis* (mesquite) accessions. Exp Agric 19:187–198

FFN (1991) Spotlight on species: *P. cineraria*. Farm Forestry News 4(3)

Ffolliot PF, Thames JL (1983) Collection, handling, storage and pre-treatment of *Prosopis* seeds in Latin America. FAO, Rome, Italy

Gates PJ, Brown K (1988) *Acacia tortilis* and *Prosopis cineraria:* leguminous trees for arid areas. Outlook on Agriculture 17:61–64

Goyal Y, Arya HC (1981) Differentiation in cultures of *Prosopis cineraria* Linn. Curr Sci 50: 468–469

Goyal Y, Arya HC (1984) Tissue culture of desert trees: I. Clonal multiplication of *Prosopis cineraria* by bud culture. J Plant Physiol 115:183–189

Green B, Tabone T, Felker P (1990) A comparison of amide and ureide nitrogen sources in tissue culture of tree legume *Prosopis alba* clone B2V50. Plant Cell Tissue Organ Cult 21:83–86

Gurumurti K, Raturi DP, Bhandari HCS (1984) Biomass production in energy plantations of *Prosopis juliflora*. Indian Forester 110:879–894

Harris PJC, Pasieczink NM, Bradbury M, Ver-Cruz MT (1996) Comparative physiology, field performance and propagation of *Prosopis*. Overseas Development Administration, London, UK, ODA Project R4733, Final Report

Hartwell JL (1967–1971) Plants used against cancer. A survey. Lloydia 30–34

Jordon M (1987) In vitro culture of *Prosopis* species. In: Bonga JM, Durzan DJ (eds) Cell and tissue culture in forestry, vol. 3. Case histories: Gymnosperms, Angiosperms and Palms. Nijhoff/Junk, The Netherlands, pp 370–384

Jordan M (1988) In vitro culture of *Prosopis* species. In: Bonga JM, Durzan DJ (eds) Cell and tissue culture in forestry, vol 3. Nijhoff, Boston, pp 370–384

Jordan M, Cortes I, Goreux A (1987) Potentialities of cell and callus tissue culture to regenerate two mesquite species (*Prosopis tamarugo* and *P. chilensis*). Gartenbauwissenschaft 52: 166–169

Kackar NL, Solanki KR, Singh M, Vyas SC (1991) Micropropagation of *Prosopis cineraria*. Indian J Exp Biol 29:65–67

Kackar NL, Vyas SC, Singh M, Solanki KR (1992) In vitro regeneration of *Prosopis cineraria* (L.) Druce using root as explants. Indian J Exp Biol 30:429–430

Kant U, Ramani V (1990) Insect induced plant galls in tissue culture. Proc Indian Acad Sci Animal Sci 99:257–265

Kaul ON (1956) Propagating mesquite (*Prosopis juliflora*) by root and shoot cuttings. Indian Forester 82:569–572

Khoshoo TN, Subrahmanyam GV (1989) Eco-development of arid lands in India with non-agricultural economic plants – a holistic approach. In: Wickens GE, Goodin JR, Field DV (eds) Plants for arid lands. Unwin Hyman, Boston, pp 243–265

Killian SE (1990) A study on the germinative behavior of the seeds of some *Prosopis* species. In: Habit MA Saavedra JC (eds) The current state of knowledge on *Prosopis juliflora*. FAO, Rome

Leakey RRB, Last FT (1980) Biology and potential of *Prosopis* species in and environments with particular reference to *P. cineraria*. J Arid Environ 3:9–24

Leakey RRB, Mesen JF, Tchoundjeu Z, Longman KA, Dick JM, Newton A, Matin A, Grace J, Munro RC, Muthoka PN (1990) Low-technology techniques for the vegetative propagation of tropical trees. Commonw For Rev 69:247–257

Lesney MS, Felker P (1995) Two field and greenhouse diseases of *Prosopis* (mesquite). J Arid Environ 30:417–422

Lewis WH, Elvin-Lewis MPF (1977) Medical botany. Wiley, New York

Lima PCF (1990) *Prosopis* vegetative propagation through cuttings. In: Habit MA, Saavedra JC (eds) The current state of knowledge on *Prosopis juliflora*. FAO, Rome, pp 223–227

Lima PCF, Haji FNP (1993) Occurrence of caterpillar defoliation in mesquite (*Prosopis juliflora*) in the semi-arid tropics of Brazil. Bol Pesqui Florestal 26/27:57–59

Luna RK (1996) *Prosopis juliflora* (Swartz) DC. In: Plantation trees. International Book Distributors, Delhi

Mahoney D (1990) Trees of Somalia – a field guide for development workers. Oxfam/HDRA, Oxford

Manga VK, Sen DN (1995) Influence of seed traits on germination in *Prosopis cineraria* (L.) MacBride. J Arid Environ 31:371–375

Mann HS, Shankarnarayan KA (1980) The role of *Prosopis cineraria* in an agropastoral system in Western Rajasthan. In: LeHouerou HN (ed) Browse in Africa. The current state of knowledge. International Livestock Centre for Africa, Addis Ababa, Ethiopia, pp 437–442

Murashige T (1974) Plant propagation through tissue culture. Annu Rev Plant Physiol 25:135–166

Muthana KD (1988) Technology for propagation and by-products utilization of *Prosopis chilensis*. Indian Farming 38:11–13, 21

Muthana KD, Arora GD (1983) *Prosopis juliflora* (Swartz) D.C. A fast growing tree to bloom the desert. CAZRI, Jodhpur, India, CAZRI Monograph No.22

Mwangi E, Swallow B (2005) Invasion of *Prosopis juliflora* and local livelihoods: case study from the lake Baringo area of Kenya. ICRAF Working Paper no. 3. World Agroforestry Centre, Nairobi

Nandwani D (1990) Tissue culture of forest plants of Rajasthan I. *Prosopis* spp. PhD thesis, University of Jodhpur, Jodhpur, pp 1–101

Nandwani D, Ramawat KG (1989) In vitro growth of three species of *Prosopis*. In: Tandon P (ed) Proc Tissue Culture Conference VII, NEHU, Shillong

Nandwani D, Ramawat KG (1991) Callus culture and plantlets formation from nodal explants of *Prosopis juliflora* (Swartz). DC Indian J Exp Biol 29:523–527

Nandwani D, Ramawat KG (1992a) In vitro regenerative potential of explants and callus morphogenesis in *Prosopis tamarugo*. Gartenbauwissenschaft 57:106–111
Nandwani D, Ramawat KG (1992b) High frequency plantlets regeneration from seedling explants of *Prosopis tamarugo*. Plant Cell Tissue Organ Cult 29:173–178
Nandwani D, Ramawat KG (1993) In vitro plantlets formation through juvenile and mature explants in *Prosopis cineraria*. Indian J Exp Biol 31:156–160
Nandwani D, Purohit SD, Jain SM, Ramawat KG (2005) Propagation technique in woody plants with special reference to arid and semiarid zone trees. In: Shanmughavel P, Ignacimuthu SJ (eds) Tree improvement and Biotechnology. Pointer, Jaipur, India, pp 16–52
Pareek OP (1998) National Research Centre for Arid Horticulture, In: Ghosh SP, Bhatanaga PS (eds) 50 years of horticulture research. Division of Horticulture, ICAR, New Delhi, pp 95–99
Pareek OP, Purohit AK (2002) Patch budding in *khejri (Prosopis cineraria)*. Indian J Hortic 59:89–94
Pareek OP, Purohit AK, Samadhia DK (2006) Graft compatibility among the Asian and *Algarobia Prosopis* species. J Arid Hortic 1:1–5
Parihar DR (1993) Insect fauna of *Khejri*, *Prosopis cineraria* of arid zone. Indian J For 16:132–137
Parihar DR (1994) Galls and gall-makers in *Khejri* (*Prosopis cineraria* Linn. Druce) of arid zone of India. Ann Arid Zone 33:313–317
Paroda RS (1979) Plant resource of Indian arid zones for industrial uses. In: Goodin JR, Northington KD (eds) Arid land plant resources. Inter CASLAS, Texas, pp 261–281
Pasiecznik N (1999) *Prosopis* – pest or providence, weed or wonder tree? Eur Trop For Res Network Newslett 28:12–14
Pasiecznik NM, Felker P, Harris PJC, Harsh LN, Cruz G, Tewari JC, Cadoret K, Maldonado LJ (2001) The *Prosopis juliflora–Prosopis pallida* complex: a monograph. HDRA, Coventry, UK
Pasiecznik NM, Harris PJC, Tavares J de P, Cassama M (1998) Pretreatment of *Prosopis* seeds to break dormancy. Int Tree Crops J 9:187–193
Pimentel MdeL (1982) Extração de sementes de algarobeira (*P.juliflora* (SW) DC) atraves de processo quimico. In: I Simposio Brasileiro sobre Algarobeira, vol 1. Natal, Brazil, pp 330–335
Puri S, Jain M, Sharma P (1992) In vitro regeneration of *Prosopis cineraria*. Nitrogen Fixing Tree Res Rep 10:189
Raj Bhansali R (2001) Biotechnology in arid agriculture and forestry. In: Verma SK, Nahar NM, Raj Bhansali R, Jindal SK, Satya Vir (eds) Priorities of research for development of arid regions. Scientific, Jodhpur, pp 113–127
Raj Bhansali R (2003) In vitro multiplication of important desert trees. In: Narain P, Amal Kar, Kathju S, Singh MP, Kumar P (eds) Human impact on desert environment. Arid Zone Research Association / Scientific, Jodhpur, India, pp 405–410
Raj Bhansali R (2008) Butt rots of *Prosopis cineraria*. In: Narain P, Singh MP, Amal Kar, Kathju S, Kumar P (eds) Diversification of arid farming systems. Arid Zone Research Association / Scientific, Jodhpur, pp 361–367
Raj Bhansali R, Jindal SK (1997) Diseases of desert trees and their management. In: Gupta JP, Sharma BM (eds) Agroforestry in sustained productivity in arid regions. Scientific, Jodhpur, India, pp 135–145
Raj Bhansali R, Jindal SK (2000) Role of farmers in promotion of eco-friendly multipurpose trees of arid zone. In: Chaudhary V, Singh K, Kakralya BL (eds) Environmental protection (forwarded by Sunder Lal Bhuguna). Pointer, Jaipur, pp 92–101
Raj Bhansali R, Singh M (2003) Micropropagation of arid zone fruit trees of India. In: Jain SM, Ishii K (eds) Micropropagation of woody trees and fruits. Kluwer, Dordrecht, pp 381–432
Ramani V, Kant U (1993) In vitro "gall" formation on the leaf rachis of *Prosopis cineraria* (Linn.) Druce (Mimosaceae) by *Lobopteromyia prosopidis mani* (Diptera: Cecidomyiidae). Phytophaga 5:7–11
Ramawat KG, Nandwani D (1991a) Arid-land afforestation by tissue culture. Bionature 11:103–109
Ramawat KG, Nandwani D (1991b) Propagation of *Prosopis* species: problem perseverance and perspectives. Ann Arid Zone 30:247–258

Rao HS (1953) Vegetative propagation and forest tree improvement. Indian For 79:176–183
Rawat TS, Sharma HC, Pundir JPS (1982–1983) Air layering – a success in propagation of Khejri (*Prosopis cineraria* L.). Udanika 5:43–44
Roonwal ML (1975) Field and other observations on the harvester termite, *Anacanthotermes macrocephalus* (Hodotermitidae) from Indian desert. Z Angew Entomol 78:424–440
Sandys-Winsch DC, Harris PJC (1991) A simple method for the vegetative propagation of *Prosopis juliflora*. Nitrogen Fixing Tree Res Rep 9:117–118
Saxena SK, Khan WA (1974) A quick method for obtaining clean seeds of *Prosopis juliflora*. Ann Arid Zone 13:269–272
Sharma K, Lodha P, Kant U (1995) Histopathology of stem gall of *Prosopis cineraria* (Linn.) Druce induced by an unknown chalcid. J Indian Bot Soc 74:129–133
Shekhawat NS, Kackar A (1987) Protoplasts of *Prosopis cineraria*: isolation and culture. Nitrogen Fixing Tree Res Rep 5:51–53
Shekhawat NS, Rathore TS, Singh RP, Deora NS, Rao SR (1993) Factors affecting in vitro clonal propagation of *Prosopis cineraria*. Plant Growth Regul 12:273–280
Singh MP (1998) Injurious hexapoda associated with *Prosopis*. In: Tewari JC, Pasiecznik NM, Harsh LN, Harris PJC (eds) *Prosopis* species in the arid and semi-arid zones of India. *Prosopis* Society of India and the Henry Doubleday Research Association, Coventry, UK, pp 105–108
Solanki KR, Kackar NL, Jindal SK (1984) Propagation in *Prosopis cineraria* (L.) MacBride by air layering. Curr Sci 53:1166–1187
Solanki KR, Kackar NL, Jindal SK (1986) Propagation in *Prosopis cineraria* (L.) MacBride by air layering. Indian Forester 112:202–207
Srivastava KK, Mishra DK (1998) Diseases of *Prosopis juliflora* in Rajasthan. In: Tewari JC, Pasiecznik NM, Harsh LN, Harris PJC (eds) *Prosopis* species in the arid and semi-arid zones of India. Society of India and the Henry Doubleday Research Association, Coventry, UK, pp 109–110
Tabone TJ, Felker P, Bingham RL, Reyes I, Loughrey S (1986) Techniques in the shoot multiplication [tissue culture] of the leguminous tree *Prosopis alba* clone B2V50. For Ecol Manage 16:191–200
Tewari JC, Harsh LN (1998) Forestry research in arid tract of India. In: Faroda AS, Singh M (eds) Fifty years of arid zone research in India. CAZRI, Jodhpur, pp 307–322
Tewari JC, Pasiecznik NM, Harsh LN, Harris PJC (1998) *Prosopis* species in the arid and semi-arid zones of India. In: Proceedings of a conference held at the Central Arid Zone Research Institute, Jodhpur, Rajasthan, India, 21–23 November 1993. The *Prosopis* Society of India and the Henry Doubleday Research Association, UK
Tewari JC, Harris PJC, Harsh LN, Cadoret K, Pasiecznik NM (2000) Managing *Prosopis juliflora* (Vilayati Babul): a technical manual. CAZRI, Jodhpur, India and HDRA, Coventry, UK
Valdivia SV (1972) El algarrobo. Una especie forestall prometedora para los trópicos áridos. Boletin de Divulgacion No. 32. Ministry of Agriculture, Lima, Peru
Verma, SK (1985) The leaf beetle *Clytria succincta* on *Prosopis cineraria*. FAO Plant Protection Bulletin 33:123–124
Wainright H, England N (1987) The micropropagation of *Prosopis juliflora* (Swartz) D.C.: establishment in vitro. Acta Hortic 212:49–53
Walton T, Harris PJC, Batchelor CA (1990) Comparative rooting response of shoot tips from six *Prosopis* species. Nitrogen Fixing Tree Res Rep 8:154–155
Wilson JRC, Munro K, Ingleby J, Dick McP, Mason PA, Jefwa J, Muthoka PN, Newton AC, Leakey RRB (1990) Agroforestry and mycorrhizal research for semi-arid lands of East Africa. Institute of Terrestrial Ecology, Penicuik, Scotland, UK
Wojtusik T, Felker P (1993) Interspecific graft incompatibility in *Prosopis*. For Ecol Manage 59:329–340
Wojtusik T, Felker P, Russell EJ, Benge MD (1993) Cloning of erect, thornless, non-browsed nitrogen-fixing trees of Haiti's principal fuelwood species (*Prosopis juliflora*). Agrofor Syst 21:293–300

Yao D, Batchelor CA, Koechler JJ, Harris PJC (1989) In vitro regeneration of *Prosopis* species (*P. cineraria* and *P. juliflora*) from nodal explants. Chin J Bot 1:89–97

Yousuf M, Gaur M (1993) Some noteworthy insect pests of *Prosopis juliflora* from Rajasthan, India. In: Tewari JC, Pasiecznik NM, Harsh LN, Harris PJC (eds) *Prosopis* species in the arid and semi-arid zones of India. Conference Proceedings, 21–23 November 1993, CAZRI, Jodhpur, Rajasthan, India

Chapter 19
Biotechnology Advances in Jojoba (*Simmondsia chinensis*)

Muppala P. Reddy and Jitendra Chikara

Abstract Wax esters have important applications in medicine, and in the cosmetics and food industries, besides their more traditional usage as lubricants. The value of the wax from sperm whales was one of the factors responsible for this animals being hunted to near extinction, which prompted the eventual ban on harvesting and the search for alternative sources. Recognition of jojoba oil as an alternative to sperm whale oil has led to a surge of interest in jojoba across the globe. The hardiness of this plant, which is amenable to cultivation even on water-deficient wastelands, has led to it being cultivated as a crop in several semi-arid and arid regions of the world. In addition, oil from the seed de-oiled cake is rich in protein and can be used as livestock feed and as a source of commercial enzymes. The plant is dioecious, and exhibits tremendous variability in male:female ratio in a given population, with male plants generally outnumbering female plants, leading to low yields as expected due to heterogeneity in the population. High yielding genotypes have been selected from experimental plantations, and vegetative propagation methods have been developed to provide genetically uniform, known sex plants to boost yields. Due to limited production, jojoba waxes are not available for a number of applications in spite of high demand. The advent of genetic engineering has provided novel opportunities to tailor the composition of plant lipids and also engineer agronomically suitable oilseed crops to produce high levels of wax esters in the seed oil. This chapter discusses efforts made towards the domestication, genetic improvements for yield and oil content, detoxification of cake for use as a live stock feed, and aspects of micropropagation of this species.

M.P. Reddy (✉)
Plant Stress Genomics and Technology Center, 4700 King Abdullah University of Science and Technology, Mail Box # 2053, Thuwal 23955-6900, Kingdom of Saudi Arabia
e-mail: reddymuppala@gmail.com

J. Chikara
Central Salt and Marine Chemicals Research Institute (CSIR), G.B. Marg, Bhavnagar, Gujarat 364 021, India

K.G. Ramawat (ed.), *Desert Plants*,
DOI 10.1007/978-3-642-02550-1_19,

19.1 Introduction

Jojoba [*Simmondsia chinensis*, (Link) Schneider] is an evergreen, perennial dioecious shrub native to the Sonoran desert, north-western Mexico and Baja California (Benzioni 1995). The seeds contain a unique oil (ca. 50–55%), commonly known as liquid wax, that is used commercially in the cosmetic, pharmaceutical and lubricant industries. This oil is unique in its molecular simplicity, stability under pressure and high temperature, and unsaturation characteristics, and it can be stored without going rancid. Although this plant has attracted the attention of scientists and agriculturists the world over as a new industrial crop for marginal lands (Gentry 1958), the area under plantation is decreasing significantly due to several constraints (Tobares et al. 2004; Coates et al. 2006). Being dioecious in nature and an obligate cross-pollinated crop, Jojoba exhibits tremendous variability in both morphological and yield parameters. Multiple recombination in the population gene pool has led to a huge range of possible gene combinations for selection and improvement to make cultivation more attractive and commercially viable.

19.2 Genetic Improvement

Worldwide, seed-raised jojoba populations in different countries have exhibited substantial variability in experimental as well as commercial plantations. In Israel, cultivation of jojoba was initiated during the mid-1960s in Negev by the Practical Research Institute at Ben Gurion University. Since then, significant contributions have been made in selection, genetic improvement, agronomy, pollen biology and micropropagation, etc. Both male and female plants were selected on the basis of desirable characters such as unique oil characteristics, natural resistance to pests, high harvest potential and an efficient plant structure. In Argentina, Tobares et al. (2004) conducted experiments to identify superior genotypes and establish proper agronomy practices, and observed significant differences in growth, seed yield and oil content. Purcell and Purcell (1988) also observed large variations in jojoba plantations raised in Australia and made attempts to improve the species for higher yields. Ash et al. (2005) described the agronomy and management of jojoba cultivation and evaluated the influence of fertiliser on growth, flower bud survival, seed yield, and oil content.

In the late 1970s, the Central Salt and Marine Chemicals Research Institute (CSMCRI) introduced jojoba to coastal sand dunes under a typical semi-arid climate with 300–500 mm annual rainfall. Plantations raised from seeds showed tremendous variability in phenotypic characters and yield; high and consistently yielding genotypes were selected for further propagation (Kumari et al. 1991; Chikara and Kumari 1991).

Table 19.1 Variability in morphological and phenological characters of selected male plants of Jojoba, *CV* Coefficient of variation

No.	Character		Erect type	Semi-erect type	Prostrate type
1	Plant height (m)	Range	1.34–2.70	1.34–2.41	0.62–1.59
		Mean	2.07	2.31	1.00
		CV (%)	17.87	18.61	36.00
2	Plant spread (m)	Range	2.15–2.72	3.72–6.22	2.00–5.52
		Mean	2.47	3.58	4.25
		CV (%)	9.72	22.11	39.38
3	Plant volume (m^3)	Range	5.60–12.92	3.11–32.92	6.89–86.50
		Mean	9.12	3.24	11.52
		CV (%)	30.15	63.11	41.05
4	Internode length (cm)	Range	2.25–2.97	1.70–3.82	2.42–3.02
		Mean	2.58	2.83	2.64
		CV (%)	10.08	15.55	6.44
5	Pollen fertility (%)	Range	78.57–98.00	70.59–97.22	89.89–97.44
		Mean	87.20	86.57	86.17
		CV (%)	9.86	6.79	10.27
6	Pollen grain size (μm)	Range	33.09–44.23	26.13–43.42	31.66–39.27
		Mean	37.52	35.70	34.23
		CV (%)	10.93	12.77	11.95
7	Flowering duration (days)	Range	25.00–33.00	10.00–37.00	32.00–37.00
		Mean	27.25	23.82	29.57
		CV (%)	20.81	23.28	24.45

19.2.1 Selection of Male Plants

Being wind pollinated, Jojoba male plants, with their large variation in blooming period, quantity of pollen produced, and pollen viability, greatly influence seed yield. Significant differences in pollen fertility, size of pollen and blooming period (Table 19.1) were noted in male plants (Kumari et al. 1991). Vaknin et al. (2003) also reported variations in pollen production and in vitro germination of pollen from different male plants. These characters are suggested as promising traits for selection of potential male cultivars.

19.2.2 Selection of Female Plants

Due to wide variations in seed yield, selection of female plants is most important for increasing yields in future populations. CSMCRI has made efforts in selecting female plants, which have continuously and consistently performed better over more than two decades, i.e. 1982–2007. It is evident from Table 19.2 that these plants exhibited significant variations in the percentage of productive plants as well as yield per plant.

As the plantation was raised on coastal sand dunes, the lower yields can be attributed to the poor soil conditions. However, under continuous selection, seed

Table 19.2 Average seed yield per plant (g) in mother population of Jojoba

No.	Year	Total no. of female plants	Plants bearing fruits (%)	Mean seed yield per plant (g)
1	1982	758	34.30	6.76
2	1983	754	50.27	10.12
3	1984	746	37.0	16.33
4	1985	741	32.25	18.93
5	1986	733	62.35	63.29
6	1987	733	32.20	22.30
7	1988	735	35.10	08.63
8	1989	735	3.13	07.26
9	1990	735	0.26	15.78
10	1991	725	10.07	42.35
11	1992	720	29.03	43.30
12	1993	705	24.39	112.15
13	1994	660	37.72	48.27
14	1995	620	33.38	35.75
15	1996	575	17.21	43.81
16	1997	520	37.11	94.64
17	1998	503	28.42	151.20
18	1999	480	23.12	65.97
19	2000	480	44.40	251.57
20	2001	456	41.01	267.98
21	2002	444	27.88	156.77
22	2003	444	32.65	234.89
23	2004	416	23.83	211.43
24	2005	347	25.67	98.56
25	2006	339	20.12	55.61
26	2007	324	30.44	49.68

Table 19.3 Correlation coefficients between various plant attributes in female plants of Jojoba

Plant height (cm)	Plant spread (cm)	Plant volume	Inter nodal length (cm)	Seed size (cm)	Seed yield per plant (gm)	No. of seeds per plant	100-Seed weight (gm)
	0.558**	0.820**	–0.044	–0.187	0.360**	0.308**	0.315**
		1.000**	0.055	–0.221	0.512**	0.505**	–0.031
			0.0006	0.127	0.460*	0.413*	0.121
				0.124	–0.088	–0.069	0.077
					–0.079	–0.068	–0.002
						0.984**	–0.143
							–0.254

* $P < 0.05$, ** $P < 0.001$

yield of individual female plant was recorded and maintained separately to determine yield potential. Consistent and high yielding female plants were identified from the above population. Besides high yield, the selected plants also had higher seed weight. The relationship between seed yield and other component characters was studied, and correlation coefficients calculated for all possible combinations (Table 19.3); significant positive correlations between seed yield and morphological characters were observed (Chikara and Kumari 1991).

Table 19.4 Comparison of yield per plant (g) for the mother plantation, progeny plants and selected mother plants

Year	Mother population (100 plants)	Progeny plants	Selected mother plants (20 plants)
2003	234.89	396.87	683.04
2004	211.43	180.15	271.66
2005	98.56	68.19	235.53
2006	55.61	57.32	492.07
2007	49.68	60.86	640.28

Forti et al. (1985) reported inverse correlation between plant volume and fruit yield in jojoba. Whereas Yermanos (1983), Abramovich et al. (1976) and Ramonet and Morales (1985) reported that none of the morphological characters had a significant correlation with total yield. Ramonet (1988) reported a high correlation between the number of seeds per plant and seed yield.

A jojoba plantation was developed in the experimental plot of CSMCRI in Gujarat, India, and selection of superior female plants was made on the basis of seed yield. Later, the progeny of a few plants was also raised separately in a plant to row design. When seed yield of the 20-year-old mother population was compared with that of progeny plants and selected plants, it was observed that, even during the last 5 years (Table 19.4) seed yield of selected plants was found to be significantly higher than that of the mother population. In contrast, there was no significant difference between the yield of mother plants and progeny plants (half sib families). Forti et al. (1985) reported considerable similarity in several phenological, morphological and reproductive characters in the progenies as well as mother shrubs. Earlier, Ayerza (1985), Benzioni and Dunstone (1985) and Ramonet (1988) have also suggested that selection can significantly increase the yield in jojoba.

19.2.3 Breeding

Jojoba being dioecious in nature, it is of considerable interest to plant breeders. The open pollination among male and female plants leads to the production of a half sib population, which is expected to have sufficient variability. So far, no systematic and successful efforts have been undertaken to improve seed yield, oil and fatty acid content in jojoba through artificial crossing. CSMCRI have made attempts in this direction and significant differences in fruit set percentage were observed in various crosses according to the source of pollen or male parent. A close perusal of the data revealed that most female parent plants almost breed true to type in the F_1 generation for seed characters in various crosses. Among such crosses, the ability of female plants to combine with male plants varied significantly (CSMCRI, personal communication). Vaknin et al. (2003) found that fertilisation of female flowers with pollen from several pollen donors resulted in relatively high fruit set and emphasised the need to select male plants with high quality pollen for fertilisation of females blooming at different periods.

Lloyd and Webb (1996) were unable to generate superior seed producers in the F_1 generation even after a decade of selection and hybridisation. Ramonet and Morales (1985) made crosses between five male and five female 9-year-old jojoba plants to evaluate the differential capability of male pollen to fertilise female flowers, and observed significant differences in the ability of male plants to set fruit but they could not increase the percentage of fruit set. Benzioni and Vaknin (2002) and Purcell and Purcell (1988) also made unsuccessful attempts to increase fruit set in jojoba through artificial pollination.

Ayerza (1996) studied the fatty acid profile of jojoba wax and revealed variations in fatty acid composition between clones. Kondra and Thomas (1975) tried to modify the fatty acid components of different plants and concluded that the alternation/modification of fatty acid traits is controlled by a single gene. However, none of these studies showed any possibility to obtain such alternation through breeding. Hybridisation between selected male and female plants could result in significant differences in F_1 progeny seeds, particularly for $C_{20:1}$, $C_{22:1}$ and $C_{18:1}$ acids. When crossed with different male plants, the same female plant exhibited significant alteration in fatty acid composition, indicating that the male genotype does play a significant role (CSMCRI, personal communication). Tobares et al. (2004) studied agronomical and chemical traits as descriptors for discrimination and selection of jojoba clones using seed weight, yield and wax content as a selection index, and revealed large genetic variation among jojoba clones, which could permit improvement by selection and breeding.

19.3 Seed Oil Content and De-oiled Cake Applications

Being dimorphic, Jojoba shows tremendous variability in seed characteristics. The oil content does not vary significantly in the population. Clarke and Yermanos (1980) studied oil content in ca. 1,156 native jojoba plants and found that approximately 53.2% and 34% of samples exceeded 53%. This variation can be attributed mainly to variations in the environmental conditions at the time of seed setting. Gentry and McGill (1979) observed a correlation between large seeds and high wax content.

The meal remaining after oil extraction is rich in protein, carbohydrates, and fibre (Abbott et al. 1991); however, use of meal as food or feed is limited due to the presence of anti-nutritional factors such as simmondsin and its derivatives (Elliger et al. 1973), heat labile trypsin inhibitors (Samac and Storey 1981), and condensed tannins with antitryptic activity (Sanchez and Price 1988). Attempts have been made by a number of workers to remove the anti-nutritional factors in order to be able to utilise the meal for animal consumption (Verbiscar et al. 1981; Verbiscar and Banigan 1982; Wolf et al. 1994; Flo et al. 1997). Shrestha et al. (2002) isolated major proteins and demonstrated their in vitro digestibility. The biochemical parameters of rats fed on a diet containing 15% of the protein isolate were normal; the isolate caused no change in liver and kidney functions and was recommended as a safe food grade ingredient (El-Anany 2007).

19.4 Sex Identification

In most dioecious plants, pistils and stamens develop on separate individuals, which are distinguished as "pistillate plants" (female plants) and "staminate plants" (male plants), respectively (Parker 1990). In such plants, gender influences economic values (Coates and Ayreza 2008). Identification of the sex of the plant prior to its transplantation to the field will help to achieve the desired ratio of male and female plants, reduce cultivation costs and boost yields by increasing fruit-bearing plants per unit area. Attempts were made to identify male/female plants by quantification of metabolic constituents such as proteins, carbohydrate and nucleic acids; indeed, significant differences were found between male and female plants (Prasad and Iyengar 1982, 1985). Agrawal et al. (2007) developed sex-specific random amplified polymorphic DNA (RAPD) markers at an early stage of development of the taxon, which will help in screening of plants at the juvenile stage, in systematic plantation of the desired male:female ratio, and in further breeding and marker-assisted selection.

19.5 Molecular Aspects of Oil Synthesis

The liquid wax from jojoba seed oil has a carbon chain length of C_{38} to C_{44} and is composed mainly of $C_{20:1}$ fatty acids and $C_{22:1}$ fatty alcohols (Miwa 1971). It has a multitude of important applications in medicine, in the cosmetics industry, and as a lubricant. The high cost and limited production of jojoba waxes restrict their application in spite of high demand. However, the advent of genetic engineering has provided novel opportunities to increase the availability of plant-derived lipids (Kalscheuer et al. 2006). Enzymes involved in wax synthesis in developing jojoba seeds have been characterised (Metz et al. 2000) and the corresponding cDNAs cloned (Metz and Lassner 1994). Overexpression of these genes has resulted in production of jojoba wax in the seeds of transgenic model plants (Lassner et al. 1996; Lardizabal et al. 2000). Kalscheuer et al. (2006) demonstrated the possibility of de novo synthesis of wax esters in a recombinant microorganism by co-expression of a fatty alcohol-producing bifunctional acyl-coenzyme. A reductase from the jojoba plant and a bacterial wax ester synthase from *Acinetobacter baylyi* strain ADP1 were identified to provide a basis for the biotechnological production of these industrial relevant compounds from inexpensive and renewable substrates.

19.6 Vegetative Propagation

Jojoba being dioecious, it does not reproduce true-to-type plants by sexual propagation and high heterogeneity has been observed in seed-raised plantations (Yermanos et al. 1967; Purcell and Purcell 1988; Chikara and Kumari 1991). In order to maintain a proper sex ratio in the plantation, as well as desirable characters, selected best accessions only should be asexually propagated. To circumvent these impediments vegetative propagation techniques have been developed. Vegetative propagation by layering (Alcaraz and Ayla-Rocha 1982; Reddy 2003; Bashir et al. 2005), grafting (Bashir et al. 2006), and cuttings (Hogan et al. 1978; Feldman et al. 1980; Lee and Palzkill 1984, Benzioni 1995; Singh et al. 2003; Cao and Gao 2003; Bashir et al. 2007a) have established with satisfactory results using elite genotypes. However, a major limitation of these techniques is the availability of a maximum number of possible propagules. Also, propagation is seasonal.

19.6.1 Micropropagation

Micropropagation is an alternative method of vegetative propagation, which is well suited to the multiplication of elite clones, offers many advantages, is not limited by the number of selected elite genotype, produces pathogen-free plants, and can provide a commercial production within a limited time frame and space. The techniques can also be used for genetic improvement of the species. Micropropagation is accomplished by several means, i.e. multiplication of shoots from different explants such as shoot tips or axillary buds; or direct formation of adventitious shoots or somatic embryos from tissues, organs or zygotic embryos. In vitro culture – a key tools of plant biotechnology – exploits the totipotent nature of plant cells; such techniques have been exploited successfully for raising in vitro plants to the point where micropropagation has reached a commercial level in many plants in recent years (Chandra and Mishra 2003).

Several investigators have attempted to propagate jojoba in vitro on various culture media containing different concentrations of plant growth regulators with varying levels of success (Mandani et al. 1978; Birnbaum et al. 1984; Lee and Palzkill 1984; Jacoboni and Standardi 1987; Letouzé 1987; Scaramuzzi and D'ambrosio 1988; Chaturvedi and Sharma 1989; Mills and Benzioni 1992; Kacker et al. 1993; Benzioni 1995; Mills et al. 1997; Elhag et al. 1998; Llorente and Apostolo 1998; Sardana and Batra 1998; Agrawal et al. 1999; Khanam et al. 1999; Agrawal et al. 2002; Roussos et al. 1999; Gao and Cao 2001; Prakash et al. 2002, 2003; Tyagi and Prakash 2001, 2004; Bashir et al. 2007b, 2007c, 2007d; Bashir et al. 2008; Singh et al. 2008). In most such experiments, cytokinins (BA, kinetin and zeatin) have been used in combination with auxins (NAA, IAA and IBA). While working with different genotypes of jojoba, Bashir et al. (2007b, 2008) showed that BA alone was better than kinetin (alone) or BA + kinetin for in vitro shoot initiation whereas Singh et al. (2008), working with different mature (20 years old) genotypes of jojoba, reported that BAP in combination with adenine proved best for shoot induction and multiplication, and that an increase in KNO_3 concentration in the medium improved shoot multiplication rate and in vitro flowering in 20% of male cultures. The number of shoots produced in vitro may depend upon the source of explant (Gao and Cao 2001; Agrawal et al. 2002), the type of explant (Hassan 2003), the composition of media, the nature of growth regulators and their concentrations, plant genotype (Llorente and Apostolo 1998; Elhag et al. 1998; Prakash et al. 2003; Tyagi and Prakash 2004), and type of vessels and cultural conditions (Benzioni et al. 2003; Mills et al. 2004). Explants of male and female shoots even exhibit differential morphogenic behaviour under the influence of various adjuvants (Prakash et al. 2003). However, Singh et al. (2008) observed similar responses when working with mature genotypes of both male and female plants.

Experiments involving rooting on media with either auxin alone or in combination with other hormones showed significant differences (Chaturvedi and Sharma 1989; Tyagi and Prakash 2004; Bashir et al. 2008; Singh et al. 2008). Tyagi and Prakash (2004) observed differences in the rooting behaviour of male and female plants, whereas Singh et al. (2008) did not observe any variation in the rooting response of male and female genotypes. Clonal differences in rooting and subsequent acclimatisation were also recorded (Apostolo and Llorente 2000; Apostolo et al. 2001; Bashir et al. 2007d, 2008).

Meyghani et al. (2005) reported peat and perlite at a ratio of 1:1 or 1:2 (v/v) as the most suitable media for transplanting or adaptation of jojoba plantlets, whereas Singh et al. (2008) reported the best survival rate when sand alone was used as a substrate. The survival rate was higher for plantlets, which developed roots in vitro in response to IBA. However, no differences were observed in field establishment (Bashir et al. 2008).

19.6.2 Somatic Embryogenesis

Somatic embryos, which are bipolar structures, arise from individual cells and have no vascular connection with the maternal tissue of the explant. Embryos may develop directly from somatic cells (direct embryogenesis), or the development of recognisable embryogenic structures can be preceded by numerous, organised, non-embryogenic mitotic cycles (indirect embryogenesis). Somatic embryogenesis has a great potential for clonal multiplication. In addition, under controlled environmental conditions, somatic embryos germinate readily, similar to their seedling counterparts.

Besides being useful for clonal multiplication, Jojoba regeneration through somatic embryogenesis may also be useful for genetic transformation (Kim and Liu 1999) and to develop new products from jojoba oil (Benzioni 1995). Reports on somatic embryogenesis in jojoba are sparse, are based on indications from

Table 19.5 Somatic embryogenesis in Jojoba (*Simmondsia chinensis*)

No.	Explant	Morphogenetic response	Reference
1	Zygotic embryo	Embryogenesis	Lee and Thomas 1985
2	Zygotic embryo	Embryogenesis/somatic embryo	Wang and Janick 1986a
3	Leaf/zygotic embryo	Embryogenesis/somatic embryo	Wang and Janick 1986b
4	Leaf	Embryogenesis/somatic embryo	Hamama et al. 2001
5	Leaf/zygotic embryo	Embryogenesis/somatic embryo	Mohammed et al. 2008

immature zygotic embryos, and have focussed on in vitro wax production (Lee and Thomas 1985; Wang and Janick 1986a, 1986b). Hamama et al. (2001) developed a protocol for the induction, maturation and germination of somatic embryos from leaf tissue of jojoba. Direct somatic embryogenesis was observed with some zygotic embryo explants, whereas leaf-derived embryogenic calli did not mature on any of the maturation/germination media examined for up to 4 weeks of culture (Mohammed et al. 2008). The comprehensive efforts made towards developing suitable protocols for somatic embryogenesis/direct organogenesis in jojoba are summarised in Table 19.5.

19.7 Conclusions and Prospects

Jojobas is dioecious in nature, and thus exhibits tremendous variability in half sib populations at inter- and intra-clonal levels in morphological as well as yield-contributing characters. Experiments conducted by various researchers indicate the necessity of selecting desirable male and female plants, and their most compatible combinations, to enhance yields.

To ensure further development of jojoba as a commercial crop, it is crucial to identify the factors that contribute to the extreme variability observed in different genotypes. Identification of genotypes capable of ensuring both profitable yield and wide genetic variability will be a challenging task that will require a complete set of information in order to understand how a given phenotype is constituted at the molecular, biochemical, reproductive and agronomic level, thereby facilitating the rapid identification of molecular and metabolic markers that are important to define a required phenotype. A multidisciplinary approach based on molecular genetics, functional genomics, plant reproductive biology, biochemistry and agronomy will provide accurate information with which to identify genotypes with stabilised yields in various production systems.

Creating a monoecious jojoba (Yermanos 1981) would enable further improvements in seed yield, and to a greater extent, making the crop economically viable with predictable yields. Enormous scope exists for genetic improvement of jojoba through in vitro techniques adopting biotechnological tools, including development of transgenic crops to synthesise jojoba wax in conventional oil seeds, and even employing recombinant microorganisms.

References

Abbott TP, Nakamura LK, Buchholz G, Wolf WJ, Palmer DM, Gasdorf HJ, Nelsen TC, Kleiman R (1991) Processes for making animal feed and protein isolates from jojoba meal. J Agric Food Chem 39:1488–1493

Abramovich R, Tal M, Forti M (1976) Selection and improvement of *Simmondsia*. In: Memorias de la II Conference Internacional Sobre la jojoba y su Aprovenchamiento. Ensenada, Baja California, Mexico, pp 89–91

Agrawal V, Prakash S, Izhar S (1999) Differential hormonal requirements for clonal propagation of male and female jojoba plants. In: Altman A, Ziv M (eds) Plant biotechnology and in vitro biology in the 21st century. Proceedings of the 9th International Congress of the International Association of Plant Tissue Culture Biotechnology, Jerusalem, Israel, 14–19 June 1998, pp 25–28

Agrawal V, Prakash S, Gupta SC (2002) Effective protocol for in vitro shoot production through nodal explants of *Simmondsia chinensis*. Biol Plant 45:449–453

Agrawal V, Sharma K, Gupta S, Kumar R, Prasad M (2007) Identification of sex in *Simmondsia chinensis* (Jojoba) using RAPD markers. Plant Biotechnol Rep 1:207–210

Alcaraz ML, Ayla-Rocha B (1982) Asexual reproduction of Jojoba by layering. Hortic Sci 17:893–896

Apóstolo NM, Llorente B (2000) Anatomy of normal and hyperhydric leaves and shoots of in vitro grown *Simmondsia chinesis* (link) schn. In Vitro Cell Dev Biol Plant 36:414–418

Apostolo NM, Brutti C, Ferrarotti SA, Llorente BE, Krymkiewicz NL (2001) Stimulation of root development with cyclodextrins on jojoba shoots in vitro. In Vitro Cell Dev Biol Plant 37:414–418

Ash GJ, Albiston A, Cother EJ (2005) Aspects of Jojoba agronomy and management aspects. Econ Bot 12:261–295

Ayerza R (1985) Index of Jojoba (*Simmondsia chinensis* L.). Production without irrigation in Villa Dolores, Córdoba, Argentina. In: Elias-Cesnik A (ed) VI International Conference on Jojoba and Its Uses, Ben Gurion University of The Negev, Beer-Sheva, Israel, pp 7–14

Ayerza R (1996) Evaluation of eight jojoba clones for rooting capacity, plant volume, seed yield, and wax quality and quantity. In: Princen LH, Rossi C (eds) Proceedings of the Ninth International Conference on Jojoba and Its Uses. Peoria, IL, pp 1–3

Bashir MA, Ahmad M, Anjum MA (2005) Response of six promising jojoba strains to air layering. Biosci Res 3:172–177

Bashir MA, Ahmad M, Anjum MA (2006) Propagation of six promising jojoba strains through veneer grafting. Int J Agric Biol 8:482–484

Bashir MA, Ahmad M, Anjum MA (2007a) Effect of various potting media on growth of rooted jojoba (*Simmondsia chinensis*) cuttings. Int J Agric Biol 9:147–151

Bashir MA, Rashid H, Anjum MA (2007b) In vitro shoot initiation from nodal explants of jojoba (*Simmondsia chinensis*) strains. Biotechnology 6:165–174

Bashir MA, Rashid H, Anjum MA (2007c) In vitro shoot multiplication of six promising strains of jojoba (*Simmondsia chinensis*) Biotechnology 6:309–315

Bashir MA, Anjum MA, Rashid H (2007d) In vitro root formation in micropropagated shoots of jojoba (*Simmondsia chinensis*). Biotechnology 6:465–472

Bashir MA, Muhammad AA, Hamid R (2008) In vitro propagation of some promising genotypes of jojoba (*Simmondsia chinensis*). Afr J Biotechnol 7:3878–3886

Benzioni A (1995) Jojoba domestication and commercialization in Israel. Hortic Rev 17:234–266

Benzioni A, Dunstone RL (1985) Environmental and hormonal effects on flowering of jojoba. In: Wisniak J, Zabicky J (eds) Proceedings of the Sixth International Conference on Jojoba and its Uses. Beer-Sheva, Israel, pp 171–178

Benzioni A, Vaknin Y (2002) Effect of female and male genotypes and environment on wax composition of jojoba. J Am Oil Chem Soc 70:297–302

Benzioni A, Mills D, Wenkart S, Zhou Y (2003) Effects of ventilation on the performance of jojoba (*Simmondsia chinensis*) clones: multiplication stage. Acta Hortic 616:135–138
Birnbaum E, Matias S, Wenkart S (1984) Vegetative propagation of jojoba by tissue culture. In: Wisniak J, Zabicky J (eds) Proceedings of the Sixth International Conference on Jojoba and its Uses. Beer-Sheva, Israel, Ben Gurion University, pp 233–241
Cao B, Gao HD (2003) Technology of cutting propagation of *Simmondsia chinensis* (Link) Schneider (in Chinese). J Nanjing For Univ 27:62–66
Chandra R, Mishra M (2003) Comprehensive micropropagation of horticultural crops. International Book Distributing Company, Lucknow UP, India
Chaturvedi HC, Sharma M (1989) In vitro production of cloned plants of jojoba [*Simmondsia chinensis* (Link) Schneider] through shoot proliferation in long-term culture. Plant Sci 63:199–207
Chikara J, Kumari A (1991) Evaluation of yield potential in jojoba (*Simmondsia chinensis*) under Indian conditions. Proc Natl Acad Sci India 61:481–485
Clarke JA, Yermanos DM (1980) Jojoba – variability in oil content and composition in a collection of 1156 native plants. J Am Oil Chem Soc 57:176–178
Coates W, Ayerza A (2008) Supplemental pollination-increasing Jojoba (*Simmondsia chinensis* L. schneider) seed yields in the Arid Chaco environment. Ind Crops Prod 27:364–370
Coates W, Ayerza A, Palzkill D (2006) Supplemental pollination of jojoba – a means to increase yields. Ind Crops Prod 24:41–45
El-Anany AM (2007) Nutritional, biochemical and histopathological studies on Jojoba protein isolate. Braz J Food Technol 10:199–211
Elhag H, El-Olemy MM, Mossa JS, Tag-El-Din SS, Al-Zoghet MF, Al-Alsheikh AMA (1998) In vitro propagation of jojoba. Program Abstracts of the Annual Conference on New Crops and New Uses: Biodiversity and Sustainability, Phoenix, AZ
Elliger CA, Waiss AC, Lundin RE (1973) Simmondsin, an unusual 2-cyanomethylene cyclohexyl glucoside from *Simmondsia chinensis*. J Chem Soc Perkin Trans 19:2209–2212
Feldman WR, Hogan L, Palzkill DA (1980) Factors affecting growth of jojoba cuttings in the liner stage. In: Puebla M (ed) Memorias: IV Reunion Internacional de la Jojoba, Consejo Internacional de la jojoba. Hermosillo, Sonora, Mexico, pp 121–129
Flo G, Abbott T, Vermaut S, Van Boven M, Daenes P, Decuypere E, Pederson M, Cokelaere M (1997) Growth performance of rats fed jojoba proteins. Possible correlations with trypsin inhibitory activity in jojoba proteins. J Agric Food Chem 45:4384–4387
Forti MA, Nerd A, Benzioni A (1985) Effect of genetic background on flowering pattern, growth and yield of Jojoba. In: Wisnaik J, Zabicky J (eds) Proceedings of Sixth Int Conf on Jojoba and its uses, Ben-Gurion University, Negev, Beer Sheva, Israel, pp 293–298
Gao HD, Cao B (2001) Study on technology of tissue culture of *Simmondsia chinensis* (Link) Schneider (in Chinese). J Jiangsu For Sci Technol 28:12–14
Gentry HS (1958) The natural history of jojoba (*Simmondsia chinensis*) and its cultural aspects. Econ Bot 12:261–295
Gentry HS, McGill LA (1979) Jojoba survey and germplasm collection – 1977. In: Yermanos DM (ed) Proceedings of the 3rd International Conference on Jojoba and its uses, Riverside, CA, pp 63–70
Hamama L, Baaziz M, Letouzé R (2001) Somatic embryogenesis and plant regeneration from leaf tissue of jojoba. Plant Cell Tissue Organ Cult 65:109–113
Hassan NS (2003) In vitro propagation of jojoba (*Simmondsia chinensis* L) through alginate-encapsulated shoot apical and axillary buds. Int J Agric Biol 5:513–516
Hogan L, Lee CW, Palzkill AD, Feldman RW (1978) Recent progress in the propagation of jojoba by stem cutting. In: Yermanos DM (ed) Proceedings of the 3rd International Conference on Jojoba, Riverside, CA, pp 1–4
Jacoboni A, Standardi A (1987) Tissue culture of jojoba (*Simmondsia chinensis* Link). Acta Hortic 212:557–560

Kacker NL, Joshi SP, Singh M, Solanki KR (1993) In vitro regeneration of female plants of *Simmondsia chinensis* (Link) Schneider (Jojoba) using coppice shoots. Ann Arid Zone 32:175–177

Kalscheuer R, Stöveken T, Luftmann H, Malkus U, Reichelt R, Steinbüchel A (2006) Neutral lipid biosynthesis in engineered *Escherichia coli*: jojoba oil-like wax esters and fatty acid butyl esters. Appl Environ Microbial 72:1373–1379

Khanam A, Rao YBN, Farook SA (1999) Standard in vitro protocol for high frequency mass micropropagation of jojoba [*Simmondsia chinensis* (Link) Schneider]. Adv Plant Sci 12:361–366

Kim SW, Liu JR (1999) Somatic embryogenesis and plant regeneration in zygotic embryo cultures of balloon flower. Plant Cell Tissue Organ Cult 58:227–230

Kondra ZP, Thomas PM (1975) Inheritence of oleic, linoleic and linolenic acid in seed oil of rapeseed (*Brassica napus*). Can J Plant Sci 55:205–210

Kumari A, Chikara J, Iyenger ERR (1991) Variability in *Simmondsia chinensis* under semi arid conditions. Indian J Genet 51:85–88

Lardizabal KD, Metz JG, Sakamoto T, Hutton WC, Pollard MR, Lassner MW (2000) Purification of a jojoba embryo wax synthase, cloning of its cDNA, and production of high levels of wax in seeds of transgenic *Arabidopsis*. Plant Physiol 122:645–655

Lassner MW, Lardizabal K, Metz JG (1996) A jojoba β-ketoacyl-CoA synthase cDNA complements the canola fatty acid elongation mutation in transgenic plants. Plant Cell 8:281–292

Lee CW, Palzkill DA (1984) Propagation of jojoba by single node cuttings. Hortic Sci 19: 841–842

Lee CW, Thomas JC (1985) Jojoba embryo culture and oil production. Hortic Sci 20:762–764

Letouzé R (1987) Biotechnology of jojoba. Int Ind Biotechnol 1987:277–280

Llorente B, Apostolo NM (1998) Effect of different growth regulators and genotype on in vitro propagation of jojoba. N Z J Crop Hortic Sci 26:55–62

Llorente BE, Juarez ML, Apóstolo MN (2007) Exogenous trehalose affects morphogenesis in vitro of jojoba. Plant Cell Tissue Organ Cult 89:193–201

Lloyd DG, Webb CJ (1996) The avoidance of interference between the presentation of pollen and stigma in angiosperms: dichogamy. N Z J Bot 24:135–162

Mandani A, Lee CW, Hogan L (1978) In vitro propagation of *Simmondsia chinensis* via shoot tip culture. Hortic Sci 13:35–37

Metz JG, Lassner MW (1994) Fatty acyl reductases. United States Patent 5,370,996. Issued December 6

Metz JG, Pollard MR, Anderson L, Hayes TR, Lassner M (2000) Purification of a jojoba embryo fatty acyl-coenzyme A reductase and expression of its cDNA in high erucic acid rape seed. Plant Physiol 122:635–644

Meyghani H, Ghazvini RF, Hamidoghli Y (2005) Micropropagation from stem segments of salt tolerant jojoba seedlings. J Korean Soc Hortic Sci 46:83–187

Mills D, Benzioni A (1992) The effect of NaCl salinity on growth and development of jojoba clones: II. Nodal segments grown in vitro. J Plant Physiol 139:737–741

Mills D, Wenkart S, Benzioni A (1997) Micropropagation of *Simmondsia chinensis* (Jojoba). In: Bajaj YPS (ed) Biotechnology in agriculture and forestry, vol 40. High-tech and micropropagation VI. Springer, Berlin, pp 370–393

Mills D, Yanqing Z, Benzioni A (2004) Improvement of jojoba shoot multiplication in vitro by ventilation. In Vitro Cell Dev Biol Plant 40:396–402

Miwa TK (1971) Jojoba oil wax esters and derived fatty acids and alcohols:gas chromatographic analyses. J Am Oil Chem Soc 48:259–264

Mohammed AM, Aly Essam A, Amer Wasef A, Al-Zayadneh AE, Negm E (2008) Growth regulators influence the fatty acid profiles of in vitro induced jojoba somatic embryos. Plant Cell Tissue Organ Cult 93:107–114

Parker JS (1990) Sex chromosomes and sexual differentiation in flowering plants. Chromosome Today 10:187–198

Prakash S, Agrawal V, Gupta SC (2003) Influence of some adjuvants on in vitro clonal propagation of male and female jojoba plants. In Vitro Cell Dev Biol Plant 39:217–222
Prakash V, Agrawal S, Gupta SC (2002) Effective protocol for in vitro shoot production through nodal explants of *Simmondsia chinensis*. Biol Plant 45:449–453
Prasad VV, Iyengar ERR (1982) Physiological differences in male and female plants of Jojoba (*Simmondsia chinensis* L.). Curr Sci 51:1039–1040
Prasad VV, Iyengar ERR (1985) Phenological and biochemical changes in male and female shrubs of Jojoba (*Simmondsia chinensis* Link) during different seasons. Proc India Acad Sci 95:203–211
Purcell HC, Purcell II HC (1988) Jojoba crop improvement through genetics. J Am Oil 65:1–13
Ramonet RR (1988) Selection criteria and evaluation procedures for Jojoba plant improvement. In: Baldwin AR (ed) Proceedings of the 7th International Conference on Jojoba and Its Uses, Phoenix, AZ, pp 60–68
Ramonet R, Morales AM (1985) Seed yield variability and selection criterion of jojoba clones in Mexico. In: Wisniak J, Zabicky J (eds) Proceedings of the Sixth International Conference on Jojoba and Its Uses. Ben Gurion University of The Negev, Beer-Sheva, Israel, pp 279–285
Reddy YN (2003) Effect of different concentration of auxins on rooting and root characters of air and ground layers of jojoba [*Simmondsia chinensis* (Link) CK Schneider]. Ethiop J Sci 26:155–159
Roussos PA, Tolia-Marioli A, Pontikis CA, Kotsias D (1999) Rapid multiplication of Jojoba seedlings by in vitro culture. Plant Cell Tissue Organ Cult 57:133–137
Samac D, Storey R (1981) Proteolytic and trypsin inhibitor activity in germinating jojoba seeds (*Simmondsia chinensis*). Plant Physiol 68:1339–1344
Sanchez LM, Price RL (1988) Effect of jojoba tannins on jojoba albumins and globulins and on trypsin and chymotrypsin. In: Baldwin AR (ed) Proceedings of the 7th International Conference on Jojoba and its Uses. American Oil Chemists' Society, Champaign, IL, pp 375–385
Sardana J, Batra A (1998) In vitro regeneration of jojoba (*Simmondsia chinensis*): a plant of high potential. Adv Plant Sci 11:143–146
Scaramuzzi F, D'Ambrosio A (1988) Organogenesis and propagation in vitro of *Simmondsia chinensis* (Link) Schn (jojoba) from vegetative fragments. Acta Hortic 227:411–413
Shrestha MK, Peri I, Smirnoff P, Birk Y, Goldhirsh A (2002) Jojoba seed meal proteins associated with proteolytic and protease inhibitory activities. J Agric Food Chem 50:5670–5675
Singh A, Reddy MP, Patolia JS (2008) An improved protocol for micropropagation of elite genotypes of *Simmondsia chinensis* (Link) Schneider. Biol Plant 52:538–542
Singh KJ, Nayyar H, Dutta A, Dhir KK (2003) Rhizogenetic studies of jojoba: hormone effect, rooting medium and seasonal variation. Ind For 129:1405–1411
Tobares L, Frati M, Guzmánc C, Maestri D (2004) Agronomical and chemical traits as descriptors for discrimination and selection of jojoba (*Simmondsia chinensis*) clones. Ind Crops Prod 19:107–111
Tyagi RK, Prakash S (2001) Clonal propagation and in vitro conservation of jojoba [*Simmondsia chinensis* (Link) Schneider]. Indian J Plant Genet Resour 14:298–300
Tyagi RK, Prakash S (2004) Genotypes and sex specific protocols for in vitro micropropagation and medium term conservation of jojoba. Biol Plant 48:119–123
Vaknin Y, Mills D, Benzioni A (2003) Pollen production and pollen viability in male jojoba plants. Ind Crops Prod 18:117–123
Verbiscar AJ, Banigan TF (1982) Jojoba seed meal as an animal feed. NSF final report on Grant AER 76–23895. Anver Bioscience Design, Sierra Madre, CA
Verbiscar AJ, Banigan TF, Weber CW, Reid BL, Swingle RS, Trei JE, Nelson EA (1981) Detoxification of jojoba meal by lactobacilli. J Agric Food Chem 29:296–302
Wang YC, Janick J (1986a) Somatic embryogenesis in Jojoba. J Am Soc Hortic Sci 111:281–287
Wang YC, Janick J (1986b) In vitro production of jojoba liquid wax by zygotic and somatic embryos. J Am Soc Hortic Sci 111:798–806

Wolf WJ, Schaer ML, Abbott TP (1994) Nonprotein nitrogen content of defatted jojoba meals. J Sci Food Agric 65:277–288

Yermanos DM (1981) Monoecious jojoba. In: Everett H, Pryde L, Princen H, Deb Mukherjee K (eds) New sources of fats and oils. American Oil Chemists Society, Champaign, IL, pp 247–254

Yermanos DM (1983) Performance of jojoba under cultivation between 1973–1982. In: Elias-Cesnik A (ed) Proceedings of the 5th International Conference on Jojoba and Its Uses through 1982. Univ. of Arizona, Tucson pp 197–211

Yermanos DM, François LE, Tammadoni T (1967) Effects of soil salinity on the development of jojoba. Econ Bot 21:69–80

Chapter 20
Date Palm Cultivation in the Changing Scenario of Indian Arid Zones: Challenges and Prospects

R. Raj Bhansali

Abstract The arid zone of Rajasthan or Great Indian Thar Desert, popularly known as Thar, is a vast tract of dry land of about 2.34 million square kilometres. The whole tract is distinguished by low and erratic rainfall, low humidity, high solar radiation, strong dust-raising winds, scant vegetation and a dry, sand-dune-dominated landscape. During the past five decades, strenuous efforts have been made towards control of desertification, ecological regeneration and restoration of the Thar desert in order to reclaim the productivity of this vast unproductive desert land. For this purpose, a 649-km-long man-made canal was constructed to bring Himalayan water into the water-starved desert. Although the Thar desert becomes lush green with the slightest precipitation, its natural sustainable regeneration has been very slow due to intense biotic (overgrazing, extraction of fodder and fuel wood) and abiotic pressures. Over-exploitation of fodder and fuel wood, the two basic necessities of life for desert people, leads to destruction of desert ecosystems and enhances desertification. Perhaps the over-exploitation of land, water and biological resources since the earliest times have made the Rajasthan desert a 'man-maintained' if not 'man-made' landscape. Introduction of fast-growing exotic species of trees and grasses from isoclimatic regions of the world in attempts to stabilise shifting sand dunes, creating 'microclimates' through shelterbelt plantations, and 'fencing and enclosures' for regeneration of indigenous species have proved highly successful in the control of desertification. Various practices of date palm cultivation for the control of desertification in arid zones have been elaborated over the last three decades. The climatic features existing in the Indian arid zone are compatible with the requirements of successful date palm plant production. Weather data and date palm growth parameters go hand-in-hand from planting to production level. Introduction of conventional offshoot and hi-tech methods of tissue culture to supply superior planting material have been described. Planners,

R. Raj Bhansali
Division of Plant Sciences and Biotechnology, Central Arid Zone Research Institute, Jodhpur 342033, India
e-mail: anjalirikhab@gmail.com

K.G. Ramawat (ed.), *Desert Plants*,
DOI 10.1007/978-3-642-02550-1_20,

researchers and modern farmers are making serious efforts at converting this vast mass of arid land into profitable green land, thereby changing the economic status of the region.

20.1 Introduction

20.1.1 Indian Arid Zone

The Great Indian Desert, known as the Thar Desert, in Western Rajasthan is the biggest desert in India. It encompasses about 70% (208,110 km^2) of the total landmass of Rajasthan – hence the "Desert State of India". Most of the Thar Desert is spread across the Western part of Rajasthan in India and South-Eastern Pakistan, stretching between 22° 30″ to 32° 05″ N and from 68° 05″ to 75° 45″E, covering Western Rajasthan (19.6 million ha), Gujarat (6.22 million ha) and South Western parts of Haryana and Punjab (2.75 million ha; Kar 1996; Faroda and Harsh 1999), and embracing the districts of Jaisalmer, Barmer, Bikaner and Jodhpur. Thus this huge stretch of barren land extends into the Southern part of Haryana, Punjab and Northern Gujarat and Sind province of Pakistan. The desert in Rajasthan is bounded by the Sutlej River in the northwest, the Aravali mountains in the east, the salty marshland of the Rann of Kachchh in the South and the Indus River in the west (Anonymous 1965). The Aravali hills also demarcate the state of Rajasthan into two distinct climatic regions i.e. semi-arid to the east of the Aravalies, and an arid region lying to the west.

The Thar desert has sand dunes of various types, sizes and structures; some are tall (70–120 m) and run in chains several kilometers in length. The desert is characterised by these massive rolling sand dunes, excessive heat (50°C in May and June, with surface sand temperatures rising to 70°C; Krishnan and Rao 1978). Dust storms and dust-raising winds often blow with high velocities of 140–150 km h^{-1}, and are common in May and June. From March onwards, when the surface is dry and temperatures soar, the summer winds associated with the South-West monsoon reach a maximum speed of 20 km h^{-1}. During the dry period in May is the ideal time for winds to carry or move sands from place to place. The moving sand during this period can be termed "dynamic sand" (Kar et al. 1994; Kar 1996). The arrival of the monsoon in July puts an end to this activity. Mirage (naturally occurring optical phenomenon) is also a distinctive feature of the desert areas, falsely suggesting the presence of large areas of water adjacent to the arid land.

The annual average rainfall is less than 10 inches (250 mm); 90% of this rain occurs during the Monsoon or "rainy season", falling between July and September. Water is scarce, but is found at a depth of 100–200 feet (~30–60 m) below ground. Arid zones are characterised by low and highly variable precipitation compared to high evaporative demand. Therefore, moisture stress is the characterising feature of arid zones (Krishnan 1974).

20.1.2 Vegetation

The vegetation cover in arid regions of India falls under the category of "Thorn Forest Type" (xerophytic nature), i.e. extremely slow growing and thinly populated. The population of many plant species in the Indian arid zone has declined to an alarming level due to environmental changes and over-exploitation of natural resources in the past (Raj Bhansali and Jindal 2000). The desert climatic, soil and water conditions threaten the establishment and propagation of many economically important plants including fruit and forest trees. Furthermore, pests and diseases also limit growth and cultivation under the harsh climatic conditions of the desert. Natural regeneration in some desert plant species is extremely poor. Moreover, methods for vegetative propagation of these woody trees have limited success for rapid multiplication of selected plant species (Abrol and Venkateswarlu 1995; Harsh and Tewari 2007).

20.1.3 Climatic Features

The arid zone is characterised by sub-tropical climatic conditions. Great uncertainty exists in the prediction of the response of arid ecosystems to elevated CO_2 and global warming. The complexities of changes in precipitation, vegetation-climate feedback, and the direct physiological effects of CO_2 on vegetation present particular challenges for climate change modelling of arid regions, which are expected to undergo significant changes under a scenario of climate warming (Lioubimtseva 2004). Rainfall in the arid zone is scant, varying from <100 to >400 mm with an average of 250 mm rainfall per annum (Table 20.1). The usual dates of the arrival and withdrawal of monsoon are 1 July and 15 September, respectively. The contribution of this seasonal rainfall to annual rainfall is very high (91–96%) over the South and central parts of Western Rajasthan and Gujarat (Rao 2009). Another interesting feature of the rainfall is the high coefficient of variability in annual rainfall, which often exceeds 50% in the North-Western Indian arid zone and can be higher than 70% in the extreme Western regions of arid Rajasthan, where annual rainfall is very low (Ramakrishna and Rao 1992; Rao 2009).

Table 20.1 Climatological features of potential date palm growing areas of the Indian arid zone. *PE* Potential evapotranspiration, *HSU* Heat summation unit

Region	PE (mm)	HSU (> base 18°C)	Fruiting period (days)	Rainy days	Humidity (%)	Rainfall (mm)	Temperature [range (mean); °C]
Kachachh	1,897	2,900–3,056	165	17	52	350	20–33 (26)
Jaisalmer	2,069	2,900–3,240	145	13	64	215	19–34 (26)
Bikaner	1,772	2,900–3,200	145	19	58	305	18–33 (26)
Jodhpur	1,843	2,900–3,200	145	20	65	350	20–33 (26)
Ganganagar	1,662	2,400–2,900	122	16	56	296	17–33 (25)
Hissar	1,615	2,400–2,513	122	25	64	446	17–33 (25)
Ferozpur	1,362	2,400–2,600	122	19	65	300	16–31 (24)

Sandy soils associated with dunes and inter-dunes occupy about 31% area of the arid zone. These soils are characterised by 85–90% sand, low water retention capacity (150–200 mm m^{-1}) and low fertility status, moderate to severe wind erosion, surface crusting and high water infiltration. High salinity in soil and ground water are associated with these soils (Dhir 2003). The soils of the Thar desert of Rajasthan have been classified into six types, viz. dune soil, sandy plain soil, brown light loam soil, grey brown loam soil, hardpan soil and indus alluvial soil (Dhir and Singh 1985; Venkateswarlu and Kar 1996), with dune soil and sandy plain soils occurring most commonly. Natural underground water resources, as well as man-made tube wells, hydrological reservoirs, and canals [e.g. the Indira Gandhi Nahar Project (IGNP) canal discussed below] are ideal ways to boost agriculture production and improve the desert economy.

20.1.4 Project IGNP

The IGNP is one of the gigantic canal projects in India aimed at transforming the desert wasteland into agriculturally productive land. The IGNP starts from the Harike Barrage, a few kilometres below the confluence of the Sutlej and Beas rivers in Punjab State. It runs South-West in Punjab and Haryana but mainly in Rajasthan for a total of 650 km and ends near Jaisalmer, in Rajasthan. It uses water released from the Pong dam and will provide irrigation facilities to the North-Western region of Rajasthan, i.e. a part of the Thar Desert. It consists of the Rajasthan feeder canal (with the first 167 km in Punjab and Haryana and the remaining 37 km in Rajasthan), with the 445 km main canal being located entirely within Rajasthan. Seven districts of Rajasthan are covered: Barmer, Bikaner, Churu, Hanumangarh, Jaisalmer, Jodhpur, and Sriganganagar. The IGNP project will enhance the living standards of the people of the state. A tree-planting programme for greening the desert in IGNP areas was started in 1965 that involved planting of shelterbelts along roads and canals, block plantations and sand dune stabilisation to check the spread of desert. Tree species used for planting were *Dalbergia sissoo*, *Eucalyptus terticornis*, *Eucalyptus camaldulensis*, *Morus alba*, *Tecomella undulata*, *Acacia tortilis*, *Azadirachta indica*, *Albizia lebbeck*, *Cassia fistula*, *Popular ciliata*, *Melia azedarch*, and *Acacia nilotica*.

20.2 Requirements for Date Palm Cultivation

20.2.1 Agro Climatic Zones

Date palms require very specific climatic conditions for successful cultivation, and their ability to produce abundant fruit at extremely low humidity is possible only if the supply of groundwater/canal irrigation is sufficient. A relatively dry rainy season is required to attain the Pind stage (when the fruit is fully ripe and dark reddish and the fruit is soft). At the time of fruit ripening, rainfall adversely affects the quality of fruits. Late onset of monsoon, lower total rainfall and fewer rainy days are the primary requirements of date palm cultivation as these conditions produce a high quantity of good quality fully ripened soft dates (Pind Khajoor). The climatic parameters governing the success of date palm cultivation in arid regions depends on the availability of heat units with the pattern of rainfall. Date palms in the Mediterranean region have a rainless summer from February to September. This enables complete ripening (up to Pind stage) of fruit on the tree (Chih Cheng and Krueger 2007). In the Thar Desert, rainless summer occurs only rarely. The June–September monsoon coincides with the fruit ripening season. This shortens the period of date ripening from 180–200 days to 100–170 days (Chandra et al. 1992). Therefore, cultivation of date palm varies with rainfall and rainfall pattern. Thus, on the basis of the main climatic parameters, the potential of date palm cultivation in arid zones can be categorised into four zones (Fig. 20.1):

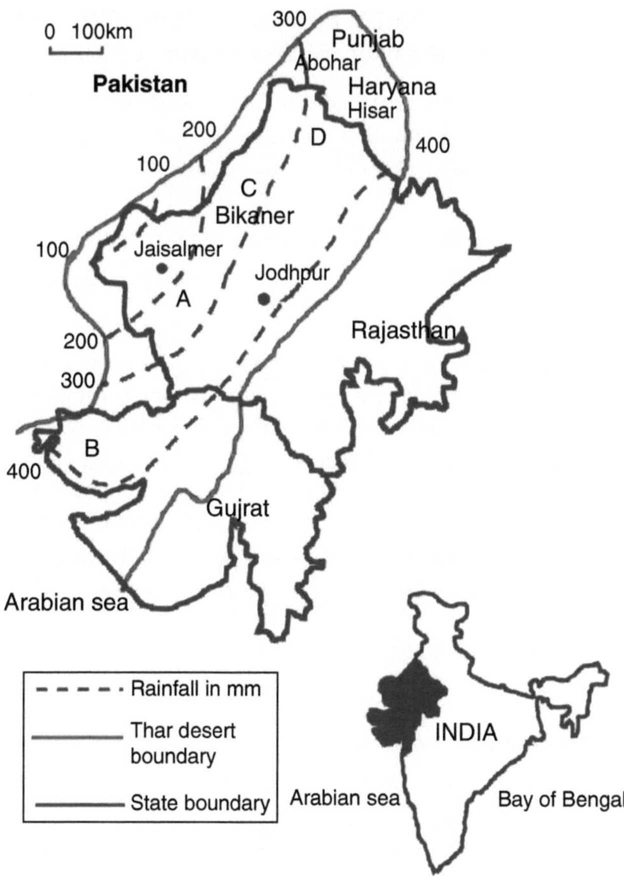

Fig. 20.1 Potential date palm growing zones in Indian arid zone

1. Extremely dry areas with lowest rainfall and highest heat summation units (HSU), viz. Jaisalmer and Western parts of Barmer, Bikaner and Jodhpur districts. This zone has the highest potential.
2. Arid coastal areas of Kachchh and part of Saurashtra. Date palm regions of Kachchh can be divided into three zones: (i) Area (Mundra, Dhrub and Zarpara) close to the coast, having deep coastal soil and a very high water table (1–3 m). This is a major date palm growing area where little irrigation is required; (ii) deep sandy soil away from coast villages (Anjar and Khedoi), with a low water table (10–20 m); and (iii) sides of rivulets, rivers and where water accumulates near Kera, Tera and Vada. The area has a high water table (1–2 m). Maturity of fruits is earlier than in zones (1) and (ii). At present, in the arid Kachchh region, 1.9 million date palm plants produce 95,000 Mt date palm fruits. Forward-looking farmers in this region have planted "Barhee date palm" in 10,000 acres of land.
3. Dry areas with storm-type rains, comprising Jodhpur, Bikaner, eastern part of Barmer, western part of Nagaur, Churu and Ganganagar district.
4. Dry areas with rains followed by cloudy weather, viz. Abohar, Sirsa, Ganganagar, eastern part of Churu and Western part of Sikar districts.

Accordingly, fruits can be harvested at Doka stage in zone 4, at Doka or Dang stage in zones 2 and 3 and at Dang or early Pind stage in zone 1. Rainfall accompanied with high relative humidity spoils the fruit due to rotting and splitting.

20.2.2 Soils

Dates grow in various types of soil – light, medium and heavy – but require good drainage and air penetration into the deep loamy soils. Date palms are the crop plants most tolerant to high pH, as well as being resistant to alkaline and saline conditions (Chundawat 1990; Chandra et al. 1992). In view of the large investment required to bring a date plantation into being and to maintain profitable production, a sandy loam soil, 2–3 m deep with good water holding capacity and drainage is most desirable. Dates can survive well in soil with a salt concentration of 4%, provided the root system does not come into contact with a stratum of soil where the sodicity is more than 1%.

20.2.3 Temperature

High temperatures (higher than early in the fruiting season) are essential at the time of date palm ripening. Prolonged, hot and dry summers and a moderate winter temperature are the primary prerequisites. The principal climatic parameters, viz. temperature and rainfall (total amount and distribution pattern) therefore primarily govern date palm performance. The ideal mean temperature during flowering and fruit ripening varies from 25°C to 40°C, depending on the cultivar. The rate with which the fruit reaches maturity, and the development of its quality depend upon the pattern of maximum temperatures. The effect of temperature is generally evaluated on the basis of 'Heat Units' (Zaid and de Wet 2002a).

The term 'Heat Summation Units' is a means of expressing this energy. At Jaisalmer and Bikaner, the accumulation of sufficient HSU between the flowering to fruit ripening period is required for the date fruits to attain the "Dang" stage while still on the trees. Chandra and Chaudhary (1990) have observed that the requirement of HSUs differs from one cultivar to the next in reaching the "colour turning" and "Pind" stages. To reach the colour turning stage from spathe initiation, 1,951 HSU are required in Halawy, as compared to 2,411 HSU in cultivar Shamran and 2,648 HSU in cultivar Medjool. To attain Pind stage, cultivar Gangangar requires most HSU (3,843) with the least (3,101 HSU) in Halawy. Due to variations in climate from one place to another, the time of maturity as well as the quality of fruit produced in different date palm growing regions differs considerably.

HSU during the period March to August, work out at 2,000–2,400 above 18°C and 3,500–4,000 above 10°C. On the basis of a base temperature of 10°C, the HSU requirement for date palm varies in the range of 1,950–3,650 depending upon the cultivar. It is obvious, therefore, that the Thar desert meets this requirement (Chundawat 1990; Chandra et al. 1992). However, the rate of maturity of the fruit, and the development of its quality depend on the pattern of daily maximum temperature and HSU during the fruit-ripening period. Thus, the time of maturity as well as the quality of fruit produced in different regions varies.

20.3 Horticultural Aspects of Date Palm

20.3.1 Date Palm (Phoenix dactylifera L.)

Date palm is the tree of life ('*Nakhla*' in Arabic), and is the oldest of all cultivated fruit trees. It is a monocotyledonous, dioecious plant belonging to the family Palmaceae. It has large number of secondary roots with smaller lateral roots, which arise from the base of the stem. The larger air passages in the stele and extra-stellar regions of roots indicate its requirements of plentiful water and air supply. The genus *Phoenix* has characteristic upward and lengthwise folding of pinnae. It produces furrowed seeds yielding either male or female plants. Date palm may reach an age of over 100 years and a height of over 20 m. There are about 12 species native to tropical and subtropical parts of Africa and Asia (Chih Cheng and Krueger 2007). The date palm most probably originated from the area around northeastern Africa (the Nile delta), northern Arabia, Iraq and western Iran (Nixon 1950; Nixon and Carpenter 1978). This area is known as the "Fertile Crescent" (ancient Mesopotamia), where Old World agriculture is thought to have arisen.

Indeed, the date palm has been cultivated in this area from ancient times, possibly being one of the first domesticated crops. The tree is considered as a blessed tree, whose history has been linked to the Arabian Region since ancient times. It is believed to have been established as early as 4000 B.C. in Mesopotomia (South Iraq; Zohary and Hopf 1993). The tree was used in the construction of the Temple of the Moon God in Ur (Iraq) some 4,000–5,000 years ago. Over several millennia, date palm culture spread to other parts of the world, including Western India.

20.3.2 Production Status

Worldwide date production has increased tremendously from 1,809,091 t in 1962 to 6,924,975 t in 2005 (FAO 2006). The top ten date-producing countries in 2005 were Egypt, Saudi Arabia, Iran, United Arab Emirates (UAE), Pakistan, Algeria, Sudan, Oman, Libyan Arab Jamahiriya and Tunisia. The top five date-importing countries in 2004 were India, Pakistan, Yemen, Morocco, and UAE. At present, India imports 60,000 t dry and soft dates every year. To help meet this demand, various arid regions of India congenial for date palm cultivation are being considered. The United States (California and Arizona) produced 16,500 t dates in 2005 and exported 4,202 t. The largest production of dates is in Egypt, where production has increased from 439,539 t in 1982 to 1,170,000 t in 2005 (16.9% of worldwide production). In UAE, there were about 1.5 million date palms producing 8,400 t when the country was founded in 1971. This has increased more than 90-fold to an estimated 18 million date palms producing 760,000 tons in 2005. Date palm plantations are distributed throughout the Middle East, North Africa and South Sahel, areas of East and South Africa, the southwestern United States, Central and South America and even in Southern Europe (notably in Spain and Italy) – the total number of date palm trees in the world is approximately 105 million, covering an area of 800,000 ha (FAO 2006). Date palms have yet to be developed in other suitable areas of the globe in which dry climates are experienced, and where there is a desperate need to stabilise and create new sustainable macro- and micro-environments.

20.3.3 Nutritional Status

Date palm fruits are eaten as raw dates (fresh fruits), dry dates (Chuhhara) and soft dates (Pind Khajoor). Dates are highly nutritious and delicious, containing sugars, proteins, fats and minerals. The fully ripe fruits contain 75–80% sugars (glucose and fructose). Dates are also good sources of iron, potassium, calcium, magnesium, sulphur, copper and phosphorus, along with various vitamins, including thiamine, riboflavin, biotin, folic and ascorbic acid. The date fruits are relatively easy to store due to the high sugar content. The fruit plays an important role in the daily nutrition of human populations in regions where date palms are grown. Its additional use as a livestock feed supplement gives the tree much added value. The secondary products generated from the fruits include syrups, jams, ice-creams, baby foods, alcoholic beverages and soft drinks. In addition to producing a valuable dessert fruit, important by-products, such as building materials and versatile starting materials for handicrafts, can be derived from date palm leaves and trunks, making it an important multi-purpose tree and a significant earner of revenue for both small and large farmers. The date palm also makes a significant contribution towards the creation of equable micro-climates within oasis ecosystems, thus enabling agricultural development to be sustained in many drought- and saline-affected regions. This is reflected in its widely acknowledged sustainability value in social, economic and ecological terms (Zohary and Hopf 1993). Planting of date palm has been considered suitable in the states of Rajasthan, Gujarat, Haryana and Punjab (Manohar and Chandra 1995).

20.3.4 Promising Cultivars

Date varieties have been developed over thousands of years of selection of seedlings, propagating only those possessing desirable characteristics through offshoots. Several date specialists have attempted to list and to describe the varieties botanically (Srivastava and Dhavan 1981; Kalra and Jawanda 1992; Pundir and Porwal 1998). An estimated 3,000 cultivars of date palm are available worldwide. Various superior cultivars have been selected and developed as having desirable traits, i.e. high quality, pest resistant and vibrant coloured fruits (Fig. 20.2). Prominent date palm cultivars suitable for the arid climatic conditions of India are listed in Table 20.2, and include varieties such as Halawy, Shamran, Khadrawy, Medjool, Barhee, Zagloul, Hayani, Zahidi, Khalas and Sewi. The wild species of date palm, viz. *P. sylvestris* (L. Roxb), which is found growing in almost all states of India, produces inferior quality fruit with little flesh. This palm is used for production of gur and a drink known as neera.

20.3.5 Irrigation

Date palms require large quantities of water. The shape of the leaves also influences the evaporation rate. Dates grow in hot climates with high levels of radiation, and the evaporation rate is high. Water consumption per hectare is high during the hottest months (April–June). The golden rule is to ensure that the greatest diameter of the bulb of the plant is at the same level as the soil surface after transplanting, and to ensure that water does not go over the top of the date plant. The fully grown date palm is known as a drought-resistant fruit tree and is able to survive for long periods

Fig. 20.2 Date palm cultivars having different types of fruits (*insets*)

Table 20.2 Promising date palm cultivars for the Indian arid zone

Cultivar	Colour of fruit	Fruit size (cm)	Time of maturity	Yield/ Doka stage (kg)
Halawy	Light orange with yellow	3.56 x 2.10	Two weeks in July	100
Shamran	Yellow with slightly pink	3.50 x 2.18	End of July	100
Barhee	Golden yellow	2.90 x 2.30	Mid August	100
Khalas	Yellow	3.2 x 2.4	Mid July	75
Medjool	Yellow	3.90 x 2.80	Mid August	75
Sewi	Yellowish green	2.9 x 1.90	Early August	50
Kuneizi	Red	3.50 x 2.20	Mid July	40
Zaagloul	Red	3.90 x 2.20	Mid July	150
Zahidi	Yellow	3.0 x 2.20	Mid August	125
Khadrawy	Greenish yellow	3.10 x 2.10	End of July	40

without irrigation. However, continuous drought conditions will retard the growth of the plant.

20.3.6 Fertiliser

Nutrient management in date palm trees is essential for optimum growth and production. Date palm trees should be provided with organic manure as well as inorganic fertiliser. Well rotted farmyard manure (20–40 kg per tree) should be applied to date-bearing trees during the period September–December. Besides organics, each date-bearing tree should be given 0.5–1.0 kg nitrogen, 0.5–1.0 kg phosphorus and 0.25–0.5 kg potash per year. Ammonium sulphate (1–2 kg per palm) should be added along with manure. The entire quantity of phosphorus and potassium and 60% of nitrogen should be given 3 weeks before flowering, with the remaining nitrogen being applied during the months of March and April after fruit set. Fertiliser is applied according to the size and age of the tree, in a ratio of 2:1:3:1. It is recommended to test for microelement deficiencies, and to spray the foliage when necessary with S, Cu, Fe, Mg, Mn. Farmyard manure is also applied during December–January. The nitrogen dose should be given 2 weeks before flowering, i.e. in the 1st week of February. However, nutritional studies conducted on 30-year-old palms using 500–2,000 g N, 500 g P_2O_5 and 500 g K did not show any significant response under a canal irrigation system at Abohar (Chandra et al. 1992; Pareek and Nath 1996).

20.3.7 Intercropping

Intercropping date palm with other suitable crops bring a good income and also improves the fertility of the soil. During the first few years, intercropping can be practiced with no shortage of irrigation. Intercrops such as cluster bean, cowpea, moong, moth bean, mustard and gram can be sown during summer and winter in rain-fed or irrigated fields (Chandra et al. 1992; Pareek and Nath 1996).

20.3.8 Pruning and Training

Being monocotyledonous, date palm trees have a single stem, therefore very little pruning and training is involved in the production of fruits. Trees are usually pruned once a year. In some growing regions trees are pruned after harvesting, while elsewhere trees are pruned in the spring, before the clusters are covered with sacks. The best period for pruning is during the winter months (November and December). Removal of one-third of the central strands after fruit set leads to better development of fruits and hastens ripening.

20.3.9 Flowering and Pollination

The dioecious nature of date palm means that male and female flowers are borne on separate trees (Fig. 20.3). Spathe initiates in the axillary branch of crown leaves in February in arid zones. Female plants require artificial pollination for good fruit setting as pollination through natural means such as wind and insects is negligible.

Fig. 20.3 Female flowers produced in tissue-cultured date palm

Pollination is a cumbersome and expensive practice due to the pattern of flowering of palm trees and the requirement of climbing several times to the crown. Hand pollination is performed by dusting pollen through cotton balls onto freshly opened female spathes in the early morning for 2–3 days (Zaid and de Wet 2002c). Pollen should be collected from mature male plants and dried (6 h in sunlight followed by 18 h in shade), and can be stored in air-tight glass vials at ambient temperature for 8 weeks or in a refrigerator at 9°C for a year. Pollination of the female spathe just after cracking gave higher fruit set percentage. The pollen of *P. dactylifera*, and of other *Phoenix* species, has been found to exert a direct influence on the size, shape and colour of the seed, and also on the size of the fruit, the speed of development of the fruit, and the time of fruit ripening of vegetatively propagated female varieties of date palm (Zaid and de Wet 2002c).

Spathe emergence occurs from the 2nd week of January to the 3rd week of March in Jodhpur and Bikaner, in the last 2 weeks of April in Abohar, and in February in Kachchh (Pareek and Nath 1996). The time required from flowering to different stages of fruit production depends on climatic conditions, date variety and management practices (Table 20.3).

20.3.10 Fruit Thinning

Excess fruit load may cause shrivelling of berries, breaking of spathe stalks, more damage due to rain and humidity, and delayed ripening. It also reduces the size and

Table 20.3 Flowering characteristics of important date palm cultivars in arid climates

Cultivar	Spathe Emergence Month	Days	Opening Days	Length of spathe (cm)	Number of bunches	Strands per bunch	Length of strand (cm)	Days to fruit maturity
Halawy	January–February	28–46	24–37	33	6–9	43.6	37.0	112
Shamran	February	30–48	23–36	28	5–8	34.3	32.6	136
Khadrawy	January–February	32–43	29–35	33	3–5	28.2	27.6	126
Zagloul	Early February	17–22	11–18	30	3–8	31.2	25.2	129
Zahidi	February	26–41	20–29	30	3–6	36.1	30.7	129
Medjool	February	19–34	21–30	30	4–6	40.6	40.1	130

Table 20.4 Fruitt characteristics of important date palm cultivars in arid climate. *TSS* Total soluble solids

Cultivar	Colour (Doka)	Shape	Pulp:stone ratio	TSS	Taste	Fruit skin	Maturity
Halawy	Yellow	Oblong	9.0	42	Very sweet	Thin	Early
Shamran	Reddish yellow	Obovate	6.7	36	Astringent	Thin	Late
Khadrawy	Light yellow	Oval	7.3	35	Astringent	Thin	Medium
Zagloul	Purple red	Obovate	7.3	71	Sweet	Medium thick	Early
Muscut–2	Red	Oblong	6.8	38	Sweet	Thin	Medium
Zahidi	Yellow	Obovate	7.5	34	Slightly sweet	Medium thick	Medium
Medjool	Orange	Oblong round	6.38	35	Astringent	–	Late

quality of the fruit. It is therefore necessary to keep the optimum quantity of fruit and thin out the rest. This is usually accomplished either by reducing the number of fruit on each bunch and/or by removing some bunches (Table 20.4). The number of fruit that a palm can safely carry depends on the cultivar, age, size and vigour of the palm, and number of green leaves it has. Under normal conditions, 1–2 bunches in the 4th year and 3–4 bunches in the 5th year may be left. Normally 8–10 bunches per palm are retained in India.

In short-stranded varieties like Khadrawy, the strands are generally cut back to even up the bunch from the top. Most fruit thinning is done by removing one-half to two-third of the strands from the centre. In long-stranded varieties like Deglet Noor, one-third to one-half of strands are cut in similar way as in Khadrawy. In addition, strands are also cut back to remove about one-third of the flowers. The optimal number of fruits to be left is between 1,300 and 1,600 per palm depending on the variety. The percent thinning is generally 40–50 in Khadrawy, 50–55 in Hallawi, and 50–60 in Zahidi and Barhee.

20.3.11 Fruit Development

Fruit set takes place after fertilisation, which is characterised by the loss of two unfertilised carpels. The fruit goes through four developmental stages, viz. Gandora or Kimiri (fruit is still hard and green); Doka or Khalal (fruits are fully grown but remain hard; their colour becomes yellow or red); Dang or Rutab (softening of fruits start from tips and finally the whole fruit is softened), and Pind or tamer (fruit is fully ripened; fruit weight decreases as a result of fruit dehydration). The fruits become edible from the Doka stage onwards, except for varieties that are highly astringent at that stage. Pre-harvest application of ethrel (2-chloroethyl phosphonic acid) at 1,000 ppm on fruit bunches at the colour break stage (when fruits start changing their colour from green to yellow or red) hastened ripening of fruits and also increased the size and weight of the fruit. Ethephon also advances the ripening of fruit. Application of ethephon (1,000 ppm) and gibberellic acid (GA_3; 100 ppm) at the fruit colour turning stage induced earlier ripening by 1 week (Manohar and Chandra 1995).

20.3.12 Harvesting

Bunches are harvested at full Doka stage for fresh date varieties edible at this stage. The fruits are separated from strands and graded. Any shrivelled, diseased and undersized fruits are discarded to make the produce more attractive for the market. For the preparation of dry dates, fruits are harvested at full Doka, and, for soft dates (Pind Khajoor), partial-to-full-Dang fruits are harvested. Date palm trees bear commercial fruit 6 years after planting. Initially, yields are low but increase with the age of the trees. Fruit yield differs from variety to variety and also depends upon the age of tree and orchard management. On average, 50 kg Doka fruits per year can be harvested from trees aged up to 15 years and thereafter 75 kg can be harvested per tree per year by using a recommended package of practices (Chundawat 1990).

20.3.13 Post Harvest and Storage

Fresh Doka (Khalal stage) fruit cannot be stored at room temperature for more than 4 days, whereas they can be stored up to 30 days under refrigeration and up to 50 days at freezing temperatures. Fresh dates are washed and packed in cardboard boxes. Fruits of Halawy turn to Dang stage in the refrigerator. For the preparation of dry dates, full Doka fruits are washed and dipped in boiling water for 10 or 20 min, sulfiltated in 1,500 ppm potassium metabisulphite and dried either in an air-circulating oven at 48–52°C for 70–95 h or through sun drying for 80–120 h if the weather is dry. Halawy is the most suitable variety for making high quality dry dates by immersing the fruit in boiling water for 20 min and subsequently drying in an air-circulating tray drier at

45°C for 60–65 h. Fully Dang fruits can be converted into soft dates simply by drying them in air-circulating oven (Manohar and Chandra 1995).

20.3.14 Diseases and Pests

Date palm plants are affected by various diseases like Bayoud disease (*Fusarium oxysporum* f. sp. *albedinis*; Zaid et al. 2002), Black scorch – also called medjnoon or fool's disease – (*Ceratocystis paradoxa*), Diplodia disease (*Diplodia phoenicum*) and Graphiola leaf spot (*Graphiola phoenicis*). In deserts, these plants are also affected by scale insects (*Parlatoria blanchardi* Targ.), termites (*Odontotermes obesus*; Chandra et al. 1992; Manohar and Chandra 1995) and birds.

20.3.15 Propagation

Date palm can be propagated naturally in only two ways: by seed and by the offshoots or suckers that spring up around the base, or sometimes on the stem, of the palm until it attains an age of 10–20 years (Fig. 20.4). The use of tissue culture techniques is a recently developed third method, which has now been adopted by advanced commercial laboratories.

Fig. 20.4 Development of offshoots from mother plant of different ages

20.3.15.1 Seed Germination

Propagation by seed is sexual method of propagation. Seed is the most convenient material with which to propagate date palm: seeds can be stored for years, they germinate easily and are available in large numbers. However, for several reasons, this method cannot be used commercially for propagating desired date palm cultivars in a true-to-type manner.

20.3.15.2 Offshoots

For commercial plantations, date palm is always propagated through offshoots (suckers). This is the only commercial method of vegetative clonal propagation used to multiply the best varieties. Offshoots are produced from axillary buds situated on the base of the trunk during the juvenile life of the palm. However, they develop slowly and the numbers produced per tree are limited. Furthermore, suckers are produced only within a certain period in the mother palm's life. Thus, during the lifetime of an individual plant, only a low number of transplantable offshoots is available. Offshoot production varies from 10 to 30 depending on the cultivar and the cultivation practices used. No field-based method is as yet available with which to increase the numbers of offshoots produced by each tree (Zaid and de Wet 2002b). Offshoots have to be large enough (i.e. 8–15 kg) to survive when transplanted in the field.

Offshoot propagation, which represents asexual or vegetative propagation, offers the following advantages:

1. Offshoot plants are true-to-type to the parent palm. The offshoots develop from axillary buds on the trunk of the mother plant and consequently the fruit produced will be of the same quality as the mother palm, thus ensuring product uniformity.
2. The offshoot plant will bear fruits 2–3 years earlier than seedlings. The life span of the date palm is divided into two distinct developmental phases: vegetative, in which buds forming in the leaf axils develop into offshoots; and generative, in which buds form inflorescences and offshoots cease. The axillary bud of a leaf differentiates into an offshoot. Offshoots are separated for planting after growing for up to 3–5 years (Nixon and Carpenter 1978). Their curved form distinguishes offshoots while mother plants have a straight form. Zahidi, Berim and Hayani varieties are known to produce large numbers of offshoots, while Mektoum and Barhee varieties produce relatively low numbers of offshoots.

20.3.15.3 Transplantation Methods

Selection

Ground offshoots (suckers), which have reached at least 8–15 kg average weight and have a well-developed root system, are selected for planting. Aerial offshoots are not used for planting, as these are either devoid of roots or have very poor root development (Zaid and de Wet 2002b). The best time for the removal of offshoots and transplanting into the nursery for rooting is after the soil begins to warm up in the spring and early summer.

Rooting

Two types of offshoots occur on a date palm tree: basal and upper offshoots. Basal offshoots are more physiologically active than upper ones, growing faster as the number of leaves produced increases with age. Numerous factors should be considered for rooting of offshoots, including the size of the offshoot (often expressed in weight), type (upper or lower), origin of the offshoot, the method of removal and preparation for planting, as well as treatment of the offshoot after planting (Nixon and Carpenter 1978). To promote rooting, the base of the offshoot should be in contact with moist soil for at least 12 months before removal.

Detachment

Around 4–5 days before the offshoots are to be removed from the mother palm, their inner leaves should be cut back to one-half and the outer leaves to two-thirds of their length. For the production of offshoots, no green leaves should be removed from an offshoot until it is cut from the mother palm. If leaves interfere with cultivation, they may be tied together. After 3–5 years of attachment to the parent palm, offshoots will form their own roots and start producing a second generation of offshoots (Nixon 1936; Nixon and Carpenter 1978). Offshoots are removed from the mother palm tree using a sharp chisel.

Planting

Offshoots can be planted in a nursery bed of 1 m^3 for rooting for 1–2 years. A mixture of 10–15 kg well-rotted farmyard manure, fertile topsoil, 50 g captan and 50–100 g insecticidal dusts are used for refilling the pits. Moreover, the offshoots are dipped in a solution of carbendazim and chloropyrophos or endosulfan. In addition, the root initials of offshoots may be dipped in IBA solution (1,000 ppm) for 2–5 min to ensure better root development.

Spacing and Density

Date palm offshoots are planted in the field in a square system of planting at a distance of 8–10 m depending on soil fertility and irrigation regime. A total of 156 suckers can be planted in a 1-ha area at 8 m planting distance in such a square system. Normally, dates are grown at a density of about 120–200 trees per hectare on average (Chandra et al. 1992). Young offshoots and tissue culture-derived plants should be protected from harsh climatic conditions.

20.4 Tissue Culture

Micropropagation of plants through plant tissue culture techniques has already attained the dimensions of a full plant-based industry in several countries. The advantages of this technology over conventional propagation are: (1) only a small laboratory space is required; (2) a large number of plants can be produced in short duration; (3) cloning of selected material is possible; (4) there is no seasonal effect on plants because they can be multiplied under controlled conditions in the laboratory throughout the year; (5) genetically uniform plants are produced; (6) easy and fast exchange of plant material between different regions of a country or between countries is ensured without any risk of the spread of diseases and pests; (7) large-scale production is economically reliable; and (8) the plantlets are easy to handle and transport, and do not requires phyto-sanitary regulation.

Three methods of micropropagation are currently been used depending upon the objectives and plant species: (1) organogenesis (shoot, root and plantlet formation from callus), (2) bud proliferation, and (3) somatic embryogenesis (direct or indirect, with the possibility of scale-up of the technology through cell culture in bioreactors; Sangwan and Harada 1975; Welander 1976; Reynolds 1979; Tisserat 1982; Raj Bhansali 1990, 1993; Raj Bhansali et al. 1991).

Date palm trees have benefited greatly from the application of plant tissue culture techniques, since, for large-scale propagation, the offshoots growing at the base of the mother tree constitute the only source of explants used for the initiation stage. Because of the low success level at this initial stage, a large number of offshoots are often needed. Therefore, over the past more than 30 years, various in vitro techniques for the production of planting materials have been developed in different parts of the world (Welander 1976; Reynolds 1979; Tisserat 1981a, 1981b; Zaid and Tisserat 1983; Sharma et al. 1986; Raj Bhansali and Kaul 1991; Loutfi and El Hadrami 2007; Al-Khayri 2007). Growth and development of tissue-cultured date palm plants involves several steps: (1) the best quality germplasm from around the world is used; (2) the actively growing tissues are taken from the desired plant under the highest hygienic conditions; (3) explants are then planted on suitable media in the laboratory; (4) direct organogenesis to plantlets from the developing somatic embryos takes place under laboratory-controlled temperature, humidity and light regimes. (5) the regenerated plants are then transferred to the greenhouse, where they gradually become hardened and accustomed to the normal environment in which they will be planted in the field. This process takes almost 3 months. The nursery period takes 3–6 months, during which the plant develops further and becomes accustomed to the actual temperatures and conditions it will experience in the field. This process is closely monitored; (6) after this period the plants are ready to be planted anywhere.

In most cases, females plant are preferred for date palm commercialisation, but there are many cases where excellent male plants are also indispensable, exhibiting a metaxenia effect. An Al Ain city male is a unique date palm with many interesting and exceptional metaxenia effects on inflorescences, but no offshoots are available. In such a plant, the only source of primary explants for the large-scale propagation of Al Ain city male is that specimen only. In all such cases, female/male healthy date palm offshoots of about 2 to 4 years old are used as sources of explants from desired cultivars. The suckers/inflorescences are obtained from adult palms after careful elimination of all leaves and adjoining tissues, and the shoot tips are excised from the offshoots (Fig. 20.5). These explants are surface sterilised in 2% sodium hypochlorite. Plantlets are obtained directly from shoot tip explants of date palm in culture through shoot development. Both somatic embryogenesis and organogenesis are used in clonal propagation of date palm plants on a commercial scale. Callus and morphogenesis in culture have been induced from different explants including zygotic embryos, roots (Sharma et al. 1980), young leaves (Sharma et al. 1984), shoot tips (Zaid and Tisserat 1983; Raj Bhansali et al. 1988), inflorescence (Drira and Benbadis 1985; Bhaskaran and Smith 1992; Loutfi and Chlyah 1998) and axillary buds (Bouguedoura et al. 1990).

MS (Murashige and Skoog 1962) tissue culture medium is the most extensively used nutritive media in date palm propagation as it helps to release the formation of new buds and somatic embryos. The following chemicals are also added to this basal nutrient MS solution: amino acids (arginine, asparagine, glycine, adenine and glutamine); vitamins (inositol, biotin, pyrodoxine, nicotinic acid and thiamine), which are important in enhancing the formation of new buds; and other organic materials, e.g. sucrose (30–40 g/l). Activated charcoal (0.3–3 g/l)/polyvinylpyrrolidone (PVP; 2 g/l) reduces the phenolic toxicity produced by explants and increases growth of living explants and formation of date palm organs (Tisserat 1979). Agar is used at a concentration of 8 g/l. The various types of auxins and cytokinins and their concentrations (according to explant quality, physiological status and developmental stage of the cultivated explants) are added. This basal medium is augmented with various combinations of auxins and cytokinins. The different media contain 2,4-dichlorophenoxy acetic acid (2,4-D), indole acetic

Fig. 20.5 Apical shoot tip extracted from the offshoot for tissue culture

Table 20.5 Date palm explants and their in vitro responses

Explant	Availability	Morphogenetic potential	Remark
Apical tip	One per tree/ offshoot	Very high	Sacrifice whole plant or offshoot
Inflorescence	Abundant	Immature–high Mature–low	Dependent on flowering time
Lateral bud	Several	Variable Young–high Old–low	Sacrifice the plant to obtain explant
Zygotic embryo	Abundant	Variable Immature–high Mature–low	Non-clonal, specific genotype
Leaf	Abundant	Cultures developed but with no roots	Clonal but difficult to culture
Stem	Abundant	Failed	Clonal but difficult to culture
Root	Abundant	No shoots	Regeneration not reported

acid (IAA), naphthoxy acetic acid (NOA), or naphthalene acetic acid (NAA) as auxins, while the cytokinins used include 6-benzylaminopurine (BAP), kinetin or N6-isopentenenyladenine (2-ip; Al-Khayri 2001, Al-Khayri and Al-Bahrany 2001). Auxin concentrations range from 0 to 10 mg/l while cytokinins range from 0 to 5 mg/l. Before adding the agar and the activated charcoal, the pH is fixed at 5.8 ± 0.1. The solution is then poured either into a tube (2.5 x 20–25 cm) at 15–17 ml per tube, or into a 100–150 ml flask. These nutrient media are autoclaved at 121°C for 20 min. After cultivating the explants in a suitable medium, they are incubated in a growth room, in complete darkness at the early stages, and in specific light conditions in the advanced stages. The cultures are kept at 26°C in the dark for 8 h; and at 28°C during illumination for 16 h, and are subcultured at 30 day intervals. Several explant sources and types of media that have been used to obtain morphogenetic responses in vitro are given in Table 20.5.

20.4.1 Somatic Embryogenesis

Asexual embryogenesis has been achieved from excised zygotic embryos (Ammar and Benbadis 1977; Reynolds and Murashige 1979) and also from somatic tissues (Tisserat et al. 1979; Mater 1986; Sharma et al. 1984, 1986; Raj Bhansali et al. 1988; Daquin and Letouze 1988; Dass et al. 1989; Raj Bhansali and Kaul 1991; Bhaskaran and Smith 1992; Sudhersan et al. 1993; Al-Khayri 2003). Callus tissues are induced in various date palm cultivars (e.g. Muscut, Medjool, Sayar, Samran, Jaglool and Khadrawy) when explants are incubated in complete darkness for 3–6 months at 25°C (Sharma et al. 1984, 1986; Raj Bhansali et al. 1988; Yadav et al. 1998; Bhargava et al. 2003).

During growth of embryogenic callus, date palm explants release excessive browning substances, which causes serious problems with this technique. These substances (phenols) have profound physiological effects on the establishment and growth of embryogenic callus. Browning of explant tissues and culture medium is due to oxidation of polyphenols and formation of quinones. These are highly reactive and toxic to the tissues. The inhibitory effects may result from the bonding of phenols with proteins and their subsequent oxidation into quinones. Presoaking of explants in ascorbic acid and citric acid solutions, and adding these compounds to the culture medium, helps to curtail the oxidation of phenols. Incorporation of polyvinylpyrrolidone (PVP), cysteine-HCl and ascorbic acid also minimised browning problems in several date palm species (Dass et al. 1989). Zaid and Tisserat (1983) suggested soaking date palm explants in an antioxidant solution (150 mg/l citric acid and 100 mg/l ascorbic acid) prior to surface sterilisation treatment. Raj Bhansali and Kaul (1991) also used these antioxidant solutions for 30–60 min in cold storage (0–4°C). Furthermore, use of nutritionally balanced media containing activated charcoal (3 g/l) has significantly checked the browning problems in date palm explants. Raj Bhansali et al. (1988) found that shoot tips and lateral bud cultures grew successfully if transferred frequently (after periods of incubation of 7–15 days) to fresh medium. The influence of physical conditions, nutrient medium and carbon source has been studied in various varieties (Tisserat 1979, 1981a, 1981b; Wangkaew et al. 1991; Veramendi and Navarro 1996, 1997).

Complete protocols (callus initiation, diagnosis of somatic embryos, nodular callus, embryo development and multiplication) for propagation of date palm via somatic embryogenesis have been reported by various workers (Sharma et al. 1986; Raj Bhansali et al. 1988; Sudhersan et al. 1993; El Hadarami et al. 1995; Figs. 20.6–20.10). At the Central Arid Zone Research Institute (CAZRI), Jodhpur, a method for the clonal propagation of date palm through repetitive somatic embryogenesis (RSE) was developed during 1986–1989. Nowadays, the principles involved in multiplication of date palm plantlets are adequate to produce large numbers of plantlets through somatic embryogenesis (Fig. 20.11). Free-living plants have been established in pots and in the field at CAZRI (Fig. 20.12).

Various workers have now field-tested tissue-culture raised and zygotic seedlings. Tissue culture plants raised from explants taken from female mother plants produce female flowers (Fig. 20.13, 20.14), whereas zygotic seedlings produced male/female flowers with different plant characteristics and flowering behaviour. Flowers are pollinated as usual by conventional methods, and fruit setting and development of the fruits are normal (Figs. 20.15, 20.16). This indicated the true-to-type behaviour of plantlets developed from the RSE process, which has now been developed sufficiently with certain varieties to be highly efficient in raising date palm from tissue culture (Raj Bhansali et al. 1988; Raj Bhansali and Singh 2000, 2003). Comparison of the quality of fruits of Hilali cultivar grown from tissue culture and from offshoots revealed no significant differences in the chemical and physical characteristics of fruits, indicating clearly that tissue culture techniques are now a viable method of date palm propagation.

Fig. 20.6 Development of callus from apical shoot tip for induction of somatic embryogenesis

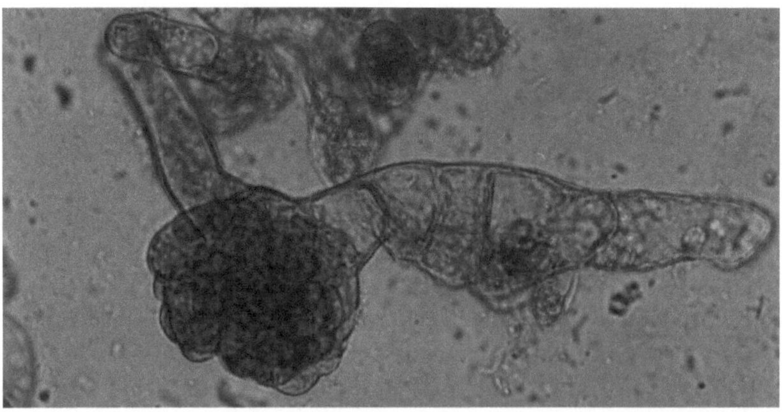

Fig. 20.7 Early diagnosis of embryogenic cultures through staining technique, indicating globular embryo

20.4.2 Suspension Culture

Several workers have also attempted to establish suspension culture of date palm friable callus for rapid embryogenesis (Sharma et al. 1986; Bhaskaran and Smith 1992). Embryogenic callus tissues are cut with a sterile scalpel into as small pieces as possible and then transferred to 50 ml liquid medium in 250 ml Erlenmeyer flasks. The flask contents are filtered using a sieve (500 μm diameter); the filtrate

Fig. 20.8 Initiation of somatic embryos from callus

Fig. 20.9 Germination of somatic embryos producing leafy shoots

obtained is incubated on a rotary shaker (100 rpm) at 25°C under the same light conditions. The liquid MS culture medium is diluted to half strength and supplemented with 2,4-D (0.1 mg/l), BAP (0.5 mg/l) and sucrose (3%). The pro-embryo

Fig. 20.10 Complete germinated somatic seedling

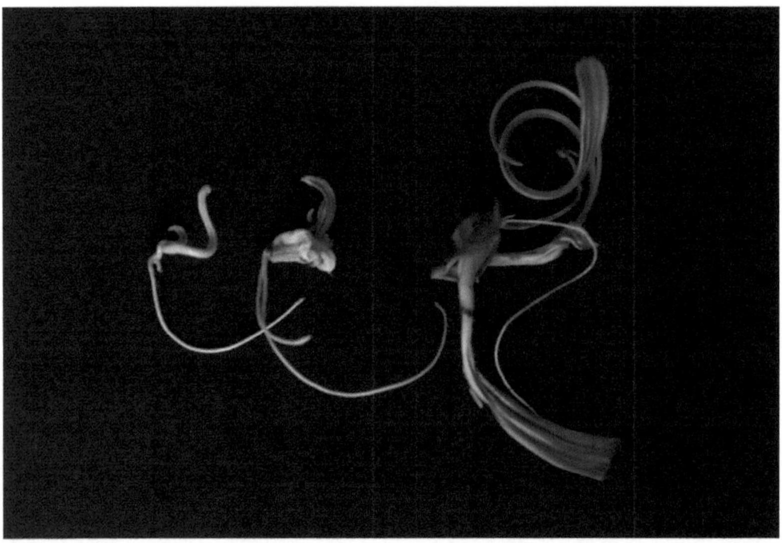

Fig. 20.11 Various stages of germination of somatic seedlings

masses develop into embryos after passing through several sequential and distinct embryo developmental phases. Hundreds of embryos can be developed from suspension culture within 3 weeks. These embryos need to be subcultured for 1 month to promote further growth. Approximately 1,000 embryos can be obtained from 200 mg embryogenic friable callus cultured per vessel (Bhaskaran and Smith 1992). Up to 40% of these embryos germinated into normal plantlets upon plating

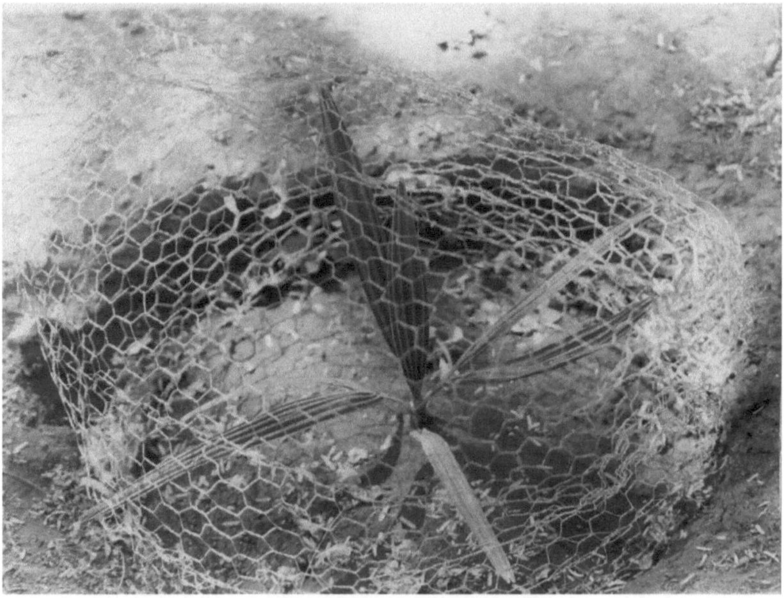

Fig. 20.12 Free-living tissue cultured date palm somatic seedling

on solid medium. The somatic embryogenesis method in date palm will help in studies on developmental embryogenesis and could be used for encapsulation of embryos for long-term storage and shipment for export.

20.4.3 Somaclonal Variation

The procedures developed initially for in vitro multiplication of date palm employed somatic embryogenesis, which requires explants to first produce callus from which somatic embryos are derived, and from which the final plantlets are subsequently developed. This method is still widely used with tremendous success (Tisserat 1982; Ammar and Benbadis 1997; Daquin and Letouze 1988; El Hadrami and Baaziz 1995). However, concerns regarding the fidelity of some plants obtained through somatic embryogenesis have emerged, with reference to the mother plant. Some abnormal vegetative traits as well as unusual flowering behaviour and fruiting habits have been reported (McCubbin et al. 2000). This is most probably due to somaclonal variations encountered as a result of callus formation and/or maintaining a long callus phase prior to embryogenesis. Corniquel and Mercier (1994) used restriction fragment length polymorphism (RFLP) and randomly amplified polymorphic DNA (RAPD) methods for cultivar identification of date palm. Off-types are quite common among tissue culture-produced date palm trees of the cultivar

Fig. 20.13 Emergence of flowering from date palm established from repetitive somatic embryogenesis (RSE) tissue culture method

'Barhee' (Cohen et al. 2004) – characterised mostly as a low fruit setting phenotype. Most flowers in such trees turn into parthenocarpic fruitlets having three carpels. Other flower abnormalities include distortions of carpels and stigmas. About 50% of trees reverted to normal within 10 years of planting. Eshraghi et al. (2005) reported that the genetic similarity between the mother plant and the somatic embryogenesis-derived regenerates ranged between 94% (for R1, R2) and 83% (for R5) when RAPD markers were used. Molecular techniques (RAPD markers/CTAB

Fig. 20.14 Female flower developed from somatic date palm seedling

method) have been employed in more than 50 cultivars in order to confirm varietal authenticity in the case of date palm micropropagated through somatic embryogenesis, with very promising results (Javouhey et al. 2000).

20.4.4 Direct Organogenesis

Date palm can also be propagated successfully by organogenesis tissue culture methods. Serious efforts have been made to develop date palm tissue culture as a means of mass producing high-yielding, disease resistant clones in large numbers for plantation (Reuveni 1979). Initially, several problems occurred in date palm tissue culture for indirect and direct organogenesis. These included browning of media and explants, resulting in premature death of tissues. The growing tip produced roots, whereas the callus was short-lived and could not survive in subculture.

Refinements in culture conditions, type of media, addition of charcoal and antioxidising chemicals, type of growth hormones, and media composition have now been improved significantly by various workers (Zaid and de Wet 2002c; Sudhersan et al. 1993; Al-Khateeb 2006, 2008). During the first 2 months in culture, the explants enlarge on the medium and, in some cases, readily initiate again into shoots. In other cases, growth subsequently slowed down, the explants became

Fig. 20.15 Fruit setting and developments of fruits – in vitro raised date palm

whitish and later adventitious buds could be seen. However, browning is also seen on some explants. Some media promoted the development of axillary buds while others did not. The new meristems or buds thus produced developed further into shoots. Well-developed shoot growth is first observed after about 6 months in culture. The shoots subsequently elongate in the following months, and other shoots begin to appear from the same explant. The initial number of shoots produced is low, between two and five per explant but, over time, these shoots are harvested as they grow and new ones form. The shoots are transferred to a rooting medium,

Fig. 20.16 Fully matured date palm fruits developed on Muscut-2 variety raised from RSE tissue culture method

which is the same basal medium but containing NAA. In an earlier report of direct organogenesis in date palm, Tisserat (1984) obtained few recoverable shoots. A similar observation was also made by Varughese (2000). Beauchesne (1983) reported many more adventive buds and shoots from explants in culture from direct organogenesis. Induction of vegetative bud development from shoot-tip or axillary bud explants in culture has been reported (Tisserat 1984; Bouguedoura et al. 1990; Raj Bhansali and Kaul 1991). Recently, MS medium with 20 mg/l adenine sulphate, BAP (1.0 mg/l), IBA (0.1–6.0 mg/l), 30 g/l sucrose, 3 g/l activated charcoal, myo-inositol (100 mg/l) and thiamine HCl (0.4 mg/l) has been employed for shoot initiation (Eshraghi et al. 2005). The cultures are initially kept under complete darkness at 25°C and then transferred to a multiplication media containing 2-isopentyladenine (2iP; 1.0 mg/l) and thidiazuron (TDZ; 0.1–0.5 mg/l). Low cytokinin and auxin concentrations supported bud and shoot multiplication in date palm cv. Sukry (Al-Khateeb 2006, 2008). Similarly, Taha et al. (2001) developed a rapid method of in vitro multiplication of date palm from shoot tips on MS medium supplemented with 2iP (2 mg/l) and NAA (1 mg/l). Shoot bud proliferation was strongly enhanced when cultured on MS medium containing 2iP (3 mg/l) and NAA (0.5 mg/l). Culturing on full-strength MS medium showed a higher multiplication rate compared with half-strength MS medium. Khierallah and Bader (2007) developed a stepwise method through direct organogenesis. The best combinations of plant growth regulators and other conditions in order to achieve organogenesis and multiplication directly from shoot tips without callus formation have been reported. MS modified medium supplemented with 2 mg/l 2iP, 1 mg/l BAP, 1 mg/l NAA and 1 mg/l naphthoxy acetic acid (NOA) supported bud formation from shoot tips after 16 weeks (6.2 bud/explant). Subculturing the formed buds on agitated liquid multiplication medium supplemented with 4 mg/l 2iP, 2 mg/L BAP, 1 mg/l NAA and 1.0 mg/l NOA gave the optimum average bud number (12.6 buds). In the elongation stage, MS medium with 0.5 mg/l GA_3 and 0.1 mg/l NAA enhanced plantlet length to 5.3 cm. Optimum rooting percentage (90%) was achieved when shoots were transferred to medium containing 1 mg/l NAA. The average root number after 8 weeks was 5.4, with a length of 9 cm. Rooted shoots (plantlets) were transplanted into small pots containing a mixture of peat moss and Perlite (2:1) and placed in plastic tunnels or in a greenhouse. The survival percentage was 85% after 3 months when the plants were transferred to bigger pots (Raj Bhansali et al. 1988). These results define a successful protocol for the in vitro propagation of date palm cv. Maktoom. Belal and El Deeb (1997) have also reported direct organogenesis in Egyptian cultivars (Zaghloul and Samani) on MS medium supplemented with a combination of auxins and cytokinins for shoot initiation, differentiation, shoot growth and shoot rooting. The influence of different carbon sources and concentrations on in vitro shoot multiplication of date palm cv. Khuneizi has been investigated. However, the number of buds produced per explant tissues is limited in comparison to somatic embryogenesis. Zaid et al. (2007) successfully developed a method for mass propagation through organogenesis (initiation, multiplication, elongation and rooting) from inflorescence tissues of a rare and unique male.

20.4.5 Advantages of Direct Organogenesis

In date palm, clonal propagation through suckers is the normal way of propagation but a more efficient method of cloning is highly desirable for large-scale production. There is also a need for further experiments to seek alternative ways of in vitro propagation that could reduce culture-induced variation in tissue culture plants. One of the methods currently advocated is the direct organogenesis method of plant propagation in vitro. Direct organogenesis has emerged as the most promising potential tissue culture method due to the lower occurrence of variation in regenerants. This technique has resulted from serious efforts aimed at finding ways to control abnormalities in regenerated plants, to maintain the quality of the plants, and to test their conformity with their parental genotype and phenotype during callus cultures due to the possibility of somaclonal variation. The successful development of this new technique is expected to greatly reduce the number of steps in culture, and possibly also the length of culture time. Gains in these two factors should help minimise the risk of somaclonal variation. There have been ongoing experiments all over world to develop in vitro propagation procedures for the date palm without somaclonal variation. The direct organogenesis method has the advantage of omitting the callus and embryo phases and significantly reducing the total number of stages in culture by forming new shoots direct from the explants. Various date varieties have been developed through direct organogenesis. In addition, unlike the procedure in some aspects of somatic embryogenesis, direct organogenesis occurs in the presence of light.

The current yield of plantlets is relatively low and further experiments are needed to substantially increase the number of plantlets recoverable per explant. The number of plants obtainable is important, first because large numbers of plants are required for the establishment of large plantations, and secondly because the incidence of contamination in culture can sometimes lead to substantial losses of young plants. Many more plants could be recovered by the somatic organogenesis method but the use of high auxin concentrations and the length of time of callus culture have been cited as possible reasons for tissue-culture-related variation (McCubbin et al. 2000). Planting of the material produced, relatively longer generation cycles, and high investment in the initial cost of plantation are needed to produce true-to-type progeny. The direct organogenesis method has the advantage of completely eliminating both the use of high auxin concentrations and the callus phase from the in vitro culture programme.

20.4.6 Molecular Characterisation

Improvement of date palm is very difficult due to its long life cycle, strongly heterozygous nature and non-availability of a method to determine sex at an early stage of development. Most earlier studies on genetic characterisation, detection of genetic variation and gene mutation have been conducted on the basis of variation in chromosome number, isoenzyme polymorphism and biochemical diversity (Booij et al. 1995). However, the chromosomes are numerous and small, and mitotic examination of tissue-culture-derived palm plants is unreliable. Recently, various workers have demonstrated the utility of RAPD markers for the analysis of genetic diversity among cultivars and within plant populations. Talaat and Al-Qaradawi (2009) have studied the genetic diversity among 15 different cultivars of date palm using simple sequence repeat (SSR) markers to determine the genetic similarity and/or diversity among the well-known Qatari date palm cultivars. Similarly, molecular techniques including mitochondrial plasmid-like DNAs markers have been used for molecular characterisation of Tunisian date palm varieties and to develop preventive procedures to protect them against Bayoud disease (*Fusarium oxysporum albedinis*), which causes vascular fusariosis. Recently, clonal plants of date palm were regenerated from juvenile leaves on 2,4-D-containing medium by producing adventitious shoot buds directly from the basal part of leaves as well as excessive calli. RAPD profiles were used to test for the somaclonal variation in plantlets that can sometimes be induced by 2,4-D during recurrent somatic embryogenesis. Nine arbitrary 10-mer primers were used to amplify DNA from 180 plantlets. The RAPD patterns of the plantlets were identical to those of the original mother plant, indicating that 2,4-D did not induce somaclonal variation that can be detected by the RAPD technique (Ahmed et al. 2009).

20.5 Conclusion

There is evidence that the climate has changed repeatedly during Earth's history and, in the past, regions that are now deserts were not always so arid, so devoid of life and vegetation (Singh et al. 1974).

Land-use changes continue to influence the microclimate, rainfall patterns and fauna of the Thar region. The introduction of irrigation to the Thar Desert via the construction of the IGNP has resulted in substantial man-induced changes in the microclimate, flora and fauna of the area due to the conversion of grasslands into irrigated cultivated lands. Land-use changes continue to influence the microclimate, rainfall patterns and fauna of the Thar region. In this changing scenario, date palm trees can represent essential and integral components of arid and semi-arid farming systems in dry and desert regions. The tremendous advantage of the date palm come from its various qualities such as resilience, requirement for limited inputs, long-term productivity and multi-purpose attributes. At the present time the yield of offshoots is almost as valuable as that of fruit, and growers therefore desire to secure as many offshoots of their best varieties as possible.

The demand for date palm planting material can be met by using in vitro technology. Plant tissue culture – considered a separate subject associated with plant biotechnology – is a rapidly expanding area of biology that has tremendous potential applications in hi-tech horticulture. Significant advances have been made in plant cell and tissue culture, and regeneration of whole plants is now a routine procedure for many plant species. These can subsequently been exploited as 'model' species for genetic transformation studies. These technologies have already led to many breakthroughs in the developed world, particularly in the United States, Europe, the United Kingdom, Germany and France, in making the agricultural cultivation of certain plants a commercially viable business. Currently, UAE, Israel, the United Kingdom and France have modern, very well developed date palm tissue culture facilities, where several million date palms are produced annually to meet the high demand for offshoot production (Date Palm Tissue Culture Laboratory 2006). Plant tissue culture has already opened up vast commercial possibilities, especially in the propagation of elite forest trees, horticulture, and ornamental and plantation crops. In the Indian arid zone context, tissue culture techniques may be even more useful in solving relevant and intractable problems. Building up trained manpower for commercial exploitation of developed technology, particularly for date palm, will be essential. Appropriate mechanisms should be evolved to integrate date palm tissue culture technology with conventional methods of date palm propagation. Planting date palm in Western Rajasthan can help check desertification and can strengthen rural economies by generation of employment besides providing net monetary income to farmers. Considering the various agriculture advances and recent technological innovations in various fields aimed at regenerating and revitalising the arid zone biosphere, it is not hard to visualise Great Indian Thar Desert full of lush green forest, orchards and crop plants.

References

Abrol IP, Venkateswarlu J (1995) Sustainable development of arid areas in India with particular reference to Western Rajasthan. In: Sen AK, Amal K (eds) Land degradation and desertification in Asia and the Pacific Region. Scientific, Jodhpur, pp 135–153

Ahmed O, Chokri B, Drira N, Mohamed M, Mokhtar T (2009) Regeneration and molecular analysis of date palm (*Phoenix dactylifera* L.) plantlets using RAPD markers. Afr J Biotechnol 8:813–820

Al-Khateeb AA (2006) Role of cytokinin and auxin on the multiplication stage of date palm. Biotechnology 5:349–352

Al-Khateeb AA (2008) Regulation of in vitro bud formation of date palm (*Phoenix dactylifera* L.) cv. Khanezi by different carbon sources. Bioresour Technol 99:6550–6555

Al-Khayri JM (2001) Optimization of biotin and thiamine requirements for somatic embryogenesis of date palm (*Phoenix dactylifera* L.). In Vitro Cell Dev Biol Plant 37:453–456

Al-Khayri JM (2003) In vitro germination of somatic embryos in date palm: effect of auxin concentration and strength of MS salts. Curr Sci 84:101–104

Al-Khayri JM (2007) Date palm *Phoenix dactylifera* L. micropropagation. In: Jain SM, Haggman H (eds) Protocols for micropropagation of woody trees and fruits. Springer, Heidelberg, pp 509–526

Al-Khayri JM, Al-Bahrany AM (2001) Silver nitrate and 2-isopentyladenine promote somatic embryogenesis in date palm (*Phoenix dactylifera* L.). Sci Hortic 89:291–298

Ammar S, Benbadis A (1997) Multiplication vegetative du palmier dattier (*Phoenix dactylifera* L.) par la culture de tissues des jeunes plantes des semis. C R Acad Sci 284:1789–1792

Anonymous (1965) Soil conservation in the Rajasthan Desert – work of the Desert Afforestation Research station, Jodhpur. Ministry of Food and Agriculture booklet, Government of India

Beauchesne G (1983) Vegetative propagation of date palm (*Phoenix dactylifera*) by in vitro culture. In: Proceedings of 1st International Symposium of Date Palm. Riyadd, Saudi Arabia, pp 698–699

Belal AH, El Deeb MD (1997) Direct organogenesis of date palm (*Phoenix dactylifera* L.) in vitro. Assiut J Agric Sci 28:67–77

Bhargava SC, Saxena SN, Sharma R (2003) In vitro multiplication of *Phoenix dactylifera* (L). J Plant Biochem Biotechnol 12:43–47

Bhaskaran S, Smith RH (1992) Somatic embryogenesis from shoot tip and immature inflorescence of *Phoenix dactylifera* CV. Barhee. Plant Cell Rep 12:22–25

Booij I, Monfort S, Ferry M (1995) Characterization of thirteen date palm (*Phoenix dactyliera* L.) cultivars by enzyme electrophoresis using the phastsystem. J Plant Phys 145:62–66

Bouguedoura N, Michaux-Ferriere N, Bompard JL (1990) Comportment in vitro de bourgeons axillaries de type inde'termine' du palmier dattier (*Phoenix dactylifera* L.). Can J Bot 68:2004–2009

Pareek OP, Nath V (eds) (1996) CFR – Indian arid zone. Coordinated fruit research in Indian arid zone – a two decades profile (1976–1995)., NRC for Arid Horticulture, Bikaner, pp 55–66

Chandra A, Chaudhary NL (1990) Performance of date palm cultivars in Thar Desert. Part II. Curr Agric 14:4

Chandra A, Chandra A, Gupta IC (1992) Date palm research in Thar Desert. Scientific, Jodhpur

Chih Cheng TC, Krueger RR (2007) The date palm (*Phoenix dactylifera* L.): overview of biology, uses, and cultivation. HortScience 42:1077–1082

Chundawat BS (1990) Arid fruit culture. Oxford & IBH, New Delhi
Cohen Y, Korchinsky R, Tripler E (2004) Flower abnormalities cause abnormal fruit setting in tissue culture-propagated date palm (*Phoenix dactylifera* L.). J Hortic Sci Biotechnol 79:1007–1013
Corniquel B, Mercier L (1994) Date palm (*Phoenix dactylifera* L.) cultivar identification by RFLP and RAPD. Plant Sci 101:163–172
Daquin F, Letouze R (1988) Regeneration du palmier dattier (*Phoenix dactylifera* L.) par embryogenese somatique: amelioration de efficacite par passage en milieu liquide agite. Fruits 3:191–194
Dass HC, Kaul RK, Joshi SP, Raj Bhansali R (1989) In vitro propagation of *Phoenix dactylifera* L. Curr Sci 58:22–24
Date Palm Tissue Culture Laboratory (2006) http://datepalm.uaeu.ac.ae/subpages/Laboratory.html, accessed 4 April 2009
Dhir RP (2003) Thar Desert in retrospect and prospect. Proc Indian Natl Sci Acad 69:167–184
Dhir RP, Singh KS (1985) Soils of India and their management, FAI, New Delhi, pp 343–364
Drira N, Benbadis A (1985) Multiplication vegetative du palmier dattier (*Phoenix dactylifera* L.) par reversion on culture in vitro debauches florales de pieds femelles. J Plant Physiol 119:227–235
El Hadrami I, Cheikh R, Baaziz M (1995) Somatic embryogenesis and plant regeneration from shoot-tip explants in *Phoenix dactylifera* L. Biol Planta 37:197–203
El Hadrami L, Baaziz M (1995) Somatic embryogenesis and analysis of peroxidases in *Phoenix dactylifera* L. Biol Planta 37:205–211
Eshraghi P, Zarghami R, Ofoghi H (2005) Genetic stability of micropropagated plantlets in date palm. J Sci (Iran) 16:311–315
FAO (2006) 2005 worldwide date production statistics. 19 Mar 2006. Food and Agriculture Organisation of the United Nations, Rome
Faroda AS, Harsh LN (1999) From an empty bowl to self-reliance: success of farmers of desert villages. Environ News 3(June–July):14
Harsh LN, Tewari JC (2007) Agroforestry systems in arid regions of India, In: Puri S, Pamwar P (eds) Agroforestry: systems and practices. New India Publishing, New Delhi, pp 175–189
Javouhey M, Dagin F, Letouze R (2000) Somatic embryogenesis, an efficient tool for date palm (*Phoenix dactylifera* L.) industrial micropropagation. Characterization and genetic stability of original offshoots and regenerated plantlets by RAPD markers. In: Proceedings of International symposium on methods and markers for quality assurance in micropropagation. ISHS Acta Hortic 530:237–241
Kalra SK, Jawanda JS (1992) Fruit characters and quality assessment of some promising date varieties at Abohar. Punjab Hortic J 12:39–43
Kar A (1996) Desertification processes in arid Western India. In: Miyazaki T, Tsunekawa A (eds) Towards solving the global desertification problem, vol 4. National Institute for Environmental Studies, Tsukuba, pp 20–29
Kar A, Singh N, Kumar S (1994) Wind erosion control measures for gas pipeline between Gamnowala Tar and Ramgarh Jaisalmer district). CAZRI, Jodhpur
Khierallah HSM, Bader SM (2007) Micropropagation of date palm (*Phoenix dactylifera* l.) var. Maktoom through direct organogenesis. Acta Hortic 736:213–224
Krishnan A (1974) Proceedings of ICAR Summer Institute on Desert Ecosystem and its Improvement. CAZRI, Jodhpur, pp 8–27
Krishnan A, Rao GGSN (1978) Soil temperature regime in the arid zone of India. Theor Appl Climatol 27:15–22
Lioubimtseva E (2004) Climate changes in arid environments: revisiting the past to understand the future. Prog Phys Geogr 28:502–530
Loutfi K, Chlyah H (1998) Vegetative multiplication of date palm from the in vitro cultured inflorescence: effect of some growth regulator combinations and organogenetic potential of various cultivars. Agronomie 18:573–580

Loutfi K, El Hadrami I (2007) *Phoenix dactylifera* date palm. In: Litz RE (ed), Biotechnology of fruits and nut crops. CABI, Wallingford, pp 144–156
Manohar MS, Chandra N (1995) Date palm culture in Rajasthan. Directorate of Research, Rajasthan Agriculture University, Bikaner
Mater AA (1986) In vitro propagation of *Phoenix dactylifera* L. Date Palm J 4:137–151
McCubbin MJ, van Staden J, Zaid A (2000) A southern African survey conducted for off-types on date palms produced using somatic embryogenesis. Proceedings of Date Palm International Symposium, Windhoek, Namibia
Murashige T, Skoog F (1962) A revised medium for rapid growth and bioassay with tobacco tissue cultures. Physiol Plant 15:473–497
Nixon RW (1936) Metaxenia and interspecific pollinations in *Phoenix*. Proc Am Soc Hortic Sci 33:21–26
Nixon RW (1950) Imported cultivars of dates in the United States. U.S. Department of Agriculture Circular no. 834, US Department of Agriculture, Washington DC
Nixon RW, Carpenter JB (1978) Growing dates in the United States. United States Department of Agriculture Bulletin no. 207, US Department of Agriculture, Washington DC
Pareek OP, Nath V (eds) (1996) CFR – Indian arid zone. Coordinated fruit research in Indian arid zone – a two decades profile (1976–1995). NRC for Arid Horticulture, Bikaner, pp 55–66
Pundir JPS, Porwal R (1998) Performance of different date palm cultivars under hyper arid-supplementary irrigated Western plains of Rajasthan (India). In: Proceedings of First International Conference on Date Palms, Al-Ain, UAE, 8–10 March 1998, pp 329–336
Raj Bhansali R (1990) Somatic embryogenesis and regeneration of plantlets in Pomegranate. Ann Bot 66:249–253
Raj Bhansali R (1993) Bud culture of shoot multiplication and plantlet formation of *Tecomella undulata* (Rohida) woody tree of arid zone. Trop Sci 33:1–8
Raj Bhansali R, Kaul RK (1991) Into future – date through tissue culture. Indian Hortic 36: 6–10
Raj Bhansali R, Singh M (2000) Somatic embryogenesis in fruit and forest trees of arid zone. In: Jain SM, Gupta PK, Newton RJ (eds) Somatic embryogenesis in woody plants, vol 6. Kluwer, Dordrecht, pp 141–168
Raj Bhansali R, Singh M (2003) Micropropagation of arid zone fruit trees of India. In: Jain SM, Katsuaki I (eds), Micropropagation of woody trees and fruits. Kluwer, Dordrecht, pp 381–432
Raj Bhansali R, Kaul RK, Dass HC (1988) Mass cloning of date palm plantlets through repetitive somatic embryogenesis. J Plant Anat Morphol 5:73–79
Raj Bhansali R, Driver JA, Durzan DJ (1991) Adventitious embryogenesis and plant regeneration from rescued peach embryos. Indian J Exp Biol 29:334–337
Raj Bhansali R, Jindal SK (2000) Role of farmers in promotion of eco-friendly multipurpose trees of arid zone. In: Chaudhary V, Singh K, Kakralya B L (eds) Environmental protection, forwarded by Sunder Lal Bhuguna. Pointer, Jaipur, pp 92–101
Ramakrishna YS, Rao AS (1992) Climatic features of the Indira Gandhi Canal Region. In: Abrol IP, Venkateswarlu J (eds) Prospects of Indira Gandhi Canal project. ICAR, New Delhi, pp 1–10
Rao AS (2009) Climate and microclimate changes influencing the fauna of the Hot Indian arid zone. In: Sivaperuman C, Baqri QH, Ramaswamy G, Naseema M (eds) Faunal ecology and conservation of the Great Indian Desert, pp 13–23
Reuveni O (1979) Embryogenesis and plantlets growth of date palm (*Phoenix dactylifera* L.) derived from callus tissues. Plant Physiol 63:138
Reynolds JF (1979) Morphogenesis of palms in vitro. In Vitro 15:210
Reynolds JF, Murashige T (1979) Asexual embryogenesis in callus cultures of palms. In vitro Cell Dev Biol 15:383–387
Sangwan RS, Harada H (1975) Chemical regulation of cell growth, organogenesis, plant regeneration and somatic embryogenesis in *Antirrhinum majus* tissue and cell culture. J Exp Bot 26:868–881

Sharma DR, Kumari R, Chowdhuri JB (1980) In vitro culture of female date palm (*Phoenix dactylifera* L.) tissue. Euphytica 29:169–174
Sharma DR, Dawra S, Chowdhuri JB (1984) Somatic embryogenesis and plant regeneration date palm (*Phoenix dactylifera* L.) cv. Khadrawy through tissue culture. Indian J Exp Biol 22:763–766
Sharma DR, Deepak S, Chowdhary JB (1986) Regeneration of plantlets from somatic tissues of date palm (*Phoenix dactylifera* Linn.). Indian J Exp Biol 24:763–766
Singh G, Joshi RD, Chopra SK, Singh AB (1974) Late quaternary history of vegetation and climate of Rajasthan desert, India. Philos Trans R Soc Lond 267B:467–501
Srivastava HC, Dhavan S (1981) Performance of some date varieties in Haryana. Agric Sci Digest 1:76–78
Sudhersan C, Abo El-Nil, Al-Baiz A (1993) Occurrence of direct somatic embryogenesis on the sword leaf of in vitro plantlets of *Phoenix dactylifera* L. Cultivar Barhee. Curr Sci 65:887–888
Taha HS, Bekheet SA, Saker MM (2001) Factors affecting in vitro multiplication of date palm. Biol Planta 44:431–433
Talaat AA, Al-Qaradawi AY (2009) Molecular phylogeny of Qatari genotypes of date palm using simple sequence repeats markers. Biotechnology 8:126–131
Tisserat B (1979) Propagation of date palm (*Phoenix dactylifera* L.) in vitro. J Exp Bot 30:1275–1283
Tisserat B (1981a) Production of free-living palms through tissue culture. Date Palm J 1:43–54
Tisserat B (1981b) Date palm tissue cultures. USDA/ARS Advances in Agricultural Technology, Western series, No.17, Agricultural Research Series. USDA, Oakland, CA
Tisserat B (1982) Factors involved in the production of plantlets from date palm cultures. Euphytica 31:201–214
Tisserat B (1984) Propagation of date palms by shoot tip cultures. Hortic Sci 19:230–231
Tisserat B, Foster G, De Mason D (1979) Plantlet production in vitro from *Phoenix dactylifera* L. Dates Growers Rep 54:19–23
Varughese A (2000) Mass propagation of date palm through tissue culture: an efficient method by SAPAD. In: Proceedings of the Date Palm International Symposium, Windhoek, Namibia
Venkateswarlu J, Kar A (1996) Wind erosion and its control in arid north-west India. Ann Arid Zone 35:85–99
Veramendi J, Navarro L (1996) Influence of physical conditions of nutrient medium and sucrose on somatic embryogenesis of date palm. Plant Cell Tissue Organ Cult 45:159–164
Veramendi J, Navarro L (1997) Influence of explant sources of adult date palm (*Phoenix dactylifera* L.) on embryogenic callus formation. J Hortic Sci 72:665–671
Wangkaew P, Pienngarm B, Polthampitak T (1991) Tissue culture of date palm. Kaenkaset Khon Kaen Agric J 19:191–200
Welander T (1976) Effects of nitrogen, sucrose, IAA and kinetin on explants of *Beta vulgaris* grown in vitro. Physiol Plant 36:7–10
Yadav NR, Singh J, Yadav RC, Chowdhury VK, Chowdhury JB (1998) Somatic embryogenesis and plant regeneration from cell suspension cultures of *Phoenix dactylifera* L. cv. Khadrawy. Physiol Mol Biol Plants 4:135–138
Zaid A, Tisserat B (1983) In vitro shoot tip differentiation in *Phoenix dactylifera* L. Date Palm J 2:163–182
Zaid A, de Wet PF (2002a) Climatic requirements of date palm. In: Zaid A (ed) Date palm cultivation. FAO Plant Production and Protection. FAO, Rome, pp 57–72
Zaid A, de Wet PF (2002b) Date palm propagation. In: Zaid A (ed) Date palm cultivation. FAO Plant Production and Protection. FAO, Rome, pp 73–105
Zaid A, de Wet PF (2002c) Pollination and bunch management. In: Zaid A (ed) Date palm cultivation. FAO Plant Production and Protection. FAO, Rome, pp 145–175
Zaid A, de Wet PF, Djerbi M, Oihabi A (2002) Diseases and pests of date palm. In: Zaid A (ed) Date palm cultivation. FAO Plant Production and Protection. FAO, Rome, pp 227–281

Zaid A, Al Kaabi HH, El Korchi B (2007) Large scale in vitro propagation of a rare and unique male date palm (*Phoenix dactylifera* L.) using inflorescences technique. ISHS Acta Hortic) 736:234–254

Zohary D, Hopf M (1993) Date palm *Phoenix dactylifera*. Domestication of plants in the Old World, 2nd edn. Clarendon, Oxford

Chapter 21
Runoff-Rainwater for Sustainable Desert Farming

Ulrich Lüttge

Abstract A historical overview shows that in arid and desert regions in Palestine, Arabia and North Africa, agriculture has been practiced by ancient populations using sustainable rainwater resources. Methods of such desert farming include forming terraces in secondary and tertiary wadis, building desert farms with runoff water flooding from large catchment areas, constructing canal systems and micro catchments. Israeli scientists have surveyed the historical remnants of ancient desert farms in the Negev desert. They have reconstructed several such farms in the Negev to study the mechanisms behind the techniques of desert farming and to demonstrate that crops of substantial quality and yield can be obtained. As desertification continuously reduces the arable land available on our globe, as water increasingly becomes a rare resource, and as large populations suffer hunger, techniques allowing fully sustainable desert farming using runoff-rainwater should become a valuable option, not for profit but for supporting people in local communities. The applicability of this approach depends on geo-morphological and soil conditions but is possible in many arid regions of the world.

21.1 Introduction

21.1.1 The Scenario of Precipitation in a Stern Desert

The most dominant single constraint to plant growth and farming in deserts is precipitation. The rainfall map of Israel and Palestine shows that the ancient Nabatean town of Avdat is located just below the isohyet of the very low average annual rainfall of 100 mm (Fig. 21.1). Annual precipitation shows some rhythmicity

U. Lüttge
Institute of Botany, Department of Biology, Technical University of Darmstadt, Schnittspahn-Strasse 3–5, 64287 Darmstadt, Germany
e-mail: luettge@bio.tu-darmstadt.de

K.G. Ramawat (ed.), *Desert Plants*,
DOI 10.1007/978-3-642-02550-1_21,

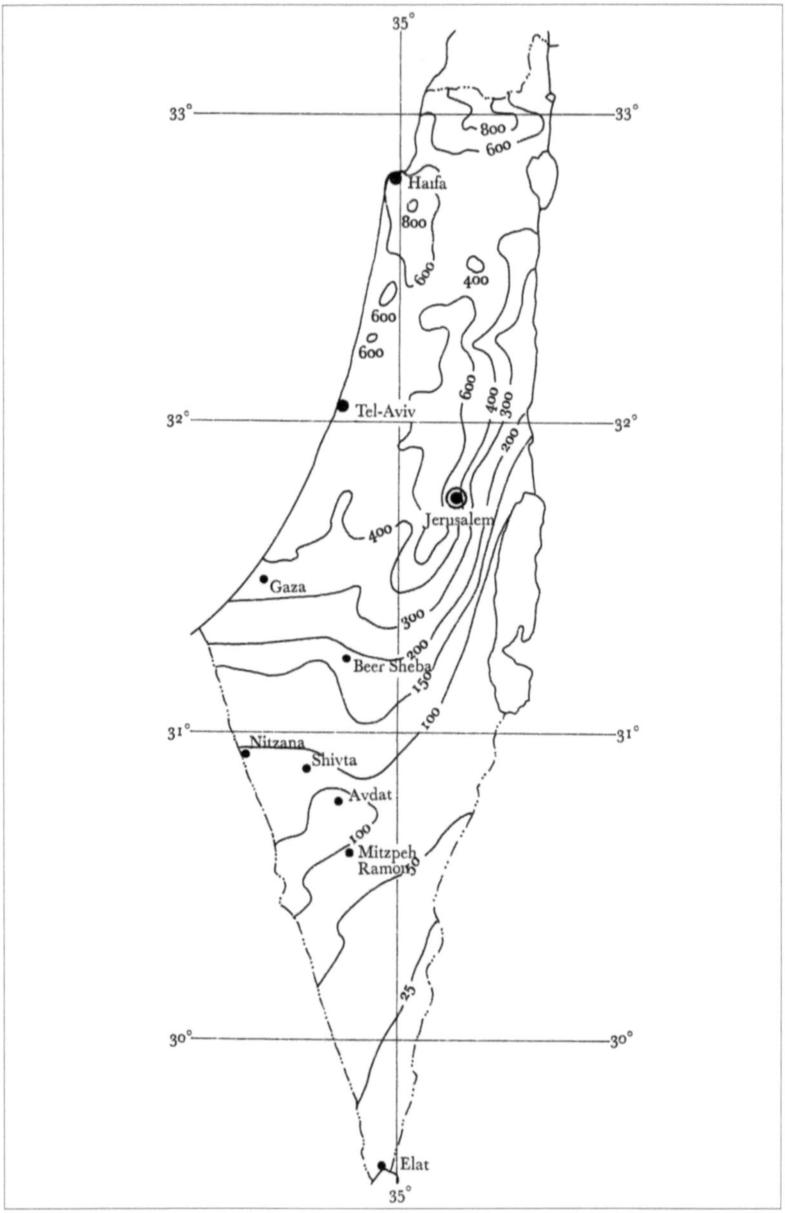

Fig. 21.1 Isohyets of average annual rainfall (mm) in Palestine (from Evenari 1982)

over the years, corresponding to the biblical seven fat and seven meagre years (Fig. 21.2; Shanan et al. 1967). Nevertheless, it is well known that agriculture was performed in historical times at this very location. Reconstruction of historical

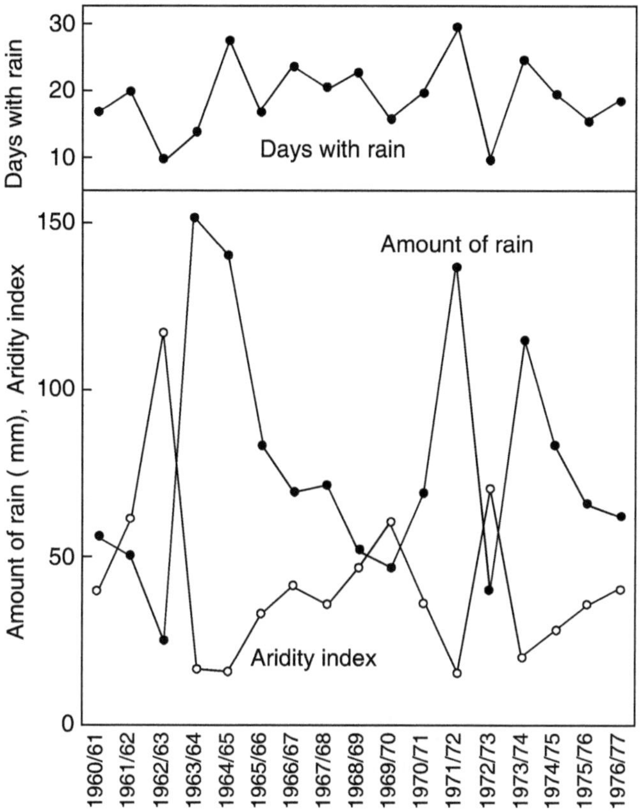

Fig. 21.2 Number of rainy days, annual precipitation and aridity index as given by the ratio of evaporation and precipitation in the Negev desert in the years 1960–1977

desert farms has demonstrated that lush agriculture solely supported by rainwater in runoff systems is possible without any artificial irrigation. It is completely sustainable without any net consumption of natural resources.

In this chapter, I shall describe the historical background of farming in the Negev desert, explain the reconstruction of desert farms in the twentieth century, and give a brief outlook of applicability for sustained management aimed at supporting at least local communities in arid environments. (This chapter draws largely on a book by Evenari 1982; see also Evenari 1987, 1989).

21.1.2 Historical Reminiscence

The history of development and desert farming in areas of Palestine, Arabia and North Africa is summarised in Table 21.1 (Evenari and Koller 1956; Evenari et al.

Table 21.1 Brief historical survey of development and desert farming in areas of Palestine, Arabia and North Africa

Period: years BC	Negev / South-Transjordania / Sinai	Southern Arabia	Southern Algeria / Tunisia
Middle Palaeolithic: 60,000–30,000	Mediterranean climate, moist steppes		
Upper Palaeolithic: 30,000–15,000	Arid climate		
Epi-Palaeolithic: 15,000–8,000	Mediterranean climate, moist		
Neolithic: 8,000–4,000	Arid climate, settlement of the Negev, already desert agriculture at Jericho		
4,000–2,000	Runoff water technique introduced from Southern Arabia, Negev densely settled	Runoff water technique well known; development of irrigation agriculture	Runoff farming introduced by the Canaanites
2,000–1,000	No traces of settlements		
1,000–700	Kings of Judah (Solomon); King Uzziah (779–738 BC): Chronicles 2, 26:10, Old Testament	Wealthy country, dam of Marib	Phoenicians, runoff farming practiced without interruption to date
600	Israelitic period of Negev settlement terminated; the area is abandoned		
300	Immigration of the Nabateans from Southern Arabia into the Negev, foundation of the town of Avdat		
Period: years AD	**Negev**	**Southern Arabia**	**Northern Africa**
106	Rabbel II, the last Nabatean king, occupation of the Nabatean empire by the Roman emperor Trajan		Cornucopia of the Roman empire
300–630	Byzantine period, highest development of Negev agriculture		
575		Destruction of the dam of Marib by a large flood (34th sura, 16–17, of the Koran)	
637–641	Arabs conquer the Negev, decline and emigration of the inhabitants, pasture grounds of Bedouins		

(*continued*)

Table 21.1 (continued)

Period: years AD	Negev	Southern Arabia	Northern Africa
1954–1958	Exploration of ancient terraces with stone walls, canals and fields with the aid of aerial photography and jeep excursions by Michael Evenari and collaborators		
1959	Reconstruction of the desert farm at Avdat and later also at Mashash		
1960	First harvest of barley		

1958, 1961, 1966, 1971, 1975; Aharoni et al. 1960; Evenari 1964a, 1964b, 1983; Shanan et al. 1969). There was some interaction between these three major areas. In Palaeolithic times, the climate in the Negev desert of Palestine (considered in geographical terms of incorporating the political units of both Israel and Palestine) changed several times between more arid and moister Mediterranean periods. Runoff-water farming developed between 4000 and 2000 BC. Desert farming became rather active at the time of the kings of Judah, where the Old Testament (in Chronicles 2, 26:10) says of king Uzziah:

"and he built towers in the wilderness and cut out many cisterns, for he had large herds, both in the Shephelah and in the plain, and he had farmers and vinedressers in the hills and in the fertile lands, for he loved the soil."

Later, the Nabateans and then the Byzantines followed, until the area was abandoned in the seventh century mostly for political reasons and not due to changes of climate.

Thirteen hundred years later, in the middle of the twentieth century, the Israeli plant scientist Michael Evenari discovered strange heaps and walls of stones on the hills of the Negev desert that quite obviously had nothing to do with building settlements. He advanced the hypothesis that they were traces of extinct ancient historical agriculture in the area. He then proved this hypothesis by actual reconstruction of a runoff-water desert farm at Avdat and other farms at different locations in the Negev, producing a variety of crops and fruits without any artificial irrigation.

21.2 Methods of Sustainable Water Use by Desert Farming

Various methods of sustainable water use by desert farming are summarised in Table 21.2 and are described as follows.

Table 21.2 Methods of sustainable water use for desert farming

Source of water	Method	Comments
Groundwater	Kanat	System of underground canals distributing water from mostly hillside or mountain sources to agricultural sinks
Runoff rain-water	Terraces in secondary and tertiary wadis	Primary wadis cannot be used because the floods would be too strong
Runoff rain-water	Desert farms with runoff water-flooding from large catchment areas	Collection of stones piled up in heaps and walls to reduce infiltration
Runoff rain-water	Canal systems	Canals and including small wadis to direct runoff water towards the fields
Runoff rain-water	Micro catchments	Small plots, often for individual trees

21.2.1 Kanats

Although not within the scope of the present chapter, it is worth drawing attention to kanats (or qanats). These are chain-well systems and aquifers established for a source-to-sink distribution of groundwater. Such systems are still used in large areas of arid regions, for example in Iran, Afghanistan and India.

21.2.2 Terraces in Secondary and Tertiary Wadis

Michael Evenari and collaborators found archaeological remnants of terrace structures in secondary and tertiary wadis, which have served historic desert agriculture (Fig. 21.3) (Shanan et al. 1959). Small stone walls crossed the wadis at distances of between 12 and 15 m; these walls were 60–80 cm high above the lower terrace levels and 10–20 cm above upper terrace levels. Each wall functioned as a dam, holding back part of the floodwater running down the wadis after rain. Thus, the water could penetrate and saturate the loess soil allowing cultivation of crops. Such terraced wadis have kept their function up to the present time, and are still used occasionally by Bedouins for the cultivation of barley. Primary wadis cannot be used in this way because the floods would be too strong and destructive.

21.2.3 Desert Farms with Runoff Water-Flooding from Large Catchment Areas

In contrast to the terrace farms within the wadis, where the floods of the wadis serve directly as sources of water for the fields, runoff water-farming in the strict sense

Fig. 21.3 Terraced wadi – how it may have looked in ancient historical times (drawing by P. Treitel, from Evenari 1982)

collects water from large catchment areas of the hills and mountains surrounding valleys with farms (Tadmor et al. 1960; Shanan et al. 1961; Evenari et al. 1968). The amount of runoff water, R, obtained in a farm can be described by:

$$R = P - I - D - F \qquad (21.1)$$

where P is the amount of precipitation. As the equation shows, retainment of water from the farm is composed of three contributions, where I is retainment of water due to interception by vegetation; D is retainment of water in depressions of the ground; and F, retainment of water due to infiltration of the soil. Thus, due to retainment, R is in the range of 20–30% of P. With an average annual rainfall in the Negev desert near Avdat of 86.4 mm m^{-2} (see Fig. 21.1) and a catchment area that is 20–30 times larger than the farm, the farm can receive an amount of water of 340–750 mm m^{-2} annually (Table 21.3). This is sufficient for agriculture since the farm soil is made up of loess to a depth of 3 m. The runoff water saturates the loess

Table 21.3 Relationship between rainfall, runoff water, catchment area and water received by a typical desert farm in the Negev

Parameter	Amount of water
Amount of rain, *P*, average of 26 years	86.4 mm m^{-2}
Runoff, *R* (20–30%)	17–25 mm m^{-2}
Catchment area 20–30 ha (= 200,000–300,000 m^2), i.e. 20–30 times the size of the farm	3.4–7.5 10^6 mm
Farm 1 ha (= 10,000 m^2)	340–750 mm m^{-2}

to full water capacity, evaporation is reduced because of the depth of the soil, and the water-holding force of the loess and the plants can gradually acquire the water from the loess during their growth.

The amount of infiltration given by the parameter *F* also explains the intriguing observation of mounds and small strips of walls of stones and gravel found on the hills of the catchment area around farms (Fig. 21.4; Tadmor et al. 1958). Infiltration is greatly supported by small depressions, cracks and capillary forces around stones on the ground. Stones also prevent the destruction of porous aggregates of small soil particles by the impact of falling raindrops. The rate of infiltration, *f*, is given by:

$$f = f_t + (f_0 - f_t)e^{-ct} \tag{21.2}$$

where f_0 is the initial infiltration rate at the beginning of a rain event, and f_t the rate at time *t* (h); *c* is a constant. At the beginning of a rain, the rate of infiltration, f_0, is high, i.e. 17–18 mm h^{-1} for the loess soil of Avdat. The infiltration rate then declines rapidly to more or less constant values, f_t, of about 2–3 mm h^{-1} for loamy soils and as low as 0.2–0.5 mm h^{-1} for loess soils, which, by colloidal binding of the water, can rapidly form almost impermeable superficial soil crusts when wetted. This leads to a considerable decrease of infiltration, and hence increased runoff (see Eq. 21.1). The constant *c* depends on the soil type and varies within a range of 2–20. It is high for loess soils forming crusts.

Hence, Michael Evenari and colleagues advanced the hypothesis that historical desert farmers collected stones from the ground, and piled them up in small heaps and walls that are still found all around the catchment area. They checked this by comparing runoff from slopes where they removed all stones with those where the stones were left in place (Fig. 21.5). Depending on the inclination of the slopes, runoff was increased up to 2.8-fold by removing the stones. Runoff is lower from steeper slopes because they have less dense and coarser soil layers, and hence reduced formation of crust than slopes with lower inclination.

21.2.4 Canal Systems

Field observations and surveying historical remnants in the Negev desert revealed that, in addition to preparing the catchment slopes for runoff by removing stones,

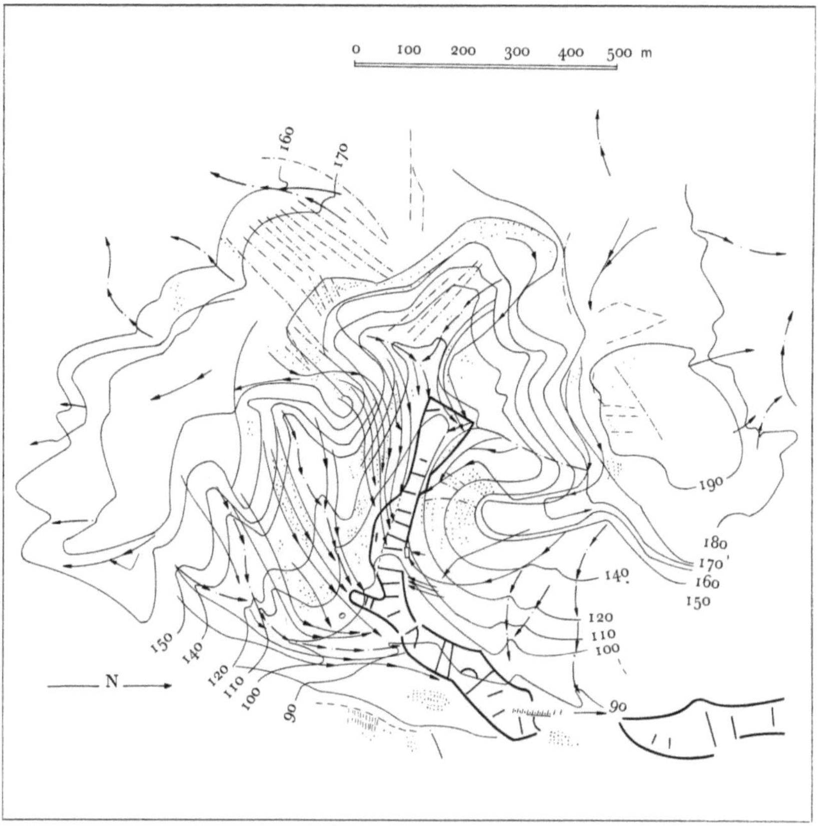

Fig. 21.4 Map obtained by measuring and surveying remnants of a historical desert farm in the Negev. *Thick lines* Terraces of the farm, *thin solid lines with arrowheads* catchment area with canals directing the runoff towards the farm, *dashed lines with arrowheads* small wadis incorporated in the runoff system, *dotted lines without arrowheads / small dots* walls and heaps of stones collected from the catchment area and piled up. Isoclines are in metres of altitude

directing runoff towards farms was also supported in places by digging extended systems of canals and incorporating smaller wadis into the runoff system (Figs. 21.4, 21.6). This allowed farms of much larger sizes than shown in Table 21.3 to be supported by extension of the catchment areas from several hundred hectares to several square kilometres. An example of the results of surveying is shown in Fig. 21.7. A diversion dam (a) served to channel the water from the large wadi (c) into a canal (b). Historically, the farm went through three distinct periods. In phase I (d, and dotted lines at the bottom of the map) terrace flood-water was obtained from the wadi. This phase dates from the Israelitic period. Erosion and alluvial material raised the level of the terraces, and this part of the farm had to be abandoned. In phase II, the farm consisted of parts (e) and (f), where the water came from the canal attached to the wadi via the diversion dam (a). Again, erosion and deposition of alluvial material led to abandonment of part (e). In phase III (f) only

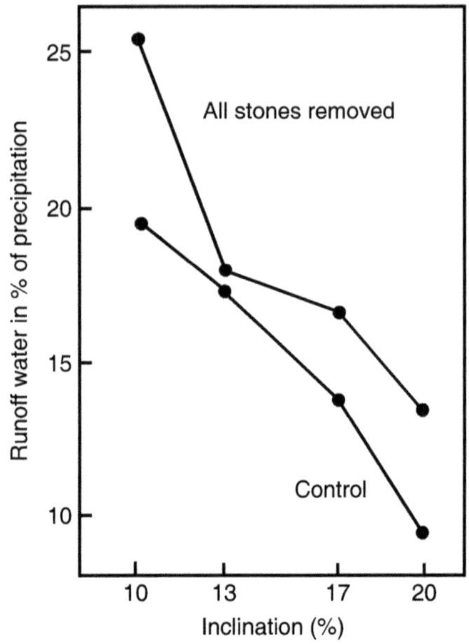

Fig. 21.5 Runoff water (y-axis) from slopes of increasing inclination (x-axis) in a catchment area with stones removed or left in place

Fig. 21.6 Farm with cultivated terraces, canals and a farmhouse as it may have looked in ancient historical times (drawing by P. Treitel, from Evenari 1982)

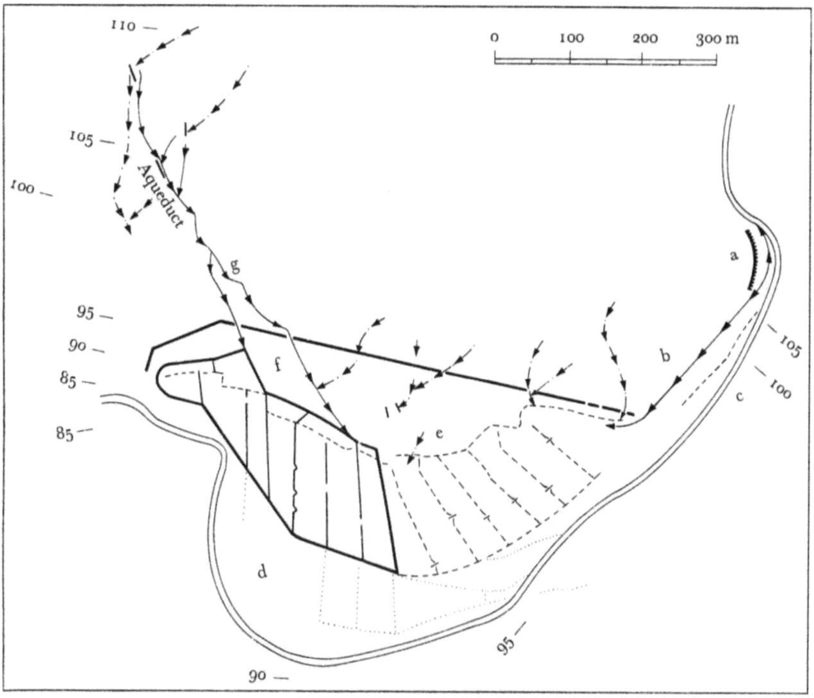

Fig. 21.7 Map obtained by measuring and surveying remnants of a historical desert farm in the Negev. *a* Diversion dam; *b* large canal; *c* large wadi; *d–f* phases I, II and III of the farm, respectively; g canals of phase III of the farm. Isoclines are in metres of altitude

seven farm terraces still received water from an aqueduct with attached small wadis (g, upper left of the map).

21.2.5 Micro Catchments

In the arid regions of Southern Tunisia, farmers cultivate olive trees and other fruit trees in micro catchments. Similar structures were established in the experimental farms in the Negev (Figs. 21.8, 21.9; Shanan et al. 1970). It was found that, under the conditions of the Negev, the optimum sizes of micro catchments were 200–250 m^2 for fruit trees and 150–200 m^2 for grape vines. The micro catchments can be considered as mini-runoff farms. Their advantage, besides the smaller costs for construction and maintenance, is that they are more effective. Due to the shorter distance that the water needs to flow to reach the cultivated plot, a rainfall that will not lead to a flood in a large catchment area can already cause a flood in a smaller one (Table 21.4).

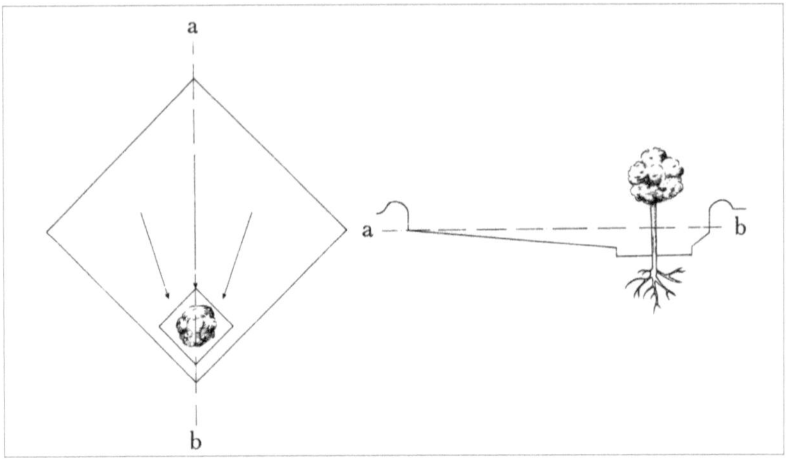

Fig. 21.8 Scheme of a micro catchment. *Arrows* indicate the direction of floods of runoff water

Fig. 21.9 Micro catchments established in the Negev desert at Avdat (original photographs by Otto L. Lange)

Table 21.4 Relationship between catchment area and amount of runoff water explaining the higher efficiency and relative advantage of micro catchments (example of a rain and flood in the Negev on 10 December 1963)

Catchment area	Runoff water produced	Ratio
340 ha	1,238 m^3	3.6 m^3 ha^{-1}
3 ha	156 m^3	52 m^3 ha^{-1}

Fig. 21.10 Reconstructed runoff-rainwater farm at Avdat in the Negev desert after a rain in the spring of 1971 (original photographs Detlef Schulze)

21.3 Reconstruction of Runoff-Rainwater Farms in the Negev Desert

To test their theories about the functioning and the actual effectiveness of runoff-rainwater farms, Michael Evenari and collaborators reconstructed several such farms in the Negev desert, some exactly the way in which, according to survey findings, they must have been in historical times, and some with modifications for experiments to optimise the output. Figure 21.10 shows the canals, dams and flooded plantations of the reconstructed farm at Avdat during a rainfall flood in the spring of 1971. Figure 21.11 shows the farm flooded in the spring of 1971 and dry in the autumn of 1978 (Cohen et al. 1968; Tadmor et al. 1970, 1971; Spiegel-Roy et al. 1971; Evenari 1984). Major crops grown successfully are listed in Table 21.5. Cherries, plums and apples were also grown but the yields were less successful. Crop yields are listed in Table 21.6.

Fig. 21.11 Reconstructed runoff-rainwater farm at Avdat in the Negev desert after a rain in the spring of 1971 (*top*) and in the autumn of 1978 (*bottom*). Remnants of the historical Byzantine town of Avdat can be seen on top of the hill in the background (original photographs, *top* Detlef Schulze, *bottom* U.L.)

Table 21.5 Crops successfully grown in reconstructed runoff-rainwater farms in the Negev desert

Barley	Wheat	
Apricots	Peaches	
Grape vines		
Almonds	Pistachios	Olives
Peas	Chick peas	Onions
Artichokes	Asparagus	
Sunflower		
Pasture grasses		

Table 21.6 Crop y yields of runoff-rainwater farming at Avdat in the Negev desert. Ranges given for trees depend on the cultivar used

Crop	Average yield (1969–1978)	Average yield of the best year
Almonds	5.0–8.9 kg/tree	8.2–20.8 kg/tree
Pistachio	6.3–6.8 kg/tree	10.1–13.9 kg/tree
Peaches	21.5–35.3 kg/tree	48.5–134.2 kg/tree
Grape vine	7.8–12.3 kg/plant	18.5–20.8 kg/plant
Barley	2.67 t/ha	4.8 t/ha
Wheat	3.55 t/ha	4.4 t/ha
Sunflower	1.91 t/ha	2.7 t/ha
Peas	5.45 t/ha	6.9 t/ha

21.4 Conclusions and Outlook: Applicability of Runoff-Rainwater Farming for Sustained Management to Support Local Communities in Arid Environments

Desert soils are usually fertile and the major limiting factor for plant growth generally is water, i.e. water is the very severe constraint determining the vegetation of deserts. Over the decades, promising technologies for achieving a sustainable water supply to arid lands (National Academy of Sciences 1974) have remained an important issue of high priority. The various methods of runoff-rainwater farming described above require only a limited level of technological advancement for establishment and are completely sustainable in terms of the use of natural resources. Their crop efficiency and potential monetary profit cannot be compared with that of advanced high-tech agriculture in highly developed areas. However, the quality and return of crops of runoff-rainwater farms is high, and is substantial enough to support local communities in arid environments. Figure 21.12 shows runoff-rainwater farming by the Djessur nomads in Tunisia, where the technique has been used continuously since Phoenician times (Table 21.1).

In a world where galloping desertification continues to reduce the available arable land, where water is becoming an increasingly rare resource, and where large populations suffer hunger, the techniques of sustainable desert farming using

Fig. 21.12 Runoff-rainwater farming by the Djessur nomads near Chenini, Tunisia (original photograph U.L.)

Table 21.7 Areas or countries where the methods of runoff-rainwater farming could be practised and in fact are already practised partially today

Syria – Jordan – Yemen – Southern Arabia – Sinai – Northern Sahara/Tunisia
Afghanistan – Pakistan
Southern Spain
Kenya – Mali – Nigeria
North and South America
Australia

runoff-rainwater should become a valuable option, not for profit but to support the people of local communities. As described in this chapter, the applicability of these techniques depends on geo-morphological conditions and soil types. Table 21.7 lists areas and countries where these methods could be practised and in fact are already practised to a certain extent today.

References

Aharoni Y, Evenari M, Shanan L, Tadmor NH (1960) The ancient desert agriculture of the Negev. V. An Israelite agricultural settlement at Ramat Matred. Isr Explor J 10:23–36, 97–111

Cohen OP, Evenari M, Shanan L, Tadmor NH (1968) Runoff farming in the desert. II. Moisture use by young apricot and peach trees. Agron J 60:33–38

Evenari M (1964a) Die Wüste Negev soll wieder fruchtbar gemacht werden. I. Archäologische Befunde. Umsch Wiss Tech 1964:463–466

Evenari M (1964b) Die Wüste Negev soll wieder fruchtbar gemacht werden. II. Gegenwart und Zukunft. Umsch Wiss Tech 1964:481–485

Evenari M (1982) Ökologisch-landwirtschaftliche Forschungen im Negev. Analyse eines Wüsten-Ökosystems. Technische Hochschule Darmstadt, Darmstadt

Evenari M (1983) Die Nabatäer im Negev (1983). In: Lindner M (ed) Pertra und das Königreich der Nabatäer. Delp, Nürnberg, pp 118–138

Evenari M (1984) Bewässerung wie zu Salomos Zeiten. Die Wüste blüht. Bild Wissensch 9:91–102

Evenari M (1987) Und die Wüste trage Frucht. Ein Lebensbericht. Bleicher, Gerlingen

Evenari M (1989) The awakening desert. Autobiography of an Israeli scientist. Springer, Heidelberg

Evenari M, Koller D (1956) Ancient masters of the desert. Sci Am 194:39–45

Evenari M, Aharoni Y, Shanon L, Tadmor NH (1958) The ancient desert agriculture of the Negev. III. Early beginnings. Isr Explor J 8:231–268

Evenari M, Shanan L, Tadmor N, Aharoni Y (1961) Ancient agriculture in the Negev. Science 133:979–996

Evenari M, Shanan L, Tadmor NH (1966) Die Landwirtschaft der Negev-Wüste in Vergangenheit und Gegenwart. Nova Acta Leopold N.F. 31(176):149–169

Evenari M, Shanan L, Tadmor NH (1968) Runoff farming in the desert. I. Experimental layout. Agron J 60:29–32

Evenari M, Shanan L, Tadmor NH (1971) Runoff agriculture in the Negev desert of Israel. In: McGinnies WG, Goldman BJ, Paylore P (eds) Food, fiber and the arid lands. The University of Arizona Press, Tucson, pp 311–322

Evenari M, Nessler U, Rogel A, Schenk O (1975) Antike Technik im Dienste der Landwirtschaft in ariden Gebieten. Tropenlandwirt 76:11–21

National Academy of Sciences (1974) More water for arid lands. Promising technologies and research opportunities. Report of an Ad Hoc Panel of the Advisory Committee on Technology Innovation. Board on Science and Technology for International Development. Commission on International Relations. National Academy of Sciences, Washington DC, pp 1–153

Shanan L, Tadmor NH, Evenari M (1959) The ancient desert agriculture of the Negev. II. Utilization of runoff from small watersheds in the Abde (Ovdat) region. Ktavim, Isr J Agric Res 9:107–129

Shanan L, Tadmor NH, Evenari M (1961) The ancient desert agriculture of the Negev. VII. Exploitation of runoff from large watersheds. Ktavim, Isr J Agric Res 11:9–31

Shanan L, Evenari M, Tadmor NH (1967) Rainfall patterns in the Central Negev desert. Isr Explor J 17:163–184

Shanan L, Evenari M, Tadmor NH (1969) Ancient technology and modern science applied to desert agriculture. Endeavour 28:68–72

Shanan L, Tadmor NH, Evenari M, Reiniger P (1970) Runoff farming in the desert. III. Microcatchments for improvement of desert range. Agron J 62:445–449

Spiegel-Roy P, Evenari M, Mazig D (1971) Performance and moisture use of apricot trees under runoff farming. J Am Soc Hortic Sci 96: 696–701

Tadmor NH, Evenari M, Shanan L, Hillel D (1958) The ancient desert agriculture of the Negev. I. Gravel mounds and gravel strips near Shivta (Sbeita) Ktavim, Isr J Agric Res 8:127–151

Tadmor NH, Shanan L, Evenari M (1960) The ancient desert agriculture of the Negev. VI. The ratio of catchment to cultivated area. Ktavim, Isr J Agric Res 10:193–206

Tadmor NH, Evenari M, Shanan L (1970) Runoff farming in the desert. IV. Survival and yields of perennial range plants. Agron J 62:695–699

Tadmor NH, Shanan L, Evenari M (1971) Runoff farming in the desert. V. Persistence and yields of annual range species. Agron J 63:91–95

Chapter 22
Biotechnological Approaches to Aphrodisiac Plants of Rajasthan, India

Neelam Jain, Shaily Goyal, and Kishan Gopal Ramawat

Abstract The human quest for sexual enhancers from natural substances is as old as civilisation itself. Ancient history in most cultures helped society in its desire to improve the sexual experience as evident by writings in holy texts and sculptures in Hindu temples. Many so-called aphrodisiac plants or herbal formulations are currently in use in different countries worldwide. In Asian countries, this is a lucrative, multi-million dollar business fuelled by the desire for enhanced performance, the secretive approach of patients, and exploitation by herbalists. As erectile dysfunction can result from many causes, there is no single panacea; the problem and its solution vary from person to person. Although many herbal drugs have been used traditionally for sexual enhancement, almost nothing is known about their side effects, and their efficacy has seldom been proved scientifically and/or clinically. The recent surge of interest in the clinical evaluation of herbal drugs has strengthened the folklore usage of herbal medicines. This chapter presents an introduction to the aphrodisiac plants of Rajasthan, describing the bioactive molecules known in these plants and their pharmacological evaluation by modern tests, herbal formulations containing extracts of these plants available on the market, and biotechnological approaches towards amelioration of such plants.

22.1 Introduction

Plants are an important source of medicines and play a key role in the health of the world's population. In almost all regions and cultures of the world, from ancient times up to today, plants have been used as medicines (Ramawat et al. 2009). Nowadays, the recent worldwide shift back towards herbal preparations over

N. Jain, S. Goyal, and K.G. Ramawat (✉)
Laboratory of Bio-Molecular Technology, Department of Botany, M.L. Sukhadia University, Udaipur 313001, India
e-mail: kg_ramawat@yahoo.com

K.G. Ramawat (ed.), *Desert Plants*,
DOI 10.1007/978-3-642-02550-1_22, © Springer-Verlag Berlin Heidelberg 2010

synthetic pharmaceuticals has emphasised the importance of research focussed on medicinal plants (Ramawat and Goyal 2008). Today, medicinal plants are important to the global economy, as approximately 80% of traditional medicine preparations involve the use of plants or plant extracts (Dhyani and Kala 2005). The use of herbs is very common in developing countries, particularly in rural settings. However, during the last decade, an increase in the use of plants has been observed in metropolitan areas of developed countries (Harnack et al. 2001). In India, the use of herbal drugs is the basic principle of the Indian system of medicine, Ayurveda, and has been in practice since ancient times (Ramawat and Goyal 2008). Plants are used extensively to relieve sexual dysfunction. For example, ginseng is an essential constituent in Chinese medicines (Murphy and Ferraro 2009), and at least 6 million Americans use the roots of this slow-growing perennial (Nocerino et al. 2000). Another root known as Maca (*Lepidium meyenii*) is traditionally used by Peruvian inhabitants living at high altitudes, as a nutrient, energiser, aphrodisiac and fertility enhancing agent. It has been proved to be effective in improving sexual desire in men, and sexual behaviour in male rats and mice. Thus, we can see that every society has evolved its own formulations based on local plants to meet their requirement for aphrodisiac substances (Yakubu et al. 2007; Shah 2002).

Sexual dysfunction affects not only sexual relationships, but also overall quality of life. Man has always sought to attain high performance, and several plants and their formulations have been available since ancient times to satisfy their desire. Use of herbal treatments is advantageous because of the readily availability of the medicinal plants and their low cost. Such formulations were very popular among Indian kings, and several formulations are still labeled with the prefix ‘Rajsahi’ meaning used by kings.

22.2 Erectile Dysfunction

The first report discussing sexual pathological disorders dates back 3,000–4,000 years, when medical information was passed from one generation to the next through oral poems among the Hindu population (Herman 1969). These poems were finally inscribed around 2000–1000 BC, and one of them (Samhita of Sushruta) already reported the pursuit of substances to enhance sexual experience (Sandroni 2001).

Ayurveda defines erectile dysfunction (ED) as follows (in Sanskrit language):

Sankalpapravano nityam priyaam vashyaamapi sthreeyam ||

na yaathi lingashaithilyaath kadaachidyaathi vaa yadi |

Shwaasaarthaha swinnagaatrshcha moghasankalpacheshtitaha ||

mlaanashishnashcha nirbeejaha syodetat klaibyalaxanam |

Translated, this means that even though a man has a strong desire to perform a sexual act with a cooperative partner, he cannot perform the sexual act because of looseness (absence of erection) of his phallus (penis). Even if he performs the sexual act with determined efforts, he does not get an erection and becomes afflicted with tiredness, perspiration and frustration when attempting to perform sex (Suri 2007).

Erectile dysfunction, sometimes called "impotence", is the repeated inability to get or keep an erection firm enough for sexual intercourse. Erectile dysfunction can mean a total inability to achieve erection, an inconsistent ability to do so, or a tendency to sustain only a brief erection (Suri 2007). Many people believe that impotence is a disorder associated with modern civilisation. However, man's preoccupation with potency or the lack thereof, has been present throughout the ages (Shah 2002).

Although ED does not affect life expectancy, it can have a significant negative impact on an individual's well-being and quality of life (Wagner et al. 2000). There are many causes of ED but the most important is the fact that this is an age-related phenomenon. The most reliable evidence for this comes from the Male Massachusetts Aging Study (Feldman et al. 1994). This study reported that the probability of ED increased from 5.1% at 40 years of age to 15% at 70 years (Feldman et al. 1994). In most men, ED is thought to be the consequence of organic deterioration, especially circulatory insufficiency (Krane et al. 1989). The usual origin is an organic factor or disease, such as hypertension, hypercholesterolaemia, vascular and heart diseases, diabetes mellitus, surgery and trauma, side effects of medications or neurodegenerative disorders, and the use of tobacco. Psychological problems also make an important contribution to the weakening of sexual performance, diminishing self-esteem and disrupting personal relationships (Feldman et al. 1994).

It is estimated that ED affects 15–30 million men worldwide (Shah 2002). According to the National Ambulatory Medical Care Survey (NAMCS), for every 1,000 men in the United States, 7.7 physician office visits were made for ED. Almost 15 years later, that figure had increased three-fold to 22.3 in 1999 (Wagner et al. 2000).

22.3 Aphrodisiacs

Throughout the ages, men and women have incessantly pursued every means to increase, preserve and recapture their sexual capacity or to stimulate the sexual desire of selected members of the opposite sex. One of the most recurrent methods has been the use of aphrodisiacs. Herbal medicines are a major source of aphrodisiacs and have been used worldwide for thousands of years by different cultures and civilisations. Traditional herbs represent an extraordinary reservoir of active ingredients, which are still present in about 25% of all prescriptions of modern "western" medicines (Zanolari 2003).

In 1996, the United States Food and Drug Administration defined an aphrodisiac as "A food, drink, drug, scent or device that can arouse or increase sexual desire or libido. A broader definition includes products that improve sexual performance".

The Oxford English Dictionary defines an aphrodisiac as "A drug or preparation inducing venereal desire". Venereal desire is often described as sexual appetite and can be understood as a desire for sexual stimulation.

The name aphrodisiac comes from the Greek goddess of sensuality and love "Aphrodite". The Greeks referred to sexual pleasure as "aphrodisia". Aphrodisiacs have been used throughout history in an attempt to increase sexual desire and pleasure. Various substances of animal and plant origin have been used as aphrodisiacs in ancient medicines. Many herbs can enhance sexual desire and sex drive in men and women, so that longer-lasting erections, female lubrication, and more intense orgasms could be achieved, resulting in satisfactory sex.

There seems to be no limit to the belief of man in his pursuit of aphrodisiacs, and it has been possible to compile a catalogue of over 500 animals, vegetables, and mineral substances that have, at some time, been evocated for their aphrodisiac properties (Taberner 1985). References to such substances have crept into holy texts from the Kama Sutra and the Bible to the Koran, and literature from Shakespeare and Ovid to Gilbert and Sullivan plays in the twentieth century (Wedeck 1963). Aphrodisiac plants and herbs have been used throughout history to enhance libido, improve potency and fertility, increase sex drive and endurance, as sexual enhancers for the purposes of seduction, and for increased energy.

In Hindu tradition, in Sanskrit literature on the subject of love, the most famous is still the Anunga Runga, written by the poet Kullianmull in the fifteenth century, and the Kama Sutra, composed by Vatsyayana sometime between the first and fourth centuries AD. There is a short chapter in the Kama Sutra that explains methods for "attracting others to oneself", including magic rituals, natural remedies of different potencies and general advice (Vatsyayana et al. 2000). Erotic sculptures in Hindu temples (Fig. 22.1) helped society to understand sexual behaviour.

Aphrodisiacs may be categorised into three groups according to their mode of action (Sandroni 2001): (1) those increasing libido (i.e. sexual desire) by affecting the central nervous system (CNS) by altering specific neurotransmitters or specific

Fig. 22.1 Sculpture in a 11th centuary Hindu temple near Udaipur on the subject of love

sex hormone concentrations, e.g. *Brassica rapa*, *Prunus amygdalus* and *Zingiber officinale*; (2) those increasing potency (i.e. effectiveness of erection) by inducing vasodilation resulting in sustained erection, e.g. *Panax ginseng*; and (3) those increasing sexual pleasure by causing irritation of genital mucosa and enhancement of sensory experience during coitus, e.g. *Coryanthe yohimbe*, *Crocus sativus*. Modern chemical drugs, such as *Viagra*, *Levitra* and *Cialis* improve erectile enlargement and erection, but they are not considered aphrodisiacs. These drugs do not have any direct effect on the brain or sensation of genital organs and libido, although these drugs enhance the ability to achieve erection and thus sexual pleasure. Plants listed in the literature for their use as aphrodisiac include *Abrus precatorius*, *Abuttilon indicum*, *Asparagus adscendens*, *Chlorophytum borivilianum*, *Bauhinia vahlii*, *Bombax ceiba*, *Butea monosperma*, *Chloroxylon swietenia*, *Pedalium murex*, *Peganum harmala*, *Pimpinella bracteata*, *Phyllanthus amarus*, *Tinospora cordifolia*, *Withania somnifera*. Other plants invoked for this purpose are *Aristolochia indica*, *Alpinia galonga*, *Coryanthe yohimbe*, *Crocus sativus*, *Datura stramonium*, *Myristica fragrans*, etc. However, neither have all been used in a single preparation nor has much phytochemical/pharmacological or tissue culture work been carried out (Ramawat et al. 1998, 2004).

22.3.1 *Butea monosperma Lamk. (syn. B. frondosa Koen. ex Roxb., Fabaceae)*

Butea monosperma is a medium-sized tree found throughout India except in extreme arid regions. It bears large orange-red flowers. Mainly seeds and bark are used, but flowers, leaves and gum are also collected. Several γ-pyrone and furan derivatives and related compounds (flavones, butein, palasonin; Fig. 22.2) have been isolated (Dev 2006). *Butea monosperma* is traditionally used in Ayurvedic medicines for many gynaecological problems and the extract is used in many herbal formulations for male impotency including ED. The bark extract of *Butea frondosa* is reported to reduce significantly mount latency, intromission latency, ejaculation latency and post-ejaculatory interval, and to increase significantly mount frequency, intromission frequency and ejaculation frequency in sexually active and inactive male rats (Ramachandran et al. 2004). A powder made from the underground tubers of *Butea superba* is used for the treatment of ED in mature males (Cherdshewasart et al. 2008). Furthermore, these tubers have androgenic effects on the reproductive organs of intact and ovariectomised rats and exhibit anti-estrogenic activity on leutinising hormone (LH) secretion in ovariectomised rats (Malaivijitnond et al. 2009).

22.3.2 *Chlorophytum borivilianum Sant. et Fernand. (Liliaceae)*

Chlorophytum borivilianum, commonly known as 'Safed musli' due to its white tuberous roots, is a traditional medicinal plant (Fig. 22.3). It is distributed mainly in

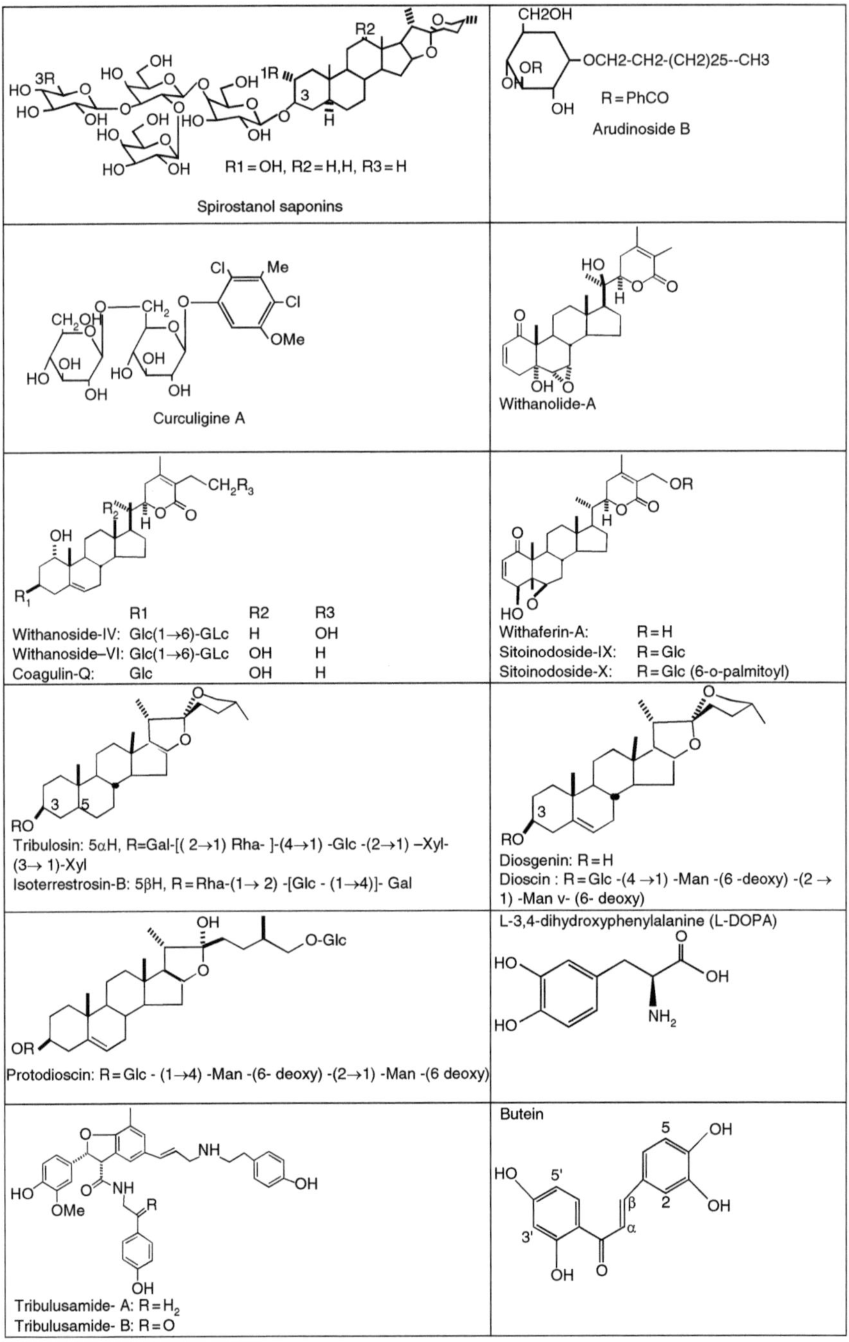

Fig. 22.2 Bioactive molecules of selected aphrodisiac plants

Fig. 22.3 *Chlorophytum borivilianum*; flowers and root tubers

Southern Rajasthan, Northern Gujarat and Western Madhya Pradesh (Maiti and Geetha 2005).

Chlorophytum species are known to contain steroidal saponins and several related compounds (Arora et al. 2004; Tandon and Shukla 1995). The ethanolic extract of *C. borivilianum* roots showed profound anabolic and spermatogenic effects in treated albino rats as evidenced by weight gains of body and reproductive organs, and the sexual behaviour of the animals as reflected in the reduction of mount, ejaculation, post ejaculatory and intromission latency. This was associated with an increase in mount frequency and attractiveness towards females (Thakur et al. 2009; Kenjale et al. 2008). The extract prepared from dried root tubers inhibited uptake of ^{3}H-dopamine in triatal syneptosomes, which could thereby lead to enhanced dopaminergic tone in the CNS. This has a beneficial effect on the brain and, in the human body, results in increased alertness, mental ability and sexual and maternal characters (Sarangdevot 2006).

Seeds of *C. borivilianum* have low viability and seed-raised plants required 3 years to produce commercially useful roots. Therefore, attempts have been made to raise plants through biotechnological methods by producing plantlets through organogenesis (Dave et al. 2003, 2004; Purohit et al. 2003; Mathur at al. 2008; Joshi et al. 2009) and somatic embryogenesis (Jain et al. 1997; Suri et al. 1999; Arora et al. 1999). Regenerants were evaluated for their fidelity using molecular markers (Arora et al. 2006; Lattoo et al. 2006; Mathur at al. 2008).

22.3.3 *Curculigo orchioides* Gaertn. (Hypoxidaceae)

The tuberous roots of *Curculigo orchioides*, also called Kali musli (Fig. 22.4), are used widely as a tonic for health, vigour and vitality due to the presence of flavanone glycosides and other steroidal saponins (Rastogi and Mehrotra 1993). Several bioactive compounds isolated from the plant include flavones, glycosides, steroids, saponins, and triterpenoids (Tandon and Shukla 1995). Two new curculigosides

Fig. 22.4 *Curculigo orchioides* plant and flower (*inset*)

were isolated from in-vitro-grown tubers of *C. orchioides* (Valls et al. 2006). The medicinal properties of this herb are attributed mainly to curculigosides and curculigo saponins (Xu et al. 1992). One study reported the effect of an ethanolic extract of the rhizomes on sexual behaviour in rats. Increased activity was indicated by reduction in mount latency, increase in mount frequency, increased penile erection index and enhanced attractability towards females (Thakur et al. 2009; Chauhan et al. 2007).

Biotechnological approaches towards producing this plant include development of micropropagation methods in static cultures (Suri et al. 1998a, 1998b; Wala and Jasrai 2003; Thomas 2007), bulbils formation in agitated medium in shake flasks (Suri et al. 2000; Nema et al. 2008), anther culture (Augustine et al. 2008), encapsulation (Nagesh et al. 2009) and, finally, plantlet regeneration on a mass scale (Suri et al. 2000).

22.3.4 *Mucuna pruriens* (L.) DC (Fabaceae)

Mucuna pruriens is a climbing herb found in several parts of India and also cultivated. It is an important source of L-DOPA, which is present in the seeds up to 4% (Dev 2006). Mishra (2004) reported increased sexual activity of male albino rats upon administration of seed powder at the dose of 75 mg kg^{-1} day^{-1}. The test drug affected different components of copulatory behaviour, including mount frequency, mount latency, intromission frequency, and intromission latency. Seed extracts of this plant are also known to increase spermatogenesis and the weight of male sexual organs.

No citable clinical studies for this plant showing curative effects for male sexual dysfunction are available (Mishra 2004). The plant contains L-DOPA, which is reported to arouse sexual desire in patients with Parkinson's disease (Kumar et al. 1994; Misra and Wagner 2008).

22.3.5 *Tribulus terrestris L. (Zygophyllaceae)*

Tribulus terrestris is an annual creeping herb distributed throughout drier parts of India (Fig. 22.5). The plant contains several steroidal saponins, which have been isolated from almost all parts of the plant. Genins derived from these saponins belong to both spirostanol (e.g. diosgenin, hecogenin, tigogenin) and furostanol (e. g. genins from tribol, tribulosaponin-A, B) classes (Kostova and Dinchev 2005; Dev 2006). Other important constituents include flavonoids, lignanamides, cinnamic amides and alkaloids. It is used as a diuretic, cardio tonic, aphrodisiac, emollient, appetiser, expectorant and laxative, etc., in the Indian system of medicine. It increases athletic and sexual performance and activity in both men and women by supporting the body's hormonal feedback system, acting as a nutrient to the glands and stimulating hormone release (Dev 2006). It is long been used in traditional Chinese and Indian medicine for the treatment of various ailments, and is popularly claimed to improve sexual function in man. Several researchers have presented evidence that protodioscin (PTN), a compound present in the *T. terrestris* extract, can improve sexual desire and enhance erection (Adaikan et al. 2000; Adimoelja 2000). *T. terrestris* extract increased testosterone levels

Fig. 22.5 *Tribulus terrestris*; plant with fruits, a flower and a fruit

and spermatogenesis in male animals (Georgiev et al. 1988). The aphrodisiac activity of the plant has been clinically proved (Miller 1988).

Daily treatment with *T. terrestris* extract (2.5–10.0 mg/kg body weight) for 8 weeks resulted in an increase in the relaxation of penile tissue isolated from treated animals as compared with that in controls. The relaxant response to electric field simulation, acetyl-choline and nitroglycerin in nor-adrenaline-treated precoxified corpus cavernosal tissue from the treated group was found to be higher compared with that in the control group. It was concluded that the enhanced relaxant effect observed was probably due to an increase in the release of nitric oxide (NO) from the endothelium and nitrinergic nerve endings, which may account for its claimed aphrodisiac properties (Adaikan et al. 2000). A similar investigation concluded that the proerectile aphrodisiac property of *T. terrestris* might possibly be the result of an increase in androgen and subsequent release of NO from the nerve endings innervating the corpus cavernosum (Gauthaman et al. 2003).

In another study, sexual behaviour and intracavernous pressure (ICP) were compared in both normal and castrated rats in an attempt to understand the role of *T. terrestris* containing PTN as an aphrodisiac. The authors concluded that *T. terrestris* extract appears to possess aphrodisiac activity (Gauthaman et al. 2002). They further studied rats chronically treated with *T. terrestris*, which increased the number of nicotinamide adenine dinucleotide phosphate-diaphorase (NADPH-d)-positive neurons and androgen receptor (AR) immunoreactivity in the para ventricular region of the brain. Androgens are known to increase both AR and NADPH-d positive neurons either directly or via conversion to oestrogen. The mechanism underlying the observed increase in AR and NADPH-d positive neurons can probably be ascribed to the androgen increasing properties of *T. terrestris*, thus supporting the aphrodisiac claims of *T. terrestris* (Gauthaman and Adaikan 2005). Studies conducted on primates, rabbits and rats using *T. terrestris* extracts showed increased levels of sex hormones, presumably due to the presence of protodioscin, which may be useful in mild-to-moderate cases of ED (Gauthaman and Adaikan 2008). Another species of *Tribulus* namely, *T. alatus*, also possesses aphrodisiac activity. In male rats, it was found that *T. alatus* extract increased free serum testosterone levels significantly. It was concluded that *T. alatus* possesses aphrodisiac activity due to its androgen-increasing properties (El-Tantawy et al. 2007).

In contrast to the above findings, Neychev and Mitev reported the influence of *T. terrestris* extract on androgen metabolism in young men and concluded that *T. terrestris* steroid saponins possess neither direct nor indirect androgen-increasing properties (Neychev and Mitev 2005).

Tissue culture studies on *T. terrestris* include callus formation (Zafar and Haque 1990) and micropropagation (Ali et al. 1997). Based on folklore usage and recent findings, several herbal formulations are available on the Asian market (Table 22.1, Fig. 22.6) and patents have been obtained (Reyes 2004; Hessel 2006; Palpu et al. 2008; Bombardelli et al. 2008). These formulations and patents contain several herbs and other ingredients.

Table 22.1 Aphrodisiac herbal formulations available on the market

Product	Ingredients	Contact
SA1	Korean red ginseng, fermented soybean, *Tribulus terrestris*, *Fructus rubi*, *Fructus lycii*, *Semen cusctae*, *Dioscorea* rhizome, *Fructus cornii*, *Fructus crataegi*	Park et al. 2006
Herbagra	*Withania somnifera*, *Asparagus racemosus*, *Mucuna pruriens*, *Argyria nervosa*, *Tribulus terrestris*, *Nardostachys jatamansi*, *Pueraria tuberosa*, *Sida cordifolia* and *Anacyclus pyrethrum*	http://www.wikimensinfo.com
Passion Rx	acetylcarnitine, *Withania somnifera*, catuaba, choline, Cistanches, *Cnidium monnieri*, *Coleus forskohlii*, dimethyl-amino-ethanol (DMAE), horny goat weed, maca, *Mucuna pruriens*, *Muira puama*, passion flower, *Pfaffia paniculata*, rehmannia, rhodiola, shilajit, *Tribulus terrestris*, and Tongkat ali	http://www.raysahelian.com
Vita ex-gold	*Withania somnifera*, shilajit, abhrak bhasma, *Crocus sativus*, kauncha beej, *Chlorophytum borivilianum*, Makardhwaj, *Myristica fragrans*, Samel, *Asparagus racemosus*, *Bacopa monnieri*, chandan, safed mulithi and juice of pan (betel leaves)	http://www.allayurveda.com
Vuka 2000	*Tribulus terrestris*, *Muira puama*, *Ginkgo biloba* extract, ginseng, amino acids arginine and histidine, zinc and vitamin E	http://www.anyvitamins.com/vuka2000.htm
VigRX Plus	*Epimedium*, *Cuscuta* seed extract, *Ginkgo biloba*, Asian Red Ginseng, *Muira pauma*, Bioperine, Damiana and *Tribulus terrestris*	http://www.real-natural-cures.com
Tentex royal	*Prunus amygdalus*, *Crocus sativus* and *Tribulus terrestris*	Gopumadhavan et al. 2003
VigoPower	*Eurycoma longifolia* Jack (Tongkat Ali), *Butea superba*, Horny Goat Weed, *Tribulus* fruit and *Smilax myosotiflora*	http://www.vigopower.com/
Libilov	Purified extracts of *Tribulus terrestris* and *Ginkgo biloba* and L-arginine	http://www.nutrica.com/en/products/libilov_info.htm

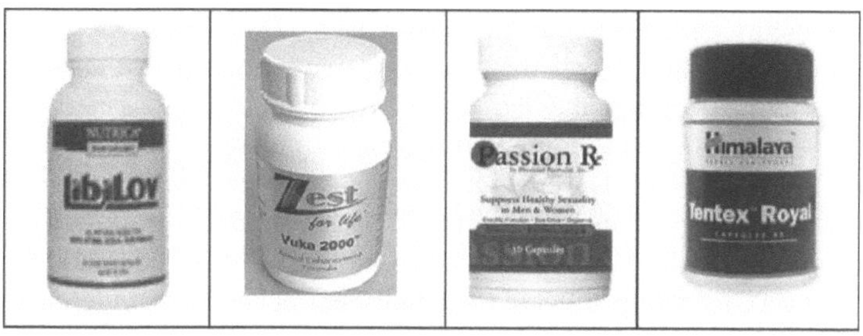

Fig. 22.6 Some aphrodisiac herbal formulations available on the market

Fig. 22.7 *Withania somnifera*; plant with fruits

22.3.6 Withania somnifera (L.) Dunal. (Solanaceae)

Withania somnifera is commonly known as 'ashwagandha', meaning smells like a horse (Fig. 22.7). Its medicinally valuable part is the root. Characteristic compounds present in the plant are steroids with an ergostane skeleton, which have been named withanoloides. More than 45 withanoloides have been isolated from the leaves, fruits and roots of *W. somnifera*. This plant has several geno- and chemotypes and hence the nature and percentage of secondary metabolites occurring in plant materials of different origin varies greatly (Bandyopadhyay and Jha 2003; Dev 2006). The plant is cultivated for its roots in Rajasthan and Madhya Pradesh (India). It is highly valued in Ayurveda as an alterative, restorative and as an anabolic agent. It has diuretic, antidepressant, and cardio-protective activities and is used in the treatment of skin diseases, CNS disorders, memory deficit, gastrointestinal problems, cough, asthma and hypertension (Dev 2006).

Recent studies have shown that the aqueous extract of *Withania somnifera* increased spermatogenesis and the weight of male sexual organs in immature rats. However, serum testosterone and follicle stimulating hormone (FSH) levels were low, whereas LH levels were high in treated animals. It was concluded that the extract has a direct spermatogenic influence on the seminiferous tubules of immature rats, presumably by exerting a testosterone-like effect (Mishra 2004). Another study showed a contrasting effect of methanolic extract of *W. somnifera* roots on the sexual competence of male rats. It was concluded that the root extract induced a marked impairment in libido, sexual performance, sexual vigour and penile ED;

therefore, use of *W. somnifera* roots may be detrimental to male sexual competence (Ilayperuma et al. 2002).

As mentioned above, *W. somnifera* is cultivated and micropropagation is not desirable. However, some reports on micropropagation (Sen and Sharma 1991; Sabir et al. 2007; Sangwan et al. 2007) and hairy root culture for the production of useful metabolites (Kumar et al. 2005; Ray et al. 1996; Banerjee et al. 1994) of the plant are available.

As listed in Table 22.1, aphrodisiac formulations contain mixtures of several herbs, but scientific validation of the aphrodisiac properties of many of these plants is still awaited. However, some recent studies describe the aphrodisiac properties of other folklore plants like *Blepharis sindica* (Mathur and Sundaramoorthy 2006) and *Asparagus racemosus* (Thakur et al. 2009) on the basis of a few selected parameters.

22.4 Conclusion

In Asian countries, ED is not discussed publicly, explaining why patients keep it secret and consume formulations available in the market (no prescription is required in several countries). At the same time, herbalists do not disclose their formulations, and these herbs and herbal formulations are quite popular among the Asian population. This herb-based business is booming due to the large population, the desire for enhanced performance, and the low cost of drugs without known side effects. In recent years, some of these traditionally used herbs have been evaluated by modern scientific methods, which has supported their traditional uses. It is well established that the effect of combined herbs differs from that of an individual herb or its isolated pure compound. Therefore, there is a need to evaluate marketed herbal formulations rather than individual plants, extracts or isolated compounds. This type of herbal treatment is officially recognised (and reimbursed) in India and several other countries.

Acknowledgements This work was supported by financial assistance from UGC-DRS under the special assistance programme for medicinal plant research, and the DST-FIST programme for infrastructure development to K.G.R. Financial assistance in the form of JRF and RA from UGC (N.J.) and CSIR (S.G.) is also acknowledged.

References

Adaikan PG, Gauthaman K, Prasad RN (2000) Pro-erectile pharmacological effects of *Tribulus terrestris* extract on rabbit corpus cavernosum. Ann Acad Med Singapore 29:22–26

Adimoelja A (2000) Phytochemicals and the breakthrough of traditional herbs in the management of sexual dysfunctions. Int J Androl 23:82–84

Ali G, Mughal MH, Srivastava PS, Iqbal M (1997) Micropropagation of *Tribulus terrestris* L., an important medicinal plant. J Plant Biol 40:202–205

Arora DK, Suri SS, Ramawat KG, Merillon JM (1999) Factors affecting somatic embryogenesis in long term callus cultures of 'Safed musli'(*Chlorophytum borivilianum*), an endangered wonder herb. Indian J Exp Biol 37:75–82

Arora DK, Jain AK, Ramawat KG, Merillon JM (2004) *Chlorophytum borivilianum*: an endangered aphrodisiac herb. In: Ramawat KG (ed) Biotechnology of medicinal plants: vitalizer and therapeutic. Science, Enfield, pp 111–128

Arora DK, Suri SS, Ramawat KG (2006) Assessment of genetic variability in regenerants from long-term cultures of Safed Musli (*Chlorophytum borivilianum*). Indian J Biotechnol 5:527–534

Augustine AC, Nivas S, D'Souza L (2008) Induction of embryos and plantlets from anthers of *Curculigo orchioides* Gaertn. – an endangered medicinal herb. Indian J Biotechnol 7:541–546

Bandyopadhyay M, Jha S (2003) *Withania* species. Review. J Trop Med Plants 4:273–284

Banerjee S, Naqvi AA, Mandal S, Ahuja PS (1994) Transformation of *Withania somnifera* (L.) Dunal. by *Agrobacterium rhizogenes*: infectivity and phytochemical studies. Phytother Res 8:452–455

Bombardelli E, Morazzoni P, Riva A, Seghizzi R (2008) Formulation useful in the treatment of male and female impotence. US Patent No. 7438934

Chauhan NS, Rao ChV, Dixit VK (2007) Effect of *Curculigo orchioides* rhizomes on sexual behaviour of male rats. Fitoterapia 78:530–534

Cherdshewasart W, Bhuntaku P, Panriansaen R, Dahlan W, Malaivijitnond S (2008) Androgen disruption and toxicity tests of *Butea superba* Roxb., a traditional herb used for the treatment of erectile dysfunction, in male rats. Maturitas 60:131–137

Dave A, Purohit SD, Bilochi G, Joshi N (2003) Scaling-up production and field performance of micro propagated medicinal herb 'Safed Musli' (*Chlorophytum borivilianum*). In Vitro Cell Dev Biol Plant 39:419–425

Dave A, Sharma P, Purohit SD (2004) In vitro propagation of *Chlorophytum borivilianum* using encapsulated shoot buds. Eur J Hortic Sci 69:37–42

Dev S (2006) A selection of prime Ayurvedic plant drugs, ancient–modern concordance. Anamaya, New Delhi

Dhyani PP, Kala CP (2005) Current research on medicinal plants: five lesser known but valuable aspects. Curr Sci 88:335

El-Tantawy WH, Temraz A, El-Gindi O (2007) Free serum testosterone level in male rats treated with *Tribulus alatus* extracts. Int Braz J Urol 33:554–559

Feldman HA, Goldstein I, Hatzichristou DG, Krane RJ, McKinlay JB (1994) Impotence and its medical and psychosocial correlates: results of the Massachusetts Male Aging Study. J Urol 151:54–61

Gauthaman K, Adaikan PG (2005) Effect of *Tribulus terrestris* on nicotinamide adenine dinucleotide phosphate-diaphorase activity and androgen receptors in rat brain. J Ethnopharmacol 96:127–132

Gauthaman K, Adaikan PG (2008) The hormonal effects of *Tribulus terrestris* and its role in the management of male erectile-dysfunction – an evaluation using primates, rabbit and rat. Phytomedicine 15:44–54

Gauthaman K, Adaikan PG, Prasad RN (2002) Aphrodisiac properties of *Tribulus terrestris* extract (Protodioscin) in normal and castrated rats. Life Sci 71:1385–1396

Gauthaman K, Ganesan AP, Prasad RN (2003) Sexual effects of puncturevine (*Tribulus terrestris*) extract (protodioscin): an evaluation using a rat model. J Altern Complement Med 9:257–65

Georgiev P, Dimitrov M, Vitanov S (1988) Effects of Tribestan (from *Tribulus terrestris*) on plasma testosterone and spermatogenesis in male lambs and rams. Veterinarna Sbirka 86:20

Gopumadhavan S, Mohamed R, Venkataranganna MV, Kala SK, Mitra SK (2003) Assessment of 'Tentex royal' for sexual activity in an experimental model. Indian J Clin Pract 10:23–26

Harnack LJ, Rydell SA, Stang J (2001) Prevalance of use of herbal products by adults in the Minneapolis/St Paul, Minn, Metropolitan area. Mayo Clin Proc 76:688–694

Herman JR (1969) Impotencia throughout the ages. J Am Soc Psychosom Dent Med 16:93–99

Hessel LL (2006) Aphrodisiac herbal composition for a female comprising *Epimedium grandiflorum*, *Turnera diffusa* var. *aphrodisiaca*, *Ilex paraguariensis* and *Smilax* spp. Int patent no. PCT/IB2005/052768

Ilayperuma I, Ratnasooriya WD, Weerasooriya TR (2002) Effect of *Withania somnifera* root extract on the sexual behaviour of male rats. Asian J Androl 4:295–8

Jain S, Ramawat KG, Sonie KC (1997) Somatic embryogenesis in *Chlorophytum borivilianum* – a medicinal plant of Aravalli Hills. In: Ravishankar GA, Venkataraman LV (eds) Biotechnological applications of plant tissue and cell culture, Oxford/IBH, New Delhi, pp 199–203

Joshi N, Dave A, Vyas S, Purohit SD (2009) Growth and shoot proliferation in *Chlorophytum borivilianum* Sant. et Fernand. in vitro under different carbon dioxide environment. Indian J Biotechnol 8:323–327

Kenjale R, Shah R, Sathave S (2008) Effects of *Chlorophytum borivilianum* on sexual behaviour and sperm count in male rats. Phytother Res 22:796–801

Kostova I, Dinchev D (2005) Saponins in *Tribulus terrestris* – chemistry and bioactivity. Phytochem Rev 4:111–137

Krane RJ, Goldstein I, Saenz de Tejada I (1989) Impotence. N Engl J Med 321:1648–1659

Kumar KVA, Srinivasan KK, Shanbhag T, Rao SG (1994) Aphrodisiac activity of the seeds of *Mucuna pruriens*. Indian Drugs 31:321–327

Kumar V, Murthy KNC, Bhamid S, Sudha CG, Ravishankar GA (2005) Genetically modified hairy roots of *Withania somnifera* Dunal: a potent source of rejuvenating principles. Rejuvenation Res 8: 37–45

Lattoo SK, Bamotra S, Dhar RS, Khan S, Dhar AK (2006) Rapid plant regeneration and analysis of genetic fidelity of in vitro derived plants of *Chlorophytum arundinaceum* Baker – an endangered medicinal herb. Plant Cell Rep 25:499–506

Maiti S, Geetha KA, (2005) Characterization, genetic diversity and cultivation of *Chlorophytum borivilianum* – an important medicinal plant of India. Plant Genet Resour 3:264–272

Malaivijitnond S, Ketsuwan A, Watanabe G, Taya K, Cherdshewasart W (2009) Androgenic activity of the Thai traditional male potency herb, *Butea superba* Roxb., in female rats. J Ethnopharmacol 121:123–129

Mathur A, Mathur AK, Verma P, Yadav S, Gupta ML, Darokar MP (2008) Biological hardening and genetic fidelity testing of micro-cloned progeny of *Chlorophytum borivilianum* Sant. et Fernand. Afr J Biotechnol 7:1046–1053

Mathur M, Sundaramoorthy S (2006) Ecological and aphrodisiac properties of *Blepharis sindica*. Niger J Nat Prod Med 10:17–25

Miller WL (1988) Molecular biology of steroid hormone synthesis. Endocr Rev 9:295–318

Mishra LC (2004) Scientific basis for Ayurvedic therapies. CRC, New York

Misra L, Wagner H (2008) Extraction of bioactive principle from *Mucuna pruriens* seeds. Phytother Res 22:796–801

Murphy LL, Ferraro JS (2009) Ginseng and male sexual behavior. In: Ramawat KG (ed) Herbal drugs: ethnomedicine to modern medicine. Springer, Heidelberg, pp 57–66

Nagesh KS, Shanthamma C, Bhagyalxmi N (2009) Role of polarity in de novo shoot bud initiation from stem disc explants of *Curculigo orchioides* Gaertn. and its encapsulation and storability. Acta Physiol Plant 31:699–704

Nema RK, Ramawat KG, Merillon JM (2008) Rapid micropropagation of *Curculigo orchioides* in shake flask culture. Pharmacogn Mag 4:315–319

Neychev VK, Mitev VI (2005) The aphrodisiac herb *Tribulus terrestris* does not influence the androgen production in young men. J Ethnopharmacol 101:319–323

Nocerino E, Amato M, Izzo AA (2000) The aphrodisiac and adaptogenic properties of Ginseng. Fitoterapia 71:s1–s5

Palpu P, Rao CV, Govindarajan R, Rawat AKS, Srivastava SK, Deb B, Subramanian S (2008) Functional aphrodisiac rolled herbal bidis and cigarettes. US Patent no. 20070042054

Park SW, Lee CH, Shin DH, Bang NS, Lee SM (2006) Effect of SA1, a herbal formulation, on sexual behavior and penile erection. Biol Pharm Bull 7:1383–1386

Purohit SD, Dave A, Bilochi G, Joshi N (2003) Factors influencing in vitro culture of *Chlorophytum borivilianum*. Oikoassay 15:19–27

Ramachandran S, Sridhar Y, Sam SK, Saravanan M, Leonard JT, Anbalagan N, Sridhar SK (2004) Aphrodisiac activity of *Butea frondosa* Koen. ex Roxb. extract in male rats. Phytomedicine 11:165–168

Ramawat KG, Goyal S (2008) The Indian herbal drugs scenario in global perspectives. In: Ramawat KG, Merrilon JM (eds) Bioactive molecules and medicinal plants. Springer, Heidelberg, pp 325–347

Ramawat KG, Jain S, Suri SS, Arora DK (1998) Aphrodisiac plants of Aravalli hills with special reference to Safed Musli. In: Khan I, Khanum A (eds) Role of biotechnology in medicinal and aromatic plants, vol 1. Ukaaz Pub, Hyderabad, pp 210–223

Ramawat KG, Sonie KC, Sharma MC (2004) Therapeutic potential of medicinal plants. In: Ramawat KG (ed) Biotechnology of medicinal plants: vitalizer and therapeutic. Science, Enfield, pp 1–18

Ramawat KG, Dass S, Mathur M (2009) The chemical diversity of bioactive molecules and therapeutic potential of medicinal plants. In: Ramawat KG (ed) Herbal drugs: ethnomedicine to modern medicine, Springer, Heidelberg, pp 7–32

Rastogi RP, Mehrotra BN (1993) Compendium of Indian medicinal plants. Central Drug and Research Institute Lucknow, PID and CSIR, New Delhi

Ray S, Ghosh B, Sen S, Jha S (1996) Withanolide production by root cultures of *Withania somnifera* transformed with *Agrobacterium rhizogenes*. Planta Med 62:571–573

Reyes J (2004) Composition to boost libido. US Patent no. 6803060

Sabir F, Sangwan NS, Chaurasiya ND, Misra LN, Tuli R, Sangwan RS (2007) Rapid micropropagation of *Withania somnifera* L. accessions from axillary meristems. J Herbs Spices Med Plants 13:118–128

Sandroni P (2001) Aphrodisiacs past and present: a historical review. Clin Auton Res 11:303–307

Sangwan RS, Chaurasiya ND, Lal P, Misra LN, Uniyal GC, Tuli R, Sangwan NS (2007) Withanolide A biogeneration in in vitro shoot cultures of Ashwagandha (*Withania somnifera* DUNAL), a main medicinal plant in Ayurveda. Chem Pharm Bull 55:1371–1375

Sarangdevot YS (2006) Development of biotechnology for medicinal plants using callus and cell cultures. PhD Thesis, ML Sukhadia University, Udaipur, India

Sen J, Sharma AK (1991) Micropropagation of *Withania somnifera* from germinating seeds and shoot tips. Plant Cell Tissue Organ Cult 26:71–73

Shah J (2002) Erectile dysfunction through the ages. Br J Urol Int 90:433–441

Suri S (2007) (http://www.ayurhelp.com)

Suri SS, Jain S, Ramawat KG (1998a) Plantlet regeneration and bulbil formation in vitro from leaf and stem explant of *Curculigo orchioides* – an endangered medicinal plant. Sci Hortic 1210:1–8

Suri SS, Arora DK, Sharma R, Ramawat KG (1998b) Rapid micropropagation through direct somatic embryogenesis and bulbil formation from leaf explants in *Curculigo orchiodes*. Indian J Exp Biol 36:1130–1135

Suri SS, Jain S, Arora DK, Ramawat KG (1999) In vitro high frequency regeneration of plantlets and tuberous root formation in *Chlorophytum borivilianum*. Gartenbau 64:106–110

Suri SS, Arora DK, Sharma R, Ramawat KG (2000) A method for large-scale multiplication of *Curculigo orchioides* through bulbil formation from leaf explant in shake flask culture. Indian J Exp Biol 38:145–148

Taberner PV (1985) Sex and drugs – Aphrodite's legacy. Trends Pharmacol Sci 6:49–54

Tandon M, Shukla YN (1995) Phytoconstitutes of *Asparagus adscendens, Chlorophytum arundinaceum* and *Curculigo orchioides*: a review. Curr Res Med Arom Plants 17:42–50

Thakur M, Chauhan NS, Bhargava S, Dixit VK (2009) A comparative study on aphrodisiac activity of some ayurvedic herbs in male albino rats. Arch Sex Behav (in press). doi: 10.1007/s1050800894448

Thomas DT (2007) High-frequency, direct bulblet induction from rhizome explants of *Curculigo orchioides* Gaertn., an endangered medicinal herb. In Vitro Cell Dev Biol Plant 43:442–448

Valls J, Richard T, Larronde F, Leblais V, Muller B, Delaunay JC, Monti JP, Ramawat KG, Merillon JM (2006) Two new benzylbenzoate glucosides from *Curculigo orchioides*. Fitoterapia 77:416–419
Vatsyayana M, Burton R, Johnson A (2000) Kama Sutra. Selection du Reader's Digest, Paris
Wagner G, Fugl-Meyer KS, Fugl-Meyer AR (2000) Impact of erectile dysfunction on quality of life: patient and partner perspectives. Int J Impot Res 12:144–146
Wala BB, Jasrai YT (2003) Micropropagation of an endangered medicinal plant: *Curculigo orchioides* Gaertn. Plant Tissue Cult 13:13–19
Wedeck HE (1963) Love potions through the ages, a study of amatory devices and mores. Philosophical Library, New York
Xu JP, Xu RS, Li XY (1992) Cycloartane type sapogenins and their glycosides from *Curculig orchioides*. Phytochemistry 31:2455–2458
Yakubu MT, Akanji MA, Oladiji AT (2007) Male sexual dysfunction and methods used in assessing medicinal plants with aphrodisiac potentials. Pharmacogn Rev 1:49–56
Zafar R, Haque J (1990) Tissue culture studies on *Tribulus terrestris* linn. Indian J Pharma Sci 52:102–103
Zanolari B (2003) Natural aphrodisiacs. Studies of commercially-available herbal recipes, and phytochemical investigation of *Erythroxylum vacciniifolium* Mart. (Erythroxylaceae) from Brazil. PhD thesis. Institute of Phytochemistry and Pharmacognosy of the University of Lausanne, Lausanne, pp 1–263

Index

MIX
Papier aus verantwortungsvollen Quellen
Paper from responsible sources
FSC® C105338

If you have any concerns about our products,
you can contact us on
ProductSafety@springernature.com

In case Publisher is established outside the EU,
the EU authorized representative is:
Springer Nature Customer Service Center GmbH
Europaplatz 3, 69115 Heidelberg, Germany

Printed by Libri Plureos GmbH
in Hamburg, Germany